AF870363

COMMON PROBLEMS AND TRENDS OF MODERN PHYSICS

Proceedings of the 7th Winter School on Hadronic Physics

COMMON PROBLEMS AND TRENDS OF MODERN PHYSICS

Folgaria (Trento) Italy
February 10 – 15, 1992

Editors

T. Bressani
Università di Torino

S. Marcello
INFN, Sezione di Torino

A. Zenoni
INFN, Sezione di Pavia

World Scientific
Singapore • New Jersey • London • Hong Kong

Published by

World Scientific Publishing Co Pte Ltd
5 Toh Tuck Link, Singapore 596224
USA office: 27 Warren Street, Suite 401-402, Hackensack, NJ 07601
UK office: 57 Shelton Street, Covent Garden, London WC2H 9HE

British Library Cataloguing-in-Publication Data
A catalogue record for this book is available from the British Library.

Proceedings of the 7th Winter School on Hadronic Physics
Common Problems and Trends of Modern Physics

Copyright © 1992 by World Scientific Publishing Co Pte Ltd

All rights reserved. This book, or parts thereof, may not be reproduced in any form or by any means, electronic or mechanical, including photocopying, recording or any information storage and retrieval system now known or to be invented, without written permission from the publisher.

For photocopying of material in this volume, please paye a copying fee through the Copyright Clearance Center, Inc., 222 Rosewood Drive, Danvers, MA 01923, USA. In this case permission to photocopy is not required from the publisher.

ISBN-13 978-981-02-1203-2
ISBN-10 981-02-1203-8

PREFACE

The 7th Winter School on "Hadronic Physics" was held in Folgaria (Trento), Italy from February 10 to February 15, 1992. The Course, organized by the Sezione di Torino of the Istituto Nazionale di Fisica Nucleare (I.N.F.N.), was attended by 86 participants.

The main goal of the School was to present to young physicists some actual open problems on hadronic physics at intermediate energy. The course was organized in the following way: senior lecturers illustrated the general framework of a topic, leaving then to younger lecturers the task of developing in detail specific aspects. Particular attention was devoted to physics at DAΦNE, the ϕ-factory at Frascati recently approved by I.N.F.N. and that will be operational in 1995-96.

The book contains most of the lectures given at the School in the following fields:

- Theoretical physics
- Subnuclear physics
- Nuclear physics
- Physics at DAΦNE
- Experimental techniques

Some papers reflect the tutorial level of the Lectures given at the Course.

We are strongly indebted to Prof. N.Cabibbo, President of I.N.F.N., for the financial support to the School. Acknowledgements are due to Miss L.Bonafini, Miss F.Gaudino and Mrs. C.Nuncibello for their precious support in organizing the School. Mr. R.Veronesi is acknowledged for his valuable technical assistance.

Thanks are due to the Folgaria Tourism Agency and to the Folgaria Meeting Committee, in particular to Dr. G.Dorigati, Mr. F.Demozzi and Miss A.Ciech for their assistance in the organization.

T. Bressani
S. Marcello
A. Zenoni

CONTENTS

II. SUBNUCLEAR PHYSICS

III. NUCLEAR PHYSICS

IV. PHYSICS AT DAΦNE

V. EXPERIMENTAL TECHNIQUES

I. THEORETICAL PHYSICS

Common Problems and Trends of Modern Physics, pp. 3–62
ed. T. Bressani, S. Marcello and A. Zenoni
©1992 World Scientific Publishing Company

THE RELATIONSHIP BETWEEN RELATIVISTIC QCD-LIKE FIELD THEORIES AND THE NON RELATIVISTIC QUARK MODEL

G.MORPURGO

Istituto di Fisica dell'Universita' di Genova

Istituto Nazionale di Fisica Nucleare-Sezione di Genova

Via Dodecaneso 33-16146 Genova (Italy)

ABSTRACT

Starting from a QCD or QCD-like Lagrangian, an exact parametrization in spin-flavor space of several physical quantities (baryon magnetic moments, hadron masses,....) is presented. Such a parametrization turns out to be unexpectedly simple; its structure is similar to that obtained in the non relativistic quark model.
A comparison between the exact general parametrization and the results of the non relativistic quark model clarifies the reason of the quantitative success of the latter in many cases. Moreover the parametrization explains why certain "classical" formulas, like Gell Mann-Okubo mass formula and Coleman-Glashow's one, work so well; it also suggests a new octet-decuplet mass formula which is satisfied to a part per thousand.

1. Introduction: The Problem

It is known since long time that the non relativistic quark model (NRQM)[1] gives a fair *quantitative* account of many hadron properties; think for instance of the baryon magnetic moments and masses, meson masses and mixing angles, the baryon electromagnetic transition matrix elements $(\underset{\sim}{10})\to(\underset{\sim}{8})+\gamma$, the meson $V\to P\gamma$ decays, the semileptonic decays of baryons and mesons. The above list includes only quantities relevant to the lowest "light" baryon and meson multiplets, but of course many more spectroscopical data both of light and heavy hadrons are explained by the model. *The problem to which these lectures are addressed is to clarify the reason for this quantitative agreement; in other words to understand why the NRQM works so well.* We confine to light hadrons, for two reasons: a) that the success of the model for light hadrons is generally regarded as misterious, whereas for charmonium or bottomonium the original[1a] (v/c)<1 justification is usually invoked;b)that in the lowest multiplets of light hadrons several physical quantities are known accurately for many members of a multiplet;as we shall see this helps in analyzing the reason of the agreement of the model with facts. These lectures will be structured as follows:I exemplify and illustrate in some detail the problem stated above on two specific cases,namely baryon magnetic moments and baryon masses[2a].Subsequently I will survey the applications of the method to many other properties of interest referring for the details to the original papers[2].

2.Two Examples

2.1)Magnetic Moments Of The Octet Baryons

The prediction of non relativistic quark model (NRQM)[1] for the magnetic moments of the baryons is compared with the data in table I.

The fit is based on assigning to the constituent quarks $\mathcal{P}$ and $\mathcal{N}$ the magnetic moments $+(2/3)\mu\underset{\sim}{\sigma}^{\mathcal{P}}$ and $-(1/3)\mu\underset{\sim}{\sigma}^{\mathcal{N}}$ (in the same ratio as the charges) and to the strange quark λ a magnetic moment $-(a/3)\mu\underset{\sim}{\sigma}^{\lambda}$ where a<1 may be due to the λ quark mass being larger than that of $\mathcal{P}$ and $\mathcal{N}$; the magnetic moment operator of a baryon is then written as the following two parameter expression:

$$\underset{\sim}{M}= \mu\sum\left[(2/3)\underset{\sim}{\sigma}^{\mathcal{P}} -(1/3)\underset{\sim}{\sigma}^{\mathcal{N}} -(1/3)\underset{\sim}{\sigma}^{\lambda}\right]+ \frac{A}{3}\sum\underset{\sim}{\sigma}^{\lambda} \qquad (1)$$

We used in Eq.1,instead of a,the parameter $A=\mu(1-a)$ to separate (last

TABLE I
The experimental values of the magnetic moments of the 8 baryons compared with the NRQM result (in Bohr proton magnetons)

	Experimental	Calculated NRQM
N	-1.913	-1.86
P	+2.793	input
Λ	-0.613±0.004	input
Σ(-)	-1.16±0.025	-1.04
Σ(+)	+2.42±0.05	+2.68
Ξ(-)	-0.65±0.03	-0.50
Ξ(o)	-1.25±0.014	-1.43
	$\lvert\mu(\Sigma\to\Lambda)\rvert$=1.61±0.09	$\mu(\Sigma\to\Lambda)$=-1.61

term) the effect of the flavor violation by the λ quark.To obtain the magnetic moments of the octet baryons one calculates the expectation value of $\underset{\sim}{M}$ using the wave functions of the NRQM; the spin-flavor factor of these functions has the SU_6 structure; this is a consequence of the orbital angular momentum L=0 and space symmetric wave function in the NRQM-not an additional assumption. For instance the spin-flavor wave function for a proton, W_p is:

$$W_p = NS[\alpha_1(\alpha_2\beta_3 - \alpha_3\beta_2)\mathcal{P}_1\mathcal{P}_2\mathcal{N}_3] \qquad (2)$$

where N is a normalization constant and S means symmetrization over 1,2,3; the spin-flavor wave function W_B for all octet baryons are listed in Ref.1b.

Clearly the magnetic moments calculated from (1) for octet baryons depend only from the two parameters μ and A (or, if one prefers μ and a); the fit in table I is with μ=2.79, a=0.65.

The problem is to compare Eq.1 [used in conjunction with the L=0 NRQM wave functions] with what one gets from a relativistic field theory of quarks and gluons.

A calculation of magnetic moments from field theory, say from QCD, is hard. Here we shall: a) Give a general procedure to parametrize in a relativistic field theory the results of calculations of physical quantities (e.g. magnetic moments or masses) in a way leading to "nuclear physics" formulas typical of the NRQM; b) Show that if the field theory satisfies two properties, these "general parametrizations" become rather simple; the number of terms in them is much smaller than one might have expected. The two properties are: 1) The electromagnetic current is carried only by the quarks; 2) The only flavor matrix λ in the strong Lagrangian is λ_8 (from the mass terms). These properties are satisfied by QCD, although they are more general than QCD. Note that for the above properties to hold it is necessary, for instance, that the pion field does not appear explicitely in the Lagrangian of the underlying field theory.

For a theory with the above properties, the most general expression of the octet baryon magnetic moments turns out to be a sum of 12 terms. If flavor breaking is taken into account only to first order, these reduce to 7;we get the following expression for the magnetic moments of octet baryons (in Eq.(3) S=strangeness,Q=charge, $\underset{\sim}{J}$=total angular momentum):

$$\underset{\sim}{M}=(\mu+KS)\sum\left[(2/3)\underset{\sim}{\sigma}^{\mathcal{P}}-(1/3)\underset{\sim}{\sigma}^{\mathcal{N}}-(1/3)\underset{\sim}{\sigma}^{\lambda}\right]+\frac{A}{3}\sum\underset{\sim}{\sigma}^{\lambda}+$$
$$+F(2\underset{\sim}{J})Q+H(2\underset{\sim}{J})S+\frac{L}{3}Q\sum\underset{\sim}{\sigma}^{\lambda}+G(2\underset{\sim}{J})QS \qquad (3)$$

A fit with (3) of the magnetic moments of the seven baryons $P,N,\Lambda,\Sigma^{+,-}$, $\Xi^{o,-}$ produces-(compare Sect.4.1)-the following values for μ,A,K,H,G,L (all in proton magnetons):μ=2.869,A=+1.005,K=+0.289,F=-0.076,H =+0.086, G=-0.15, L=-0.17.It appears that all the new terms F,K,H,G,L present in the general parametrization are smaller than A;this is the reason for the approximate validity of the NRQM formula (Eq.1) and shows that the agreement of the NRQM is not due to chance.At the same time it shows that there is no ground to expect Eq.1 to give a very accurate fit to the magnetic moments.

Another example of parametrization,starting from the basic field theory is that of the baryon masses.Also in this case (and in fact in all cases) it turns out that the NRQM is just a simplified version of the general parametrization obtained from field theory.For the baryon masses,our next example,the general parametrization is extremely simple,even simpler than the above one for magnetic moments.

2.2)Masses Of The Octet And Decuplet Baryons

Above I presented the general parametrization for magnetic moments correct only up to 1st-order in flavor breaking.Now,for the $\underset{\sim}{8}$ and $\underset{\sim}{10}$ masses,I will display the exact general parametrization correct to all orders in flavor breaking.It is:

$$\tilde{H}=M_o+B\sum_i P_i^{\lambda}+C\sum_{i>k}(\underset{\sim}{\sigma}_i\cdot\underset{\sim}{\sigma}_k)+D\sum_{i>k}(\underset{\sim}{\sigma}_i\cdot\underset{\sim}{\sigma}_k)(P_i^{\lambda}+P_k^{\lambda})+E\sum_{\substack{i\neq k\neq j\\(i>k)}}(\underset{\sim}{\sigma}_i\cdot\underset{\sim}{\sigma}_k)P_j^{\lambda}+$$
$$+a\sum_{i>k}P_i^{\lambda}P_k^{\lambda}+b\sum_{i>k}(\underset{\sim}{\sigma}_i\cdot\underset{\sim}{\sigma}_k)P_i^{\lambda}P_k^{\lambda}+c\sum_{\substack{i\neq k\neq j\\(i>k)}}(\underset{\sim}{\sigma}_i\cdot\underset{\sim}{\sigma}_k)(P_i^{\lambda}+P_k^{\lambda})P_j^{\lambda}+d\,P_1^{\lambda}P_2^{\lambda}P_3^{\lambda} \qquad (4)$$

Here M_o,B,C,D,E,a,b,c,d are 9 real parameters;the $\underset{\sim}{\sigma}_i$'s are spin Pauli matrices; and:

$$P^{\lambda}=(1-\lambda_8)/3 \qquad (5)$$

is the projection operator on the strange λ quark; P^{λ} is related to flavor breaking; if the mass of the strange quark were equal to that of the non strange ones, all P^{λ} terms would be absent in (4). The quark indices i,k,j in (4) and the sums extend from 1 to 3; the mass M_B of a baryon B is obtained taking the expectation value of $\tilde{H}$ (Eq.4) in the spin-flavor states W_B mentioned above (Eq.1).

$$M_B=\langle W_B|\tilde{H}|W_B\rangle \qquad (6)$$

We stress again that in spite of its non relativistic, NRQM like, appearance, Eqs.4 and 6 are exact, relativistic results; in fact they are the general result of a calculation based on relativistic field theory. The surprising simplicity of Eq.4, that is its rather small number of parameters (surprising precisely because Eq.4 is exact), is largely due to the fact that the underlying field theory is QCD-like, as we shall see later when deriving Eq.4 . We postpone, as stated, this derivation and also the detailed comparison of Eq.6 with the results of the NRQM; but, to exemplify the use of the general parametrization (Eq.4) we anticipate some applications.

1) Limiting to first order flavor breaking, that is neglecting in Eq.4 all terms containing more than one P^{λ} (compare however Sect.3.5 for the correspondence between order in flavor breaking and order in P^{λ}), we remain with five parameters M_0 ,B,C,D,E; because the $\underset{\sim}{8}$ and $\underset{\sim}{10}$ masses are eight, we obtain three relationships among the masses; as expected they are the Gell Mann-Okubo formula for the octet and the equal spacing rule for the decuplet:

$$(N+\Xi)/2=(3\Lambda+\Sigma)/4 \quad , \qquad \Omega-\Xi^*=\Xi^*-\Sigma^*=\Sigma^*-\Delta \qquad (7)$$

2) Keeping now also the terms quadratic in P^{λ}, but omitting the cubic term, three additional parameters, a,b,c enter; but, because a and b enter only in the combination (a+b) in the masses of Ξ,Ξ^*,Ω, (the only masses in which they appear), we remain with 7 parameters and still have one relationship among the eight masses. It is an old relation first written by Okubo[3] correct to second order in flavor breaking:

$$\Omega-\Delta=3(\Xi^*-\Sigma^*) \tag{8}$$

3)The general parametrization Eq.4 can also be approximated in an other way,in addition to the above ones (both based only on the order in P^λ);as we shall see,it may be reasonable to assume that the 3-quark terms quadratic in P^λ are negligible with respect to all other terms; thus we are going to neglect in (4) the terms that,besides being at least quadratic in P^λ,contain simultaneously the variables of three different quarks.This approximation is expressed by:

$$c=d=0 \tag{9}$$

With (9),we remain with 6 parameters and obtain,besides the Okubo formula (8), another new relationship that connects the $\underset{\sim}{8}$ and $\underset{\sim}{10}$ masses,the following:

$$(1/2)(N + \Xi) + T = (1/4)(3\Lambda + \Sigma) \tag{10}$$

with:

$$T=\Xi^*- (1/2)(\Omega + \Sigma^*) \tag{11}$$

This relationship is satisfied at such an accuracy that,before comparing it with the data,it is necessary to rewrite it so that it refers just to the strong part of the masses or,equivalently,that it is unaffected by the electromagnetic contributions.How to do this will be stated in Sect.4.3;at this stage we give only the result (now of course the charge labels appear explicitely;N means Neutron-not Nucleon):

$$(1/2)(N + \Xi^o) + T = (1/4)(3\Lambda + \hat{\Sigma}^+) \tag{12}$$

(1132.4±0.8) (1133.9±0.1)

where:

$$T=\Xi^{*-}- (1/2)(\Omega + \Sigma^{*-}) \tag{13}$$

and $\hat{\Sigma}^+$ stays for:

$$\hat{\Sigma}^+= 2\Sigma^+- \Sigma^o+ 2(N - P) \tag{14}$$

The agreement (shown by the numbers (in MeV) below left and right sides of Eq.(12)) is rather striking.For comparison,the standard Gell-Mann Okubo mass formula (also made free from electromagnetic effects) is the same as Eq.(12) without the T term;its two sides are (left)=1127.2±0.3 and (right)=1133.9±0.1.

Note that Eq.12 was obtained[4],except for electromagnetic effects, long time ago, from the NRQM, neglecting three quark potentials.

One final remark: the procedure that we shall illustrate for the magnetic moments or masses can be extended to other cases. It turns out that the treatment of many problems in the NRQM is just a simplified parametrization obtained omitting certain terms in the exact general parametrization obtained from field theory. In particular we conclude that: a) the language of the NRQM is reobtained quite generally from QCD-like relativistic field theories. This language does not imply that $(v/c)\ll 1$ or that $q\bar{q}$ virtual pairs and virtual gluons are neglected; all that matters is the existence of a correspondence between the exact state of the hadrons and the non relativistic states with a fixed number of quarks of an appropriately constructed "model" hamiltonian; b) As to the quantitative results of the NRQM (e.g. the fair fit (2) to the magnetic moments; or the above mass formula obtained neglecting three quark interactions), we will suggest (Sect.9) the reason why the terms that enter the NRQM description are, in the cases examined, more important than the neglected terms of the general parametrization.

In the next section the general parametrization will be derived focussing the attention on the magnetic moments.

3. The Parametrization Procedure

3.1. The Transformation V From The Model States To The Exact States

Because the procedure to derive from field theory the general parametrization is similar in all cases, it is convenient to illustrate it on the specific case of baryon magnetic moments.

Call H the exact Hamiltonian of the underlying relativistic field theory of quarks and gluons and let $|\Psi_B\rangle$ be its eigenstates for the octet baryons B ($B=P,N,\Lambda,\Sigma^{\pm,0},\Xi^{-,0}$) at rest; we thus have:

$$H|\Psi_B\rangle=M_B|\Psi_B\rangle \tag{15}$$

where M_B is the mass of the baryon. The problem is to calculate the expectation value in $|\Psi_B\rangle$ of the magnetic moment operator $\boldsymbol{\mathcal{M}}$:

$$\boldsymbol{\mathcal{M}} = \frac{1}{2}\int d^3\mathbf{r}\; \mathbf{j}(\mathbf{r})\times \mathbf{r} \tag{16}$$

For the current $\mathbf{j}(\mathbf{r})$ in (16) we assume the canonical form:

$$j_\mu(x) = e\left[\frac{2}{3}\,\bar{u}_R(x)\gamma_\mu u_R(x) - \frac{1}{3}\,\bar{d}_R(x)\gamma_\mu d_R(x) - \frac{1}{3}\,\bar{s}_R(x)\gamma_\mu s_R(x)\right] \tag{17}$$

where the fields $u_R(x), d_R(x), s_R(x)$ are the renormalized fields of the respective quarks [we omitted a color index on the quark fields and a sum over colors in the current because unimportant for the following]. In what follows the values of the renormalized (or "physical") masses of the quarks will not intervene explicitely, except to ensure that it is reasonable to treat to first order the flavor breaking (due to λ-$\mathcal{P}$ [or $\mathcal{N}$] mass difference) and except when comparing the results with the NRQM. But the mass renormalization implies that the "physical" (or better the "effective") masses of the quarks appearing in the quark propagator for calculations in the low k^2 regime are not the bare masses (governing the high momentum transfer regime and determined to be a few MeV); they must be identified with the constituent quark masses, possibly in the range of some hundreds of MeV for the light quarks; in fact we identify the constituent quarks $\mathcal{P}, \mathcal{N}$ and λ -the symbols used in (1)- with the fields of the renormalized quarks:

$$\mathcal{P} \equiv u_R(x), \qquad \mathcal{N} \equiv d_R(x), \qquad \lambda \equiv s_R(x) \tag{18}$$

The exact hamiltonian H is, as stated, flavor independent, except for the only breaking due to the mass difference between the λ quark renormalized mass $(m+\Delta m)$ and the common value (m) of the $\mathcal{P}, \mathcal{N}$ renormalized masses; thus <u>the only flavor breaking part of H</u> is:

$$m\int d^3\mathbf{r}\,\left[\bar{u}_R(x)u_R(x)+\bar{d}_R(x)d_R(x)+\bar{s}_R(x)s_R(x)\right] + \Delta m\int d^3\mathbf{r}\;\bar{s}_R(x)s_R(x) \tag{19}$$

The magnetic moment of a baryon is:

$$\mathbf{M}_B = \langle\Psi_B|\boldsymbol{\mathcal{M}}|\Psi_B\rangle \tag{20}$$

Below we will often omit the index B, restoring it when needed; to avoid confusion the index B will be used only to specify the baryon.

It is useful to read the equation (20) as follows: $\mathbf{M}$ is the expectation value of a simple operator [the current $j_\mu(x)$ in (16) has indeed the simple structure (17)] on the complicated state $|\Psi\rangle$. In fact the state $|\Psi\rangle$ in (15), being the exact state of a strongly interacting system of quarks and gluons, is extremely complicated; it can be displayed as a superposition of an infinite number of Fock states, starting with 3 quarks; it includes 4 quarks and one antiquark, 3 quarks plus one gluon and so on. Schematically:

$$|\Psi\rangle = |qqq\rangle + |qqqq\bar{q}\rangle + |qqqG\rangle + |qqqGG\rangle \ldots \qquad (21)$$

where the dots stay for the sum of an infinite number of additional states; the amplitudes that multiply each state in (21) (depending on momenta, spins, flavors, colors of the intervening quarks, antiquarks and gluons) have been left understood.

We introduce now an auxiliary hamiltonian $\mathcal{H}$, non relativistic or semirelativistic (by this we mean that the "kinetic" energy of a particle with mass m can be written $p^2/(2m)$ or $[(p^2+m^2)^{1/2} - m]$, or also differently) that operates, by construction, only in the 3-quark sector ; the operator $\mathcal{H}$ -to be called the "model hamiltonian"- has the only purpose of providing a set of 3-quark states that we call the 3-quark "model" states $|\phi_B\rangle$. In principle we can write the exact state $|\Psi_B\rangle$ in (15) as:

$$|\Psi_B\rangle = V|\phi_B\rangle \qquad (22)$$

where V is some very complicated unitary operator; again in principle V can be expressed in terms of H and $\mathcal{H}$ using, for instance, the adiabatic construction of the bound states[5] (compare for this the Appendix, Sect. 11). With (22) the expression (15) for $\underset{\sim}{M}_B$ becomes:

$$\underset{\sim}{M}_B = \langle\phi_B|V^{\dagger}\underset{\sim}{\mathcal{M}}\, V|\phi_B\rangle \qquad (23)$$

The difference between (15) and (23) is that in (15) $\underset{\sim}{\mathcal{M}}$ is simple but the states $|\Psi\rangle$ very complicated; in (23), on the other hand, the states $|\phi\rangle$ are simple and all the complications are transferred to $V^{\dagger}\underset{\sim}{\mathcal{M}}\, V$.

This has an advantage: $V^{\dagger}\underset{\sim}{\mathcal{M}}\, V$ is indeed a complicated field operator, but since it has to act only on the coordinates (space, spin, flavor, color) of the <u>three</u> quarks present in the state $|\phi\rangle$, it must be [after contraction of all the field operators] necessarily a function of these coordinates only. In the following the three quarks in $|\phi\rangle$ will be numbered 1,2,3. Thus, after the elimination of all the creation and destruction operators, $V^{\dagger}\underset{\sim}{\mathcal{M}}\, V$ in (23) behaves as a three body operator acting on 1,2,3.

In (15) $\boldsymbol{\mathcal{M}}$ transforms as an axial vector (since it is a magnetic moment) and the same is true for $V^{\dagger}\boldsymbol{\mathcal{M}}V$ in (23) because V is a rotationally invariant operator,being expressed in terms of H and $\mathcal{H}$ which are both rotationally invariant.As to $|\phi\rangle$ in (23),it depends on how we select the model Hamiltonian $\mathcal{H}$. The main point is that choosing $\mathcal{H}$ as the simplest most naive,most unrefined non relativistic quark model hamiltonian,the parametrization of $\langle\phi|V^{\dagger}\boldsymbol{\mathcal{M}}V|\phi\rangle$ will be seen to be very simple.We select $\mathcal{H}$ so that for the lowest octet (and decuplet) baryons the wave functions ϕ_B of the baryons states $|\phi_B\rangle$ have a non relativistic space-spin structure with the following properties:

1)ϕ_B is the product of a space part $X(\mathbf{r}_1,\mathbf{r}_2,\mathbf{r}_3)$ symmetrical in $\mathbf{r}_1,\mathbf{r}_2,\mathbf{r}_3$, times a spin-flavor part $W_B(1,2,3)$ times a colour singlet factor $S_c(1,2,3)$.

2)the space part has orbital momentum $\mathbf{L}$=0;thus $X\equiv X_{L=0}(\mathbf{r}_1,\mathbf{r}_2,\mathbf{r}_3)$.

Altogether for baryon B:

$$\phi_B = X_{L=0}(\mathbf{r}_1,\mathbf{r}_2,\mathbf{r}_3)\cdot W_B(1,2,3)\cdot S_c(1,2,3) \tag{24}$$

The assumption $\mathbf{L}$=0 and the symmetry of the space wave function imply automatically that the spin factors $W_B(1,2,3)$ of the wave function have the SU_6 structure.Note that $|\phi_B\rangle$,a |3q,no gluon state> in the Fock space,should be thought as written in terms of creation operators a_i^+ of the three quarks applied to the Fock vacuum of non interacting (mass renormalized) quarks;schematically this writing is,in a compact but obvious notation:

$$|\phi_B\rangle=\sum_{1,2,3}\phi_{B,L=0}(1,2,3)\ a_1^+a_2^+a_3^+|0\rangle \tag{25}$$

In the exact state $|\Psi\rangle$ (21) the |qqq> part includes all kinds of configuration mixing;whereas the model state $|\phi_B\rangle$ (25) has been chosen as a purely L=0 |qqq> state.In summary the V operator transforms the pure 3-quark model L=0 state $|\phi_B\rangle$ into the exact state $|\Psi_B\rangle$,a superposition of infinite Fock states of quarks, antiquarks and gluons; V includes configuration mixing. Of course V is constructed in a given frame (here the rest frame), but the results are fully relativistic if the Lagrangian of the underlying field theory is such.

The following additional remarks on the model Hamiltonian $\mathcal{H}$ -hence on the model wave functions ϕ - are in order:

a) Because the operator V in (22) is written in terms of creation and destruction operators of Dirac particles, the spinors appearing in $W_B(1,2,3)$ must be four component spinors; otherwise the operation of V on $|\phi_B\rangle$ would not be defined. This is achieved by simply completing the Pauli spinors of the NR wave function with two zeros in the lower components. This is entirely compatible with our non relativistic choice of the model hamiltonian $\mathcal{H}$. In fact $\mathcal{H}$ operates on two component Pauli spinors, but it is formally possible to enlarge the space of such spinors to that of four component spinors, provided that $\mathcal{H}$ is extended so that it does not connect the spaces of the upper and lower components and gives zero when operating on the latter.

b) the $X_{L=0}(\mathbf{r}_1,\mathbf{r}_2,\mathbf{r}_3)$ part of the wave function (24) does not carry the baryon index B; this is because we have chosen $\mathcal{H}$ to be flavor independent [in particular the masses of the quarks <u>in $\mathcal{H}$</u> are taken equal, all with the value $m+(\Delta m/2)$, in the notation of (19)]. All the flavor breaking is left to V: recall that V depends on the exact Hamiltonian H of the theory and H contains the flavor breaking term (19). We might have made a different choice, assuming $\mathcal{H}$ to be flavor dependent, but, at least for the problem at hand, it is convenient to proceed as indicated.

The operator $V^{\dagger}\mathbf{M}\,V$ has a role only when inserted between 3-quark states; this is because V in (22) contains $\mathcal{H}$ that was constructed specifically for the 3-quark states. But of course one can introduce auxiliary Hamiltonians for states different from the 3-quark ones, for instance the $q\bar{q}$ states of the non exotic mesons or the $qq\bar{q}\bar{q}$ states corresponding to exotic mesons (if these are bound); then V would operate also in sectors additional to the 3-quarks one.

Finally we wish to clarify in more detail why calculating the expectation value of the field operator $V^{\dagger}\mathcal{M}V$ in the baryon 3-quark state $|\phi_B\rangle$ amounts to calculate the expectation value of a certain quantum mechanical three body operator $\tilde{\mathcal{M}}$ on the wave function ϕ_B in <u>ordinary three body quantum mechanics</u>; this is so because, since we need only the expectation value of $V^{\dagger}\mathcal{M}V$ on a three quark-no gluon state $|\phi_B\rangle$, the only relevant part of $V^{\dagger}\mathcal{M}V$ is its projection in the $|3q,\text{no }G\rangle$ Fock sector:

$$\tilde{\mathcal{M}} = \sum_{3q,3q'}\sum |3q\rangle\langle 3q|V^{\dagger}\mathcal{M}V|3q'\rangle\langle 3q'| \qquad (26)$$

where the sums in (26) are on all the possible 3-quark, no gluon states. After normal ordering of all creation and destruction operators in $\tilde{\mathcal{M}}$ and their contraction with those in $\langle\phi_B|$ and $|\phi_B\rangle$, the operator $\tilde{\mathcal{M}}$ becomes a function only of the spin-flavor-space variables of the three quarks in ϕ_B; thus parametrizing $\tilde{\mathcal{M}}$ reduces to construct the most general 3-quark axial operator of the $\boldsymbol{\sigma}_i$'s, λ_i's and $\mathbf{r}_i$'s of the three quarks ($i=1,2,3$) including in it, of course, only those terms, that have a non vanishing expectation value in ϕ_B. Some steps in this construction will be given in the next section.

3.2. The Most General Three Body Operator $\tilde{\mathcal{M}}$

After contraction of all creation and destruction operators (as stated in the past section) the field operator $\tilde{\mathcal{M}}=V^{\dagger}\mathcal{M}V$ or, better, its projection in the 3-body sector $\tilde{\mathcal{M}} = \sum_{3q,3q'}\sum |3q\rangle\langle 3q|V^{\dagger}\mathcal{M}V|3q'\rangle\langle 3q'|$, is only a function of the space $\mathbf{r}_i$, spin $\boldsymbol{\sigma}_i$, flavor f_i and color operators of the three quarks (we shall, in the following, suppress the color variables); we must write the most general expression for such an operator, that we continue to call $\tilde{\mathcal{M}}$ [to be precise we should have adopted a different symbol because $\tilde{\mathcal{M}}$ (Eq. 26) operates in Fock space, whereas our present $\tilde{\mathbf{M}}$ acts in the conventional three particle space of quantum mechanics]. The number of independent operators that can be constructed in the spin-flavor space of three quarks is finite;we use for them,in general,the symbol $Y_\nu(\boldsymbol{\sigma},f)$ where the index ν specifies to which operator we refer.Thus *the most general operator of the space and spin-flavor variables has necessarily the form*:

$$\sum_\nu R_\nu(\mathbf{r},\mathbf{r}')Y_\nu(\boldsymbol{\sigma},f) \qquad (27)$$

where $R_\nu(\mathbf{r},\mathbf{r}')$ is an operator (not necessarily local) acting in the coordinate space $\mathbf{r}\equiv(\mathbf{r}_1,\mathbf{r}_2,\mathbf{r}_3)$ of the 3 quarks.Because to calculate a physical quantity such as a magnetic moment one has to take the expectation of (27) on the wave function $\phi_B = X_{L=0}(\mathbf{r}_1,\mathbf{r}_2,\mathbf{r}_3)\cdot W_B(1,2,3)$ (Eq.23),the factorizability of ϕ_B implies for the expectation value the following structure:

$$\sum_\nu \langle\phi_B(\mathbf{r})|\ R_\nu(\mathbf{r},\mathbf{r}')\,|\phi_B(\mathbf{r}')\rangle\langle W_B|Y_\nu(\boldsymbol{\sigma},f)\,|W_B\rangle \qquad (28)$$

that we can also rewrite as:

$$\sum_\nu g_\nu\langle W_B|Y_\nu(\boldsymbol{\sigma},f)\,|W_B\rangle \qquad (29)$$

having set:

$$g_\nu = \langle\phi_B(\mathbf{r})|\ R_\nu(\mathbf{r},\mathbf{r}')\,|\phi_B(\mathbf{r}')\rangle \qquad (30)$$

For the 8 and 10 baryons ϕ_B has,by construction,L=0;therefore the operators $R_\nu(\mathbf{r},\mathbf{r}')$ in (27) must be scalar operators;because the magnetic moment $\tilde{\mathbf{M}}$ is an axial vector,it follows that the $Y_\nu(\boldsymbol{\sigma},f)$ that enter in the parametrization of magnetic moments must be axial vector operators,to be constructed only in terms of the spins $\boldsymbol{\sigma}_i$ of the three quarks; we will indicate by $\mathbf{G}_\nu(\boldsymbol{\sigma},f)$ such axial vector operators;hence the parametrization of the magnetic moment operator $\tilde{\mathbf{M}}$ is written as:

$$\tilde{\mathbf{M}} = \sum_\nu g_\nu\,\mathbf{G}_\nu(\boldsymbol{\sigma},f) \qquad (31)$$

and the magnetic moments $\mathbf{M}_B$ are:

$$\mathbf{M}_B = \langle \phi_B | \tilde{\mathbf{M}} | \phi_B \rangle = \sum_\nu g_\nu \langle \phi_B | \mathbf{G}_\nu(\boldsymbol{\sigma}, f) | \phi_B \rangle \qquad (32)$$

Our next problem (to be treated in the next section) is to list all the axial vectors $\mathbf{G}_\nu(\boldsymbol{\sigma},f)$ formed with the spins and flavors of the three quarks.

Note incidentally that when we shall deal with the parametrization of the the masses, the $Y_\nu(\boldsymbol{\sigma},f)$'s in (27) will be scalars under rotations; indeed for the masses one must parametrize the expectation value of the Hamiltonian H, which is a scalar.

3.3. The Spin Dependence Of The Most General Axial Vector $\mathbf{G}_\nu(\boldsymbol{\sigma},f)$ Formed With The Spin And Flavors Of 3 Quarks

We prove in this section the following theorem: The most general hermitian axial operator $\mathbf{G}_\nu(\boldsymbol{\sigma},f)$ to be inserted in (31) constructed with the spins and flavors of 3 quarks has necessarily the following structure (for states of given J):

$$\mathbf{G}(\boldsymbol{\sigma},f) = \boldsymbol{\sigma}_1 \Gamma_a^J(f) + \boldsymbol{\sigma}_2 \Gamma_b^J(f) + \boldsymbol{\sigma}_3 \Gamma_c^J(f) \qquad (33)$$

where $\boldsymbol{\sigma}_1, \boldsymbol{\sigma}_2, \boldsymbol{\sigma}_3$ are the spins of the three quarks and $\Gamma_a^J(f), \Gamma_b^J(f)$, $\Gamma_c^J(f)$ are three functions of the flavor operators of the three quarks 1,2,3. Moreover because the W_B's are symmetrical in 1,2,3, the same must be true for the right hand side in the expression (33) of $\mathbf{G}(\boldsymbol{\sigma},f)$.

The proof is based on the reality of the model spin flavor functions W_B (and on the reality of flavor operators in a QCD-like theory).

Proof: The most general hermitian axial vector formed with the spins $\boldsymbol{\sigma}_1, \boldsymbol{\sigma}_2, \boldsymbol{\sigma}_3$ of three spin 1/2 particles and depending on the flavor operators f is given by the following expression (34) plus all the terms [possibly with different coefficients] obtained performing any permutation on 1,2,3.

$$\boldsymbol{\sigma}_1 [a(f) + b(f)(\boldsymbol{\sigma}_2 \cdot \boldsymbol{\sigma}_3)] + c(f)(\boldsymbol{\sigma}_1 \times \boldsymbol{\sigma}_2) + d(f)(\boldsymbol{\sigma}_1 \times \boldsymbol{\sigma}_3) \qquad (34)$$

In (34) a(f), b(f), c(f), d(f) are hermitian operators acting on the flavor variables and having real matrix elements between real functions [we call such operators "real"].

We will show that in calculating the expectation value of the operator (34) on a spin-flavor state with a real wave function and a given value of the total angular momentum J, a) cross product terms in (34) give no contribution, and b) the term $\sigma_1(\sigma_2\cdot\sigma_3)$ can be rewritten purely in terms of σ_1 and of the c-number J. It follows that the most general hermitian axial vector operator $G(\sigma,f)$ constructed in terms of the σ_i's of the three particles (and of the flavor operators f) is, when used for performing an expectation value as specified above, a combination of $\sigma_1 r^J_a(f)$, $\sigma_2 r^J_b(f)$, $\sigma_3 r^J_c(f)$ where $r^J(f)$ are three operators depending (for a given J) only on f.

To show this consider first the term $c(f)(\sigma_1\times\sigma_2)$ in (34); the same argument obviuosly applies to $d(f)(\sigma_1\times\sigma_3)$. It is sufficient to consider a given component of $c(f)(\sigma_1\times\sigma_2)$, for instance the z one and prove the above statement for $c(f)(\sigma_1\times\sigma_2)_z$. Consider any spin-flavor state represented by a real function R(1,2,3) of the spin-flavor variables; we will show that:

$$\langle R|(\sigma_1\times\sigma_2)_z c(f)|R\rangle=0 \tag{35}$$

The most general form of a real function of the 1,2,3 spin-flavor variables is:

$$R(1,2,3) = \alpha_1\beta_2R_a + \alpha_2\beta_1R_b + \alpha_1\alpha_2R_c + \beta_1\beta_2R_d \tag{36}$$

where R_a, R_b, R_c, R_d are four real functions expressed through α_3,β_3 [the spin basis for the third particle] and through the complete flavor basis for 1,2,3. Because it is $(\sigma_1\times\sigma_2)_z = 2i[\sigma_1^+\sigma_2^- - \sigma_1^-\sigma_2^+]$ the operator $(\sigma_1\times\sigma_2)_z$ gives zero when operating on $\alpha_1\alpha_2$ and on $\beta_1\beta_2$ terms in R (36). When applied to $\alpha_1\beta_2$ and to $\alpha_2\beta_1$ it is: $(\sigma_1\times\sigma_2)_z\alpha_1\beta_2 = -2i\beta_1\alpha_2$ and $(\sigma_1\times\sigma_2)_z\alpha_2\beta_1=+2i\beta_2\alpha_1$.

It therefore follows:

$$\langle R|(\sigma_1\times\sigma_2)_z c(f)|R\rangle = -2i\ \langle\alpha_1\beta_2R_a+\beta_1\alpha_2R_b|c(f)|\beta_1\alpha_2R_a-\alpha_1\beta_2R_b\rangle =$$
$$=-2i[\langle R_b|c(f)|R_a\rangle - \langle R_a|c(f)|R_b\rangle] = 0 \qquad (37)$$

The vanishing of (37) is due to the assumed reality of c(f) and to the fact that R_a and R_b are real functions:

$$\langle R_b|c(f)|R_a\rangle = \langle R_a|c(f)|R_b\rangle^* = \langle R_a|c(f)|R_b\rangle$$

It must be stressed that the above property (37) holds for the <u>expectation value</u> of $(\sigma_1\times\sigma_2)c(f)$ over a real function,but not necessarily for a non diagonal matrix element.

Note,incidentally [we will use this later],that also the scalar:

$$(\sigma_1\times\sigma_2)\cdot\sigma_3 \qquad (38)$$

(times any hermitian real flavor dependent operator) has a vanishing expectation value on any real spin-flavor state of three particles.Indeed:

$$(\sigma_1\times\sigma_2)\cdot\sigma_3=(\sigma_1\times\sigma_2)_z\cdot\sigma_{3z}+(\sigma_1\times\sigma_2)_x\cdot\sigma_{3x}+(\sigma_1\times\sigma_2)_y\cdot\sigma_{3y}\equiv$$
$$\equiv(\sigma_1\times\sigma_2)_z\cdot\sigma_{3z}+(\sigma_3\times\sigma_1)_z\cdot\sigma_{2z}+(\sigma_2\times\sigma_3)_z\cdot\sigma_{1z}$$

The expectation value of $(\sigma_i\times\sigma_k)_z\cdot\sigma_{jz}$ $(i\neq k\neq j)$ vanishes as it is seen using (37) in which c(f) is replaced by $\sigma_{jz}c(f)$,a real operator.

Consider next the operator $\sigma_1(\sigma_2\cdot\sigma_3)$ [multiplied in (34) with a real operator b(f)] and take its z component.It is:

$$\sigma_{1z}(\sigma_2\cdot\sigma_3) = \frac{1}{2}\sigma_{1z}[(\sigma_2+\sigma_3)^2 - 6]$$

On writing:

$$2J = \sigma_1+\sigma_2+\sigma_3$$

we obtain:

$$\sigma_{1z}(\sigma_2\cdot\sigma_3) = \frac{1}{4}\sigma_{1z}[(2J-\sigma_1)^2-6] + \frac{1}{4}[(2J-\sigma_1)^2-6]\sigma_{1z}$$

and with some algebra:

$$\sigma_{1z}(\sigma_2\cdot\sigma_3) = \frac{1}{4}[(4|J|^2-7)\sigma_{1z} + \sigma_{1z}(4|J|^2-7)] -2J_z \qquad (39)$$

In calculating the expectation value of $\sigma_{1z}(\sigma_2\cdot\sigma_3)$ on a state with a given J we can write $|J|^2=J(J+1)$,a c number.

In conclusion the most general axial vector operator [*as far as its expectation value on a real spin flavor state with a given J is concerned*] is:

$$\mathbf{G}(\boldsymbol{\sigma},f) = \boldsymbol{\sigma}_1\Gamma^J_a(f) \quad \text{or} \quad = \boldsymbol{\sigma}_2\Gamma^J_b(f) \quad \text{or} \quad = \boldsymbol{\sigma}_3\Gamma^J_c(f) \tag{40}$$

where the $\Gamma^J(f)$ are real flavor operators.What we have proven,essentially, is that the most general axial vector formed with three spin 1/2 particles [under the italicized condition written above] is a combination of $\boldsymbol{\sigma}_1,\boldsymbol{\sigma}_2,\boldsymbol{\sigma}_3$ and nothing else.One might wish to see more directly why the only axial vectors are those listed in (40) since,for instance,we can multiply $\boldsymbol{\sigma}_1$ by $(\boldsymbol{\sigma}_1\cdot\boldsymbol{\sigma}_2)$ or by other scalar products of the spin matrices,and remain with an axial vector. The answer appears from the equations (39),and (41),(42) below,where we have restricted to the z components (the x and y behave similarly):

$$\sigma_{1z}(\boldsymbol{\sigma}_1\cdot\boldsymbol{\sigma}_2) = \sigma_{2z} + i(\boldsymbol{\sigma}_1\times\boldsymbol{\sigma}_2)_z \tag{41}$$

$$\sigma_{1z}[\boldsymbol{\sigma}_1\cdot(\boldsymbol{\sigma}_2\times\boldsymbol{\sigma}_3)] = (\boldsymbol{\sigma}_2\times\boldsymbol{\sigma}_3)_z + i[\boldsymbol{\sigma}_1\times(\boldsymbol{\sigma}_2\times\boldsymbol{\sigma}_3)]_z = (\boldsymbol{\sigma}_2\times\boldsymbol{\sigma}_3)_z + \\ +i(\boldsymbol{\sigma}_1\cdot\boldsymbol{\sigma}_2)\sigma_{3z} - i(\boldsymbol{\sigma}_1\cdot\boldsymbol{\sigma}_3)\sigma_{2z} \tag{42}$$

On recalling that the expectation value of $(\boldsymbol{\sigma}_i\times\boldsymbol{\sigma}_k)$ gives zero,as shown above,the equations (41,42) [in conjunction with (39)] exemplify how the multiplication of $\boldsymbol{\sigma}_i$ by scalar products of $\boldsymbol{\sigma}$'s does not create new axial vectors in addition to (40) with non zero expectation value.

3.4.Flavor Dependence And A Complete List Of the $\mathbf{G}_\nu(\boldsymbol{\sigma},f)$'s

Having clarified the spin dependence of $\mathbf{G}_\nu(\boldsymbol{\sigma},f)$,we now determine the most general flavor operator $\Gamma^J(f)$ in (40).

Call $P^{\mathcal{P}},P^{\mathcal{N}},P^{\lambda}$ the projection operators for $\mathcal{P},\mathcal{N},\lambda$ quarks;in terms of the Gell Mann matrices λ_3 ,λ_8 (λ_3=Diag[1,-1,0] and λ_8=Diag[1,1,-2]), they are of course:

$$P^{\mathcal{P}} = \frac{1}{6}(2+3\lambda_3+\lambda_8) \quad , \quad P^{\mathcal{N}} = \frac{1}{6}(2-3\lambda_3+\lambda_8) \quad , \quad P^{\lambda} = \frac{1}{3}(1-\lambda_8) \tag{43}$$

With the above projection operators and calling Ψ the column symbol u_R,d_R,s_R the current density (17) takes the form:

$$j_\mu(x) = e\,\overline{\Psi}(x)\gamma_\mu Q\,\Psi(x) \tag{44}$$

where Q stays for the combination of projection operators:

$$Q= \frac{2}{3} P^{\mathcal{P}} - \frac{1}{3} P^{\mathcal{N}} - \frac{1}{3} P^{\lambda} \quad (45)$$

[Note that in Ref.2a we used the symbol P^q instead of Q].

The mass hamiltonian (19), breaking the flavor, is rewritten:

$$m \int d^3\underline{r}\ \overline{\Psi}(\underline{r})\Psi(\underline{r}) + \Delta m \int d^3\underline{r}\ \overline{\Psi}(\underline{r})P^{\lambda}\Psi(\underline{r}) \quad (46)$$

Consider now the operator $\tilde{\underline{\mathcal{M}}}$ given by (31). No matter how complex is the calculation leading, after contraction of the creation and destruction operators of the fields, to the expression of $\tilde{\underline{\mathcal{M}}}$ in terms of the three quark variables, the result of this calculation in a QCD-like theory must be linear in Q_i (where Q is the combination (45) of the projection operators and i is a quark index, i=1,2,3).

Indeed, if in the course of calculating $\tilde{\underline{\mathcal{M}}}$, we keep all the P^{λ} 's without simplifying Q_i with P_i^{λ} (compare the eq.(47) below), the Q appearing in the current (44) remains in the final result for $\tilde{\underline{\mathcal{M}}}$; this is due to the fact that in a QCD-like theory the only flavor operators in the Hamiltonian H are the P^{λ} 's (appearing in the mass-breaking part of (46)) and the P_i^{λ} commute with the Q_k for any choice of the indices i,k. Of course in simplifying the end result one will note that the product of Q_i and P_i^{λ} with the same index i is:

$$Q_i \cdot P_i^{\lambda} = -\frac{1}{3} P_i^{\lambda} \quad (47)$$

due to $P_i^{\mathcal{P}} \cdot P_i^{\lambda} = P_i^{\mathcal{N}} \cdot P_i^{\lambda} = 0$ and $P_i^{\lambda} \cdot P_i^{\lambda} = P_i^{\lambda}$; then, due to (47), Q can disappear if multiplied by a P^{λ} with the same quark index. But below we will list the possible flavor factors in the final expression of $\tilde{\underline{\mathcal{M}}}$ before the use of (47).

Thus for all octet baryons the operator $\Gamma(f)$ appearing in (40) must have necessarily one of the following forms [the indices i,k,j have values 1,2,3]:

$$Q_i \quad , \quad Q_i \cdot P_k^{\lambda} \quad , \quad Q_i \cdot P_k^{\lambda} \cdot P_j^{\lambda} \quad (48)$$

(If we were to consider the Ω^- -in the decuplet- we should add $Q_i \cdot P_1^{\lambda} \cdot P_2^{\lambda} \cdot P_3^{\lambda}$). We stress that in general the final flavor structures might be expected to include the Casimir flavor SU_3 operators, the values of which depend on the SU_3 representation; the QCD-like property of the theories considered here imply that the flavor matrices for strong and electromagnetic interactions all commute so that the final results are independent of the flavor Casimir operators; this simplifies the parametrization very considerably.

It follows that the most general operator $\underset{\sim}{G}_\nu(\underset{\sim}{\sigma}, f)$ in (40) for the octet (to be then inserted in (31,32)) is the product of an operator chosen from the basis $\underset{\sim}{\sigma}_1, \underset{\sim}{\sigma}_2, \underset{\sim}{\sigma}_3$ in spin space and an operator from the basis (48) in flavor space.

As already stated, because the spin-flavor functions $W_B(1,2,3)$ are symmetric in 1,2,3 for the lowest baryon octet and decuplet, each $\underset{\sim}{G}_\nu(\underset{\sim}{\sigma}, f)$ in (32) must be symmetric in 1,2,3. Thus all the $\underset{\sim}{G}_\nu(\underset{\sim}{\sigma}, f)$ are obtained taking one of the operators (48) [with the index i chosen to be, say, 1 and the other indices k, j different or equal among them and from 1 in all possible ways], multiplying this operator by $\underset{\sim}{\sigma}_1$ or $\underset{\sim}{\sigma}_2$ or $\underset{\sim}{\sigma}_3$ in all possible ways and symmetrizing. By doing so the following 13 different $\underset{\sim}{G}_\nu(\underset{\sim}{\sigma}, f)$ are obtained (we list them before the simplification (47)) [we have included also $\underset{\sim}{G}_{12}$ which is non-zero only for the Ω^-]:

$$\underset{\sim}{G}_1 = \sum_i Q_i \underset{\sim}{\sigma}_i \qquad \underset{\sim}{G}_2 = \sum_i Q_i P^\lambda_i \underset{\sim}{\sigma}_i \tag{49}$$

$$\underset{\sim}{G}_3 = \sum_{i\neq k} Q_i \underset{\sim}{\sigma}_k \quad , \quad \underset{\sim}{G}_4 = \sum_{i\neq k} Q_i P^\lambda_i \underset{\sim}{\sigma}_k \quad , \quad \underset{\sim}{G}_5 = \sum_{i\neq k} Q_i P^\lambda_k \underset{\sim}{\sigma}_k \ ,$$
$$\underset{\sim}{G}_6 = \sum_{i\neq k} Q_i \underset{\sim}{\sigma}_i P^\lambda_k \quad , \quad \underset{\sim}{G}_7 = \sum_{i\neq k\neq j} Q_i P^\lambda_k \underset{\sim}{\sigma}_j \tag{50}$$

$$\underset{\sim}{G}_{7b} = \sum_{i\neq k} Q_i P^\lambda_i \underset{\sim}{\sigma}_i P^k_\lambda \ , \quad \underset{\sim}{G}_8 = \sum_{i\neq k\neq j} Q_i P^\lambda_k P^\lambda_j \underset{\sim}{\sigma}_i \quad , \quad \underset{\sim}{G}_9 = \sum_{i\neq k\neq j} Q_i \underset{\sim}{\sigma}_k P^\lambda_k P^\lambda_j \quad ,$$
$$\underset{\sim}{G}_{10} = \sum_{i\neq k\neq j} Q_i P^\lambda_i \underset{\sim}{\sigma}_k P^\lambda_j \ , \quad \underset{\sim}{G}_{11} = \sum_{i\neq k\neq j} Q_i P^\lambda_i \underset{\sim}{\sigma}_i P^\lambda_j \quad , \quad \underset{\sim}{G}_{12} = \sum_{i\neq k\neq j} Q_i P^\lambda_i \underset{\sim}{\sigma}_i P^\lambda_k P^\lambda_j \tag{51}$$

[Note: The term G_{7b} was omitted in Ref.2a. This had no consequence because in Ref.2a we only used the terms linear in P^{λ}; here we call this term G_{7b} instead of G_8 so as to leave all other indices unchanged].

Nothing can be done -without additional assumptions- with so many terms, that is parameters; below we restrict to flavor violation to first order. In this case only the G_ν's from G_1 to G_7 (the vector notation for G_ν is often suppressed from now on) intervene; from the experimental values of the seven magnetic moments of the octet baryons we then determine their coefficients and compare the result with that given by the NRQM description.

First order flavor breaking means that among the above terms we keep only those linear, at most, in the P^{λ}'s. But, because $(P_i^{\lambda})^n = P_i^{\lambda}$, the terms linear in P^{λ} already include to all orders flavor breaking contributions additive in the quarks; only flavor breaking effects of second order and higher due to terms carrying the indices of two different quarks are left out (see next section).

The structure of (49)(50)(51) simplifies with the abbreviations:

$$\underset{\sim}{\Sigma}^q \equiv \sum_i Q_i \underset{\sim}{\sigma}_i \quad , \quad \underset{\sim}{\Sigma}^{\lambda} \equiv \frac{1}{3}\sum_i P_i^{\lambda} \underset{\sim}{\sigma}_i \qquad (52)$$

as well as the charge Q, the strangeness S, the total spin $2\underset{\sim}{J}$:

$$Q \equiv \Sigma_i Q_i \quad , \quad S \equiv -\Sigma_i P_i^{\lambda} \quad , \quad 2\underset{\sim}{J} \equiv \Sigma_i \underset{\sim}{\sigma}_i \qquad (53)$$

The first seven G_ν's ($\nu=1,\ldots 7$) in terms of the quantities (52)(53) are reported below; we have used the simplification (47):

$$G_1 = \underset{\sim}{\Sigma}^q$$

$$G_2 = -\underset{\sim}{\Sigma}^{\lambda}$$

$$G_3 = Q\cdot(2\underset{\sim}{J}) - \underset{\sim}{\Sigma}^q$$

$$G_4 = \frac{1}{3} S\cdot(2\underset{\sim}{J}) + \underset{\sim}{\Sigma}^{\lambda}$$

$$G_5 = 3Q\cdot\underset{\sim}{\Sigma}^{\lambda} + \underset{\sim}{\Sigma}^{\lambda}$$

$$G_6 = -S\cdot\underset{\sim}{\Sigma}^q + \underset{\sim}{\Sigma}^{\lambda}$$

$$G_7 = -Q\cdot S\cdot(2\underset{\sim}{J}) - \frac{1}{3} S\cdot(2\underset{\sim}{J}) - 3Q\cdot\underset{\sim}{\Sigma}^{\lambda} + S\cdot\underset{\sim}{\Sigma}^q - 2\,\underset{\sim}{\Sigma}^{\lambda}$$

G_{7b} to G_{12} can be expressed similarly; they contain additional powers of S, up to the third.

Each of the seven G_ν's linear in the P^λ's is a linear combination of products of the quantities (52) and (53) containing at most the first power of S *or* of $\underset{\sim}{\Sigma}^\lambda$. Therefore the most general form for the magnetic moment $\underset{\sim}{M}$ to first order in the flavor perturbation is:

$$\underset{\sim}{M}=\sum_{1}^{7}{}_{\nu}\, g_\nu \underset{\sim}{G}_\nu \equiv \mu\, \underset{\sim}{\Sigma}^q + A\, \underset{\sim}{\Sigma}^\lambda + FQ\cdot(2\underset{\sim}{J}) + H\ S\cdot(2\underset{\sim}{J}) + L\ Q\cdot\underset{\sim}{\Sigma}^\lambda + KS\cdot\underset{\sim}{\Sigma}^q + GQ\cdot S\cdot(2\underset{\sim}{J}) \qquad (54)$$

where $g_1, g_2, g_3 \ldots g_7$ or, equivalently, μ, A, F, H, L, K, G are seven parameters. Eq. (54) is the same as the Eq. (3) of Sect. 2.1, written with the symbols introduced in Eq. (52)

Before discussing on the basis of the parametrization (54) the magnetic moments, we comment (in the next section) on the correspondence between order in flavor breaking and order in P^λ.

3.5. Order In P^λ And Order In Flavor Breaking

Clearly the omission in the parametrization of terms quadratic in P^λ implies the neglect of certain terms of 2nd-order in flavor breaking because each P^λ is multiplied by Δm (Eq. (46)).

The question is: Does the neglect of the terms quadratic in P^λ imply neglecting *all* contributions of order $(\Delta m/m)^2$ or higher? Or, similarly, does the omission of the terms cubic in P^λ imply the neglect of *all* contributions of order $(\Delta m/m)^3$ or higher?

The answer is no, as already stated in Sect. 3.4. In fact, due to $(P_k^\lambda)^n = P_k^\lambda$, the terms linear in P^λ take into account flavor breaking corrections of any order in Δm, provided that these corrections are due to the same quark. Stated differently, in a calculation (of any quantity) based on Feynman diagrams, the sum of those infinite diagrams where the line of *a given quark* is affected by any number of "perturbations" Δm simply produces a term linear in P^λ.

In conclusion "1st-order flavor breaking" is not the same as "terms linear in P^λ"; sometimes however we may loosely continue to speak of 1st or 2nd-order flavor breaking, meaning terms linear or quadratic in P^λ.

4. Comparison of the parametrized magnetic moments with experiment

4.1. The Determination Of The Parameters In Eq.(54)

To use Eq.(54) we must have the expectation values of Σ_z^q, Σ_z^λ for the 7 baryons $P, N, \Lambda, \Sigma^{+,-}, \Xi^{o,-}$; they are given in Table II:

TABLE II

The expectation values of Σ_z^q and Σ_z^λ

	$\langle\Sigma_z^q\rangle$	$\langle\Sigma_z^\lambda\rangle$
P	+1	0
N	−2/3	0
Λ	−1/3	+1/3
Σ^+	+1	−1/9
Σ^-	−1/3	−1/9
Ξ^o	−2/3	+4/9
Ξ^-	−1/3	+4/9
	$\langle\Sigma^o\uparrow\vert\Sigma_z^q\vert\Lambda\uparrow\rangle = -1/\sqrt{3}$	$\langle\Sigma^o\uparrow\vert\Sigma_z^\lambda\vert\Lambda\uparrow\rangle = 0$

The last line in the table gives the matrix elements of Σ_z^q and Σ_z^λ for the transition $\Sigma^o\to\Lambda\gamma$ to be used in next section.

Using Σ_z^q, Σ_z^λ from table II, Eq.(54) leads to the following magnetic moments (we indicate by the baryon symbol its magnetic moment in proton magnetons):

$$P = \mu + F \qquad\qquad N = -\frac{2\mu}{3}$$

$$\Lambda = -\frac{\mu}{3} + \frac{A}{3} - H + \frac{K}{3} \tag{55}$$

$$\Sigma^+ = \mu - \frac{A}{9} + F - H - \frac{L}{9} - K - G \quad , \quad \Sigma^- = -\frac{\mu}{3} - \frac{A}{9} - F - H + \frac{L}{9} + \frac{K}{3} + G$$

$$\Xi^o = -\frac{2\mu}{3} + \frac{4A}{9} - 2H + \frac{4K}{3} \quad , \quad \Xi^- = -\frac{\mu}{3} + \frac{4A}{9} - F - 2H - \frac{4L}{9} + \frac{2K}{3} + 2G$$

P and N depend only on the two parameters μ and F; this remains true also if all flavor breaking terms are included. It follows (in proton magnetons):

$$\mu = 2.869 \qquad\qquad F = -0.076 \qquad (56)$$

The $\Lambda, \Sigma^{+,-}, \Xi^{o,-}$ determine the remaining five parameters; it is:

$$A = -\frac{3}{2}(\Xi^o + N) + \frac{9}{2}\Lambda - \frac{3}{4}(\Sigma^+ + \Sigma^-)$$

$$K = +\Xi^o - \frac{1}{2}N - \frac{3}{2}\Lambda - \frac{1}{4}(\Sigma^+ + \Sigma^-) \qquad (57)$$

$$H = -\frac{1}{6}(\Xi^o + N) - \frac{1}{3}(\Sigma^+ + \Sigma^-)$$

$$G = \frac{1}{2}\Lambda + \frac{5}{12}\Sigma^- - \frac{1}{4}\Sigma^+ - \frac{1}{2}\Xi^o + \frac{1}{6}\Xi^- + \frac{5}{6}(P + N)$$

$$L = -\frac{3}{2}(\Xi^- + \Xi^o) + \frac{3}{2}P + \frac{9}{4}\Sigma^- - \frac{3}{4}\Sigma^+ + \frac{9}{2}\Lambda$$

Inserting in (57) the values P= 2.793, N= − 1.913, Σ^+= 2.48 , Σ^- = − 1.16, Λ= − 0.61 , Ξ^o= − 1.25 , Ξ^- = − 0.65 [Note: Changing to Σ^+=2.42 (Table I) implies only a minor change in (58)] we obtain:

$$A=+1.005 \;, K= + 0.289 \;, H= + 0.086 \;, G= - 0.155 \;, L= - 0.175 \qquad (58)$$

We stress that these are, to 1st-order in P^λ, the exact values of the parameters A,K,H,G,L. The fact that the magnetic moments are seven, the same number as that of first order terms in P^λ, allows the data to determine completely the parameters. They are the same values that a full calculation by the exact relativistic theory, if it satisfies the QCD-like conditions stated in the introduction, must produce.

We may also record the values of the g_ν's used in the first expression of $\underset{\sim}{M}$ in Eq. (54); one has:

$$g_1=2.79 \;; g_2=-0.94 \;; g_3=-0.076 \;; g_4=0.41 \;; g_5=0.097 \;; g_6=-0.134 \;; g_7=0.155 \qquad (59)$$

We have listed the values of the g_ν's because the G_ν's, of which they are the coefficients, are normalized, by construction, more homogeneously than $\underset{\sim}{\Sigma}^q, \underset{\sim}{\Sigma}^\lambda$, $Q\cdot(2\underset{\sim}{J}) \ldots Q\cdot S\cdot(2\underset{\sim}{J})$ that multiply μ, A, F, H, K, G, L in the second form of (54); however below we will continue to refer mainly to μ, A, F..G, L. On writing:

$$A= \mu\,(1-a) \qquad (60)$$

where a is the usual parameter a= $m_{\mathcal{P}}/m_\lambda$ appearing in the NRQM, we get:

$$a = 0.649 \qquad (61)$$

Thus the value of a≈ 0.65 used to fit the NRQM description remains valid in the complete description, in spite of the many additional degrees of freedom. The next largest parameter in (54) is K; this is not introduced usually in the NRQM description; it behaves as a correction to the magnetic moment of the individual quark linear in the strangeness, that is, essentially, linear in the mass of the hadron in which the quarks are.

Each of the remaining parameters is rather smaller than A: $|G/A| \cong 0.15$, $|L/A| \cong 0.17$, $|H/A| \cong 0.085$. In some cases, however, G, L, H have an appreciable effect; for instance the value of the combination:

$$\Lambda + 2\Sigma^- + \Sigma^+ = G - 4H + \frac{1}{9} L - F = -0.44$$

depends essentially on the small value of $H = 0.086$.

In conclusion, though affected by many corrections, the NRQM description is verified by the exact parametrization, in the following sense: μ and A, the only parameters that the NRQM description introduces, are indeed the largest ones; all other parameters are at most 15% of μ.

Of course the question remains of understanding why the correct relativistic theory (say, QCD) leads to predictions approximately confirming (in the sense just stated) the assumption of additivity. We will come back to this question in Sect.9.

4.2. The $\Sigma^0 \to \Lambda\gamma$ Transition Rate

The parametrization of sect.3.3 for an axial vector holds only for an expectation value. For a non diagonal matrix element additional terms can enter. In fact $(\boldsymbol{\sigma}_1 \times \boldsymbol{\sigma}_2) c(f)$, where $c(f)$ is a real operator acting on the flavor variables only, can have a non diagonal matrix element. In calculating the $\Sigma^0 \to \Lambda\gamma$ rate operators of this form can contribute and increase the number of parameters. We want to show, however, that in calculating the $\Sigma^0 \to \Lambda\gamma$ M1 rate, correct to 1st-order in flavor breaking, no new parameter, in addition to those in Eq.(54) appears. This is seen as follows: by Wigner-Eckart theorem we can consider only the matrix element between the $\Lambda\uparrow$ and $\Sigma^0\uparrow$ states, that is:

$$\langle \Sigma^0 \uparrow | (\boldsymbol{\sigma}_1 \times \boldsymbol{\sigma}_2)_z c(f) | \Lambda^0 \uparrow \rangle \qquad (62)$$

This is purely imaginary: in fact $(\boldsymbol{\sigma}_1 \times \boldsymbol{\sigma}_2)_z = 2i\, (\sigma_1^+ \sigma_2^- - \sigma_2^- \sigma_1^+)$ and $c(f)$ is real; because $(\boldsymbol{\sigma}_1 \times \boldsymbol{\sigma}_2)_z c(f)$ is hermitian, its coefficient (call it D) must be real if it has to be a hermitian operator. Therefore the rate $\Sigma^0\to\Lambda\gamma$ turns out to be proportional to:

$$|\langle\Sigma^0\uparrow|M_z|\Lambda^0\uparrow\rangle|^2 + D^2|\langle\Sigma^0\uparrow|(\sigma_1\times\sigma_2)_z c(f)|\Lambda^0\uparrow\rangle|^2 \qquad (63)$$

where M_z is the operator (54). In the rate no interference terms appear between the M_z matrix element and the $D\langle\Sigma^0\uparrow|(\sigma_1\times\sigma_2)_z c(f)|\Lambda^0\uparrow\rangle$ one. Now consider the zero order terms in flavor breaking; this means to select c(f) in (62) simply as Q_i. Then the transition operator has the form:

$$(\sigma_i\times\sigma_k)_z\, Q_j \quad (i\neq k\neq j) \quad \text{or} \quad (\sigma_i\times\sigma_k)_z(Q_i - Q_k) \quad (i\neq k) \qquad (64)$$

But it is:

$$\langle\Sigma^0\uparrow|(\sigma_1\times\sigma_2)_z\, Q_3|\Lambda\uparrow\rangle=0 \quad \text{and} \quad \langle\Sigma^0\uparrow|(\sigma_1\times\sigma_2)_z\,(Q_1 - Q_2)|\Lambda\uparrow\rangle=0 \qquad (65)$$

as it can be checked directly.

Note that some flavor breaking matrix elements with the factor $(\sigma_1\times\sigma_2)_z$ do not vanish, as those obtained, for instance, choosing $c(f)=Q_1\cdot P_1^\lambda$ $(=-P_1^\lambda/3)$; however, in view of (63), they contribute to the rate at the second order in the flavor breaking. Thus in calculating the rate $\Sigma^0\to\Lambda\gamma$ correct to first order in flavor breaking, only the part with the operator M_z (54) contributes; moreover the coefficients μ, A, K....L in M_z have the same values deduced in sect. 4.1 from the (diagonal) magnetic moments, because Σ^0 and Λ have the same J=1/2 [if we had to calculate a matrix element between an octet state (J=1/2) and a decuplet state (J=3/2) the coefficients, in view of (39), would not be expected to be the same]. Therefore it is:

$$\langle\Sigma^0\uparrow|M_z|\Lambda^0\uparrow\rangle= (\mu + KS)\langle\Sigma^0\uparrow|\Sigma_z^q|\Lambda^0\uparrow\rangle + A\langle\Sigma^0\uparrow|\Sigma_z^\lambda|\Lambda^0\uparrow\rangle= \qquad (66)$$

$$= -(\mu - K)\frac{1}{\sqrt{3}} = -\frac{(2.869 - 0.289)}{\sqrt{3}} = -1.49 \qquad (|\text{Exp}|= 1.61\pm0.09)$$

where use has been made of the values in the last line of the table II.

Incidentally, writing the r.h.s. of $\langle\Sigma^0\uparrow|M_z|\Lambda^0\uparrow\rangle=-(\mu-K)/\sqrt{3}$ (Eq. (66)) in terms of the baryon magnetic moments via the Eqs. (55), we reobtain a relationship derived by Okubo[3] (using 1st-order T_3^3 flavor breaking) between the $\Sigma^0\to\Lambda^0\gamma$ M1-matrix element and the magnetic moments of the octet baryons. This is a check both of the procedure and the algebra.

5. The Baryon Masses

5.1. The Possible Terms In The Parametrization Of $\langle\phi|V^{\dagger}HV|\phi\rangle$

As stated at the beginning, we now discuss in some detail the parametrization of the baryon masses. Replacing in Eq. (20) the magnetic moment operator $\boldsymbol{\mathcal{M}}$ by the exact Hamiltonian H, the same procedure to parametrize the magnetic moments leads to a parametrization of the masses of the baryons for octet and decuplet. The difference is that instead of an axial vector $\boldsymbol{\mathcal{M}}$ we must now parametrize a scalar H (under space rotations). Again the factorizability (24) of the model states reduces the effective hamiltonian $\tilde{H}$ to a combination of spin flavor operators multiplied by some coefficients, the same for all the states of $\underline{8}$ and $\underline{10}$; because L=0, terms of type (Space Vector)·(Spin Vector) do not contribute to $\tilde{H}$. As to the spin flavor structures that intervene, we recall: 1) as shown in Sect. 3.3 (Eq. (38)) the expectation value of $\boldsymbol{\sigma}_1\cdot(\boldsymbol{\sigma}_2\times\boldsymbol{\sigma}_3)$ on a real spin flavor wave function vanishes; the same is true for $\boldsymbol{\sigma}_1\cdot(\boldsymbol{\sigma}_2\times\boldsymbol{\sigma}_3)\Gamma(f)$ where $\Gamma(f)$ is any real flavor operator. The only spin structures that can appear are thus:

$$1 \quad ; \quad (\boldsymbol{\sigma}_i\cdot\boldsymbol{\sigma}_k) \tag{67}$$

2) As to the flavor structures, the charge operator Q will now be absent, of course [the electromagnetic mass differences will be considered in Sect. 6]; the only flavor operator remains P^{λ}, and only products of P^{λ}'s can enter, up to a maximum of three P^{λ}'s.

Each spin flavor operator must be symmetrical in the coordinates of the three quarks (again because the spin-flavor structure of our model states is symmetrical); thus the following 9 independent operators intervene in the expression of the masses:

$$1\ ,\ \sum_i P_i^{\lambda}\ ,\ \sum_{i>k}(\boldsymbol{\sigma}_i\cdot\boldsymbol{\sigma}_k)\ ,\ \sum_{i>k}(\boldsymbol{\sigma}_i\cdot\boldsymbol{\sigma}_k)(P_i^{\lambda}+P_k^{\lambda})\ ,\ \sum_{\substack{i\neq k\neq j\\(i>k)}}(\boldsymbol{\sigma}_i\cdot\boldsymbol{\sigma}_k)P_j^{\lambda}$$

$$\sum_{i>k}P_i^{\lambda}P_k^{\lambda}\ ,\ \sum_{i>k}(\boldsymbol{\sigma}_i\cdot\boldsymbol{\sigma}_k)P_i^{\lambda}P_k^{\lambda}\ ,\sum_{\substack{i\neq k\neq j\\(i>k)}}(\boldsymbol{\sigma}_i\cdot\boldsymbol{\sigma}_k)P_j^{\lambda}(P_i^{\lambda}+P_k^{\lambda})\ ,\ P_1^{\lambda}P_2^{\lambda}P_3^{\lambda} \tag{68}$$

Differently from the magnetic moments we will consider here all terms, of any order in flavor breaking,so that the most general $\tilde{H}$ has the form (4) already written in Sect.2.2.To prove the results listed in Sect.2.2 we will first (next section) limit the discussion to the terms linear P^{λ}, and include in Sect.5.3 the terms quadratic and cubic in P^{λ}.

Before doing this a remark is appropriate:the terms entering in the parametrization (4) are the same for the expectation value of any quantity independent of the charge Q, scalar under rotations;they are obtained,in fact,using only the factorizability of the states and the list (68) of structures in spin flavor space.Thus the parametrization Eq.(4) is true not only for the hamiltonian H,but also for its square H^2,or to H^3 or more generally any function F(H) of H.This has two consequences:

1)It shows in general (not only for the masses) that the number of independent structures [for masses those in the list (68)] has to be always larger than or equal to the number of the states;if it were not so we would have rigorous relations between masses that would be true simultaneously for the masses,their squares or for any function F(M) of the masses;clearly this is impossible except for the uninteresting case of complete mass degeneracy.

Note that similar considerations hold for the magnetic moments; we might have parametrized $H\underset{\sim}{\mathcal{M}}H$ instead of $\underset{\sim}{\mathcal{M}}$,for instance,in a way formally identical to that used for $\underset{\sim}{\mathcal{M}}$ .We did not raise this point for magnetic moments only because the situation for masses is much more familiar.

2)If we introduce an approximation,for instance that of neglecting structures quadratic or cubic in P^{λ},then one can obtain relations between the masses (the Gell Mann-Okubo mass formula will be seen to be one of them);but these relations can equally be derived for any power or function of the masses. Of course these relations do not have all the same accuracy:If it is,say,a fair approximation to treat flavor breaking to first order when dealing with H,it is certainly less justified to do the same with $H^{10}=(H_0+H_1)^{10}$.What is in this sense the best function of H,such that the effect of the neglected flavor breaking terms is minimum,is hard to say,a priori [but is hard to see why the quadratic mass formulas should be better than the linear ones, a long debated question; this is some justification for the linear mass formulas]. Of course all the above remarks apply also to the conventional group derivation of, say, the Gell Mann-Okubo mass formula.

5.2. The Parametrization Of Baryon Masses Up To Terms Linear In P^λ And The Comparison In That Approximation With The NRQM Results

The most general effective Hamiltonian $\tilde{H}$, including only terms linear in P^λ (loosely we may say to first order flavor breaking and occasionally we will say so, but as we repeat, this is not correct, see sect. 3.5) is:

$$\tilde{H}(1)=M_o+B\sum_i P_i^\lambda+C\sum_{i>k}(\boldsymbol{\sigma}_i\cdot\boldsymbol{\sigma}_k)+D\sum_{i>k}(\boldsymbol{\sigma}_i\cdot\boldsymbol{\sigma}_k)(P_i^\lambda+P_k^\lambda)+E\sum_{\substack{i\neq k\neq j\\(i>k)}}(\boldsymbol{\sigma}_i\cdot\boldsymbol{\sigma}_k)P_j^\lambda \tag{69}$$

The 1 in $\tilde{H}(1)$ recalls that (69) is correct to 1st-order in P^λ. The expression (69) contains five parameters; because we must fit 4 octet and 4 decuplet masses, we get three (well known) mass relations.

We now write the masses of octet and decuplet baryons in terms of the coefficients M_o......E of the parametrization (69). Writing again S=strangeness = $-\Sigma_i P_i^\lambda$ and noting that:

$$\sum_{i>k}(\boldsymbol{\sigma}_i\cdot\boldsymbol{\sigma}_k) = \frac{1}{2}[4J(J+1)-9] \tag{70}$$

$$\sum_{\substack{i\neq k\neq j\\(i>k)}}(\boldsymbol{\sigma}_i\cdot\boldsymbol{\sigma}_k)P_j^\lambda = (\boldsymbol{\sigma}_1\cdot\boldsymbol{\sigma}_2)P_3^\lambda+(\boldsymbol{\sigma}_1\cdot\boldsymbol{\sigma}_3)P_2^\lambda+(\boldsymbol{\sigma}_2\cdot\boldsymbol{\sigma}_3)P_1^\lambda =$$

$$= -\frac{1}{2}[4J(J+1)-9]S-\sum_{i>k}(\boldsymbol{\sigma}_i\cdot\boldsymbol{\sigma}_k)(P_i^\lambda+P_k^\lambda) \tag{71}$$

$$\sum_{i>k}(\boldsymbol{\sigma}_i\cdot\boldsymbol{\sigma}_k)(P_i^\lambda+P_k^\lambda) = \sum_i(\boldsymbol{\sigma}_i P_i^\lambda\cdot 2\mathbf{J}) + 3S \tag{72}$$

we get:

$$M=[M_o-\frac{9}{2}C]+S\left[3D+\frac{3}{2}E-B\right]+\frac{C}{2}[4J(J+1)]+[D-E]\sum_i(\boldsymbol{\sigma}_i P_i^\lambda\cdot 2\mathbf{J})-\frac{E}{2}S[4J(J+1)] \tag{73}$$

Defining:

$$\tilde{M}_o \equiv M_o - \frac{9}{2}C \quad , \beta = 3D + \frac{3}{2}E - B \quad , \gamma = \frac{C}{2} \quad , \delta = D - E \quad , \epsilon = -\frac{E}{2} \tag{74}$$

the expression (73) becomes:

$$M = \tilde{M}_o + \beta S + \gamma[4J(J+1)] + \delta \sum_i (\underset{\sim}{\sigma}_i P_i^\lambda \cdot 2\underset{\sim}{J}) + \epsilon S[4J(J+1)] \tag{75}$$

Note now that:

$$\langle\uparrow|\sum_i (\underset{\sim}{\sigma}_i P_i^\lambda \cdot 2\underset{\sim}{J})|\uparrow\rangle_{J=1/2} = 3\cdot\langle\uparrow|\sum_i \sigma_{iz} P_i^\lambda|\uparrow\rangle \tag{76}$$

$$\langle\uparrow|\sum_i (\underset{\sim}{\sigma}_i P_i^\lambda \cdot 2\underset{\sim}{J})|\uparrow\rangle_{J=3/2} = 5\cdot\langle\uparrow|\sum_i \sigma_{iz} P_i^\lambda|\uparrow\rangle \tag{77}$$

where $|\uparrow\rangle$ means a state with J_z respectively 1/2 and 3/2. The values of $\langle\uparrow|\sum_i \sigma_{iz} P_i^\lambda|\uparrow\rangle$ are listed below for 8 and 10 baryons:

N	Λ	Σ	Ξ	Δ	Σ*	Ξ*	Ω
0	1	− 1/3	4/3	0	1	2	3

From (87) we get, therefore (baryon symbols stay for baryon masses):

$$N = \tilde{M}_o + 3\gamma$$

$$\Lambda = \tilde{M}_o - \beta + 3\gamma + 3\delta - 3\epsilon \tag{78}$$

$$\Sigma = \tilde{M}_o - \beta + 3\gamma - \delta - 3\epsilon$$

$$\Xi = \tilde{M}_o - 2\beta + 3\gamma + 4\delta - 6\epsilon$$

and:

$$\Delta = \tilde{M}_o + 15\gamma$$

$$\Sigma^* = \tilde{M}_o - \beta + 15\gamma + 5\delta - 15\epsilon \tag{79}$$

$$\Xi^* = \tilde{M}_o - 2\beta + 15\gamma + 10\delta - 30\epsilon$$

$$\Omega = \tilde{M}_o - 3\beta + 15\gamma + 15\delta - 45\epsilon$$

With 5 parameters and 8 masses we get 3 mass relations: the Gell Mann Okubo formula for the octet:

$$\underset{(1127)}{\frac{N + \Xi}{2}} = \underset{(1134)}{\frac{3\Lambda + \Sigma}{4}} \tag{80}$$

and the equal spacing formula for the decuplet:

$$\underset{139}{\Omega - \Xi^*} = \underset{149}{\Xi^* - \Sigma^*} = \underset{152}{\Sigma^* - \Delta} \tag{81}$$

In preparation for the comparison with the results of the NRQM we give below the values (in MeV) of the parameters $\tilde{M}_o, \beta, \gamma, \delta, \epsilon$ in (75) obtained from the expressions (78)(79);we also display M_o,B,C,D,E,the parameters of (73).It is:

$$\tilde{M}_o=865.8 \quad , \beta= -229.9 \quad , \gamma =24.4 \quad , \delta= -19.4 \quad , \epsilon= -2.7 \qquad (82)$$

or,alternatively:

$$M_o= 1085 \quad , \quad B=187 \quad , \quad C= 48.7 \quad , \quad D= -16 \quad , \quad E= 3.4 \qquad (83)$$

One can check (as already seen from the mass formulas (80)(81)) that the formula (75)[or (69)] represents the data to a few percent,in spite of the fact that only terms linear in P^λ have been taken into account.

We can compare these results with those from a non relativistic quark model;we do this for the version of NRQM due to De Rujula,Georgi and Glashow[6] where the interactions between quarks (and,in particular the spin dependent interactions) are the Fermi Breit approximation to the QCD one gluon exchange potential.The baryon mass formula to first order in flavor breaking,is given by Eq.(6) of Ref.6,which is:$M_{DGG}=A + +B\sum_i \frac{\Delta m_i}{m_{\mathcal{P}}} +C\sum_{i>j} \mathbf{s}_i\cdot\mathbf{s}_j[1-(\Delta m_i+\Delta m_j)/m_{\mathcal{P}}]$ (here A,B,C are the coefficients of De Rujula et al.,different from ours). Comparing that formula with our mass formula in its first form (69),we see that it becomes equal to ours (to all orders in $(\Delta m/m)$) if we suppress in (69) the parameter E :

$$E=0 \qquad (84)$$

and if D and C are related by:

$$\frac{D}{C} = -\frac{m_\lambda-m_{\mathcal{P}}}{m} \equiv -\frac{\Delta m}{m} \qquad (85)$$

where the m's are the quark masses.[In particular,using (75)(84) and the value (85) of D/C,our mass formulas (78),(79) are seen to satisfy the equations (5) (11) of DGG[6],namely:eq.(5), $m_{\mathcal{P}}/m_\lambda=2(\Sigma^*-\Sigma)/(2\Sigma^*+\Sigma-3\Lambda)$; eq.(11),$\Sigma-\Lambda = \frac{2}{3}[1-(m_{\mathcal{P}}/m_\lambda)][\Delta-N]$].

We see from (83) that the value of E,the only parameter outside the NRQM description with two body potentials,is indeed small, compared to the other parameters; moreover if E=0 [the best value of D changes then slightly from (83)] (-D/C) becomes $(16.4/48.7)\cong 0.34$;this number is in good agreement with $(1-a)\cong 0.35$ obtained (eq.(61)) from the magnetic moments;in a NRQM description 1-a can be interpreted as $\cong\Delta m/m$.

5.3. The 8 And 10 Masses Including Quadratic Terms In P^{λ}; A New Mass Formula

We will show here the following: Keeping in the parametrization all terms quadratic in P^{λ} except those containing the variables of 3 quarks (we will show why such terms are expected to be small), one obtains a mass formula that improves the conventional Gell Mann-Okubo formula; this formula has been already anticipated in Sect.2.2 (Eq.12))); it is:

$$(1/2)(N + \Xi^{o}) + T = (1/4)(3\Lambda + \hat{\Sigma}^{+}) \qquad (86)$$
$$(1132.4\pm0.8) \qquad (1133.9\pm0.1)$$

where:

$$T=\Xi^{*-}-(1/2)(\Omega + \Sigma^{*-}) = 5.2\pm0.7 \text{ MeV} \qquad (87)$$

and $\hat{\Sigma}^{+}$ means: $\hat{\Sigma}^{+}=2\Sigma^{+}-\Sigma^{o}+2(N-P)$; the charge specifications and the replacement of Σ by $\hat{\Sigma}$ are due to the fact that, at this accuracy, Eq.(86) must properly treat the electromagnetic contributions; the choice of charges in (86) is such that the electromagnetic contributions to the masses (calculated at zero order in flavor breaking) cancel in Eq.(86), as we shall see. The Gell Mann Okubo formula (also made free from electromagnetic effects) is Eq.(86) without the T term; its two sides are (left)=1127.2±0.3 and (right)=1133.9±0.1.

To prove (86) recall the list (68) in the general parametrization for the masses and keep terms quadratic in P^{λ}; then the 3 following terms add to $\tilde{H}(1)$, Eq.(69):

$$\tilde{H}(2)=\tilde{H}(1)+a\sum_{i>k} P_i^{\lambda}P_k^{\lambda} +b\sum_{i>k} (\sigma_i\cdot\sigma_k)P_i^{\lambda}P_k^{\lambda} +c\sum_{\substack{i\neq k\neq j\\(i>k)}} (\sigma_i\cdot\sigma_k)(P_i^{\lambda}+P_k^{\lambda})P_j^{\lambda} \qquad (88)$$

These new terms with coefficients a,b,c, clearly affect only baryons with strangeness 2 or 3 (Ξ, in the octet and Ξ^{*}, Ω in the decuplet); the other masses are unmodified. For Ω, Ξ^{*}, Ξ the corrections due to the terms a,b,c in (88) are:

$$\Omega:\ 3(a+b+c) \quad ; \quad \Xi^{*}:\ a+b+c \quad ; \quad \Xi:\ a+b-2c \qquad (89)$$

It is clear from (89) that the equal spacing rule (81) for the decuplet no longer holds but that Okubo's relationship $\Omega-\Delta=3(\Xi^{*}-\Sigma^{*})$ remains true. Clearly with 8 masses and 7 parameters (note that a and b enter only combined as a+b) only one relationship can exist. If we included

3rd-order flavor breaking (affecting only the Ω), no relationship would remain, obviously.

Assume now that the three quark term in Eq.(88) is absent (c=0); that this approximation may be a reasonable one is suggested by the fact stated above, that the three quark term contribution (coefficient E) is already very small at 1st-order flavor breaking and the c term in (88) is expected to be further depressed being of next order in $\Delta m/m \simeq 0.34$, the flavor breaking parameter. In this approximation only the parameter (a+b) remains at 2nd-order in flavor breaking; all 8 and 10 baryon masses are then summarized by Eq.(90) below (S=strangeness):

$$M(2)=M(1)+(1/2)(a+b)(S^2+S)\equiv M(1)-T(S^2+S) \qquad (90)$$

Because we have altogether 6 parameters, two relationships now exist; they are that of Eq.(86) with $T\equiv -(a+b)/2$ (expressed in terms of the masses by Eq.(87)) and the 2nd-order Okubo equation $\Omega-\Delta^-=3(\Xi^{*-}-\Sigma^{*-})$, now rewritten with explicit charges so as to be free of electromagnetic effects–see below. Using the latter equation we can write T also as:

$$T=\Sigma^{*-}-(1/2)(\Delta^-+\Xi^{*-}) \qquad (91)$$

We prefer to evaluate T from (87) because the experimental errors on Δ's are larger than those of the other masses of the decuplet; for $\Delta^-=1233\pm\sigma$, Eq.(91) would give $T=3.2\pm0.4\pm\sigma/2$ instead of 5.2 ± 0.7 of Eq.(13); but σ is at least 3 and a more precise test of Okubo's 2nd-order formula, the equality of the T's of Eqs.(87) and (91), needs a good knowledge of the Δ masses.

Another remark: we called "new" the mass formula (86); indeed this formula, including the electromagnetic corrections, is new, as far as we know. But leaving aside the electromagnetic corrections, Eq.(86) is, as we stated, very old. A set of mass relations that, when combined linearly, lead to Eq.(86) was derived in 1966 by Federman, Rubinstein and Talmi[4] using the NRQM, neglecting three quark interactions. The new feature here is that Eq.(86) is now derived by the general parametrization; thus its validity does not depend on the potential model description. In fact we have here a new case where the general parametrization explains the quantitative success of the NRQM.

Finally we discuss electromagnetic effects. Eq.(88) parametrizes the "strong" part of the 8 and 10 masses. But the measured masses include an electromagnetic contribution δM; to obtain the strong masses one must subtract from the measured masses the calculated electromagnetic contribution. This calculation will be done in Sect.6; calling δB the electromagnetic contribution to the mass of baryon B (at zero order in flavor breaking), one can check (compare below) that in Eq.(86) the sum of the electromagnetic corrections $1/2(\delta N+\delta\Xi^{o})$ on the left is equal to the same sum on the right $(1/4)(3\delta\Lambda+\delta\hat{\Sigma}^{+})$; also (with the charge values indicated in T, Eq.(87)) it is $\delta T=0$; thus Eq.(86) written with the charges indicated and $\hat{\Sigma}^{+}$ defined above, is independent, as stated, of electromagnetic corrections at zero order in flavor breaking.

Though electromagnetic masses will be dealt in Sect.6, we sketch how their effects cancel in Eq.(86). The δB's of the 8-baryons can be seen to contain 4 parameters μ,ν,η,ρ. It is: $\delta P=\mu+(5/9)\nu+\eta+\rho$; $\delta N=(2/3)\mu$; $\delta\Lambda=(2/3)\mu+(1/9)\nu$; $\delta\Sigma^{+}=\mu+(5/9)\nu+\eta+\rho$; $\delta\Sigma^{o}=(2/3)\mu+(1/3)\nu$; $\delta\Sigma^{-}=(1/3)\mu+ +(1/9)\nu+\eta+(1/3)\rho$; $\delta\Xi^{o}=(2/3)\mu$; $\delta\Xi^{-}=(1/3)\mu+(1/9)\nu+\eta+(1/3)\rho$. One checks that: $(1/2)(\delta N+\delta\Xi^{o})=(1/4)(3\delta\Lambda+\delta\hat{\Sigma}^{+})$ where $\delta\hat{\Sigma}^{+}\equiv2\delta\Sigma^{+}-\delta\Sigma^{o}+2(\delta N-\delta P)$. The Coleman Glashow relationship $\delta N-\delta P+\delta\Xi^{-}-\delta\Xi^{o}=\delta\Sigma^{-}-\delta\Sigma^{+}$ is obviously satisfied by the above expressions of δM. As to the decuplet we find $\delta\Omega=\delta\Sigma^{-*}=\delta\Xi^{-*}=\delta\Delta^{-}$; so it is $\delta T=0$ and Okubo's decuplet formula, written as $\Omega-\Delta^{-}=3(\Xi^{*-}-\Sigma^{*-})$, is unaffected by electromagnetic effects; we repeat that the above electromagnetic δB's are all parametrized only to zero order flavor breaking but this should be sufficient at this stage.

6. More On The Baryon Masses: The Electromagnetic Masses

6.1. Parametrization Of The Electromagnetic Masses-The Coleman-Glashow Formula

Up to now I gave many details on the procedure so as to clarify the basis and technique of the general parametrization method. From now on I will more and more tend to state the results, referring to the original papers[2] for details. In this and next section and I will deal with the parametrization of the electromagnetic masses of the baryons[2f].

By a procedure identical to that illustrated for the magnetic moments and the strong masses one can write the general parametrization of the electromagnetic masses of the $\underset{\sim}{8}$ and $\underset{\sim}{10}$ baryons; once more the electromagnetic mass of a baryon B is the expectation value on the spin-flavor functions W_B of the spin-flavor operator:

$$\tilde{\Omega} = \sum_{\nu} t_{\nu}\, \Gamma_{\nu}(s,f) \tag{92}$$

where t_{ν} are real parameters. The spin flavor operator $\tilde{\Omega}$ associated to the electromagnetic masses must be, of course, rotation invariant; therefore the $\Gamma_{\nu}(s,f)$'s in (92) are the spin scalars (constructed with the Pauli spins of the quarks) already displayed (Eq.(67)) in relation with the strong masses. As to the flavor dependence, all $\Gamma_{\nu}(s,f)$'s must be quadratic in the Q_i's ($Q_i \equiv (2/3)P_i^{\mathcal{P}} - (1/3)P_i^{\mathcal{N}} - (1/3)P_i^{\lambda}$) because the electromagnetic mass is quadratic in the current density which is linear in Q_i; the $\Gamma_{\nu}(s,f)$'s will, moreover, contain a number of factors P_i^{λ} up to three, related to the order of flavor breaking up to which we parametrize. The list of all possible structures in spin-flavor space is given in Ref.2f. As an example, the list contains, at zero order in flavor breaking, six Γ_{ν}'s; after some algebra they can be written as follows:

$$\Gamma_1(8)=\Gamma_1(10)=\sum_{1}^{3}{}_i\, Q_i^2 \equiv \Gamma_1 \tag{93a}$$

$$\Gamma_2(8)=-(3/2)\Gamma_1+(3/2)\sum_{1}^{3}{}_i\, Q_i^2\sigma_{iz} \quad ; \quad \Gamma_2(10) = \sum_{1}^{3}{}_i\, Q_i^2 \tag{93b}$$

$$\Gamma_3(8)= -3\sum_{1}^{3}{}_i\, Q_i^2\sigma_{iz} \quad ; \quad \Gamma_3(10)= \sum_{1}^{3}{}_i\, Q_i^2 \tag{93c}$$

$$\Gamma_4(8)=\Gamma_4(10)=(1/2)(Q^2-\sum_{1}^{3}{}_i\, Q_i^2) \equiv \Gamma_4 \tag{93d}$$

$$\Gamma_5(8)= -(3/2)(Q^2-\sum_{1}^{3}{}_i\, Q_i^2)+3(Q\sum_{1}^{3}{}_i\, Q_i\sigma_{iz}-\sum_{1}^{3}{}_i\, Q_i^2\sigma_{iz}) \quad ; \quad \Gamma_5(10)=\Gamma_4 \tag{93e}$$

$$\Gamma_6(8) = -3\Gamma_5(8)-3\Gamma_4 \quad ; \quad \Gamma_6(10)= 2\Gamma_4 \tag{93f}$$

The above six Γ_{ν}'s are expressible (compare the Eqs.93) in terms of the four quantities $\Sigma[Q_i^2]$, Q^2, $\Sigma[Q_i^2\sigma_{iz}]$, $Q\cdot\Sigma[Q_i\sigma_{iz}]$; thus only four parameters (those called μ,ν,η,ρ at the end of the past Section) intervene at zero order in flavor breaking. In terms of such parameters we obtain the electromagnetic masses of the baryons already written at the end of the past Section; as stated there, the Coleman–Glashow formula

$$\delta P-\delta N = \delta\Sigma^+-\delta\Sigma^-+\delta\Xi^--\delta\Xi^o \tag{94}$$

$$1.29 \qquad\qquad 1.67\pm0.6 \text{ MeV}$$

(at zero order in flavor breaking) is derived immediately.

6.3. The Gal-Scheck Relationships

Using the non relativistic quark model, Gal and Scheck[7] derived long ago a set of relationships between the electromagnetic masses both for mesons and for baryons. Whereas for mesons the assumptions of Gal and Scheck are very restrictive, for baryons these assumptions amounted mostly to the neglect of three body electromagnetic interaction terms. With that neglect the NRQM leads to the following relationships that involve both the octet and the decuplet electromagnetic masses:

$$\delta\Delta^{++}-\delta\Delta^o = 2(\delta P-\delta N) + (\delta\Sigma^+-\delta\Sigma^o) + (\delta\Sigma^--\delta\Sigma^o) \tag{95}$$

$$\delta\Delta^{++}-\delta\Delta^- = 3(\delta P-\delta N) \tag{96}$$

$$\delta\Delta^+-\delta\Delta^o = \delta P-\delta N \tag{97}$$

$$(\delta\Sigma^{*+}-\delta\Sigma^{*-})+(\delta\Xi^{*-}-\delta\Xi^{*o}) = \delta P-\delta N \tag{98}$$

$$-1.2\pm0.9 \qquad\qquad -1.29 \text{ MeV}$$

$$\frac{1}{2}(\delta\Sigma^{*+}+\delta\Sigma^{*-})-\delta\Sigma^{*o} = \frac{1}{2}(\delta\Sigma^++\delta\Sigma^-)-\delta\Sigma^o \tag{99}$$

$$1.3\pm1.2 \qquad\qquad 0.85\pm0.12 \text{ MeV}$$

We did not given numbers for Eqs.(95–97) because the experimental uncertainties on the masses of Δ's are still large.

To show that the formulas above are generally derivable from a QCD-like theory (not just consequences of the non relativistic quark model) one has to calculate the electromagnetic masses of the baryons in terms of the parameters of the general parametrization. As shown in Ref.2f there are (at zero order flavor breaking) 6 parameters in play (a,b,c,d,e,f in Ref.2f) that reduce to four if three-quark terms are neglected (then c=f=0); by elimination of these four parameters the

Eqs.(95-99) are obtained.

But there is more: It is possible to show that flavor breaking does not modify the e.m. mass relationships, provided that 3-quark terms are neglected [3-quark terms in the parametrization are defined as terms that depend on the indices of three different quarks]. This means that, excluding 3-quark terms, both the Coleman Glashow and the octet-decuplet relationships are not modified by flavor breaking. That is the Coleman Glashow formula and the formulas (95) to (99) can be shown to be correct to any order in flavor breaking provided that three quark terms are omitted. A distinction however has to be made: At zero order flavor breaking the Coleman Glashow relationship is true including all terms, in particular 3-quark terms, while in the formulas (95-99) 3-quark terms must be omitted already at zero order.

We end with the following remarks:

1) That the Coleman Glashow relationship is satisfied "too" well has been known since the formula was written; that this relationship is valid (barring 3-quark terms) with inclusion of all flavor breaking corrections makes this fact more understandable. It would be of great interest to reduce the experimental error on the mass of Ξ^{0}, which is the reason for the error ±0.6 MeV on the right hand side of (94); the other masses there are known at ±0.13 MeV or less.

2) The general parametrization method allows once more to give a firm basis to the quantitative success of a NRQM calculation, as that of Gal and Scheck for the electromagnetic masses of the 10 and 8 baryons; again the basis of this success is simply that the parametrization given by the NRQM is quite similar to the parametrization that is obtained in a full calculation based on a QCD-like field theory.

3) As to the quantitative agreement of the relations (98,99) with the data (1.2±0.9 to be compared with 1.29; and 1.3±1.2 to be compared with 0.85±0.12), this confirms that 3-quark terms are negligible; this confirmation adds to that provided (Eq.(86)) by the modified Gell Mann Okubo mass formula, based on the same assumption, and satisfied, I recall, at the level of 1132.4±0.8 compared to 1133.9±0.1. That the above verification of the relations (98,99) is now possible is due to the decrease, during the years, of the experimental errors on the decuplet

masses and suggests, if feasible, further efforts in this direction; although here the difficulties (also of how the masses are defined), particularly for very broad resonances as the Δ's, are evident.

7. Parametrization of The Semileptonic Baryon Matrix Elements

7.1. The Matrix Elements In The No Flavor Breaking Approximation

We now apply the method to the semileptonic matrix elements of the baryons[2b]. We will see that: 1) In the no flavor breaking approximation the general parametrization is of course equivalent to Cabibbo's parametrization; it contains only 2 parameters (equivalent to D and F); 2) In the NRQM treatment the smaller of these parameters is absent, leading to D/F=3/2; 3) A byproduct of the present treatment is to clarify (once more, compare Ref. 1b, p. 126) why the frequent statement that the NRQM produces a value 5/3 of g_A/g_V for the nucleon is wrong.

With the weak current given of course, in terms of the quark fields $\Psi(x)$, by:

$$j^+_\mu(x) = \cos\theta\cdot\bar{\Psi}(x)\tau^+\gamma_\mu(1+\gamma_5)\Psi(x) + \sin\theta\cdot\bar{\Psi}(x)\lambda^+\gamma_\mu(1+\gamma_5)\Psi(x) \qquad (100)$$

where θ is the Cabibbo angle, $\tau^+\equiv \lambda_1+i\lambda_2$ and $\lambda^+\equiv\lambda_4+i\lambda_5$, we must parametrize the matrix elements of $\int j^+_\mu(\underset{\sim}{x},0)\exp(i\Delta\underset{\sim}{P}\cdot\underset{\sim}{x})d^3\underset{\sim}{x}$. We do this in the limit of four momentum transfer $\Delta P_\mu=0$; in particular, the initial and final baryons A and B are both at rest; the parametrization will be general and relativistic, though non covariant, exactly as that of magnetic moments. In an obvious notation the only non vanishing matrix elements (with $\Delta P_\mu=0$) are:

$$\langle 1\rangle_{\Delta S=0} = \langle B\uparrow|\int\bar{\Psi}(\underset{\sim}{x},0)\tau^+\gamma_4\Psi(\underset{\sim}{x},0)d^3\underset{\sim}{x}|A\uparrow\rangle \equiv \langle B\uparrow|T^+|A\uparrow\rangle \qquad (101)$$

$$\langle 1\rangle_{\Delta S=1} = \langle B\uparrow|\int\bar{\Psi}(\underset{\sim}{x},0)\lambda^+\gamma_4\Psi(\underset{\sim}{x},0)d^3\underset{\sim}{x}|A\uparrow\rangle \equiv \langle B\uparrow|L^+|A\uparrow\rangle \qquad (102)$$

and:

$$\langle\sigma_z\rangle_{\Delta S=0} = \langle B\uparrow|\int\bar{\Psi}(\underset{\sim}{x},0)\tau^+\gamma_4\sigma_z\Psi(\underset{\sim}{x},0)d^3\underset{\sim}{x}|A\uparrow\rangle \qquad (103)$$

$$\langle\sigma_z\rangle_{\Delta S=1} = \langle B\uparrow|\int\bar{\Psi}(\underset{\sim}{x},0)\lambda^+\gamma_4\sigma_z\Psi(\underset{\sim}{x},0)d^3\underset{\sim}{x}|A\uparrow\rangle \qquad (104)$$

Expressing once more the exact states $|A\rangle$ and $|B\rangle$ as:

$$|A\rangle = V|\phi_A\rangle \quad , \quad |B\rangle = V|\phi_B\rangle \qquad (105)$$

we obtain for Fermi matrix elements:

$$\langle 1\rangle_{\Delta S=0} = \langle \phi_B\uparrow|V^\dagger T^+ V|\phi_A\uparrow\rangle = \langle \phi_B\uparrow|\sum_{i=1}^{3} \tau_i^+|\phi_A\uparrow\rangle \tag{106}$$

where the last equality is due to the fact that V commutes with T^+ (because of isospin conservation) and $V^\dagger V=1$; for $\langle 1\rangle_{\Delta S=1}$ we obtain:

$$\langle 1\rangle_{\Delta S=1} = \langle \phi_B\uparrow|V^\dagger L^+ V|\phi_A\uparrow\rangle = \langle \phi_B\uparrow|\sum_{i=1}^{3} \lambda_i^+|\phi_A\uparrow\rangle + O[(\Delta m/m_\lambda)^2] \tag{107}$$

where the last step is due to the Ademollo Gatto theorem (Fermi matrix elements receive only 2nd-order flavor breaking corrections because L^+ is a generator of flavor symmetry).

We now parametrize the Gamow-Teller $\langle\sigma_z\rangle$ matrix elements; from (103-107) we have:

$$\langle\sigma_z\rangle_{\Delta S=0} = \langle \phi_B\uparrow|V^+\int\overline{\Psi}(\underset{\sim}{x},0)\tau^+\gamma_4\sigma_z\Psi(\underset{\sim}{x},0)d^3\underset{\sim}{x}\ V|\phi_A\uparrow\rangle \equiv \langle \phi_B\uparrow|\Gamma_z^+|\phi_A\uparrow\rangle \tag{108}$$

$$\langle\sigma_z\rangle_{\Delta S=1} = \langle \phi_B\uparrow|V^+\int\overline{\Psi}(\underset{\sim}{x},0)\lambda^+\gamma_4\sigma_z\Psi(\underset{\sim}{x},0)d^3\underset{\sim}{x}\ V|\phi_A\uparrow\rangle \equiv \langle \phi_B\uparrow|\Omega_z^+|\phi_A\uparrow\rangle \tag{109}$$

Once more the main point of the procedure is that after normal ordering and contraction of all the quark and gluon creation and destruction operators, $\underset{\sim}{\Gamma}^+$ and $\underset{\sim}{\Omega}^+$ defined in (108,109) are simple 3 body operators, expressed through the space-spin-flavor coordinates of the 3 quarks that appear in ϕ_A and ϕ_B.

To construct all $\underset{\sim}{\Gamma}^+$'s and $\underset{\sim}{\Omega}^+$'s in (108,109) we must write the most general expression of a 3-quark axial vector, linear in the τ_i^+'s, (Eq.(108)) or in the λ_i^+'s (eq.(109)), containing a product of P_i^λ 's, up to a maximum of three. Here we start with no flavor breaking (flavor symmetric case), where no P^λ appears in $\underset{\sim}{\Gamma}^+$ or $\underset{\sim}{\Omega}^+$. Later we consider first order flavor breaking (one P^λ).

In the no flavor breaking case the only axial vectors, linear in τ_i^+ or λ_i^+, that can be used to construct $\underset{\sim}{\Gamma}^+$ or $\underset{\sim}{\Omega}^+$ are the following:

$$\underset{\sim}{\Gamma}^{+}: \quad \sum_{1\ i}^{3} \tau_i^{+}\underset{\sim}{\sigma}_i \quad , \quad \sum_{1\ i\neq k}^{3} \tau_i^{+}\underset{\sim}{\sigma}_k \quad , \quad \sum_{1\ i\neq k}^{3} (\tau_i^{+}-\tau_k^{+})(\underset{\sim}{\sigma}_i \times \underset{\sim}{\sigma}_k) \tag{110}$$

$$\underset{\sim}{\Omega}^{+}: \quad \sum_{1\ i}^{3} \lambda_i^{+}\underset{\sim}{\sigma}_i \quad , \quad \sum_{1\ i\neq k}^{3} \lambda_i^{+}\underset{\sim}{\sigma}_k \quad , \quad \sum_{1\ i\neq k}^{3} (\lambda_i^{+}-\lambda_k^{+})(\underset{\sim}{\sigma}_i \times \underset{\sim}{\sigma}_k) \tag{111}$$

In principle $\underset{\sim}{\Gamma}^{+}$ can be a linear combination of the three operators (110) and $\underset{\sim}{\Omega}^{+}$ a linear combination of the three operators (111);each term in (110) and (111) is symmetrical in the quarks,again because ϕ_A and ϕ_B are symmetrical in the spin-flavor variables.This is why in (110,111) we did not include $\sum_{1\ i\neq k\neq j}^{3} \tau_j^{+}(\underset{\sim}{\sigma}_i \times \underset{\sim}{\sigma}_k)$ or $\sum_{1\ i\neq k\neq j}^{3} \lambda_j^{+}(\underset{\sim}{\sigma}_i \times \underset{\sim}{\sigma}_k)$.

We now show that the terms in (110,111) containing $(\underset{\sim}{\sigma}_i \times \underset{\sim}{\sigma}_k)$ have vanishing matrix elements.Infact,with $2\underset{\sim}{J}= \underset{\sim}{\sigma}_1+\underset{\sim}{\sigma}_2+\underset{\sim}{\sigma}_3$, it is:

$$\sum_{1\ i\neq k}^{3} (\tau_i^{+}-\tau_k^{+})(\underset{\sim}{\sigma}_i \times \underset{\sim}{\sigma}_k) = \sum_{1\ i}^{3} \tau_i^{+}\underset{\sim}{\sigma}_i \times (2\underset{\sim}{J}) - (2\underset{\sim}{J}) \times \sum_{1\ i}^{3} \tau_i^{+}\underset{\sim}{\sigma}_i \tag{112}$$

and it is straightforward to check that the matrix element of the right hand side of (112) between the two J=1/2 states ϕ_A and ϕ_B vanishes for any A,B. A similar conclusion holds of course for $\sum_{1\ i\neq k}^{3} (\lambda_i^{+}-\lambda_k^{+})(\underset{\sim}{\sigma}_i \times \underset{\sim}{\sigma}_k)$.

Thus in the flavor symmetric case the most general $\underset{\sim}{\Gamma}^{+}$ and $\underset{\sim}{\Omega}^{+}$ are:

$$\underset{\sim}{\Gamma}^{+} = a \sum_{1\ i}^{3} \tau_i^{+}\underset{\sim}{\sigma}_i + b \sum_{1\ i\neq k}^{3} \tau_i^{+}\underset{\sim}{\sigma}_k \tag{113}$$

$$\underset{\sim}{\Omega}^{+} = a \sum_{1\ i}^{3} \lambda_i^{+}\underset{\sim}{\sigma}_i + b \sum_{1\ i\neq k}^{3} \lambda_i^{+}\underset{\sim}{\sigma}_k \tag{114}$$

where a and b are two parameters.Note that the coefficients a and b in (113) and (114) <u>must be the same</u> because in the no flavor breaking approximation the operator V appearing in the (middle) expressions (108) and (109) is the same and is flavor independent;no matter how complicated, V acts in the same way on $\bar{\Psi}(\underset{\sim}{x},0)\tau^{+}\gamma_4\underset{\sim}{\sigma}\ \Psi(\underset{\sim}{x},0)$ and on $\bar{\Psi}(\underset{\sim}{x},0)\lambda^{+}\gamma_4\underset{\sim}{\sigma}\ \Psi(\underset{\sim}{x},0)$.

The above can be summarized stating that in the flavor symmetric approximation the most general expressions of $\langle 1\rangle$ and $\langle\sigma_z\rangle$ are:

$$\langle 1\rangle_{\Delta S=0} = \langle\phi_B\uparrow|\sum_{i}^{3}{}_1\,\tau_i^{+}|\phi_A\uparrow\rangle \quad ; \quad \langle 1\rangle_{\Delta S=1} = \langle\phi_B\uparrow|\sum_{i}^{3}{}_1\,\lambda_i^{+}|\phi_A\uparrow\rangle \tag{115}$$

$$\langle\sigma_z\rangle_{\Delta S=0} = \langle\phi_B\uparrow|\ a\sum_{i=1}^{3}\tau_i^{+}\sigma_{zi} + b\sum_{i\neq k=1}^{3}\tau_i^{+}\sigma_{zk}\ |\phi_A\uparrow\rangle \equiv$$

$$\equiv \langle\phi_B\uparrow|(a-b)\sum_{i=1}^{3}\tau_i^{+}\sigma_{zi} + b\sum_{i=1}^{3}\tau_i^{+}\ (2J_z)\,|\phi_A\uparrow\rangle \tag{116}$$

$$\langle\sigma_z\rangle_{\Delta S=1} = \langle\phi_B\uparrow|\ a\sum_{i=1}^{3}\lambda_i^{+}\sigma_{zi} + b\sum_{i\neq k=1}^{3}\lambda_i^{+}\sigma_{zk}\ |\phi_A\uparrow\rangle \equiv$$

$$\equiv \langle\phi_B\uparrow|(a-b)\sum_{i=1}^{3}\lambda_i^{+}\sigma_{zi} + b\sum_{i=1}^{3}\lambda_i^{+}\ (2J_z)\,|\phi_A\uparrow\rangle \tag{117}$$

Note that $(2J_z)$ in the second form of (116) and (117) can be replaced by 1 (since the functions $\phi_A\uparrow$ and $\phi_B\uparrow$ have $J_z=1/2$); we kept $(2J_z)$ only to make clear the passage from the first to the second form.

In spite of their non covariant appearance the Eqs.(114-117), are fully relativistic (in the limit $\Delta P_\mu=0$) and general (in the flavor symmetric approximation); no other term can be present.

It is straightforward to relate the parameters a and b to D and F of the Cabibbo parametrization. It is enough to consider two independent matrix elements and write them in terms of a,b or D,F. For N→P matrix elements it is:

$$\langle 1\rangle_{NP}=1 \qquad \langle\sigma_z\rangle_{NP} = \frac{5}{3}\,(a-b) + b = \frac{5}{3}\,a - \frac{2}{3}\,b = D + F \tag{118}$$

For $\Sigma^-\to\Lambda$ we have:

$$\langle 1\rangle_{\Sigma^-\Lambda} = 0 \qquad \langle\sigma_z\rangle_{\Sigma^-\Lambda} = \sqrt{2/3}\ (a-b) = \sqrt{2/3}\ D \tag{119}$$

Thus:

$$a = \frac{1}{3} D + F \quad , \quad b = F - \frac{2}{3} D \tag{120a}$$

or:

$$D = a - b \quad , \quad F = \frac{2}{3} a + \frac{1}{3} b \tag{120b}$$

The physical meaning of a and b is very direct: a is the coefficient of an expression additive in the quarks, whereas b is the coefficient of an expression containing at the same time the variables of two quarks. As already stated for magnetic moments, the NRQM, in its simplest form, amounts to keep among all possible terms, only those additive in the coordinates of the quarks; here this implies to put b=0 in (120a,b); indeed, b=0 corresponds to the well known NRQM result D/F=3/2. Note also that the NRQM approximation (b=0) implies $(g_A/g_V)_{NP}=(5/3)a$; the frequent assertion that NRQM predicts $(g_A/g_V)_{NP}= 5/3$ is incorrect; as remarked long ago[1b] 5/3 is a result of abstract SU_6, not of NRQM.

The value of D/F depends to some extent on how the experimental data are analyzed, in particular on whether the Cabibbo angle is fixed by the same hyperon semileptonic decays or is determined differently (e.g. from the $0^+ \to 0^+$ β transitions). In the first case ($\sin\theta_C = 0.231 \pm 0.003$) one gets $D=0.756\pm0.011$, $F=0.477\pm0.012$ and $D/F=1.58\pm0.04$; one has a fit to hyperon decays that does not show appreciable SU_3 violations; in the second case (with $\sin\theta_C=0.225\pm0.002$) the overall fit requires $D/F=1.74\pm0.04$ and an appreciable amount of flavor breaking is needed to have a reasonably good fit to all data. We shall discuss below the question of the flavor breaking; here we note only the values of b/a corresponding to the values of D/F stated above. It is $b/a= =[1 -(2D/3F)]/[1 +(D/3F)]$; therefore for D/F=1.58 one has b/a= −0.034, whereas for D/F=1.74 it is b/a= −0.10. In both cases b is much smaller than a, from ≈3% to ≈10%. Thus the "two quark term" (with coefficient b) in (143,144) is depressed with respect to the "additive" one quark term (with coefficient a) as it happens for the magnetic moments. Again the additive terms prevail over the two quark ones.

7.2.First Order Flavor Breaking

We now extend the above treatment (eq.(115-117) to include flavor breaking to 1st-order.Nothing changes,to 1st-order,for the vector current (<1> matrix elements).On the other hand for the axial current ($\langle\sigma_z\rangle$ matrix elements) a variety of new possible terms appear.We list them,noting that for first order flavor breaking only one P^λ operator can appear.

$\Delta S=0$: in addition to $\sum_i \tau_i^+ \underset{\sim}{\sigma}_i$,$\sum_{k\neq i} \tau_i^+ \underset{\sim}{\sigma}_k$,the matrix elements of the following operators appear in the parametrization of $\langle\sigma_z\rangle_{\Delta S=0}$ to first order:

$$1)\ \sum_i \tau_i^+ \underset{\sim}{\sigma}_i P_k^\lambda \qquad ,2)\ \sum_{i\neq k} \tau_i^+ \underset{\sim}{\sigma}_k P_k^\lambda \qquad ,3)\ \sum_{i\neq k\neq j} \tau_i^+ \underset{\sim}{\sigma}_k P_j^\lambda \tag{121}$$

$\Delta S=1$: In addition to $\sum_i \lambda_i^+ \underset{\sim}{\sigma}_i$,$\sum_{k\neq i} \lambda_i^+ \underset{\sim}{\sigma}_k$ the matrix elements of the following operators appear:

$$1)\ \sum_i \lambda_i^+ \underset{\sim}{\sigma}_i P_i^\lambda \qquad ,2)\ \sum_{i\neq k} \lambda_i^+ \underset{\sim}{\sigma}_k P_i^\lambda \qquad ,3)\ \sum_{i\neq k} \lambda_i^+ \underset{\sim}{\sigma}_i P_k^\lambda$$

$$4)\ \sum_{i\neq k} \lambda_i^+ \underset{\sim}{\sigma}_k P_k^\lambda \qquad ,5)\ \sum_{i\neq k\neq j} \lambda_i^+ \underset{\sim}{\sigma}_k P_j^\lambda \tag{122}$$

plus terms containing $(\underset{\sim}{\sigma}_i \times \underset{\sim}{\sigma}_k)$.

Clearly the number of parameters,if we take into account all terms above (with no hint as to the magnitude of their coefficients),is too large to make a general parametrization with 1st-order flavor breaking of interest;the no flavor breaking approximation is the only case where a parametrization exists with few enough parameters to have predictive power (Cabibbo parametrization).One might try to reduce the number of flavor breaking terms assuming that (compare Sect.9) the contributions of the "additive" terms among (121,122) are more important than the remaining ones,that is neglecting two quark terms; then only the term $\sum_i \lambda_i^+ \underset{\sim}{\sigma}_i P_i^\lambda$ in (122) would remain;just one more parameter would intervene, namely that (call it Δa) multiplying $\sum_i \lambda_i^+ \underset{\sim}{\sigma}_i P_i^\lambda \equiv \sum_i \lambda_i^+ \underset{\sim}{\sigma}_i$.In this case (115,116) do not change, whereas (117) changes into:

$$\langle\sigma_z\rangle_{\Delta S=1} = \langle\phi_B\uparrow|(a+\Delta a)\sum_{i}^{3}{}_{1}\lambda_i^{+}\sigma_{zi} + b\sum_{i\neq k}^{3}{}_{1}\lambda_i^{+}\sigma_{zk}\,|\phi_A\uparrow\rangle \tag{123}$$

where we expect Δa to be of the order $\approx$ $a(\Delta m/m)$.Note however that,if the flavor breaking effect that fits the data amounts,as suggested in Ref.8,to a difference between a value of $(D/F)_{\Delta S=0}$ ($\cong$1.34) and $(D/F)_{\Delta S=1}$ ($\cong$1.82) [that this is so should be checked further],then the introduction of Δa alone cannot produce such a difference.The reason is that if b=0,the value of D/F is always 3/2, independently of the value of a;now b is not zero,but is very small as seen above;thus a comparatively small change in a (by Δa) cannot produce the necessary change in D/F.In any case there is no reason why flavor breaking effects should be absent or totally negligible in this case as one finds rather frequently implied in the literature.

8.The Masses And Mixing Angles Of The Lowest Nonet Mesons

8.1.General Parametrization For Mesons

Now we apply the method used so far for baryons to the lowest mesonic nonets,find the general parametrization for their masses,and compare the results with the NRQM.We will show that the formulas obtained from field theory with the general parametrization have essentially the same structure and number of parameters as those of the NRQM.This explains the good fit that the NRQM provides in this case.As a byproduct,the mixing angle for I=0 mesons (e.g.η and η') predicted from the masses is model independent to first order in flavor breaking. We will discuss the question of the linear versus quadratic or,say, square root mixing angles.

8.2.From Field Theory To 2-Body Parametrization

We define the model states ϕ_M for mesons similarly to those ϕ_B for baryons.We choose the model Hamiltonian $\mathcal{H}$ for mesons,operating only in the subspace of Fock states with one quark and one antiquark,no gluon, as an operator having a non relativistic,exceedingly simple structure;$\mathcal{H}$ is defined in a way similar to that for baryons (Sect.3.1), where $\mathcal{H}$

acted, of course, in the subspace of Fock states with three quarks, no gluon; also now $\mathcal{H}$ is chosen to conserve the orbital angular momentum and to be spin and flavor independent. The model states of pseudoscalar and vector nonets are therefore degenerate states of $\mathcal{H}$:

$$\mathcal{H}|\phi_{\mathcal{M}i}\rangle\,|\text{no gluons}\rangle = M_o^o|\phi_{\mathcal{M}i}\rangle\,|\text{no gluons}\rangle \qquad (i=\pi,\eta,\eta',K;\rho,\omega,\phi,K^*) \quad (124)$$

In (124) $|\phi_{\mathcal{M}i}\rangle$ is a state of a meson at rest and M_o^o is the common value of the masses of the 18 model states of the 0^- and 1^-, lowest L=0 mesons. From now on we omit writing the factor $|\text{no gluon}\rangle$ in (124) and suppress the index $\mathcal{M}$ in $\phi_{\mathcal{M}i}$. Due to the spin and flavor independence of the model Hamiltonian $\mathcal{H}$, the wave function ϕ_i of a meson has the form:

$$\phi_i = S_c \cdot W_i \cdot \varphi(r) \qquad (125)$$

where S_c is the colour singlet factor for a meson (to be omitted in what follows), W_i is the $q\bar{q}$ spin-flavor wave function of the i-th meson and $\varphi(r)$ is a function of the distance r of quark and antiquark.

The exact states $|\Psi_i\rangle$ of nonet mesons are obtained applying to the $|\phi_i\rangle$'s a unitary operator V:

$$|\Psi_i\rangle = V|\phi_i\rangle \qquad (126)$$

As for baryons, V is constructed in terms of the exact field Hamiltonian H (say, the QCD Hamiltonian); we again assume H written in terms of renormalized quark masses, identified with the masses of constituent quarks, possibly in the range of some hundreds MeV for light quarks; again the precise values assigned to these masses $m \equiv m_{\mathcal{P}} = m_{\mathcal{N}}$ and m_λ will not intervene, except to ensure that the flavor breaking parameter $\Delta m/m \equiv (m_\lambda - m_{\mathcal{P}}))/m$ is decently small and can be treated perturbatively. The mass of the i-th meson is of course:

$$M_i = \langle\Psi_i|H|\Psi_i\rangle = \langle\phi_i|V^\dagger HV|\phi_i\rangle \qquad (127)$$

Because $|\phi_i\rangle$ is, by construction, a 2-body state (one quark, one antiquark) only the projection $\tilde{H}$ of the operator $V^\dagger HV$ in the subspace of these two body states enters in the calculation of (127):

$$\tilde{H} \equiv \sum |1q,1\bar{q}\rangle\langle 1q,1\bar{q}|V^{\dagger}HV|1q',1\bar{q}'\rangle\langle 1q',1\bar{q}'| \qquad (128)$$

where by $|1q,1\bar{q}\rangle$ we mean a complete set of states of one quark,one antiquark (with total three-momentum zero,since we consider a meson at rest) and the sums in (128) are over such a set of states.Once more the trivial fact that the only part of $V^{\dagger}HV$ entering the calculation of M_i (127) is the two body part $\tilde{H}$ is essential;calculating the expectation value of the field operator $V^{\dagger}HV$ in $|\phi_i\rangle$ *becomes equivalent to calculate the expectation value of a certain quantum mechanical* ***two body*** *operator* $\tilde{H}$ on the wave function ϕ_i in ordinary non relativistic two body quantum mechanics:

$$M_i = \langle\phi_i|\tilde{H}|\phi_i\rangle \qquad (129)$$

The general parametrization of meson masses amounts then to write in ordinary non relativistic quantum mechanics the most general operator $\tilde{H}$ of the relative coordinate and momentum $\underset{\sim}{r},\underset{\sim}{p}$,of the spins $\underset{\sim}{\sigma}_1$,$\underset{\sim}{\sigma}_2$ (1=quark, 2=antiquark) and of the flavor operators f(see below) invariant for translations and rotations.

8.3.The List Of The Spin-Flavor Terms In The Parametrization

The most general operator $\tilde{H}$ of the space and spin-flavour variables of the quark-antiquark pair can be written:

$$\tilde{H} = \sum_{\nu} R_{\nu}(\underset{\sim}{r},\underset{\sim}{p})G_{\nu}(\underset{\sim}{\sigma},f) \qquad (130)$$

where $G_{\nu}(\underset{\sim}{\sigma},f)$ is the set of all independent operators constructed in terms of $\underset{\sim}{\sigma}_1$,$\underset{\sim}{\sigma}_2$ and the flavor operators f of the quark and antiquark; $R_{\nu}(\underset{\sim}{r},\underset{\sim}{p})$ are operators (that we do not need to write) constructed in terms of $\underset{\sim}{r}$ and $\underset{\sim}{p}$ of the quark antiquark pair.

Performing the expectation value (131) of $\tilde{H}$,the masses M_i become:

$$M_i = \sum_{\nu} g_{\nu} \langle W_i|G_{\nu}(\underset{\sim}{\sigma},f)|W_i\rangle \qquad (131)$$

where g_{ν} stays for:

$$g_\nu = \langle\varphi(r)|R_\nu(\underset{\sim}{r},\underset{\sim}{p})|\varphi(r)\rangle \qquad (132)$$

[The symbols g_ν and G_ν in (131,132) were also used for baryons in Sect. 3.2, but there is no danger of confusion]. Having chosen $\mathcal{H}$ independent from flavor and from J, the wave function $\varphi(r)$ in (132) is independent of the meson index i; thus the g_ν's in (131) are independent of i. Therefore the mass operator (the expectation value of which must be taken only on the part W_i of the wave function) is:

$$\tilde{H} = \sum_\nu g_\nu\ G_\nu(\underset{\sim}{\sigma}, f) \qquad (133)$$

It should be noted again that this general result from field theory (as the analogous ones for baryons) appears in a notation similar to that of the NRQM. To list the $G_\nu(\underset{\sim}{\sigma}, f)$ in (133), we first discuss their flavor dependence, for a QCD-like theory. The usual $P^\lambda = \frac{1}{3}(1-\lambda_8)$ gives now zero when applied to $\mathcal{P}$ or $\mathcal{N}$ quarks or antiquarks and 1 applied to a λ quark (or antiquark). For a system of one quark and one antiquark, the list of flavor operators acting on states with isospin $I \neq 0$ is:

$$1 \quad ; \quad P_1^\lambda \quad ; \quad P_2^\lambda \qquad (134)$$

<u>In (134) and in all what follows the index 1 refers to the quark and the index 2 to the antiquark</u>. For the mesons with I=0 the list (134) is incomplete. Indeed the flavor operators in (134) connect only a $q\bar{q}$ pair to a $q\bar{q}$ of the same type (say $\mathcal{P}\bar{\mathcal{P}}$ with $\mathcal{P}\bar{\mathcal{P}}$, or $\mathcal{N}\bar{\mathcal{N}}$ with $\mathcal{N}\bar{\mathcal{N}}$ or $\mathcal{P}\bar{\lambda}$ with $\mathcal{P}\bar{\lambda}$, and so on); but the sum in Eq. (128) includes, for mesons with isospin I=0, also matrix elements of $\tilde{H}$ that connect a $\mathcal{P}\bar{\mathcal{P}}$ state with a $\mathcal{N}\bar{\mathcal{N}}$ state or a $\lambda\bar{\lambda}$ state with a $\mathcal{P}\bar{\mathcal{P}}+\mathcal{N}\bar{\mathcal{N}}$ state (in QCD they are due to the gluon annihilation diagrams). Thus there are other flavor operators in addition to (134); to list them we first introduce the I=0 flavor states:

$$|z\rangle = |\mathcal{P}_1\bar{\mathcal{P}}_2 + \mathcal{N}_1\bar{\mathcal{N}}_2 + \lambda_1\bar{\lambda}_2\rangle \quad ; \quad |w\rangle = |\lambda_1\bar{\lambda}_2\rangle \qquad (135)$$

[Note that in (135) $|z\rangle$ is not normalized; for normalization a factor $(1/\sqrt{3})$ is needed; while keeping this in mind, (135) is more convenient for calculations]. Then to the list (134) we must add the operators:

$$|z\rangle\langle z| \qquad (136a)$$

$$|z\rangle\langle w| + |w\rangle\langle z| \tag{136b}$$
$$|w\rangle\langle w| \equiv P_1^\lambda P_2^\lambda \tag{136c}$$

The first of these (136a) is a unitary singlet; the second (136b) transforms linearly in λ_8 (in fact it transforms as $P^\lambda=(1-\lambda_8)/3$). The third (136c) is quadratic in P^λ. We should have included in (136) also $i[|z\rangle\langle w|-|w\rangle\langle z|]$; we did not because the expectation value of this operator vanishes on states with a real wave function (as ϕ_i's are).

As to the $\boldsymbol{\sigma}$ dependence of the $G_\nu(\boldsymbol{\sigma},f)$'s, only the following expressions (137) are possible:

$$1 \quad ; \quad \boldsymbol{\sigma}_1\cdot\boldsymbol{\sigma}_2 \tag{137}$$

8.4. The Masses Of Mesons With I≠0

We start with the *mesons with I≠0* ($\pi,K;\rho,K^*$). For them (compare Eq. (134)) the flavor breaking part of the $G_\nu(\boldsymbol{\sigma},f)$'s contains only P_1^λ and P_2^λ. Multiplying the spin operators (137) with the flavor operators (134) and noting that the charge conjugation invariance of the Hamiltonian implies that the G_ν's must be invariant for the exchange of 1 and 2, the list of G_ν's is: 1 ; $\boldsymbol{\sigma}_1\cdot\boldsymbol{\sigma}_2$; $P_1^\lambda+P_2^\lambda$; $\boldsymbol{\sigma}_1\cdot\boldsymbol{\sigma}_2(P_1^\lambda+P_2^\lambda)$. Therefore the *most general expression* of the mass of a I≠0 meson, *correct to all orders in flavor breaking*, is:

$$M_{I\neq0} = A + B\boldsymbol{\sigma}_1\cdot\boldsymbol{\sigma}_2 + C(P_1^\lambda+P_2^\lambda) + D\boldsymbol{\sigma}_1\cdot\boldsymbol{\sigma}_2(P_1^\lambda+P_2^\lambda) \tag{138}$$

where A,B,C,D are 4 real coefficients. Because $\boldsymbol{\sigma}_1\cdot\boldsymbol{\sigma}_2=-3$ for J=0 and $\boldsymbol{\sigma}_1\cdot\boldsymbol{\sigma}_2=+1$ for J=1, the masses (indicated with the meson symbols) are:

$$\pi = A-3B\ (=138) \qquad K = A-3B+C-3D\ (=495)$$
$$\rho = A+B\ (=770) \qquad K^* = A+B+C+D\ (=894) \tag{139}$$

Eqs. (139) imply (in MeV):

$$A=612\ ,B=158\ ,C=182\ ,D=-58 \tag{140}$$

Note that (restricting to the I≠0 mesons) an equation of the form (138) holds not only for the masses, but also for their squares or any power of the masses (of course with coefficients different from (140)).

8.6.The Masses Of Mesons With I=0

For mesons with I=0 the most general mass formula is constructed adding to (138) [with coefficients A,B,C,D just determined from the masses of π,K,ρ,K^*] another part obtained multiplying the flavor operators (136) with the spin operators (137); that is:

$$M_{I=0}= A + B\sigma_1\cdot\sigma_2 + C(P_1^\lambda+P_2^\lambda)+ D\sigma_1\cdot\sigma_2(P_1^\lambda+P_2^\lambda) + \tag{141}$$
$$+(E+F\ \sigma_1\cdot\sigma_2)|z\rangle\langle z| + (H+G\ \sigma_1\cdot\sigma_2)[|z\rangle\langle w|+|w\rangle\langle z|] + (N+T\ \sigma_1\cdot\sigma_2)P_1^\lambda P_2^\lambda$$

The equation (141) contains in addition to A,B,C,D (determined (140) from $\pi,K;\rho,K^*$) six parameters E,F,H,G,N,T; one might determine them from the masses of $\eta,\eta';\omega,\phi$ plus the two mixing angles (the vector θ_V and the pseudoscalar θ_{Ps}), if the angles are known in a way free of theoretical uncertainties. Alternatively we can limit to first order flavor breaking approximation; in that case the last term in Eq.(141), of second order in flavor breaking (being proportional to $P_1^\lambda P_2^\lambda$) can be omitted (N=0,T=0). In this first order flavor breaking approximation the masses of $\eta,\eta';\omega,\phi$ fix E,F,H,G; one can then determine, to first order in flavor breaking, the mixing angles.

We restrict here to determine the parameters for pseudoscalar (P) mesons η,η'; for them $\sigma_1\cdot\sigma_2=-3$. For vector (V) mesons one proceeds similarly, simply putting $\sigma_1\cdot\sigma_2=+1$, that is replacing below A-3B with A+B; C-3D with C+D; E-3F with E+F; H-3G with H+G. As it is well known, the parametrization with A,B,C,D alone is almost sufficient for the I=0 vector mesons (the V mixing angle is nearly ideal); this means that in (141) $E+F\approx 0$ and $H+G\approx 0$; of course there would be no problem to determine the exact parameters and angles also for V mesons.

Let us proceed with P mesons; we introduce the abbreviations:

$$b=A-3B\ (=138\ \text{MeV})\ ;\quad d=C-3D\ (=357\ \text{MeV});\quad f=E-3F\ ;\quad g=H-3G \tag{142}$$

where b and d are known from (140). Thus Eq.141 for P mesons with I=0 becomes:

$$M_{I=0}(P)= b + d(P_1^\lambda+P_2^\lambda)+ f|z\rangle\langle z| + g[|z\rangle\langle w|+|w\rangle\langle z|] + O[(\Delta m/m)^2] \tag{143}$$

The diagonalizing equation:

$$\begin{vmatrix} b+f-M & f & f+g \\ f & b+f-M & f+g \\ f+g & f+g & b+2d+f+2g-M \end{vmatrix} = 0 \qquad (144)$$

relates f and g to the masses of η and η'.One solution of (144) is the π^0 mass=b≡A−3B (Eq.(139));the masses of η and η',are the two roots of:

$$(b+2f-M)(b+2d+f+2g-M)-2(f+g)^2=0 \qquad (145)$$

The solutions of (145) are $M=b+k\pm(k^2+2g^2-4fd)^{1/2}$ where $2k\equiv 3f+2d+2g$. Identifying the two solutions with the masses of η(549 MeV) and η'(958 MeV) we have $\eta'+\eta=2b+3f+2d+3g$ and $(\eta'-\eta)/2=\{2g^2-4fd+\frac{1}{4}(3f+2d+2g)^2\}^{1/2}$ where b and d have the values (142) Solving for g and f we obtain two possible solutions $(g;f)_1$ and $(g;f)_2$;in MeV they are:

$$(g;f)_1=(-133,+261) \quad ; \quad (g;f)_2=(-343,+400) \qquad (146)$$

The ratio $(g/f\sqrt{3})$ between the coefficient $g\sqrt{3}$ of the (normalized) flavor breaking term and the coefficient 3f of the (normalized) unitary singlet term is substantially smaller for the solution N.1 than for the solution N.2. Thus we must select the solution N.1 to be consistent with the assumption of neglecting terms of order higher than first in $(\Delta m/m)$.On choosing $(g;f)_1$ it is straightforward to check that the diagonalization (139) implies (in addition to $\pi^0=(1/\sqrt{2})(\mathcal{P}\bar{\mathcal{P}}-\mathcal{N}\bar{\mathcal{N}})$):

$$\eta = 0.603(\mathcal{P}\bar{\mathcal{P}}+\mathcal{N}\bar{\mathcal{N}}) - 0.522\lambda\bar{\lambda} \qquad (147)$$

$$\eta' = 0.367(\mathcal{P}\bar{\mathcal{P}}+\mathcal{N}\bar{\mathcal{N}}) + 0.854\lambda\bar{\lambda} \qquad (148)$$

With the usual definition $\eta=-\eta_1\sin\theta_P+\eta_8\cos\theta_P$ and $\eta'=\eta_1\cos\theta_P+\eta_1\sin\theta_P$, where $\eta_1=(1/\sqrt{3})(\mathcal{P}\bar{\mathcal{P}}+\mathcal{N}\bar{\mathcal{N}}+\lambda\bar{\lambda})$ and $\eta_8=(1/\sqrt{6})(\mathcal{P}\bar{\mathcal{P}}+\mathcal{N}\bar{\mathcal{N}}-2\lambda\bar{\lambda})$,we obtain:

$$\sin\theta_P \cong -0.39 \qquad (\text{that is } \theta_P \cong -23^\circ) \qquad (149)$$

8.6.Comparison Of The General Parametrization With NRQM; a) I≠0

Eq.(138) has the same form as that currently used in the NRQM to fit the L=0 meson masses (compare e.g. Ref.1b).It now emerges that (independently of the NRQM) the four parameter equation (138) is the most general expression possible for the lowest nonet meson masses in any (relativistic field) theory where the only λ_{flavor} that enters is λ_8. It is an exact equation to all orders in flavor breaking. This fact (that Eq.(138) is an exact formula, no more an approximate NRQM result) has, in our view, an important implication for the pion: The small mass of the pion appears now just an accident that depends rather critically on the values of the parameters A and B in (140); assume, for instance, that B in (138) were 100 MeV instead of 158 MeV (Eq.(140)); then the pion, if A keeps its value (140), would have a mass=312 MeV, no more so small (the ρ mass would then become 712 MeV). Now, taking e.g. QCD as the basic theory, a reduction of B in (138) by the above amount of ≈40% may possibly be obtained with an even percentually smaller variation of the quark-gluon coupling. Thus it seems difficult to attribute a profound meaning (as often done) to the smallness of the pion mass. This does not question, of course, the phenomenological treatments or results related to PCAC, insofar as they are simply based on the empirical smallness of the pion masss.

We also compare Eq.(138) with the NRQM formula for meson masses of de Rujula et al.[6] (their formula (17)) derived from QCD in the one gluon exchange approximation. A feature of that formula is that $\sigma_1 \cdot \sigma_2$ appears only multiplied by the factor $(m_1 m_2)^{-1}$ where m_1 and m_2 are the masses of the quark and antiquark; this factor $(m_1 m_2)^{-1}$ cannot be obtained by general arguments, as far as we can see. One can check, however, that the general parametrization Eq.(138) above reduces to the formula (17) of Ref.6, to all orders in $(m_\lambda - m)/m$, with the identification $(-D/B) = (m_\lambda/m) - 1$. From the experimental values of D and B (eq.140) above) one gets, as noted already[1c], $(m_\lambda - m)/m \cong 0.37$, to be compared with $(m_\lambda - m)/m \cong 0.33$ derived from baryon masses and $(m_\lambda - m)/m \cong 0.35$ from baryon magnetic moments (Sects.5.2 and 4.1).

b) *Mesons With I=0*

1. The θ_P mixing angle.-Having omitted terms of order $(\Delta m/m)^2$ in

Eq.(141),the results for I=0 mesons are correct only to 1st-order flavor breaking,as already stated;but it should be stressed that to that order they are exact:If we were able to solve the equations of the correct field theory (say,QCD) and were allowed to neglect terms of order higher than the first in $\Delta m/m$,we would obtain,in particular,the value $\theta_P \cong -23^{\circ}$ for the P mixing angle.A remark on the question of quadratic versus linear mixing angle is appropriate at this stage.We have parametrized above $M_i = \langle\phi_i|V^{\dagger}HV|\phi_i\rangle$ but,as stated in Sect.5.1,we might have parametrized as well $M_i^n = \langle\phi_i|V^{\dagger}H^nV|\phi_i\rangle$ for any real n, positive or negative,integer or fractional. The parametrization is the same independently of n;it is always an equation of the form (138) or (141);simply one has to replace on the left hand side of these equations M by M^n; the coefficients A,B,C,D,E,F,H,G,N,T side have of course different numerical values for each different n.If we do not neglect N and T (the second order terms in flavor breaking in (141)) then the exact mixing angle is always the same,independently of n;the same mixing corresponds,say, to a quadratic (M^2) or square root ($M^{1/2}$) mass formula;indeed the exact eigenstates of H^2 or $H^{1/2}$ are the same;as already noted,in the complete parametrization (that is keeping N and T) no relationship exists between mixing angle and masses;a relationship arises only neglecting N and T.Then we get different values of the mixing angle for each different choice of n;for instance for θ_P we get:

$$(M^{1/2})\ \theta_P \cong -31^{\circ} \qquad ;(M)\ \theta_P \cong -23^{\circ} \qquad ;(M^2)\ \theta_P \cong -10^{\circ}$$

Neglecting N and T,the procedure takes into account exactly (for any n) first order flavor breaking;the expansion of the mixing angle in terms of the flavor breaking parameter $\Delta m/m$ starts with a term of first order in flavor breaking:

$$\sin\theta_P^{exact} = \alpha_n \frac{\Delta m}{m} + O((\Delta m/m)^2)$$

where α_n is a coefficient depending on the selected n.Because $\Delta m/m \cong 0.37$ is not so small,terms $O((\Delta m/m)^2)$ may contribute significantly to θ_P , with a different relative weight for different n's.Which is the best angle? No theorem can be stated,in general;only solving the problem in

QCD, say, one can get the correct angle. However the following remark seems relevant: If one were to solve the problem in QCD, namely determine the mixing angle as an expansion in the flavor breaking parameter, one would never dream of using the square or square root of the QCD Hamiltonian (indeed to perform calculations with H^2 or $H^{1/2}$ the question arises, in first place, of defining properly such operators, not only via their eigenvalues M_i^2 or $M_i^{1/2}$).

Thus in performing a full field theoretical calculation, one would use H and get $\theta_P \cong -23^o$ plus terms $O((\Delta m/m)^2)$. In this sense the linear mixing angle is the most natural one; it is the choice made, without saying, in any perturbative diagonalization of an Hamiltonian.

2. Comparison with the NRQM-Compare first our parametrization for the I=0 mesons with the NRQM work of Sacharov and Zeldovich[9] (indicated as SZ). Specifically we compare our Eq.(141) with the expression for the masses proposed in SZ (their Eq.(3) $H=H_a+H_b$, with H_a and H_b given respectively by their Eqs.(4) and (1)). The structure and number of parameters of the quoted equations of SZ is identical to that of our Eq.(141) in the approximation N=T=0; the only minor difference is that SZ keep terms of second order in flavor breaking (they ignore α^2 in their Eqs.(1)(4) and replace $(1-\beta)^2$ with $1-2\beta$). With this modification the expression for the masses of SZ becomes equal to our Eq.(143) with the following correspondence of our parameters b, d, f, g with those of SZ: $b=(a_o-\frac{3}{4}c_o)_{SZ}$; $d=(b_o+\frac{3}{4}c_o\alpha)_{SZ}$; $f=(d_o)_{SZ}$; $g=-(d_o\beta)_{SZ}$. [There seems to be a numerical mistake in Eqs.(5) and (5a) of SZ for η and η'; indeed solving Eqs.(5) and (5a) of SZ (with their input parameters) we find $d_o=267$ instead of their value $d_o=580$ and $\beta=0.49$ instead of their 0.75; this is confirmed by our $(g;f)_1$ that gives $f=(d_o)_{SZ}=261$ and $\beta=-(g/f)=0.51$; the differences 261 and 267, 0.51 and 0.49 arise from slightly different input data, the α^2 term and use of $(1-\beta)^2$ instead of $(1-2\beta)$].

We next compare our general parametrization (141) for the I=0 mesons with that of De Rujula et al.[6] (DGG). The mass matrix of DGG (appearing before their Eq.(43)) differs from ours (144) because DGG neglect in their matrix the flavor breaking correction (our coefficient g). In fact if we identify our f with the β appearing in the mass matrix of DGG and include the first order flavor breaking corrections g (that DGG should have been included because the quark masses appear in the quark lines of the annihilation diagrams) our mass matrix and that of DGG coincide.

8.6. Conclusions From The Analysis Of Meson Masses

The main conclusions of the above analysis are:

1) Almost all features of mass formulas for P and V meson nonets currently used in a NRQM description are general consequences of the field theoretical parametrization, independent of the NRQM. [An exception is the $(m_1m_2)^{-1}$ factor multiplying the $\boldsymbol{\sigma}_1\cdot\boldsymbol{\sigma}_2$ term in the DGG expression for meson masses, a result related specifically to the one gluon exchange potential]. Stated differently, the meson mass formulas used in the NRQM happen to be more general than one might have thought; for mesons with $I\neq0$ they possess a number of free parameters as large as that of a full relativistic parametrization; for mesons with I=0, the same is true omitting terms quadratic in P^{λ}.

2) The procedure described above to derive the meson mass formulas clarifies the old question of the mixing angle for the I=0 meson nonets. The mixing angle (say θ_P) can be determined from the general parametrization (that is in a model independent way) knowing the masses of P mesons provided that terms of order higher than first in the flavor breaking parameter $\Delta m/m$ are negligible (and neglected). According to this procedure the linear angle is much more natural than the quadratic or square root ones; it would be the angle resulting from a QCD calculation on expanding the exact result in series of $\Delta m/m$ and neglecting terms $O((\Delta m/m)^2)$.

9. The Reasons For The Validity Of The Non Relativistic Quark Model

9.1. Introduction

One reason why the NRQM works has emerged rather clearly from the above:The NRQM is a parametrization expressed in a form very similar to that obtainable exactly from relativistic field theory.Thus,although the NRQM language is non relativistic,it represents in fact a fully relativistic (though non covariant) description of the facts.This conclusion has nothing to do with the magnitude of the velocity v of quarks inside hadrons;in other words it does not depend on v/c being or not being ≪1.

There is however one second feature of the NRQM description that,so far,we did not discuss:Additive terms in the quark variables are usually more important than two-quark terms, and these in turn more important than three quark terms (when the latter are present,that is for baryons).

As a matter of fact we saw,examining the general parametrizations of baryon masses or magnetic moments,that the data support this hierarchy typical of the NRQM.However we wish to understand what is the underlying reason for it.We will reach the following conclusion[10]: The contribution to the various quantities (masses,magnetic moments,....) due to two gluon exchange between constituent quarks in a baryon is typically reduced by a factor 5 with respect to that due to one gluon exchange.We will see that putting together this fact with the factor 3 reduction from each additional order in flavor breaking explains the hierarchy (typical of the NRQM) mentioned above.Also it allows to understand,for instance,the success,to a part per thousand,of the baryon mass formula (86).It will also explain,as we shall see,why the results of SU_3 that neglect second order flavor breaking (like Gell Mann-Okubo baryon mass formula) work much better than might be expected.

9.2.The Order Of Magnitude Of The Coefficients In The Baryon Mass Parametrization

The contents of this Section is a very short summary of Ref.10;it cannot be a substitute for it.Let us start recalling the values (in Mev) of the coefficients in Eqs.(88,69) giving the general baryon mass parametrization:

M_o=1085 ; B=187 ; C=48.7 ; D= −16 ; a+b= −11.6 ; E=3.4 ; c=1.17 (150)

We did not list the value of d (expected to be smaller than or about equal to c but) dominated by the error in the Δ mass.

We recall that M_o and B are coefficients of terms carrying zero and one quark indices (we call them α-terms);C,D,a and b are coefficients of terms carrying two quark indices (β-terms);E and c are coefficients of terms carrying three quark indices (γ-terms).

It is possible[10] to show the following:

1) In the general parametrization the various terms can be derived from Feynman diagrams (in a field theoretical calculation starting from the underlying field theory) leading,for baryons,from a three quark state to a three quark state.

2) When this is done,the α-terms correspond to diagrams that do not need the exchange of any gluon between quarks (although they may also include the exchange of any number of gluons).The β-terms correspond to Feynman diagrams that need the exchange of at least one gluon.The γ-terms need the exchange of at least two gluons.

3) As to the order in P^λ,this is immediately readable in each term counting the number of P^λ factors.

With the above correspondence between Feynman diagrams and terms in the general parametrization and paying attention on how to normalize the terms in the general parametrization,it is possible[10] to estimate the reduction factor corresponding to the exchange of one more gluon between two constituent quarks.Just to exemplify:

$$\left|\frac{E}{D}\right| = \frac{(\geq 2 \text{ Gluon exch. } |1P^\lambda)}{(\geq 1 \text{ Gluon exch. } |1P^\lambda)} \cong 0.21 \qquad (151)$$

Similarly one can determine the reduction factor due to the flavor breaking;for instance:

$$\left|\frac{D}{C}\right| = \frac{(\geq 1 \text{ Gluon exch. } |1P^\lambda)}{(\geq 1 \text{ Gluon exch. } |0P^\lambda)} \cong 0.33 \qquad (152)$$

The above Eqs.(151)(152) exemplify the assertion that each additional gluon exchange implies a reduction by a factor 5 and each P^λ a reduction by a factor 3.In Ref.10 many other examples are listed in addition to (151)(152).We conclude that Eq.(86) is so good because, when neglecting 3-quark terms of second order in P^λ, we neglect terms of order (1/25)x(1/9)=[(2 Gluon reduction)x(2P^λ reduction)]$\cong 4\cdot 10^{-3}$; this leads us into the region of accuracy of 1 MeV.

Similarly, consider a classical SU_3(flavor) formula like the Gell Mann-Okubo baryon mass formula. There one keeps terms linear in P^λ and neglects terms of type $P_i^\lambda P_k^\lambda$. The neglect of such terms does not mean just neglecting terms of order (1/3) with respect to those kept, (if this were so the Gell Mann-Okubo formula would be very poor); rather it means to neglect terms of order (1/5)x(1/3)=1/15 with respect to those kept. This is so because terms of type $P_i^\lambda P_k^\lambda$, carrying two indices, imply the exchange of at least one gluon and this carries a reduction factor of 5. Thus, when writing the Gell Mann-Okubo formula, we neglect terms of order 1/45 with respect to the gross features of the mass spectrum, and this explains its success.

10. Conclusion

For reasons of space we had to omit in these lectures a number of topics or to survey very briefly others. Among the topics entirely omitted is the question of the $\Delta\to P\gamma$ decay (compare Ref.2a); in such case the parametrization clarifies why the M1 transition matrix element given by the simple NRQM differs by about 50% from the correct matrix element. Another subject that we had not time to treat has been that of the $V\to P\gamma$ meson decays, dealt in Ref.2d. There the general parametrization method is extended to the treatment of a non diagonal matrix element in which, moreover, the meson from the decay moves relativistically. Thus, in this case, the general parametrization method solves the old question of how to apply the NRQM when one of the hadrons has a relativistic kinematics. Some $V\to P\gamma$ decays turn out to be of particular interest; for instance in the ratio $\Gamma(\omega\to\pi\gamma)/\Gamma(\rho\to\pi\gamma)$ any deviation from 9.5 [=9·1.06, where 1.06 comes from the phase space] is a measure of the gluonic annihilation contribution to $\omega\to\pi\gamma$. The present data give 9.9±1.6. Other similar examples are mentioned in Ref.2d.

We refer to[2d] for a complete discussion of $V\to P\gamma$ decays; instead we dedicate the closing remarks of these lectures to a general consequence of the general parametrization method for QCD-like theories;a consequence that,so far,we did not underline sufficiently.The point is this:The fact that only λ_8 and λ_3 intervene in the strong plus electromagnetic Hamiltonian of a QCD-like theory implies,since λ_8,λ_3 form a closed algebra,that the exact expression of the energy levels derived from such an Hamiltonian (that is the masses of hadrons) cannot depend on flavor matrices different from λ_3 and λ_8.In particular in the expression of the masses the Casimir flavor operators must be absent. This shows that any effective Lagrangian that depends explicitely on the pion fields (and thus on the $\lambda_1 \pm i\lambda_2$ flavor matrices) cannot be in general equivalent to the original QCD or QCD-like Lagrangian;it will not reproduce correctly the levels of the original Hamiltonian unless in the end result all flavor operators different from λ_8 and λ_3 (and in particular all Casimir flavor operators) disappear;but obviously in that case the introduction of such an effective Lagrangian complicates the description.This argument should remain true also if spontaneous simmetry breaking occurs,that is if the ground state of the system does not have the simmetry of the Hamiltonian.

The above is relevant to several models in which the pion degrees of freedom are introduced explicitely;models that,often, are stated to represent an approximation to QCD;for instance the skyrmion model,the chiral quark model of Manohar and Georgi[11] and various cloudy bag models with pions.Specifically in the latter case the exact general parametrization shows[12] that the magnetic moments of the various $\Delta(1232)$'s are proportional to their charges.In the cloudy bag model with pions this is exact result is not reproduced;this is an explicit demonstration that the cloudy bag model cannot be equivalent to QCD.

11.Appendix.The Gell Mann-Low construction of the transformation V

We reproduce below the Appendix of Ref.2a so as to clarify the relationship between model states $|\phi\rangle$ and exact states $|\Psi\rangle$.Moreover the

Eq.(A10) illustrates the connection between general parametrization and Feynman diagrams, a connection that we used in Sect.9.

Let H be the exact hamiltonian of the quarks and gluons; to be specific we may have in mind the H of QCD. We will indicate by $q_R(x)$ a mass renormalized quark field, with (renormalized) mass m; as stated in the text m is taken to be the mass that appears in the quark propagator $[\not{k}-im]^{-1}$ for low values of the momentum transfer k^2 ($|k|\approx R^{-1}$, with R the size of the hadron; here it is unnecessary to specify $|k|$ more precisely); we identify m with the mass of a constituent quark. We introduce the Fock states of the independent quarks and gluons with renormalized mass and, suppressing the index R, call $|qqq\rangle$ a state of three quarks and no gluon. Call $\eta = \Sigma|qqq\rangle\langle qqq|$ the projection operator into of the states of three quarks and no gluons:

$$\eta|qqq\rangle = |qqq\rangle \tag{A1}$$

$$\eta|\neq qqq\rangle = 0 \tag{A2}$$

We rewrite H identically as:

$$H = \eta H\eta + (1-\eta)H(1-\eta) + \eta H(1-\eta) + (1-\eta)H\eta \tag{A3}$$

Introduce now the model hamiltonian $\mathcal{H}$ which is a typical non relativistic quark model Hamiltonian acting only on the Fock space of the states of three quarks and no gluons. We decompose H as:

$$H = K_o + K_1 \tag{A4}$$

with:

$$K_o = \eta\mathcal{H}\eta + (1-\eta)H(1-\eta)$$

$$K_1 = \eta H(1-\eta) + (1-\eta)H\eta + \eta H\eta - \eta\mathcal{H}\eta$$

having added, to K_o, and subtracted, from K_1, the model Hamiltonian $\eta\mathcal{H}\eta$.

Referring to baryons we assume that $\eta\mathcal{H}\eta$ has degenerate eigenvalues M_o^o for all the octet and decuplet baryon states:

$$\eta\mathcal{H}\eta\ \phi_B = M_o^o\ \phi_B \qquad (B=N,\Lambda,\Sigma,\Xi,\Delta,\Sigma^*,\Xi^*,\Omega) \tag{A5}$$

where $|\phi_B\rangle|0$ gluons$\rangle$ are the L=0 model states, called simply $|\phi\rangle$ in the text. Because in the three quark sector K_o and $\mathcal{H}$ coincide, $|\phi_B\rangle$ are the degenerate eigenstates of K_o:

$$K_o|\phi_B\rangle = M_o^o|\phi_B\rangle \tag{A6}$$

In the part $\eta\mathcal{H}\eta$ of K_o the masses of the $\mathcal{P}, \mathcal{N}$ and λ quarks are taken as equal (as implied by (A5));the flavour breaking mass term appears in the term $(1-\eta)H(1-\eta)$ of K_o and in the term $\eta H\eta$ of K_1.The term $(1-\eta)H(1-\eta)$ of K_o includes,in particular,the Hamiltonian of the non interacting gluons;K_1 contains the interaction terms $\eta H(1-\eta)$ and $(1-\eta)H\eta$ of the quark gluon Hamiltonian.

We now regard K_o as an unperturbed Hamiltonian,K_1 as perturbation, imagine to insert K_1 adiabatically,and construct the true states $|\Psi\rangle$ with the procedure of Gell Mann and Low[5] [this construction procedure is of course is not compulsory,but we imagine to use it to exemplify at least one method of construction].With $K_1(t)=\exp(+iK_o t)K_1\exp(-iK_o t)$ the adiabatic $U(t,t_o)$ satisfies $i\dot{U}_\alpha(t,t_o) = \exp(-\alpha|t|)\cdot K_1(t)U_\alpha(t,t_o)$ [with $\alpha>0$, $U(t_o,t_o)=1$] and the $|\Psi\rangle$'s for the exact bound states corresponding to the lowest $|\phi\rangle$'s are:

$$|\Psi_B\rangle = \lim_{\alpha\to0} \exp(-w_B/\alpha)\cdot U_\alpha(0,-\infty)|\phi_B\rangle \tag{A7}$$

where w_B is purely imaginary ($w_B + w_B^* = 0$)[so that the factor $\exp(-w_B/\alpha)$ in front of (A7) is a pure phase factor that eliminates the singularity coming from the $\lim_{\alpha\to0}U_\alpha(0,-\infty)$; w_B in (A7) is related to the $S=U(+\infty,-\infty)$ matrix element of the $\phi_B\to\phi_B$ transition by $\lim_{\alpha\to0}\langle\phi_B|S|\phi_B\rangle = \exp(2w_B/\alpha)$].

The operator V introduced in the text can be therefore written explicitely as:

$$V = \lim_{\alpha\to0} \exp(-w_B/\alpha)\cdot U_\alpha(0,-\infty) \tag{A8}$$

Referring to magnetic moments,the formula for the magnetic moment has the form used in the text, namely:

$$\underset{\sim}{M} = \langle\phi|V^+\underset{\sim}{\mathcal{M}}\, V|\phi\rangle \tag{A9}$$

As it can be seen in a few passages,this is the same as the formula used frequently for practical calculations:

$$\underset{\sim}{M} = \frac{\langle\phi|T(\underset{\sim}{M}(0)S)|\phi\rangle}{\langle\phi|S|\phi\rangle} \equiv \langle\phi|T(\underset{\sim}{M}(0)S)|\phi\rangle_C \tag{A10}$$

where the index C means "connected".

11. References

1. G.Morpurgo: a) *Physics* 2 (1965)95 [reproduced in J.J.Kokkedee, *The Quark Model* (Benjamin, New York, 1969), p.132]; b) *Theory and Phenomenology in Particle Physics*, proceedings of the Int.School of Physics "E.Majorana" Erice, Italy, 1968 edited by A.Zichichi (Academic, New York, 1969) pp.83-217; c) *Ann.Rev.of Nuclear Science*, **20** (1970) 105.
2. G.Morpurgo: a) *Phys.Rev.* **D40** (1989)2997; b) *Phys.Rev.* **D40** (1989)3111; c) *Phys.Rev.* **D41** (1990)2865 ; d) *Phys.Rev.* **D42** (1990)1497; e) *Phys.Rev. Lett.* **68** (1992) 139; f) *Phys.Rev.* **45** (1992)1686. [In a) the method is formulated and applied to baryon magnetic moments and masses; in b) it is applied to the baryon semileptonic decays; in c) to the meson masses; in d) to the $V \to P\gamma$ decays; in e) to the derivation of a new baryon mass formula; in f) to the baryon e.m. masses].
3. S.Okubo, *Phys.Lett.* **4** (1963)14
4. P.Federman, H.Rubinstein and I.Talmi, *Phys.Lett.* **22** (1966)208; see also Ref.1b, p.163.
5. M.Gell Mann and F.Low, *Phys.Rev.* **84** (1951)181
6. A De Rujula, H.Georgi and S.Glashow, *Phys.Rev.* **D12** (1975)147
7. A.Gal and F.Scheck, *Nucl.Phys.* **B2** (1967)121; see also Ref.1b, p.165.
8. J.F.Donoghue and B.R.Holstein, *Phys.Rev.* **D25** (1982)2015
9. Ya.B.Zeldovich and A.D.Sacharov, *Yad.Phys.* **4** (1966)395; translated in *Soviet J.Nucl.Phys.* **4** (1967)283
10. G.Morpurgo, *Smallness of gluon coupling to constituent quarks in baryons and the validity of the non relativistic quark model*, INFN/GE preprint, March 2, 1992; (E) March 31, 1992 (to be published).
11. A.Manohar and H.Georgi, *Nucl.Phys.* **B234** (1984)189.
12. G.Dillon and G.Morpurgo, *Magnetic moments of the Δ's in QCD*, INFN/GE preprint, March 17, 1992 (to be published).

Common Problems and Trends of Modern Physics, pp. 63–90
ed. T. Bressani, S. Marcello and A. Zenoni
©1992 World Scientific Publishing Company

QCD AND THE QUARKS: LIGHT AND HEAVY

Giuliano Preparata (*)

INFN - Laboratori Nazionali di Frascati - Frascati (Italy)

(*) On leave of absence from Dipartimento di Fisica, Universita' di Milano, Italy.

INTRODUCTION

This year the format of the Folgaria school has been somewhat changed; it was felt that young people should have more chances to participate actively in the exchange, discussion and confrontation of new ideas. So I shall limit myself to outline in two lectures the framework in which our group in Milano is now pursuing the long range research program of understanding and analysing the physics of quarks and hadrons. Some details of what we are actually doing can be found in the lectures, reported in this book, of R. Barni, L. Gamberale, M. Gibilisco, P.G. Ratcliffe, M. Scorletti, M. Verpelli and S.S. Xue'.

1. ACD: THE RIGHT PQCD

To begin with, it is important to realize that our general knowledge of the hadron sector - $SU(3)_c$- of the Standard Model (SM), or QCD, is on a completely different footing from, say, Quantum Electro Dynamics (QED). This is due to the fact that the basic fields of QCD, quarks and gluons, unlike those of QED, electrons and photons, do not exist in a real physical sense. Technically, this is embodied in the remarkable fact that quarks and gluons do not belong to the Hilbert space of physical (finite energy) states. In the light of this unprecedented occurrence, it is not unjustified to state that the now generally accepted QCD (GAQCD) is not really a physical theory, with a sound basis of natural realism, but rather an IDEOLOGY, which rests on two assumptions (that in the course of almost twenty years have become dogmas):

- the physics of short distances (high energies) is well described by perturbative QCD (PQCD);
- the physics of long distances (low energies) can be calculated from lattice QCD (LQCD) on rather coarse lattices.

From a logical standpoint the fact that GAQCD is based on the above two assumptions, whose main motivations derive from some positive circumstantial evidence (free-field partonic behaviour for the first, and the Wilson area-law for the second), "faute de mieux" appears perfectly acceptable. However, what is not generally known or appreciated is that these two pillars of GAQCD have long been shown to be without basis.

The theoretical backing of such a preposterous sounding statement has been widely discussed in several instances, including the Folgaria Winter Courses of 1986 and 1988. In particular one should consult Refs. 1 and 2, together with the early, and only, criticism of L. Maiani et al. [3], which was rebutted in a paper of mine [4] and more recently by Castorina and Consoli in Ref. 5. Surprisingly, in spite of the evidence that has been accumulating against GAQCD, its ideology still impregnates most of the current thought on hadrons and their dynamics.

I maintain that in the absence of any serious arguments to dismiss the adverse evidence it belongs to reason and scientific integrity to finally reject GAQCD and to replace it by some other unfalsified calculational strategy. These lectures will be devoted to discuss the motivations and the main features of an approach to the physics of QCD that I first proposed in 1980 [6], which was called, for reasons that will soon become apparent, Anisotropic Chromo Dynamics (ACD), and which was later shown to realize QCD on a highly non-perturbative ground state [1], the Chromo Magnetic Liquid (CML) [7], which shall be discussed in a moment.

1.1 Why is PQCD wrong?: the (probable) QCD ground state

As I have stressed time and again the central assumption of GAQCD is that a finite mass scale exists, to be called Λ_{QCD}, that separates the physics of short distances ($\mu >> \Lambda_{QCD}$) from that of long distances ($\mu \simeq \Lambda_{QCD}$) such that for the "running" coupling constant of QCD $\alpha_s(\mu)$ one has:

$$\alpha_s(\mu) \underset{\mu \gg \Lambda_{QCD}}{\to} 0. \tag{1.1}$$

This is the well known characterization of the key notion of GAQCD: Asymptotic Freedom (AF). Eq. (1.1) has been usually taken to mean that at scales μ much larger than Λ_{QCD} (which has been plausibly determined to lie in the range 100÷200 MeV) the basic QCD interaction can be treated perturbatively, i.e. following the rules of perturbation theory (PQCD), whose algorithmic implementation is very simple, but whose physical interpretation poses to the natural philosopher more than one anguishing problem. From the purely mathematical point of view, however, the assumption that the validity of (1.1) implies the validity of PQCD at large energy scales is subject to a fundamental criticism: the asymptotic vanishing of $\alpha_s(\mu)$ is a necessary but by no means sufficient condition for the validity of PQCD.

Indeed, for the limit (1.1) to be sufficient to prove PQCD it is necessary that at the asymptotic scale μ the ground state of perturbation theory be stable, i.e. be the state of minimum energy. Lacking such a proof the foundations of PQCD are seen to be only conjectural, and people should be prepared to see this conjecture eventually falsified. This actually occurred in the years 1984-86 [1,2], but surprisingly no notice of it was taken by the community except for the early, unsubstantiated criticism already mentioned [3]. Be that as it may, in a series of theoretical developments [1,2,7] it was conclusively shown that:

(i) the perturbative ground state (PGS, characterized by the order parameter $\langle F^a_{\mu\nu} F^{a\mu\nu}\rangle = 0$, $F^a_{\mu\nu}$ are the QCD field strengths) is unstable against condensation of a constant chromomagnetic field ($\langle B^a_3\rangle = \langle F^a_{12}\rangle \neq 0$);

(ii) The size of $\langle B_3^a \rangle$, the chromomagnetic field condensate, does not scale according to the perturbative renormalization group prediction (Λ is the ultraviolet cut-off)

$$|\langle g_o B_3^a \rangle| = \Lambda^2 \exp - 24\pi^2/11g_o^2 ; \tag{1.2}$$

(iii) There exists thus no scale at which PQCD faithfully describes the dynamics of QCD;

(iv) the strenght of the condensate, being $O(\Lambda^4)$, undermines completely any attempt to solve QCD through a "low Λ-approximation", such as present day LQCD.

On the basis of these conclusions, that have found no serious challenge in the past several years, the collapse of GAQCD appears inevitable.

1.2 The chromomagnetic liquid (CML)

The study conducted on the essential instability of the PGS was focused on the dynamics of a peculiar configuration of the colour gauge-field, that I have called Savvidy state, from the theorist that first looked into its structure [8]. A Savvidy state is characterized by a quantum colour gauge-field fluctuating around the special solutions of the "classical" Yang-Mills equations that by a special gauge transformation can be reduced to a constant chromomagnetic field:

$$\vec{B}^a = B\eta^a \vec{u}, \tag{1.3}$$

and in which the energy density functional is minimum. Working in SU(2) one can show [1] that the energy density funtional on such states yields a value that is below that of the PGS, E(0), by the quantity

$$\Delta E(B) = E(B) - E(0) = g^2B^2/4\pi \left[a - 11/12\pi \ln (\Lambda^2/gB)\right] +$$
$$+ 0 \left(g^4B^2\ln(\Lambda^2/gB)\right), \tag{1.4}$$

and the constant a, that in the variational calculation of Ref. 1 could not be determined, was evaluated in Ref. 2 by a lattice Monte Carlo calculation to be for SU(2)

$$a = 10.3 \pm 0.2. \tag{1.5}$$

$\Delta E(B)$ attains its minumum for

$$gB^* \simeq 10^{-5} \Lambda^2, \tag{1.6}$$

where its value is

$$\Delta E(B^*) \simeq -1.2 \; 10^{-12} \Lambda^4. \tag{1.7}$$

The proportionality of gB^* and $\Delta E(B^*)$ to the ultraviolet cutoff Λ clearly shows the genuine character of the essential instability of the PGS, for unlike (1.2) the result (1.6) does not allow any tuning of the "bare" coupling constant g_o to produce a condensate $gB^* \sim \Lambda^2_{QCD}$, with Λ_{QCD} finite. It thus turns out that, contrary to the fundamental assumption of AF, the dynamics of the non-abelian gauge system does not exhibit a second scale Λ_{QCD} besides the cut-off scale Λ. Deprived of this important scale, the dividing line between short and long distances, PQCD looses any meaning and with it GAQCD.

We may now ask ourselves what is the physics of a Savvidy state. As I have discussed at length in Ref. 7, it can be summarized in the following terms:

(i) a fully gauge covariant magnetic field condenses with strength

$$\langle F_{ik}{}^{ab}\rangle \sim \varepsilon_{ikr}\, u_r\, \delta^{ba}\, g_o^2 B^* \sim 0(\Lambda^2); \tag{1.8}$$

(ii) the lines of force of the chromoelectric field around a colour charge are in the direction $\vec{u}$, the direction of the constant chromomagnetic field;

(iii) color is confined, i.e. an isolated color charge has an infinite energy;

(iv) the dynamics of gauge-fields is restricted to a set of finite energy "Landau states" that have the following features:

a) their spectrum is given by

$$\omega(p) = \sqrt{p^2 + \mu_g^2} \qquad \mu_g^2 = 4gB^* \exp - 8\pi^2/g^2$$

b) they are polarized in a direction perpendicular to $\vec{u}$;

c) in color space they point in a direction orthogonal to η^a;

v) rotational invariance is obviously broken by the preferred direction $\vec{u}$.

The points (i) to (iv) are evidently quite good and promising, however (v) is definitely bad. Should we then conclude that Savvidy states have nothing to do with the real QCD ground state? The real question is thus whether we can find a state with all the characteristics of a Savvidy state but (v).

A study of the mechanism by which a Savvidy state lowers its energy density with respect to the PGS shows that in order to gain the enormous energy density $\Delta E(B^*)$ of Eq. (1.7) one does not need a spatially homogeneous chromomagnetic field, like the one introduced by Savvidy, but only one that is constant within a "chromomagnetic needle", i.e. a tubular domain with area $A \simeq 2\pi/gB^* = O(1/\Lambda^2)$ and height $h \simeq 1/\mu_g$ (μ_g is the finite gluon mass), oriented along an arbitrary direction $\vec{u}$.

By breaking space into a myriad of such chromomagnetic

domains, randonly oriented in all possible directions, just like the molecules of a liquid, one can show [11] that one can gain an extra

$$\Delta E_{CML} \simeq - \mu_g^2 \; (gB^*/2). \tag{1.9}$$

Such energy gain stems from the release of the energy required to "polarize" in the same direction $\vec{u}$ all the needle-shaped chromomagnetic domains where a spatially homogeneous magnetic field B^* condenses. In this way a Chromo Magnetic Liquid (CML) arises, where the properties (i) to (iv) remain intact inside each "chromomagnetic needle", and at the same time there holds:

(v') rotational invariance is restored by the rotation of the "chromomagnetic needles".

Is this the real QCD vacuum, the actual non-perturbative ground state? I clearly possess no rigorous proof that this is the case. However the conjecture that the real QCD ground state is the CML stands a very good chance of being correct, for its energy density is by far the lowest that has been determined up to now for any configuration of the non-abelian gauge-field system. This being said, in the rest of these lectures I shall proceed as though the CML were in fact the real QCD ground state.

1.3 The Primitive World (PW): the true free-world of QCD

As I have shown in Ref. 11, in the Savvidy state of minimum energy the dynamics of colour charges is governed by the Lagrangian density ($\vec{u}$ denotes the direction of $\vec{B}^*$)

$$L_{\vec{u}}(x) = \sum_{a=1}^{8} \left[- 1/4 \; f^{\,a}_{\mu\nu,\vec{u}}(x) \; F^{\mu\nu,a}_{\vec{u}}(x) + g_{\vec{u}} J^{\mu a}_{\vec{u}}(x) \; A^{a}_{\mu\vec{u}}(x) \right], \tag{1.10}$$

where we have adopted the usual distinction, that holds in a medium, between the field intensities $F^a_{\mu\nu,\vec{u}} \equiv (\vec{E}^a_{\vec{u}}, \vec{B}^a_{\vec{u}})$ and the field magnitudes $f^a_{\mu\nu,\vec{u}} \equiv (\vec{D}^a_{\vec{u}}, \vec{H}^a_{\vec{u}})$.

Writing $n_\mu \equiv (0,\vec{u})$, the relationship between these two tensors turns out to be given by the following simple equations:

$$F^a_{\mu\nu,\vec{u}} = \partial_\mu A^a_{\nu\vec{u}} - \partial_\nu A^a_{\mu\vec{u}} + g_{\vec{u}} A^b_{\mu\vec{u}} A_\nu{}^c_{\vec{u}} f_{abc} \tag{1.11}$$

$$f^a_{\mu\nu,\vec{u}} = \varepsilon^{\alpha\beta}{}_{\mu\nu}(n) F^a_{\alpha\beta,\vec{u}} \tag{1.12}$$

where

$$\varepsilon_{\mu\nu}{}^{\alpha\beta}(n) = -(\delta^\alpha{}_\mu n_\nu n^\beta + \delta_\nu{}^\beta n_\mu n^\alpha). \tag{1.13}$$

Furthermore the $\vec{u}$-sheet charge $g_{\vec{u}}$ can be written as

$$g_{\vec{u}} = g_o N_{\vec{u}}^{-1/2}, \tag{1.14}$$

where $N_{\vec{u}}$ is the number of different "discrete" $\vec{u}$-directions. Please note this normalization, which weighs the square of the charge of a coloured particle coupled to the colour field on a given $\vec{u}$-sheet (magnetic needle) with the probability that it possesses to find itself in that particular $\vec{u}$-sheet; evidently

$$\sum_{\vec{u}} g^2_{\vec{u}} = g^2_o.$$

It is now a simple exercise to show [6] that due to the peculiar structure of $\varepsilon^{\alpha\beta}{}_{\mu\nu}(n)$ (Eq. (1.11)), the potentials $A^a_\mu(x)$ do not carry any autonomous dynamical degrees of freedom, exactly as it happens in 1+1 space-time dimensions, where the gauge field, having no transverse degrees of freedom, possesses only a two-dimensional Coulombic structure.

The CML structure now requires that in order to obtain the full theory one sum over all different configurations, i.e. the full lagrangian of the theory is obtained from summing over all lagrangians $L_n(x)$ [Eq. (1.10)], where n_μ is a generic spacelike vector with $n^2 = -1$. It is then immediate to check that in this way we obtain precisely the lagrangian of Anisotropic Chromo Dynamics (ACD) [6], which I proposed over ten years ago in order to construct a "phenomenological" gauge theory of colour confinement. The key idea of ACD was just to force upon 3+1 dimensional space-time a 1+1 dimensional gauge dymamics, which is well known to have two highly desirable properties: colour confinement and freedom at short distances. A way to do this, it turned out, was to enlarge Minkowski space-time by means of the 4-vector n_μ ($n^2 = -1$), thus introducing locally the notion of a direction $\vec{u}$, that specifies the relevant 1+1 dimensional gauge dynamics for that particular space-time event. In QCD, through the condensation of the superstrong magnetic fields of the CML, one sees that the local direction $\vec{u}$ is nothing but the direction of such fields; in this way ACD emerges as the <u>effective</u> dynamical theory of the quantum fluctuations of the quark and the (residual) color-fields around the CML. We can express the conclusions of the above discussion in the following simple way:

$$\mathrm{QCD} \underset{\mathrm{CML}}{\rightarrow} \mathrm{ACD}.$$

As discussed at length in Ref. 7, in ACD one encounters three types of fields defined on a $\vec{u}$-sheet (or equivalently inside a chromomagnetic needle oriented along the

$\vec{u}$-direction). These are:

(i) the "longitudinal" gluon fields $A^a_{\mu,\vec{u}}(x)$ (a=1,...,8);

(ii) the Dirac quark fields $q_{i,\vec{u}}(x)$ (i=1,2,3 is the color index);

(iii) the "transverse" gluons $\eta^a_{\mu,\vec{u}}(x)$ (a = 1,2,...,8).

The fate of the longitudinal fields is quite obvious. Having no autonomous dynamics they can be integrated away in the path integral to yield an effective Hamiltonian

$$H_c(t) = \sum_{\vec{u}} g^2_{\vec{u}}/2 \int d\vec{x}\, d\vec{x}'\, J^{\mu a}_{\vec{u}}(\vec{x},t)\, J^{\nu a}_{\vec{u}}(\vec{x}',t)\, D_{\vec{u}}(\vec{x}-\vec{x}') \cdot$$

$$\cdot\, (g_{\mu\nu} + n_\mu\, n_\nu), \tag{1.15}$$

with

$$D_{\vec{u}}(\vec{x}) = \lim_{\lambda\to 0} \int d^3k/(2\pi)^3 \exp(-i\,\vec{k}\vec{x}) \frac{1}{(\vec{k}\vec{u})^2+\lambda^2} \tag{1.16}$$

The Hamiltonian H_c is the crucial ingredient of ACD, it describes an instantaneous interaction between the colour currents associated to the $\vec{u}$-sheet quark ($q_{\vec{u}}$) and gluon ($\eta^a_{\mu,\vec{u}}$) fields, alluded to above. The structure of the "propagator" function $D_{\vec{u}}(\vec{x})$ [Eq.(1.16)] is infrared singular when $\lambda \to 0$, thus leading to an equally divergent self energy for any system whose colour charge is non-zero: and this is the clear manifestation of the fundamental colour-confinement of ACD.

For the systems that have a finite energy - the hadrons - the mobile structure of the CML suggests the introduction of "collective quark fields"

$$q(x) = \sum_{\vec{u}} \frac{1}{N_{\vec{u}}^{1/2}} q_{\vec{u}}(x), \tag{1.17}$$

that average out their instantaneous belonging to a particular $\vec{u}$-sheet (or equivalently a chromomagnetic needle). Note that such averaging makes a well defined physical sense only for those states whose lifetimes are longer than the periods of revolutions of the chromomagnetic needles. In the following we shall assume that this is the case for all hadronic states. Thus we may rightly conclude that colour-singlets forget about the $\vec{u}$-sheet structure. In this way we have regained a well defined contact with the basic description of the quark model, where the vacuum looks perfectly tame, and the quark fields signal the presence of a highly non-trivial vacuum structure only in their bizarre confined behaviour.

Now that we have gone so far as to obtain a solid basis for confinement in QCD, our strategy to solve this fundamental theory consists of two steps:

(I) Integrate out all $\vec{u}$-fields in the path-integral keeping the collective field fixed;

(II) Solve the resulting effective theory of the "colour degrees of freedom" (quarks and gluons) by expanding in the gauge coupling constant g and in the effective non-local interactions generated in step (I).

The main aim of step (I) is to construct the spectrum of the Primitive World (PW), the true (physical) QCD free-particle world, that is populated by zero-width (infinite lifetime) HADRONS.

The possibility to develop a new kind of perturbative

expansion in a theory, like ACD, that seems so hopelessly non-perturbative, stems from a very peculiar property of the Hamiltonian H_c [Eq. (1.15)]. Let's put ourselves in the rest-frame of the generic hadron, whose only qualification at this point is to be a finite energy eigenstate of the hamiltonian H_c. By expressing the colour-currents $j^{a\mu}_{\vec{u}}(x)$ in terms of the creation and annihilation operators of the collective quark fields (1.17) we realize at once that the only (potentially) infrared singular terms are those associated with the operators b^+b - the quark-number operator - and d^+d - the antiquark-number operator (a similar structure can be derived for transverse gluons). All other terms - b^+d, b^+b^+, ... - are associated with terms that are finite for $\lambda \to 0$. Calling $J_{\mu\vec{u}}^{(0)}$ the currents solely constructed with the b^+b and d^+d terms, by selecting in H_c $H_c^{(0)}[J_{\mu\vec{u}}^{(0)}]$, i.e. the collection of terms containing the $J^{(0)}$-currents, it is obvious that $H_c^{(0)}$ conserves <u>separately</u>, on a fixed time-hyperplane, the number-operators N_q, $N_{\bar{q}}$ and N_g.

We are now in a position to precisely define the PW as the huge Hilbert space spanned by those states that diagonalize $H_c^{(0)}$, and are therefore characterized by a <u>fixed</u> number of quarks, antiquarks and gluons that comprise colour-singlets. Adopting the notation (N_q, $N_{\bar{q}}$, N_g), normal mesons are of type (1,1,0), baryons (3,0,0), glueballs (0,0,2) or (0,0,3), exotic qq-$\bar{q}\bar{q}$ states (2,2,0) etc. Neglecting the other terms of the Hamiltonian H_c, as is done in the PW, one obtains a world of free, zero-width hadrons which should be a good approximation to QCD-ACD, if the neglected part of the Hamiltonian can indeed be treated perturbatively(*).

(*) Actually the PW is only the subspace of the Hilbert space of the eigenstates of $H_c^{(0)}$ that is one-particle irreducible, i.e. it cannot be reduced to simpler hadrons. For instance the state (2,2,0) that describes two meson states obviously does not belong to the PW.

It is interesting to note that the existence of a PW was that feature of the quark model that rendered it most puzzling vis a vis a full fledged quantum field theoretical approach, that QCD was supposed to embody. How could a non-perturbative relativistic field theory produce a set of permanently bound states - the hadrons - that could be characterized by a <u>fixed</u> number of fundamental quanta? However the success, well beyond all reasonable expectations, of the naive quark picture was the palpable illustration that "Nature knew better". And it is now very gratifying that QCD-ACD (which from now on I will call QACD) ties the puzzling successes of the naive quark model to the dominance of the $H_c^{(0)}$-hamiltonian, an occurrence that can indeed be checked numerically in the course of our perturbative expansion in H_c' , defined as

$$H'_c = H_c - H_c^{(0)}, \tag{1.18}$$

to which we are now going to turn our attention.

1.4 <u>Beyond the PW: the right PQCD</u>

In the next lecture we shall see that due to the modest value of g (or better of the "dimensionally transmuted" string tension $\mu \simeq .5$ GeV) H_c' can indeed be treated as a perturbation. The PW, the world of the naive quark model (obviously extended to glue-carrying states), gets thus perturbed by H_c' , and a new type of perturbation theory in the creation of $q\bar{q}$-pairs can be systematically developed. From our discussion it seems justified to call this perturbative theory the <u>true PQCD</u>. For instance in first order in H'_c the ρ-meson will acquire a width through the diagram

$<\pi\pi|H'|\rho> \quad =$ 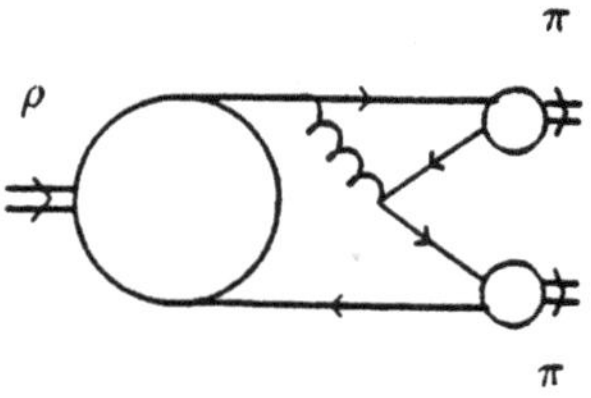

I may now conclude this Lecture by stressing the main achievement of a research program that has been carried out over a time-period of about twenty years: a solution of QCD seems to have finally emerged that apparently realizes many dreams of the hadron physicist, like

(i) confinement is well understood and described;

(ii) the simple structure of an extended quark model is retrieved;

(iii) the real world can be systematically computed by means of a well-defined and convergent perturbative approach in the creation of $q\bar{q}$-pairs.

It really looks as though the book of QCD has finally lost its seventh seal!

2. ACD AND THE QUARKS

In this second Lecture I will make a sketch of a few applications of the ACD strategy, described in the first Lecture, to a number of problems in heavy and light quark physics. Let me first briefly recall the main steps of our strategy:

i) construct the PW, i.e. diagonalize the infrared singular piece of $H_c^{(0)}$ of the current-current Hamiltonian H_c;

ii) couple the irreducible states of the PW through the infrared regular piece H'_c of the Hamiltonian H_c.

As we shall see, step (i) requires setting up the appropriate Schroedinger equations with fully relativistic kinematics in the space of fixed number of quarks and antiquarks. Step (ii) will require the careful definition and evaluation of the matrix elements of the perturbative Hamiltonian H'_c between three and more states of the PW, whose space-momentum is in general different from zero.

In order to achieve this, it will be seen, one needs construct the hadronic wave-functions for any space-momentum in agreement with relativistic invariance.

A possible solution of this important problem along the lines of Ref. 11 will be presented below, and the first relevant applications will be discussed in the talk by L. Gamberale.

2.1 The relativistic Schroedinger equation

Let's write down the QACD-Hamiltonian:

$$H_{QACD} = H_{KIN} + H_C^{(0)} + H'_c, \qquad (2.1)$$

where the kinetic Hamiltonian is given by:

$$H_{KIN} = \int d^3x \cdot \bar{q}(\vec{x},t) \left[i \vec{\gamma} \vec{\nabla} + M\right] q(\vec{x}, t), \qquad (2.2)$$

and M is the quark mass-matrix(*). According to the preceding discussion the kinetic Hamiltonian contains only the collective quark fields, i.e. those that propagate from one chromomagnetic domain to another.

(*) At this level of our discussion, where we ignore the problem of chiral symmetry breaking, the masses appearing in M are the "constituent" quark masses, which would stem from the rearrangement of the vacuum of a spontaneously violated approximate chiral symmetry. For a discussion of this problem see [9].

As for the $H_c^{(0)}$-piece of the Hamiltonian, which alone will be retained in the construction of the PW, we rewrite it as follows:

$$H_c^{(0)} = 1/2 \sum_{\vec{u}} g^2_{\vec{u}} \int d^3x \, d^3x' J^{(0)}_{\mu \vec{u},a}(\vec{x},t) \, J^{(0)}_{\nu \vec{u},a}(\vec{x}',t)$$

$$D_{\vec{u}} \, (\vec{x} - \vec{x}') \left[g_{\mu\nu} + n_\mu n_\nu \right]. \qquad (2.3)$$

On a given $\vec{u}$-sheet this Hamiltonian gives rise to the following diagrams:

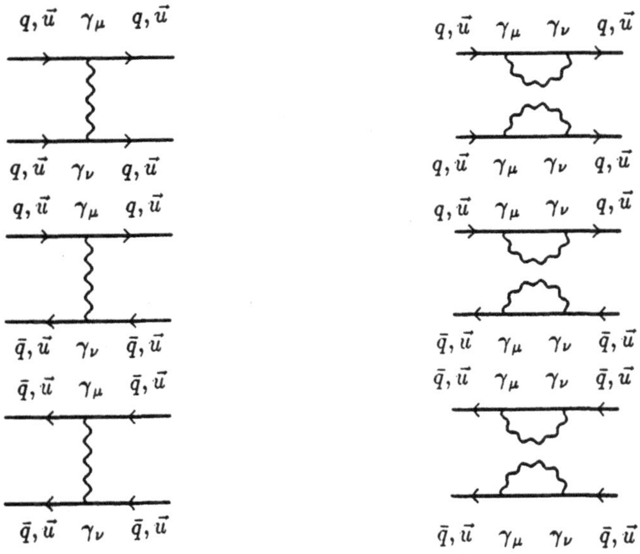

where the diagrammatical rules are

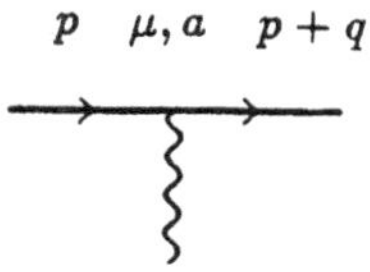

$$= \bar{u}(p+q,\vec{u}) \gamma_\mu \, \lambda^a/2 \; u(p,\vec{u})$$

for the quark vertex,

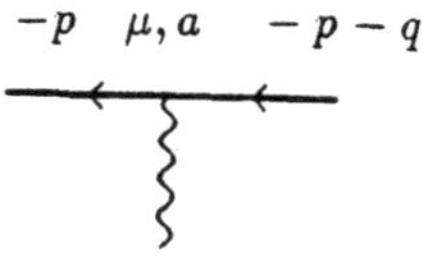

$$= \bar{v}(-p,\vec{u})\gamma_\mu \; \lambda^a/2 \; v(-p-q,\vec{u})$$

for the antiquark vertex

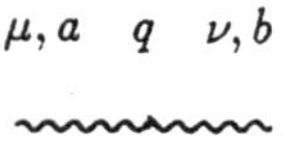

$$= \left[g_{\mu\nu} + n_\mu n_\nu\right] \frac{-i}{q^2 + \lambda^2} \cdot \delta_a^b$$

for the "longitudinal" gluon propagator.

Note that the momenta p,q, ... are one-dimensional, for they point in the $\vec{u}$-direction, reflecting the 1+1 dimensional character of QCD inside a "chromomagnetic needle". The above diagrams with their rules allow one to compute the Hamiltonian $H_c^{(0)}$ on a given $\vec{u}$-sheet.

By projecting this expressions, with specially defined projectors (*), on the collective quark (antiquark) states one can uniquely compute the POTENTIAL to be used in the spectrum equation

$$(H_{KIN} + H_c^{(0)}) \; |N_q, N_{\bar{q}}, N_g\rangle = E\,(\vec{p}) \; |N_q, N_{\bar{q}}, N_g\rangle, \tag{2.4}$$

(*) See, e.g., the contribution of M. Scorletti at this School.

which will take the form:

$$\sum_{i=q,\bar{q},g} \sqrt{\vec{k}_i^{\,2} + m_i^2} \quad \psi_{\vec{p}}(\vec{k}_1,\vec{k}_2,\ldots) +$$

$$+ \int V\,(\vec{k}_1,\vec{k}_2,\ldots;\vec{k}_1{}',\vec{k}'_2\ldots)\;\psi_{\vec{p}}(\vec{k}'_1,\vec{k}'_2\ldots) =$$

$$= E(\vec{p})\;\psi_{\vec{p}}(\vec{k}_1,\vec{k}_2,\ldots) \qquad (2.5)$$

of many-quark Schroedinger equations for the wave-functions $\psi_{\vec{p}}(\vec{k}_1,\vec{k}_2\ldots)$ with relativistic kinematics ($\sqrt{\vec{k}^2+m^2}$ instead of $\vec{k}^2/2m + m$) with a well-defined color-potential V.

These equations are central for the construction of the Primitive World, and QACD provides them free of ambiguities.

2.2 Structure of the Spectrum

So far extensive analysis has been carried out for the $q\bar{q}$-system only, i.e. in the sector (1,1,0) of the PW. The basic input parameters for the determination of the meson states are

(a) the string tension

$$\mu^2 \sim \;(gB^*)\;\; g_o^2/N_{\vec{u}} \quad \underset{\exp}{\rightarrow} \quad (.5\div.6\text{ GeV})^2 \qquad (2.6)$$

(b) the "constituent" quark masses

$$m_{u,d} = .3 \text{ GeV}$$

$$m_s = .43 \text{ GeV}$$

$$m_c = 1.58 \text{ GeV} \qquad (2.7)$$

$$m_b = 4.95 \text{ GeV}$$

$$m_{top} = ?$$

As I have mentioned above the notion of constituent mass stems from the Nambu description of spontaneous breaking of chiral symmetry. The relevant spectrum equation is in the $q\bar{q}$-meson case ($\vec{p}=0$):

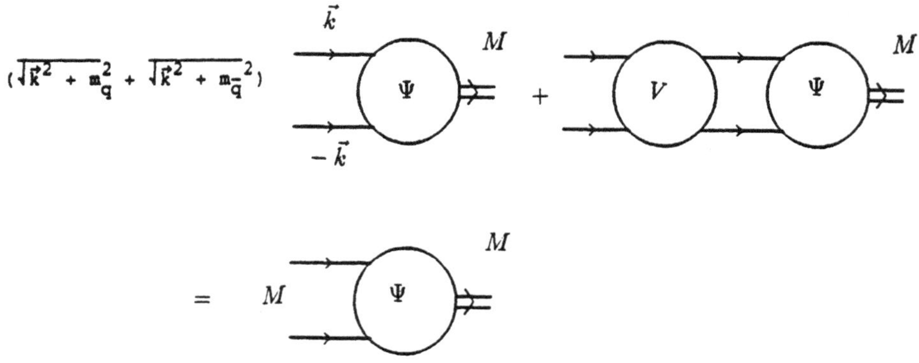

(2.8)

where the potential V, is given by the diagrams:

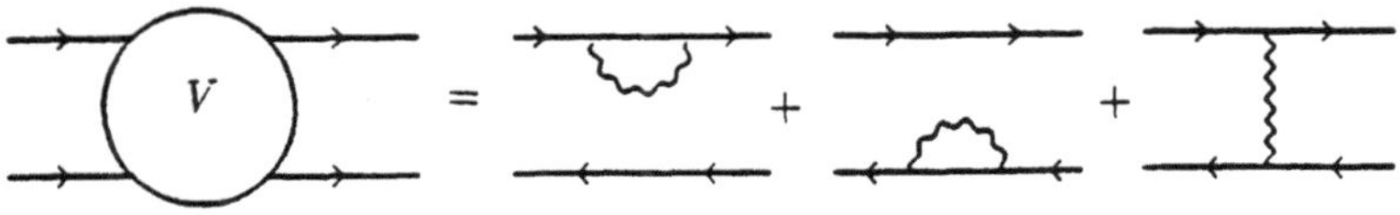

(2.9)

It is the ACD discovery that for a colour singlet [colour w.f. = $\delta_a^b\ 1/\sqrt{3}$ (a,b = 1,2,3)] the $1/\lambda$ terms of the potential cancel, leaving behind a potential term

$$V \sim \mu^2 r, \tag{2.10}$$

linear in the distance between the $q\bar{q}$-pair. This potential, which is responsible for confinement, is spin and flavor independent. The complete evaluation of the diagrams in Eq. (2.9) produces a set of other terms that are both spin and flavour dependent (through the different constituent masses) that have been evaluated in Ref. [10]. The salient features of this calculations are

(i) structure and ordering of the levels as expected by the naive quark model:

$$J^{PC} = 0^{-+},\ 1^{--},\ 0^{++},\ 1^{++},\ 2^{++},\ 1^{+-}, \ldots;$$

(ii) linear Regge trajectories;

(iii) spin-spin and spin-orbit interaction of the right size and sign;

(iv) heavy quark systems as generally expected.

2.3 Wave-functions in any frame

As already stressed, a most important problem in any Schroedinger equation approach to hadrons and their spectrum is the necessary recovery of the relativistic invariance of the full theory. This problem is especially relevant in view of the fact that the main (confining) part of the forces is instantaneous.

The strategy we have adopted for approximately solving this problem has been presented in Ref. 11. In essence the problem is to find the wave-function $\psi_{ns}(\vec{p},\vec{k})$, satisfying the $\vec{p}\neq 0$ Schroedinger equation:

$$\sqrt{(\vec{p}/2+\vec{k})^2+m_q^2}+\sqrt{(\vec{p}/2-\vec{k})^2+m_{\bar{q}}^2}\quad \psi_{sn}(\vec{p},\vec{k})+$$

$$+\int d^3k'/(2\pi)^3\; V_{ss'}(\vec{p},\vec{k},\vec{k}')\;\psi_{s'n}(\vec{p},\vec{k}')=$$

$$=\sqrt{\vec{p}^2+M_n^2}\;\psi_{sn}(\vec{p},\vec{k}), \tag{2.11}$$

once the problem at $\vec{p}=0$ has been solved, yielding a spectrum of masses M_n, and a set of corresponding wave-functions $\psi_{ns}(\vec{0},\vec{w})$.

Let's consider the $q\bar{q}$-state $|\vec{p},\vec{k}>$ describing two quarks of total momentum $\vec{p}$ and relative momentum $\vec{k}$ in a given bound state n; its relation with the corresponding state $|\vec{0},\vec{w}>$ in the rest frame of the same bound state can be expressed as some non-linear relationship

$$\vec{w}_n=\vec{w}_n(\vec{p},\vec{k}), \tag{2.12}$$

induced by the Lorentz transformation that brings the state of mass M_n to move with momentum $\vec{p}$.

In particular we may assume, as in the free case, that $\vec{w}_\perp = \vec{k}_\perp$, i.e. the momentum components transverse to the total momentum $\vec{p}$ remain identical.

Let's now consider $V_{ss'}(\vec{p},\vec{k},\vec{k}') = \langle\vec{p},\vec{k}s|\Theta_{oo}|\vec{p},\vec{k}'s'\rangle$ for $\vec{k} = \vec{k}'$. Under such Lorentz tranformation one obtains (s,s' are the quark helicities)

$$V_{ss'}(\vec{p},\vec{k},\vec{k}) = ch^2\omega \cdot V_{ss'}(\vec{0},\vec{w},\vec{w}), \qquad (2.13)$$

where $\omega=\omega(\vec{p},\vec{k})$ is the boost angle that the transformation (2.12) induces on the minkowskian space tangent to ($\vec{k}$, k_o)$^{(*)}$, and is obviously given by (we take direction of $\vec{p}$ as the 3-direction)

$$ch\omega = (\partial w_3/\partial k_3). \qquad (2.14)$$

Eq. (2.13) then follows from the observation that $V_{ss'}(\vec{p};\vec{k},\vec{k})$ is the 00-component of a second-rank tensor, and from the vanishing of its 03- and 33- components at $\vec{p}=0$ (due to the vanishing of total momentum). Setting now

$$\psi_{ns}(\vec{p},\vec{k}) = (E_n/M_n)^{1/2}\ \psi_{ns}(\vec{0},\vec{w})\ (\partial w_3/\partial k_3)^{1/2}, \qquad (2.15)$$

so as to enforce the relativistic normalization [11]

$$\int d^3k/(2\pi)^3\,|\psi_{ns}(\vec{p},\vec{k})|^2 = 2E_n, \qquad (2.16)$$

we may try to fix the structure of Eq. (2.12), by requiring that (2.11) reduce to the $\vec{p}=0$ equation:

(*) Note that k_o is the 0-component of the vector k_μ that is also non-linearly related to $\vec{w}$ and its 0-component w_o. However, for a Schroedinger equation with an instantaneous potential the dependence of the w.f. on the 0-component of the 4-vector k_μ is washed out by integration.

$$\left(\sqrt{\vec{w}^2 + m_q^2} + \sqrt{\vec{w}^2 + m^2_{\bar{q}}}\right) \psi_{ns}(\vec{0},\vec{w}) +$$

$$+ \int d^3w'/(2\pi)^3 \; V_{ss'}(\vec{0},\vec{w},\vec{w}') \; \psi_{ns'}(\vec{0},\vec{w})$$

$$= M_n \; \psi_{ns} \; (\vec{0}, \vec{w}). \qquad (2.17)$$

This impossible looking problem can be solved approximately very easily if we take advantage of the strong peaking of the confining potential for $\vec{k} = \vec{k}'$. Setting then

$$V_{ss'}(\vec{p},\vec{k},\vec{k}') \simeq (\partial w_3/\partial k_3)^{3/2} \; (\partial w_3/\partial k'_3)^{1/2} \cdot$$

$$\cdot \; V_{ss'}(\vec{0},\vec{w},\vec{w}'), \qquad (2.18)$$

which for $\vec{k} = \vec{k}'$ reduces to (2.13) and (2.14), we can rewrite (2.11) as:

$$\left(\sqrt{(\vec{p}/2 + \vec{k})^2 + m_q^2} + \sqrt{(\vec{p}/2 - \vec{k})^2 + m^2_{\bar{q}}}\right) \psi_{ns} \; (\vec{0},\vec{w})+$$

$$+ (\partial w_3/\partial k_3) \int d^3w'/(2\pi)^3 \; V_{ss'}(\vec{0};\vec{w},\vec{w}') \; \psi_{ns'}(\vec{0},\vec{w}')$$

$$= \sqrt{\vec{p}^2 + M_n^2} \;\; \psi_{ns} \; (\vec{0}, \vec{w}), \qquad (2.19)$$

which upon eliminating the potential term by means of (2.17), yields for $\partial w_3/\partial k_3$ the following simple differential equation:

$$\partial w_3/\partial k_3 = \frac{\sqrt{(\vec{p}/2+\vec{k})^2 + m_q^2} + \sqrt{(\vec{p}/2-\vec{k})^2 + m_{\bar{q}}^2} - \sqrt{\vec{p}^2 + M_n^2}}{\sqrt{\vec{w}^2 + m_q^2} + \sqrt{\vec{w}^2 + m_{\bar{q}}^2} - M_n} \tag{2.20}$$

This equation can be very easily solved, and with it our problem.

2.4 Form factors and the 3-meson vertex

Given a local current $J_\mu(\vec{x},t)$, its matrix elements between two meson states are given by the diagram:

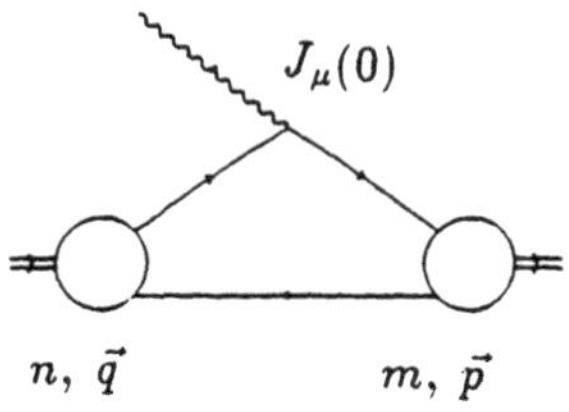

$= \langle \vec{p}, m | J_\mu(0) | \vec{q}, n \rangle$

+ antiquarks insertions

These matrix elements can now be calculated in any frame and have approximately the correct transformation properties. Due to the approximate nature of our solution of the relativistic boost problem it is convenient to prescribe a well defined reference frame for performing our calculations, which for $(p-q)^2<0$ we may choose to be the

Breit frame, i.e. $\vec{p}=-\vec{q}$ (*).

As for the most important 3-meson vertex, our main tool to go beyond the PW:

$\langle \vec{p}_2 n_2 , \vec{p}_3 n_3 | H_c'(0) | \vec{p}, n_1 \rangle =$

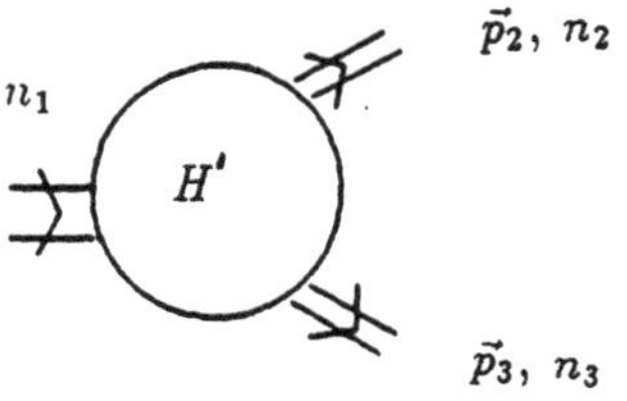

the diagrams which correspond to it are

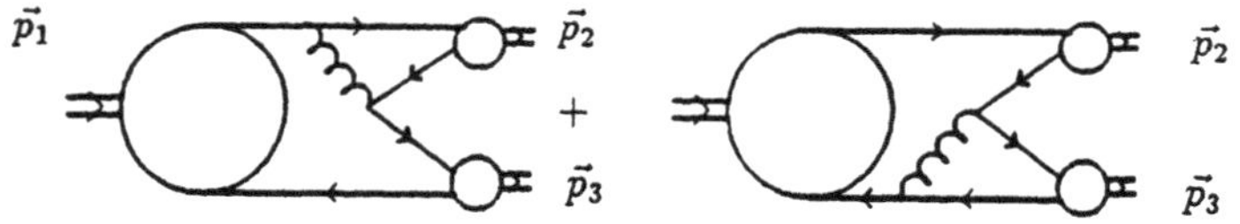

They can now be fully evaluated for any three inhabitants of the Primitive World (**). The results are quite encouraging.

(*) See the contribution of M. Gibilisco at this School.

(**) The first results for $V \to PP$ decay, such as $\rho \to \pi\pi$ are discussed in L. Gamberale's talk at this School.

CONCLUSIONS

I should like to believe that from the developments presented in these Lectures, beyond the technical details that must be acquired through an attentive study of the literature, the reader has been able to perceive that in the last few years a new non perturbative approach to QCD has reached a stage of maturity.

The need of a radically different formulation of the problems and the approximations of generally accepted QCD (GAQCD) has been throuroughly discussed in the first Lecture. It stems from a problematics that is fifteen years old [8], and has required for its correct assessment and interpretation several years of hard work both technically and psychologically.

And it was a sweet surprise that at the end of such theoretical parabola, that showed the untenability of GAQCD, a phenomenological approach to the physics of hadrons and quark confinement, Anisotropic Chromo Dynamics (ACD) [6], could be directly linked to the physical realization of QCD about a very peculiar (probable) ground state, the Chromo Magnetic Liquid (CML). In this way the destruction of the dogmas of GAQCD has been seen to contain in a very clear way the seeds of a reconstruction that looks extremely promising.

Finally I hope that through these Lectures and the communications of my pupils you have been able to fully capture both the intellectual challenge and the promise of the ideas on which we are now involved. And I also hope that this will give you a strong motivation to look seriously into these developments and to honor them with the attention of your criticism.

REFERENCES

[1] G. Preparata, Nuovo Cimento 96A, 366 (1986).

[2] L. Cosmai and G. Preparata, Phys. Rev. Lett. 57, 2613 (1986); L. Gamberale, G. Preparata, S.-S. Xue', Nuovo Cimento 105A, 309 (1992).

[3] L. Maiani, G. Martinelli, G.C. Rossi, M. Testa: Nucl. Phys. B273, 275 (1986).

[4] G. Preparata, Nuovo Cimento 96A, 394 (1986).

[5] P. Castorina and M. Consoli, Phys. Lett. B213, 493 (1988).

[6] G. Preparata, Phys. Lett. 102B, 327 (1981).

[7] G. Preparata, Nucl. Phys. B279, 235 (1987); Nuovo Cimento 103A, 1073 (1990).

[8] G.K. Savvidy, Phys. Lett. 71B, 133 (1977).

[9] P. Castorina, P. Cea, P. Colangelo, G. Nardulli and G. Preparata, Phys. Lett. 115B, 487 (1982).

[10] J.L. Basdevant, G. Preparata, Nuovo Cimento 67A, 19 (1982); J.L. Basdevant, P. Colangelo and G. Preparata, Nuovo Cim. 71A, 445 (1982).

[11] L. Cosmai, M. Pellicoro and G. Preparata, Nucl. Phys. B228, 31 (1983).

Common Problems and Trends of Modern Physics, pp. 91–97
ed. T. Bressani, S. Marcello and A. Zenoni
©1992 World Scientific Publishing Company

DETECTION OF WEAKLY INTERACTING PARTICLES BY NEARLY PERFECT SINGLE CRYSTALS WITH HIGH DEBYE TEMPERATURES

JOE WEBER
University of Maryland
College Park, MD 20742, USA

and
Center for Relativity, University of Texas
Austin, TX 78712, USA

Abstract

Nearly perfect crystals which have a well defined Mossbauer effect, have unusually large scattering cross sections for weakly interacting particles such as neutrinos and antineutrinos. This is a consequence of the fact that large momentum may be exchanged at a single site or a small number of sites, without excitation of phonons.

As in the Mossbauer effect, momentum exchanged at a nucleus results in recoil of the entire crystal as a single entity. The identity of the scattering site or sites cannot be established by subsequent measurements, and this leads to large coherence effects. Total cross sections are then proportional to the square of the number of scatterers.

1 Introduction

Consider first the diffraction of light by two slits as shown in Figure 1. For light of wavelength λ and slit separation L, the angle Θ_D giving the first diffraction minimum is given by

$$\Theta_D = \frac{\lambda}{2L} \tag{1}$$

The photons propagate in the x direction and their momentum p_x is given by

$$p_x = \frac{2\pi\hbar}{\lambda} \tag{2}$$

It is well known that the diffraction pattern will be destroyed if apparatus is provided to determine through which slit the photons pass. This occurs in the following way. The width of each slit is W. To determine through which slit the photon passes, the y component of the photon position must be determined to precision W. The uncertainty principle then implies that there is an uncertainty Δp_y in the y component of photon momentum given by

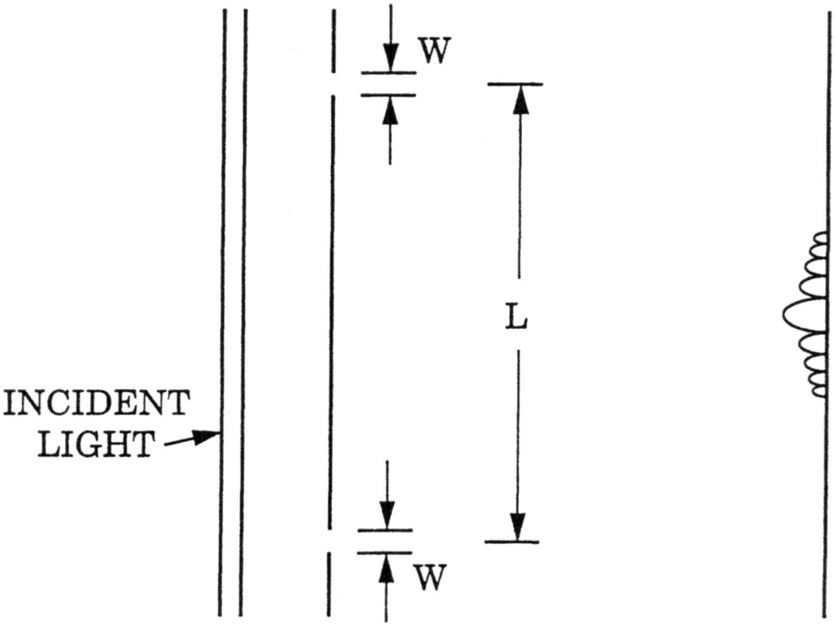

Figure 1

$$\Delta p_y \sim \frac{\hbar}{W} \tag{3}$$

(3) implies an uncertainty $\Delta\Theta_u$ in the photon direction given by

$$\Delta\Theta_u \sim \frac{\Delta p_y}{p_x} = \frac{\lambda}{2\pi W} \gg \Theta_D \tag{4}$$

The inequality (4) follows because $W \ll L$. The angular spread due to the momentum uncertainty far exceeds the diffraction angle for the first minimum. Identification of the photon scattering (diffraction) site destroys the sharp interference needed to produce a well defined diffraction pattern.

Now let us apply these ideas to the coherent scattering of neutrinos and antineutrinos. If a liquid such as perchlorethylene is employed, and a neutrino is scattered by quarks in a single nucleus, momentum is exchanged. The coordinate momentum distribution of the scatterer nucleus is changed as a result of the scattering. In principle, the momentum of each nucleus could be measured before and after scattering. Differences then reveal the scattering site.

If a solid such as lead is employed as a detector, and a neutrino is scattered by a quark in a lead nucleus, the nucleus will be displaced momentarily, then return to its permanent site, emitting phonons. Observation of these phonons again identifies the scattering site. Clearly then there will be no significant coherence effects for single nucleus scattering if a liquid is employed, or a solid in which phonons are produced by scattering.

1.1 Scattering by a Nearly Perfect Crystal With High Debye Temperature

The situation may be quite different if a single nearly perfect crystal with high Debye temperature is employed as a particle detector. Such a crystal will have a very well defined Mossbauer effect. If a nucleus emits a gamma ray, no phonons may be emitted because the crystal stiffness is very great. The entire crystal may recoil as a single entity in consequence of the exchanged momentum associated with emission of a gamma ray.

Similarly if an incident particle is scattered by quarks in a single nucleus, again the entire crystal may recoil with no phonon excitation. The identity of the scattering nucleus cannot be established by subsequent measurements. Any nucleus could have been the scatterer. The same kind of interference and coherence effects may occur as in multiple slit diffraction.

2 Theory

We assume that all volume elements containing scatterers have the interaction Hamiltonian for a weakly interacting incident particle with one scatterer,

$$H' = G \int \psi_{FS}^{*} \psi_{FP}^{*} \psi_{IS} \psi_{IP} d^{3}x \tag{5}$$

In (5) G is a coupling constant. ψ_{FS} is the final state scatterer wavefunction, ψ_{FP} is the final state wavefunction of the incident particle, ψ_{IS} is the wavefunction of the scatterer initial state, and ψ_{IP} is the wavefunction of the incident particle initial state. Let the incident particle wavefunction be

$$\psi_{IP} = \frac{1}{\sqrt{V}} e^{i\overline{p}_{IP}\cdot\overline{r}/\hbar} \tag{6}$$

before scattering and

$$\psi_{FP} = \frac{1}{\sqrt{V}} e^{i\overline{p}_{FP}\cdot\overline{r}/\hbar} \tag{7}$$

after scattering. V is the volume, p_{IP} and p_{FP} are the initial and final state momenta, respectively, of the incident particles.

Scatterers are assumed to be at points in a crystal lattice. Suppose the nth scatterer has the wavefunction product of delta functions

$$\psi_{S}^{*}\psi_{S} = \delta\left(x - n_{x}b\right)\delta\left(y - n_{y}b\right)\delta\left(z - n_{z}b\right) \tag{8}$$

Scatterers are spaced b units of length in the x, y, and z directions. Thus the nth scatterer has coordinates $x = n_{x}b$, $y = n_{y}b$, $z = n_{z}b$. If momentum $\Delta\overline{p}$ is exchanged by the n_{th} scatterer, (8) becomes

$$\psi_{FS}^{*}\psi_{IS} = \delta\left(x - n_{x}b\right)\delta\left(y - n_{y}b\right)\delta\left(z - n_{z}b\right)e^{-i\Delta\overline{p}\cdot\overline{r}/\hbar} \tag{9}$$

In (9) we have chosen $\psi_{FS} = \psi_{IS}e^{i\Delta\overline{p}\cdot\overline{r}/\hbar}$ because for ψ_{IS} having momentum p,

$$\langle p \rangle_{\substack{\text{FINAL}\\ \text{SCATTERER STATE}}} = \int \psi_{S}^{*} e^{-i\Delta\overline{p}\cdot\overline{r}/\hbar}\left(-i\hbar\nabla\right)\psi_{S} e^{i\Delta\overline{p}\cdot\overline{r}\hbar} d^{3}x \tag{10}$$

Evaluating (10) gives

$$\langle p \rangle_{\substack{\text{FINAL}\\ \text{SCATTERER STATE}}} = \overline{p}_{S} + \Delta\overline{p} \tag{11}$$

With the wavefunctions (9), the interaction Hamiltonian (5) is evaluated for the case where any scatterer could have exchanged $\Delta\overline{p}$, with the scatterer site unidentifiable by subsequent measurements.

$$H' = \frac{G}{V}\sum_{n=1}^{n=N} e^{i(\overline{p}_{IP}-\overline{p}_{FP}-\Delta\overline{p})\cdot\overline{r}_n/\hbar} \tag{12}$$

The transition probability W is given by

$$W = \frac{2\pi}{\hbar}|H'|^2\rho(E) \tag{13}$$

In (13) $\rho(E)$ is the density of final states. The formal mathematics leading to (13) also guarantees conservation of energy. We can imagine any given experiment translated in any direction with no changes in cross section. The scattering process has translational invariance. Therefore momentum is also conserved, and

$$\overline{p}_{IP} - \overline{p}_{FP} - \Delta\overline{p} = 0 \tag{14}$$

(14) implies that the final momentum of the crystal center of mass is uniquely determined by

$$\Delta\overline{p} = \overline{p}_{IP} - \overline{p}_{FP} \tag{15}$$

To each final state $\overline{p}_F$ of the scattered particle, there corresponds one and only one final state for the crystal center of mass. Therefore the total number of final states may be computed considering only the final states of the scattered particles. For zero rest mass incident particles

$$\rho(E) = \frac{p^2 V d\Omega}{8\pi^3\hbar^3 c} \tag{16}$$

In (16) $d\Omega$ is the element of solid angle into which incident particles are scattered. Expressions (12), (13), (14), (15), and (16) give

$$W = \frac{G^2N^2p_{FP}^2 d\Omega}{4\pi^2\hbar^4 cV} \tag{17}$$

If the crystal has a very high Debye temperature, $\Delta\overline{p}$ can be at least as large as $2\overline{p}$ without excitation of phonons. In this case all scattering angles are possible with the N^2 dependence given by (17). $d\Omega \to 4\pi$ and (17) becomes

$$W = \frac{G^2N^2p_{FP}^2}{\pi\hbar^4 cV} \tag{18}$$

The cross section for this process, σ, is obtained by division of (18) by the incident flux c/V,

$$\sigma = \frac{G^2N^2p_{FP}^2}{\pi\hbar^4 c^2} \tag{19}$$

2.1 Crystal Recoil Energy

If the entire crystal of mass M recoils as a single entity the recoil energy is

$$E_{RECOIL} = \frac{(\Delta p)^2}{2M} \tag{20}$$

For zero rest mass incident particles, the ratio of recoil energy to incident particle energy cp is given by

$$\frac{E_{RECOIL}}{E_{INCIDENT}} = \frac{(\Delta p)^2}{2MCP} \rightarrow \frac{(Cp)}{MC^2} \tag{21}$$

(21) is roughly the ratio of the incident particle energy to the rest energy of the entire crystal—a very small quantity. Therefore this kind of scattering is very nearly perfectly elastic.

2.2 Observable Cross Sections

An observed cross section is a summation of all possible processes which contribute to the final state of incident particles and scatterers. We must, therefore consider for the high Debye temperature crystal, additional processes in which 2, 3, 4 ... N–1 unidentifiable scatterers exchange total momentum $\Delta\overline{p}$ and scatter one neutrino.

Suppose S unidentifiable scatterers exchange $\Delta\overline{p}$, not necessarily in equal fractions with

$$\Delta p_x = \sum_{j=1}^{j=s} \frac{\Delta p_x}{v_{jx}}, \Delta p_y = \sum_{j=1}^{j=s} \frac{\Delta p_y}{v_{jy}}, \Delta p_z = \sum_{j=1}^{j=s} \frac{\Delta p_z}{v_{jz}} \tag{22}$$

For S scatterers we have a contribution to H′ which is

$$\frac{G}{V} \sum_{j=1}^{j=s} e^{i\left(\overline{p}_{IP} - \overline{p}_{FP} - \frac{\Delta\overline{p}}{v_{jx}}\right)_X Xj + i\left(\overline{p}_{IP} - \overline{p}_{FP} - \frac{\Delta\overline{p}}{v_{JY}}\right)_Y Yj + i\left(\overline{p}_{IP} - \overline{p}_{FP} - \frac{\Delta\overline{p}}{v_{jz}}\right)_Z Z_j} \tag{23}$$

Boundary conditions restrict v_{jx}, v_{jy}, v_{jz} to integers between 1 and $\frac{pL}{2\pi\hbar}$ for a cubic crystal with edge length L.

Corresponding to (12) we require

$$H' = \frac{G}{V} \sum_j \sum_n e^{i\left(\overline{p}_{IP} - \overline{p}_{FP} - \frac{\Delta\overline{p}}{v_{Jx}}\right)_X Xn/\hbar + i\left(\overline{p}_{IP} - \overline{p}_{FP} - \frac{\Delta\overline{p}}{v_{JY}}\right)_Y Yn/\hbar + i\left(\overline{p}_{IP} - \overline{p}_{FP} - \frac{\Delta\overline{p}}{v_{jz}}\right)_Z Zn/\hbar} \tag{24}$$

Clearly

$$v_{jx} = v_{jy} = v_{jz} = v_j \tag{25}$$

This corresponds to a process in which v_j scatterers collectively exchange momentum Δp with an incident particle. On average the v_{jx} terms will sum to $v_j^{\frac{1}{3}}$ and contribute a term

$$\frac{G^2 p_{FP}^2}{\pi \hbar^4 c^2} \left(\frac{N}{v_j}\right)^2 v_j \tag{26}$$

to the total cross section.

Therefore the total cross section, summing over all allowed v_j is

$$TOTAL = \frac{G^2 N^2 p_{FP}^2}{\pi \hbar^4 c^2} \left(1 + \frac{1}{2} + \frac{1}{3} + \frac{1}{4} + \dots \frac{2\pi\hbar}{p_{FP} L}\right) \tag{27}$$

(27) is based on the assumption that the Debye temperature is sufficiently high so that all of the incident particles are scattered without phonon excitation. Appropriate corrections for available materials have been discussed in the literature[1,2]

REFERENCES

1. J. Weber, *Phys. Rev.* C31, 4, 1468–1475 April 1985.

2. J. Weber, *Phys. Rev.* D38, 1, 32–39 July 1, 1988

Common Problems and Trends of Modern Physics, pp. 99–104
ed. T. Bressani, S. Marcello and A. Zenoni
©1992 World Scientific Publishing Company

GRAVITATIONAL RADIATION BAR ANTENNA WITH END COUPLED OSCILLATOR INSTRUMENTATION AND TWO COUPLED HARMONIC OSCILLATORS

JOE WEBER
Center for Relativity, University of Texas
Austin, TX 78712, USA

and

University of Maryland
College Park, MD 20742, USA

Abstract

Two coupled harmonic oscillators have two normal mode frequencies. Excitation by an external short pulse results in energy in the two normal mode frequencies after the external pulse drops to zero.

A bar gravitational radiation antenna and an end face coupled harmonic oscillator with resonant frequency close to one of the bar normal modes, is in some ways equivalent to the two harmonic oscillators. However, a short external pulse may excite phonons in the bar. These phonons remain after the external pulse drops to zero. The response is different from that of the two coupled harmonic oscillators.

I Introduction

Elastic solid and interferometer gravitational radiation antennas were invented in 1957. Theory and apparatus development have been continuing.[1,2,3,4,5,6] Supernova 1987A observations[7] and correlations with neutrino detectors have been reported. Here we begin serious study of great differences in sensitivity when changes are made in instrumentation.

About 25 years ago successful stable operation was reported with ferroelectric crystals at the center of the bar. The bar strains give output voltages of the ferroelectric crystals. Low noise amplifiers and demodulators then process the information for recording and computer analyses. (Figure 1).

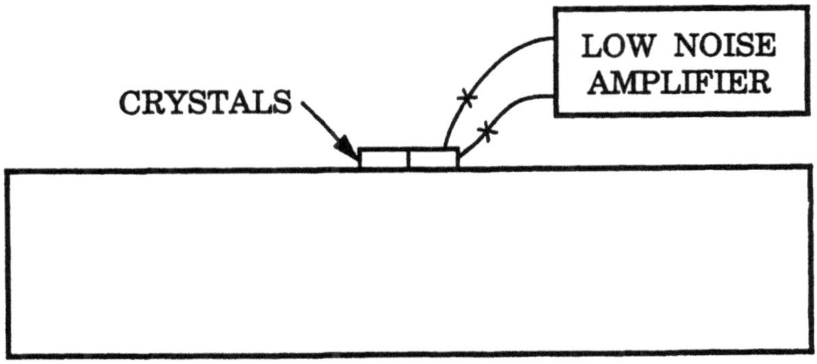

FIGURE 1. Bar With Center Ferroelectric Crystals

Many ideas have been contributed for improving sensitivity of such antennas. An early Maryland antenna employed a second degree of freedom (Figure 2). A superconducting inductance at low temperature is arranged such that the electromagnetic degree of freedom has a normal mode frequency close to the bar normal mode of interest. Energy from an external pulse may excite the electromagnetic degree of freedom and be observed.

The second degree of freedom can be a small mechanical system coupled to the end of the bar. If the bar energy can be transferred to the small mechanical system, much larger displacements result than displacements of the ends of the bar.

There are differences between the single mode center of bar crystal instrumentation and the instrumentation which employs coupled mass at the end of the bar.

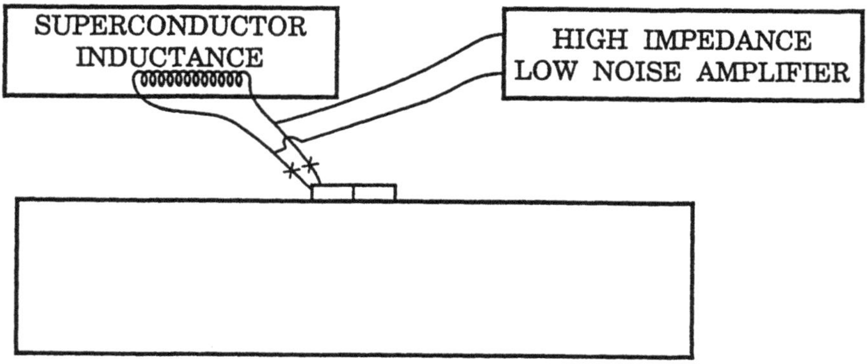

FIGURE 2. Bar With Electromagnetic Degree of Freedom Coupled to the Lowest Frequency Normal Mode

2 Two Coupled Harmonic Oscillators

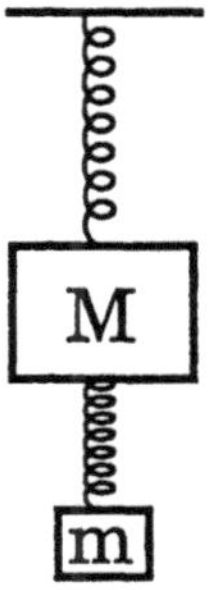

FIGURE 3. Two Coupled Oscillators

Consider the coupled harmonic oscillators shown in Figure 3. Suppose the masses are M and m, and the spring constants are K and k. Equations of motion for the coordinate X_M of M, and X_m of m are

$$M\frac{d^2X_M}{dt^2} + KX_M + k(X_M - X_m) = F \tag{1}$$

$$m\frac{d^2X_m}{dt^2} + k(X_m - X_M) = 0 \tag{2}$$

Suppose first that the force F is zero. Solutions are

$$X_M = A\sin\omega_n t \tag{3}$$

$$X_m = B\sin\omega_n t \tag{4}$$

1, 2, 3, and 4 give

$$\omega_n^2 = \frac{k}{2m} + \frac{K}{2M} + \frac{k}{2M} \pm \sqrt{\frac{k^2}{4m^2} + \frac{K^2}{4M^2} + \frac{k^2}{4M^2} - \frac{Kk}{2Mm} + \frac{k^2}{2mM} + \frac{Kk}{2M^2}} \tag{5}$$

Now we choose

$$\omega_0^2 = \frac{k}{m} = \frac{K}{M}, m \ll M \tag{6}$$

(5) and (6) give

$$\omega_n^2 = \omega_0^2 + \frac{\omega_0^2 m}{2M} \pm \omega_0^2\sqrt{\frac{m}{M} + \frac{m^2}{4M^2}} \approx \omega_0^2\left(1 \pm \sqrt{\frac{m}{M}}\right) \tag{7}$$

Let

$$\omega_1 = \omega_0\left(1 + \frac{1}{2}\sqrt{\frac{m}{M}}\right) \tag{8}$$

$$\omega_2 = \omega_0\left(1 - \frac{1}{2}\sqrt{\frac{m}{M}}\right) \tag{9}$$

These equations give

$$B = \frac{A\omega_0^2}{\omega_0^2 - \omega_n^2} \tag{10}$$

Equations (8), (9), and (10) give

$$B \approx \pm A\sqrt{\frac{M}{m}} \tag{11}$$

Now we will assume that a harmonic force $F \sin \omega t$ is applied for a short period. (1) and (2) give

$$X_M = \frac{F\left(\omega_0^2 - \omega^2\right)\sin\omega t}{M\left(\omega_0^2 - \omega^2\right)^2 - m\omega^2\omega_0^2} + A_1 \sin\omega_1 t + A_2 \sin\omega_2 t \tag{12}$$

$$X_m = \frac{F\omega_0^2 \sin\omega t}{M\left(\omega_0^2 - \omega^2\right)^2 - m\omega^2\omega_0^2} - A_1\sqrt{\frac{M}{m}}\sin\omega_1 t + A_2\sqrt{\frac{M}{m}}\sin\omega_2 t \tag{13}$$

For $X_M = 0 = X_m$ at $t = 0$ and $\dot{X}_M = 0 = \dot{X}_m$ at $t = 0$, A_1 and A_2 are evaluated with,

$$\begin{aligned} X_M \;=\; & \frac{F\left(\omega_0^2 - \omega^2\right)\sin\omega t}{M\left(\omega_0^2 - \omega^2\right)^2 - m\omega^2\omega_0^2} + \frac{F\omega\left[\omega_0^2\sqrt{\frac{m}{M}} - \left(\omega_0^2 - \omega^2\right)\right]\sin\omega_1 t}{2\left[M\left(\omega_0^2 - \omega^2\right)^2 - m\omega_0^2\omega^2\right]\omega_1} \\ & - \frac{F\omega\left[\omega_0^2\sqrt{\frac{m}{M}} + \left(\omega_0^2 - \omega^2\right)\right]\sin\omega_2 t}{2\left[M\left(\omega_0^2 - \omega^2\right)^2 - m\omega_0^2\omega^2\right]\omega_2} \end{aligned} \tag{14}$$

$$X_m = \frac{F\omega_0^2 \sin\omega t}{M\left(\omega_0^2-\omega^2\right)^2 - m\omega^2\omega_0^2} - \frac{F\omega\left[\omega_0^2-\left(\omega_0^2-\omega^2\right)\sqrt{\frac{M}{m}}\right]\sin\omega_1 t}{2\left[M\left(\omega_0^2-\omega^2\right)^2 - m\omega_0^2\omega^2\right]\omega_1}$$
$$- \frac{F\omega\left[\omega_0^2+\left(\omega_0^2-\omega^2\right)\sqrt{\frac{M}{m}}\right]\sin\omega_2 t}{2\left[M\left(\omega_0^2-\omega^2\right)^2 - m\omega_0^2\omega^2\right]\omega_2} \tag{15}$$

When the external force F drops to zero the first terms in (14) and (15) are zero. Terms containing the normal mode frequencies w_1 and w_2 remain. (14) and (15) give

$$X_M \approx \sqrt{\frac{m}{M}} X_m \tag{16}$$

This implies that the maximum energies of M and m are equal. The energy oscillates between m and M. It is most important to stress that as soon as external force F is removed, all of the energy is in the two normal modes at angular frequencies ω_1 and ω_2.

3 Bar with End Transducer

A small harmonic oscillator may be coupled to the end of the bar, as described by Professor H. J. PAIK (Figure 4). The Hamiltonian for a bar may be written as a sum over normal modes. The energy of the coupled harmonic oscillator contributes additional terms to the Hamiltonian. Constants are chosen so that the resonant frequency of the coupled harmonic oscillator is close to a bar normal mode frequency.

In some ways the coupled harmonic oscillator and close frequency bar mode are similar to the coupled oscillators described by equations 1–16.

There are differences which may be important. The quantum theory of the bar antenna would permit excitation of a single pulse of frequency much higher than the bar normal mode frequency as shown in Figure 5.

This group of phonons may propagate from end to end of the bar. Center crystal instrumentation is immediately excited by a succession of these pulses. The harmonic oscillator at the end of the bar responds to the fundamental component of the recurrent short duration signals and requires a long time for reaching its energy maximum.

Unlike the two mode operation of the pair of coupled masses of the first part of this paper, the single sine wave externally induced excitation does not drop to zero when the external source becomes zero. The external source signals continue for a very long time, and the coupled oscillator at the bar end may not absorb its maximum energy until a long period has elapsed.

FIGURE 4. Bar With End Coupled Harmonic Oscillator

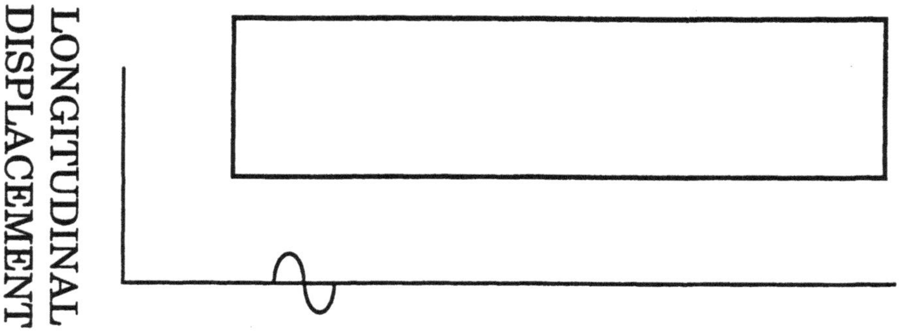

FIGURE 5. High Frequency Pulse Excitation of a Bar

REFERENCES

1. J. Weber, *Phys. Rev.* **117**, 306, (1960), General Relativity and Gravitational Waves, Interscience–John Wiley, New York, London, 1961; Chapter 8 General Relativity and Gravitation, Volume 2, A. Held (Editor) Plenum Publishing Company 1980; Chapter 1, Volume 3, Sir Arthur Eddington Centenary Symposium, World Scientific 1986.
2. Moss, Miller, Forward, Applied Optics, **10**, 2495 (1971).
3. C. Kittel, Introduction to Solid State Physics, Chapter 4, Fifth Edition, John Wiley and Sons, 1976.
4. G. Preparata, Modern Physics Letters **A1** 1990.
5. B.S. DeWitt in Gravitation, L. Witten (Editor) John Wiley and Sons 1962.
6. J. Weber, *Phys. Rev. Letter.* **18**, 478 (1967).
7. M. Aglietta Et Al, Il Nuovo Cimento **12**, 75, 1988, **14C**, 2, 1991.

Common Problems and Trends of Modern Physics, pp. 105–123
ed. T. Bressani, S. Marcello and A. Zenoni
©1992 World Scientific Publishing Company

EXCLUSIVE SEMILEPTONIC AND NON-LEPTONIC DECAYS OF HEAVY FLAVOUR MESONS IN ANISOTROPIC CHROMODYNAMICS (ACD) THEORY

MARINA GIBILISCO
Departement of Physics - University of Milan
Via Celoria 16, 20133 Milano - Italy

ABSTRACT:

Two research programs, the short-distance (SD) and the long-distance (LD) ones, can be adopted in order to shed light on the $\Delta I = \frac{1}{2}$ enhancement in the $K \to \pi\pi$ and hyperon decays and to study non-leptonic physics. In this paper I work in LD approach and I use the meson wave-functions of the ACD theory to obtain a complete, totally parameter-free description of the D, D_s, and B exclusive weak decays, either semileptonic or non-leptonic. After the calculation of the form factors parametrizing the matrix elements, I can compute the decay widths; in particular, in the non-leptonic case, I consider those decays where the final states are two pseudoscalar (PP) or a pseudoscalar and a vector (PV) mesons. The widths thus calculated are in very good agreement with the experimental data and the W-exchange contribution turns out to be very important for D° and D_s decays, thus explaining the difference existing between the lifetimes of D^+ and D°, D_s.

1. INTRODUCTION.

The problem of the non-leptonic weak decays of hadrons is a rather old one: in particular, in the last sixties, the explanation of the observed $\Delta I = \frac{1}{2}$ enhancement in the $K^\circ \to \pi^+ \pi^-$ decay was an undelayable task for any theory aiming to understand the non-leptonic physics. The effective non-leptonic Hamiltonian was written in the Wilson representation [1]:

$$H_{NL}(0) = g^2 \int d^4x\, D_W(x)\, T(J^{+\mu}\left(\frac{x}{2}\right) J_\mu\left(\frac{-x}{2}\right)) \tag{1.1}$$

where the propagation function $D_W(x)$ has an effective support $| m_W^2 x^2 | \leq 1$. The essential question in handling (1.1) is: how can we treat the current product for space-time distances $|x| \leq \frac{1}{m_W}$?

One starts by approximating the hadronic matrix element of H_{NL} introducing the well known "vacuum saturation" (VS) factorization [2] which reduces non-leptonic physics to the semileptonic one by saturating the sum over intermediate quantum states between two charged weak currents with the "vacuum" state. Then, two different ways can be followed in order to handle the current-current product: we call them the "long distance" (LD) and the "short distance" approaches. The theoretically mysterious $\Delta I = \frac{1}{2}$ rule represented the crucial matter of the comparison between these programs, that now I will briefly examine.

1.1 THE LONG DISTANCE AND THE SHORT DISTANCE PROGRAMS.

It is to try the mistery of the $\Delta I = \frac{1}{2}$ rule that in 1968 S. Nussinov and G. Preparata [3] carried out a short-distance analysis of the time-ordered product $T(J^{+\mu}\left(\frac{x}{2}\right) J_\mu\left(\frac{-x}{2}\right))$ of two charged weak currents. The conclusions of that analysis where that the absence of short-distance singularities in a theory whose short-distance behaviour is as in free-field theory strongly suggested that the physical states $|n\rangle$ between the currents could be restricted such that the momentum transfer between them and the initial and final states is $Q^2 = |(p_{i,out} - p_n)^2| \leq 1\, GeV^2$, i.e.

$$J^{+\mu}\left(\frac{x}{2}\right) J_\mu\left(\frac{-x}{2}\right) \stackrel{x\to 0}{\longrightarrow} \sum_n J^{+\mu}\left(\frac{x}{2}\right) |n\rangle\langle n| J_\mu\left(\frac{-x}{2}\right)\Big|_{|(p_{i,out}\; -\; p_n)^2|\; \leq\; 1\; GeV^2} \tag{1.1.1}$$

or, put differently, non-leptonic physics is <u>dominated by long-distance effects</u>. The outcome was very promising, including a "dynamical" explanation of the $\Delta I = \frac{1}{2}$ rule for hyperon weak decays [3] and non-leptonic K-decays [4,5].

In 1969 K. Wilson introduced his analysis of short-distance singularities - the Wilson's expansion - based upon the notion of anomalous dimensions and suggested that the very large numerical difference between $\Delta I = \frac{1}{2}$ and $\Delta I = \frac{3}{2}$ weak non-leptonic amplitudes could be explained by the different anomalous dimensions of the operators associated to the $\Delta I = \frac{1}{2}$ and $\frac{3}{2}$ channels. Thus, according to Wilson, writing (G is the Fermi constant)

$$H_{NL} \sim \frac{G}{\sqrt{2}} J^{+\mu}\left(\frac{x}{2}\right) J_\mu\left(\frac{-x}{2}\right)\Big|_{|x|\simeq\frac{1}{m_W}} = \frac{G}{\sqrt{2}} \sum_K C_K\, (m_W^{-2})\, O^K(0), \tag{1.1.2}$$

it should happen that

$$C^{\Delta I=\frac{3}{2}}(m_W^{-2}) \ll C^{\Delta I=\frac{1}{2}}(m_W^{-2}) \tag{1.1.3}$$

as a result of different SD anomalous dimensions.

With the discovery of the Asymptotic Freedom [6] of QCD Wilson's suggestion could finally be checked, with the following result [7,8]:

$$\frac{c^{\frac{1}{2}}(m_W^{-2})}{c^{\frac{3}{2}}(m_W^{-2})} \simeq \left[1 + \frac{b\alpha_s(\mu)}{6\pi}\, ln\left(\frac{m_W}{\mu}\right)\right]^{0.72} \simeq 4 \tag{1.1.4}$$

($\mu = 1\ GeV$, $b = 33 - 2n_f$) which is unable to justify the experimentally known enhancement of the $\Delta I = \frac{1}{2}$ channel. Thus the LD program, giving a satisfactory answer to this problem, sets oneself as the right approach to the non-leptonic physics.

2. NOTES ON ACD THEORY.

Anisotropic ChromoDynamics is the theory of strong interactions proposed by Giuliano Preparata in 1980, with the purpose to find an answer to an old and unsolved problem of QCD, i.e. the color-confinement. In fact, ACD was formulated with the belief that the only natural description of confinement emerges in the framework of 2-dimensional gauge theories.

In this paper I want just to recall the fundamental ideas of ACD Theory: a detailed discussion about it can be found in [9,10]. The starting point is the assumption that the dynamics of color is locally isomorphic to the dynamics of a 2-dimensional QCD (1 spatial + 1 temporal dimension). Hence, one needs a new geometry for the space-time, realized on a 7-dimensional manifold [9] which is a collection of usual Minkowskian worlds labelled by a variable n_μ, forming n-sheets. The observed isotropy of space-time is recovered by demonstrating [9] that the hadron dynamics can be globally expressed in terms of the collective quark fields, obtained by averaging over the sheet variables n_μ.

A deep investigation about the ground state of the theory showed the irremediable instability of the perturbative ground state [11,12], thus requesting a new concept to take care of this problem: the so-called "Chromomagnetic Liquid" (CML) [13], formed by a large number of needle-shaped magnetic domains, is the best candidate to be the real ground state of QCD, having all suitable properties (Lorentz and Rotational invariance, color confinement, freedom at short distances). Moreover, it is up to now the state of the chromodynamical field with lower energy content.

The first step in the construction of the hadronic world is the determination of the spectrum of the "primitive world": in fact, the confining Hamiltonian of ACD can be splitted [10] into two pieces:

$$H_C = H_C^{(0)} + H_C' \tag{2.1}$$

where $H_C^{(0)}$ is the confining, infrared singular part and $H_C^{'}$ describes pair creation and has no singularity in the infrared region. Due to this peculiar behaviour, one can treat $H_C^{'}$ perturbatively whereas under no circumstance this can be done with $H_C^{(0)}$; thus the diagonalization of the first part of the Hamiltonian produces the "primitive hadronic world" where the hadrons are stable states with a definite number of q, $\bar{q}$. After the introduction of the perturbation $H_C^{'}$, the primitive hadrons will become unstable, due to the pair creation processes: by constructing Bethe-Salpeter relativistic wave-functions, one can generate a set of effective non-local quark and gluon interactions, corresponding to the propagation of the primitive hadrons.

2.1 THE MESON WAVE-FUNCTIONS OF ACD.

The calculation of the ACD meson wave-functions in an arbitrary moving-frame has been performed in Ref. [14]: basically, such moving-frame functions are the solutions of a Schrödinger-like equation:

$$\left[\sqrt{\left(\frac{\vec{p}}{2}+\vec{k}\right)^2+m_q^2}+\sqrt{\left(\frac{\vec{p}}{2}-\vec{k}\right)^2+m_{\bar{q}}^2}\right]\Psi_{r,s,n}(\vec{p},\vec{k})+ \\ +\int\frac{d^3K'}{(2\pi)^3}\,V_{rs}^{r's'}(\vec{p},\vec{k},\vec{k}')\,\Psi_{r's'n}(\vec{p},\vec{k}') \;=\; E_n\,\Psi_{rsn}(\vec{p},\vec{k}) \tag{2.1.1}$$

where r, s denote the quark helicities and

$$V_{rs}^{r's'}(\vec{p},\vec{k},\vec{k}') \;=\; \langle p,k,r,s|H_C^{(0)}|p,k',r',s'\rangle, \tag{2.1.2}$$

$V(\vec{0},\vec{k},\vec{k}')$ has been computed [9,10] and decomposed into four distinct contributions, depending in a different way on the spin and on the flavour.

This potential is inserted in (2.1.1) at $\vec{p}\;=\;0$ in order to find the meson spectrum in terms of five parameters only, the string tension μ^2 and the quark masses m_u, m_s, m_c, m_b: the case $\vec{p}\;=\;0$ is solved in a simple way and then the problem one has to face is the determination of the potential V at $\vec{p}\;\neq\;0$. A direct calculation is not easy, involving an elaborate analysis of the structure of colour-currents for quarks which must be performed after a boost with velocity equal to $\frac{|\vec{p}|}{E}$. In order to achieve this, one uses an approximate Lorentz transformation establishing a connection between $V(\vec{p},\vec{k},\vec{k}')$ and $V(\vec{0},\vec{k},\vec{k}')$ [14]; in this way one finds the meson wave-functions in an arbitrary moving frame.

3. THE EVALUATION OF THE MATRIX ELEMENTS AND THE CALCULATION OF THE FORM FACTORS.

3.1 STUDY OF THE CONTRIBUTIONS TO NON-LEPTONIC DECAYS.

Let's consider the matrix element:

$$\langle\beta|\;H^{ab}(0)\;|a\rangle \tag{3.1.1}$$

where $|\alpha\rangle$ and $|\beta\rangle$ are two generic hadronic states and

$$H^{ab} = \frac{G}{\sqrt{2}} : J^a_\mu J^{\mu b} : \qquad (3.1.2)$$

($J^{a,b}_\mu$ are general weak currents).

From a purely topological point of view we can classify all diagrams contributing to (3.1.1) in two classes:

(a) The disconnected diagrams (D-class),

(b) The connected diagrams (C-class).

To the D-class belong those contributions for which the individual hadrons appearing in $|\alpha\rangle$ and $|\beta\rangle$ separate in two disconnected groups. These diagrams can be generated in only two ways, by inserting the vacuum between the two currents $J^a_\mu(0)$ and $J^{\mu b}(0)$ (see Fig. 1a) or by inserting the vacuum between any "Fierz rearrangement" of the product of the two currents $J^a_\mu(0)$ and $J^{\mu b}(0)$ (see Fig. 1b). These contributions can be factorized in a simple way in terms of semileptonic processes; on the other hand, the C-class comprises all the contributions which cannot be so decomposed.

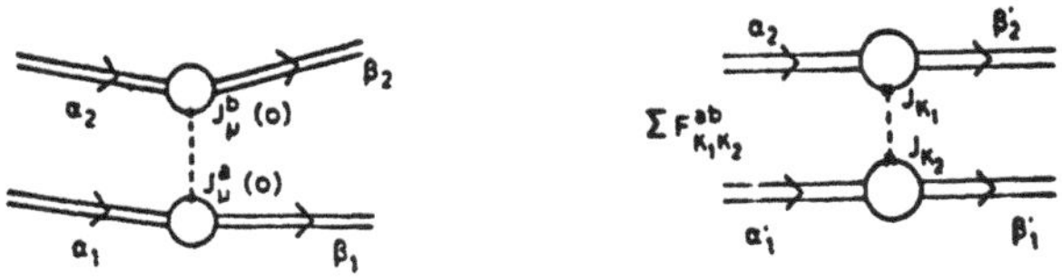

Fig. 1a: The vacuum insertion between the two currents $J^a_\mu(0)$ and $J^{\mu b}(0)$.
Fig. 1b: The vacuum insertion between any "Fierz-rearrangement" of the product of the two currents $J^a_\mu(0)$ and $J^{\mu b}(0)$.

The evaluation of the D-class diagrams requires the knowledge of the weak form factors that appear in the matrix elements: the amplitude for D (or B) decay into two pseudoscalar mesons is given by the product of the following matrix elements:

$$i)\ \langle P(p_2)|V_\mu|D(p_1)\rangle = f^+(q^2)(p_1 + p_2)_\mu + f^-(q^2)(p_1 - p_2)_\mu, \qquad (3.1.3)$$

$$ii)\ \langle P(q)|A_\mu|0\rangle = -if_P\, q_\mu. \qquad (3.1.4)$$

For a final state that comprises a pseudoscalar and a vector meson, we have two possible configurations (see fig. 2):

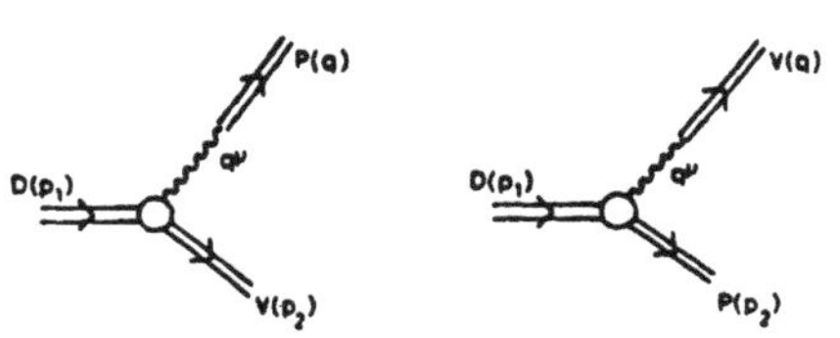

Fig. 2: The D decay into a pseudoscalar and a vector meson in the two possible configurations.

Given the relevant matrix elements:

$$iii)\ \langle V(p_2)|J_\mu|D(p_1)\rangle = i\Big[\epsilon_\mu(M_D + M_V)A_1(q^2) - \epsilon^\nu(p_1 - p_2)_\nu(p_1 + p_2)_\mu\cdot$$

$$\cdot\frac{A_2(q^2)}{(M_D + M_V)} + \epsilon^\nu(p_1 - p_2)_\nu(p_1 - p_2)_\mu\frac{2M_V(A_0(q^2) - A_3(q^2))}{q^2}\Big] +$$

$$+ \frac{2\epsilon_{\mu\nu\alpha\beta}}{(M_D + M_V)}\epsilon^\nu p_1^\alpha p_2^\beta V(q^2) \tag{3.1.5}$$

$$iv)\ \langle V(q)|V_\mu|0\rangle = f_V\epsilon_\mu \tag{3.1.6}$$

(ϵ_μ is the polarization of the vector meson), we must calculate the product of the (3.1.5) and (3.1.4) for the diagram 2a) and (3.1.6) times (3.1.3) for the diagram 2b):

A direct calculation of these matrix elements can be performed by using the ACD meson wave-functions discussed in sect. 2; the decay constant f_P, f_V for various mesons are obtained [16] in satisfactory agreement with the experimental data.

About the evaluation of (3.1.3), (3.1.5), in a first step [17] the pseudoscalar form factors $f^+(q^2)$ $f^-(q^2)$ have been calculated directly, in the Breit system, choosing the z axis of the coordinates along the direction of the spatial component of the momentum. At the present time, I perform this calculation inserting in (3.1.3) an additional current term whose origin lies in the fact that I am working in a well defined approximation (the theory is truncated by treating quark-pair creation perturbatively), that in general cannot guarantee that the "naive current" $J_\mu^{(o)}$:

$$J_\mu^{(0)} = \int\frac{d^3q\ d^3q'}{(2\pi)^3}\sqrt{\frac{m_q\ m_{\bar q}}{E_q\ E_{\bar q}}}\ \cdot$$

$$\cdot[\bar u(q)b^+(q) + \bar v(q)d(q)]\ \gamma_\mu(1-\gamma_5)\ [u(q')b(q') + v(q')d^+(q')] \tag{3.1.7}$$

is conserved (or partial-conserved) [14]. Thus, I introduce an additional term δJ_μ by the requirement:

$$\langle b|\ q^\mu\ (J_\mu^{(0)} + \delta J_\mu)\ |a\rangle = \langle b|\ D\ |a\rangle \tag{3.1.8}$$

where I have called respectively $|a\rangle$, $|b\rangle$ the generical initial and final states,

$q_\mu = (p_1 - p_2)_\mu$ and:

$$\langle b|\, D\, |a\rangle = (m_a^{(0)} - m_b^{(0)})\, \langle b|\, \bar{u}u\, |a\rangle - (m_a^{(0)} + m_b^{(0)})\, \langle b|\, \bar{u}\gamma_5 u\, |a\rangle \quad (3.1.9)$$

The additional piece δJ_μ restores the current conservation and can be evaluated in a simple way from (3.1.8) provided one knows the current quark masses $m^{(\circ)}$ in (3.1.9).

For this purpose, I use the following values [18]:

$$m_u^{(\circ)} \simeq m_d^{(\circ)} = 18\ MeV \qquad m_s^{(\circ)} = 123\ MeV \quad (3.1.10)$$

$$m_c^{(\circ)} = 1100\ MeV \qquad m_b^{(\circ)} = 4400\ MeV \quad (3.1.11)$$

3.2 THE RESULTS: FORM FACTORS

The values of the pseudoscalar form factors obtained at $q^2 = 0$ with the ACD meson wave-functions are listed in tab. 1 where one can find both theoretical and experimental results; in tabs. 2a and 2b a detailed comparison is performed with the values calculated in some different theoretical approaches.

About the results of tab. 1, note just that $f^+_{D\to K}(0)$ is in satisfactory agreement with the experimental world average [15]:

$$f^+_{D\to K}(0) = 0.69 \pm 0.04 \quad (3.2.1)$$

and with some theoretical estimates referring to QCD Sum Rules [19] or Lattice calculations [20] (see tab. 2a).

When the final pseudoscalar meson is π, I obtain in general a small value for $f^+(0)$ in comparison with other theoretical models (see tab. 2b). This fact is probably due to the uncertainty connected with the chiral role of the pion [18] which so far has not been fully taken into account in the calculation of its wave-function. Note that the additional contribution to the current in (3.1.8) is not able to increase the value of $f^+_{D\to\pi}(0)$.

Turning now to the vector mesons' form factors, in tabs. 3, 4, 5 I list the results obtained respectively for a decaying meson D, D_s and B and I compare them with the ones calculated in lattice gauge theories [20]: one can remark that these values are systematically larger than those of Lubicz-Martinelli and the difference is noteworthy especially for A_2.

A closer look to the vector form factors for $D \to K^*$ (see tab. 6) shows that my results are in agreement with most theoretical models, except for the Lattice calculation [20]. The experimental data of E691 Collaboration [21] seem to be somewhat at variance with many theoretical results; note however that the ratio $\Gamma_L(D^+ \to \bar{K}^{\circ *}\, e^+\, \nu_e)\, /\, \Gamma_T(D^+ \to \bar{K}^{\circ *}\, e^+\, \nu_e)$ derived by E691 is larger than the one obtained by WA82 [22] and MARKIII [23]; in this case, my result agrees, within the experimental errors, with all these measurements.

4. THE DETERMINATION OF THE EXCLUSIVE NON-LEPTONIC WIDTHS.

The form factors calculated in the previous section can be immediately used in order to determine the disconnected D-contributions (fig. 3a) and the essentially disconnected diagram (fig. 3b) to the non-leptonic decays: in fact, the latter is reducible to a disconnected one through a Fierz-trasformation which introduces a color-suppression factor $\xi = \frac{1}{3}$ and a minus sign.

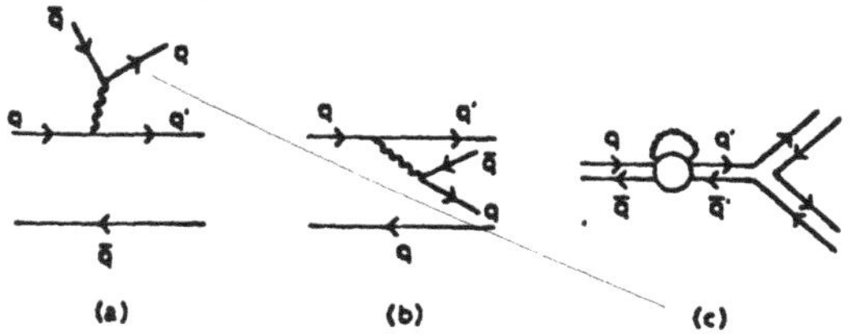

Fig. 3: The different contributions to heavy flavour meson decays.

The value of this factor is a very fundamental question concerning the study of the exclusive non-leptonic decays of heavy flavours because it affects in a remarkable way the cancellation between the different contributions to the overall amplitudes: the untenability of the value $\xi = 0$ has been discussed by G. Preparata [24].

The properly connected diagram (c) involves the exchange of a W-boson and it has been frequently considered negligible, owing to its SD helicity-suppression. In the LD program one demonstrates [25] that this is not the case: this contribution has indeed a considerable importance for D°, D_s decays, its amplitude being comparable with the disconnected one and probably producing the difference of lifetime between D^+ and D°, D_s mesons [26].

Unfortunately, one is not yet in a position to calculate the properly connected contributions, thus one must choose for them a suitable parametrization in terms of a modulus and a phase angle (see fig. 4).

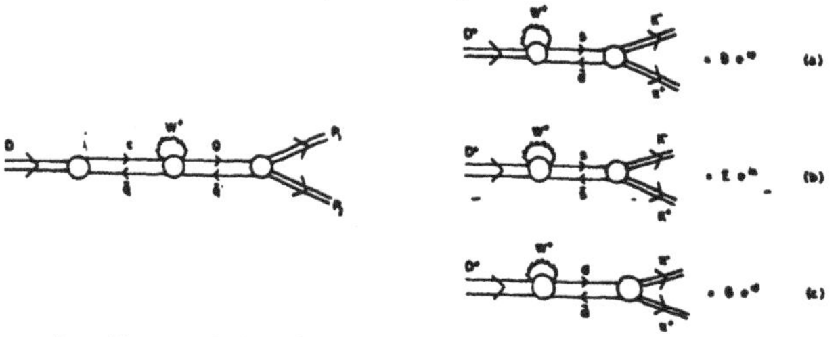

Fig. 4: The parametrization of the properly connected contributions to the decay $D^\circ \to K^-\pi^+$.

Using this parametrization, I perform a fit of those channels that have the smallest experimental errors; subsequently, I use the parameters thus obtained to calculate the remaining channels.

4.1 NON-LEPTONIC DECAYS OF D'S.

A selection of the relevant results obtained for D^+, D° and D_s exclusive non-leptonic decays is presented in tabs. 7÷12. A complete list and a detailed discussion of all calculated widths can be found in [40]; here, the following remarks will suffice:

a) One can see that this approach, in general, gives good results: I will emphasize that for the D^+ decays, the model is completely parameter-free, thus allowing, for instance, a successfully calculation of $\Gamma(D^+ \rightarrow \bar{K}^\circ\, \pi^+)$ (see tab. 7).

b) I obtain a very satisfactory agreement with the CLEO II last data for the exclusive decays $D^\circ \rightarrow PP$: after the fit of some other channels, measured by various experimental groups (see tab. 8), I can calculate the remaining widths in a completely independent way.

c) The calculated value of $\Gamma(D^+ \rightarrow \phi\pi^+)$ is completely out of the experimental range stated by E691 [27] and this failure doesn't seem to depend on the smallness of $f^+_{D\rightarrow\pi}(0)$ calculated in this approach. With regard to this problem, probably a pole diagram (see fig. 5), that I have so far neglected, this time could be important.

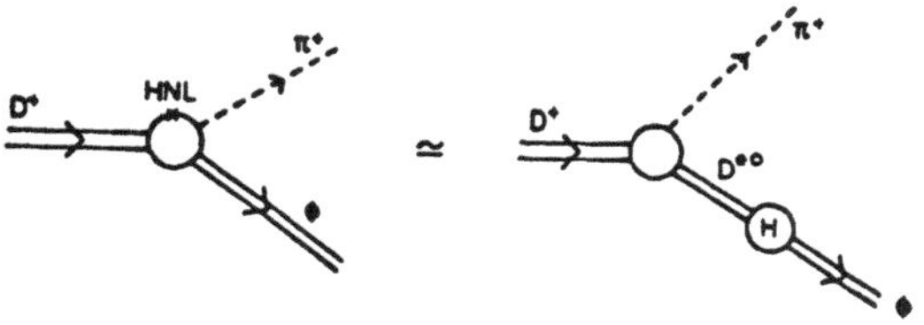

Fig. 5: A pole-model for the channel $D^+ \rightarrow \phi\, \pi^+$.

However, in order to evaluate such diagram, I need a knowledge of $\Gamma(D^{\circ *} \rightarrow D\, \pi)$, which is presently unavailable. A better determination of the experimental width is also desirable.

d) A very important result is obtained for the channel $D_s^+ \rightarrow \phi\, \pi^+$: for this decay one has no W-exchange contribution thus allowing a complete, parameter-free calculation of the amplitudes; I must note that my result is in very good agreement with all experimental data: this is a clear indication that when I are able to calculate all contributions, the model works very well.

4.2 NON-LEPTONIC DECAYS OF B'S.

The same analysis just carried out for D mesons can be successfully applied to determine the exclusive decay widths of B mesons. However, a fundamental difference does exist between D and B decays: while D mesons through the W-exchange amplitude can be mixed to well defined almost degenerate light quark resonances, B's can't, as one can easily remark from the value of their masses. Indeed, probably it is the closeness to a resonance that makes the W-exchange contribution important for D° and D_s. Thus, one expects no significant contribution from the properly connected amplitude for B and no remarkable difference between the lifetimes of $B^\pm$ and B°, as the experimental data can confirm.

In tab. 13 ÷ 15 I list my results for the exclusive widths concerning the $B \to PP$ and the $B \to PV$ channels, obtained from the evaluation of the disconnected and essentially disconnected contributions.

For the CKM matrix elements and the lifetimes, I have used [28,29]:

$$V_{cb} = 0.046 \qquad V_{ub} = 0.007 \tag{4.2.1}$$

$$\tau_{B^\pm} = 1.35 \times 10^{-12}\ sec$$
$$\tau_{B^\circ} = 1.42 \times 10^{-12}\ sec. \tag{4.2.2}$$

The values (4.2.1) are (probably) slightly high, however the error on V_{cb}, V_{ub} is not very important at this preliminary stage; in the same way, the uncertainty about the correction to vector mesons' form factors does not seem worrying at this stage of the experimental knowledge.

One must wait for more precise experimental measurements before judging these results, but I believe that one can see already a basic agreement with the presently known data.

5. THE DETERMINATION OF THE EXCLUSIVE SEMILEPTONIC WIDTH.

An important test of the form factors calculated in ACD theory is the computation of the semileptonic decay widths, which are determined by a simple evaluation of a disconnected diagram, completely known. The results obtained for the exclusive channels $D, B \to P\ e\ \nu_e$ are listed in tab. 16; the work is in progress for the case where the final meson is a vector.

The agreement with the experimental data seems good: note that I obtain:

$$\Gamma(D^+ \to \bar{K}^\circ e^+ \nu_e) = \Gamma(D^\circ \to K^- e^+ \nu_e) \tag{5.1}$$

as one can expect considering that exactly the same process and the same diagram are involved in these cases. However, this result seems to contradict the experimental data which, if confirmed, will require an indispensable theoretical explanation of this difference.

Note that the estimate of $\Gamma(B \to D\ e\ \nu_e)$ is in good agreement with the experimental data and that I can make a precise prediction for $\Gamma(B \to \pi^\circ\ e\ \nu_e)$.

AKNOWLEDGEMENT.
I desire to thank in particular Giuliano Preparata, whose contribution has been fundamental and precious for this work; I adress also a sincere thank to Giancarlo Moneti for the useful discussions about the CLEO experience and the charm physics.

REFERENCES

1 K. Wilson: *Phys. Rev.* **179**, (1969), 1499.
2 R.P. Feynman: "Consequences of SU(3) Symmetry on Weak Interactions", Proc. 1964 Int. School "E. Majorana", A. Zichichi ed. Academic Press, N.Y. (1965).
3 S. Nussinov, G. Preparata: *Phys. Rev.* **175**, (1968), 2180.
4 G. Preparata: *Phys. Lett.* **B34**, (1971), 412.
5 G. Nardulli, G. Preparata, D. Rotondi: *Phys. Rev.* **D27**, (1983), 557.
6 M. K. Gaillard, B. W. Lee: *Phys. Rev. Lett.* **33**, (1974),108.
7 G. Altarelli, L. Maiani: *Phys. Lett.* **B52**, (1974), 351.
8 J. Ellis, M.K. Gaillard, D.V. Nanopoulos: *Nucl. Phys.* **B100**, (1975), 313.
9 J. L. Basdevant, G. Preparata: *Nuovo Cimento* **A67**, (1982), 19.
10 J. L. Basdevant, P. Colangelo, G. Preparata: *Nuovo Cimento* **A71**, (1982), 445.
11 G. Preparata: *Phys. Lett.* **B201**, (1988), 139.
12 G. Preparata: *Nucl. Phys.* **B279**, (1987), 235.
13 G. Preparata: *Nuovo Cimento* **A103**, (1990), 1073.
14 L. Cosmai, M. Pellicoro, G. Preparata: *Nucl. Phys.* **B228**, (1983), 31; and *Phys. Lett.* **B121**, (1983), 272.
15 M. Aguilar-Benitez et al., Particle Data Group, *Phys. Lett* **B239** April 1990.
16 P. Cea, P. Colangelo, G. Nardulli, G. Preparata: *Phys. Lett.* **B115**, (1982), 310.
17 M. Gibilisco, G. Preparata: *Phys. Lett.* **B264** (1991), 207.
18 P. Castorina, P. Cea, P. Colangelo, G. Nardulli, G. Preparata: *Phys. Lett.* **B115**, (1982), 487.
19 C. Dominguez, N. Paver: *Phys. Lett.* **B207**, (1988), 499;
T. M. Aliev, V.L. Eletskii, Y.I. Kogan, *Sov. J. Nucl. Phys.* **40**, (1984), 527.
20 M. Crisafulli, G. Martinelli, C.T. Sachrajda: *Phys. Lett.* **B223**, (1989), 90;
V. Lubicz, G. Martinelli, C.T. Sachrajda: Rome Preprint 748 (1990).
21 J. C. Anjos et al.: *Phys. Rev. Lett.* **65**, (1990), 2630.
22 M. Adamovich et al., WA82 Coll.: *Phys. Lett.* **B268**, (1991), 142.
23 Z. Bai et al., MARKIII Coll.: *Phys. Rev. Lett.* **66**, (1991), 1011.
24 G. Preparata: *Phys. Lett.* **B193**, (1987), 312.
25 G. Preparata: "Topics on the weak decays of heavy flavoured mesons". In "Topics on Heavy Flavours", San Miniato, (1987), 403.

26 L. Angelini, L. Nitti, M. Pellicoro, G. Preparata: *Nuovo Cimento* **A99**, (1988), 45.
27 J.C. Anjos et al.: Preprint FERMILAB-PUB 87/203-E (Pub. on Phys. Rev. Lett.).
28 L3 Coll.: "Measurement of the average lifetime of b-hadrons and determination of matrix element $|V_{cb}|$" L3 Preprint, June 1991.
29 P. Abreu et al., DELPHI Coll.: "Measurement of the average lifetime of B". CERN-PPE/91-131, subm. to Zeit. Phys. C.
30 J.C. Anjos et al. *Phys. Rev. Lett.* **62**, 1587 (1989);
J. Adler et al.: *Phys. Rev. Lett.* **62**, 1821 (1989).
31 A.A. Ovchinnikov, V.A. Slobodenyuk: *Z Phys.* **C44**, 433 (1989).
32 C. Dominguez, N. Paver: *Z Phys.* **C41**, 217 (1988).
33 M. Bauer, B. Stech, M. Wirbel: *Z Phys.* **C34**, 103 (1987); ibid. **C29**, 637 (1985).
34 N. Isgur, D. Scora: *Phys. Rev.* **D40**, 1491 (1989).
35 M. Bauer, M. Wirbel: *Z Phys.* **C42**, 671 (1989).
36 F.J. Gilman, R.L. Singleton: *Phys. Rev.* **D41**, 142 (1990).
37 G. Korner, G.A. Schuler: *Z Phys.* **C38**, 511 (1988).
38 P. Ball, V. M. Braun, H. G. Dosch, M. Neubert: *Phys. Lett.* **B259**, 481 (1991).
39 R.H. Schindler: "Physics of a High Luminosity Collider operated near Charm and Tau pair Thresholds", SLAC-PUB-4996, June 1989.
40 M. Gibilisco, G. Preparata: "Theory of the Non-leptonic Decays of Heavy Flavours", Preprint MITH-92/9.

Tab. 1: Pseudoscalar form factors (after the correction).

CASE	$f^+(0)$	$f^-(0)$	REMARKS
$D \to \eta$	0.85	-0.20	
$D \to K$	0.61	0.04	$f^+_{D\to K}(0)_{exp} = 0.69 \pm 0.04$ [30]
$D \to \pi$	0.44	-0.09	
$D_s \to \eta$	0.57	0.05	
$D_s \to K$	0.55	0.03	
$B \to D$	0.76	-0.93	$f^+_{B\to D}(0)_{th} = 1.0 \pm 0.2$ [31]
$B \to K$	0.62	-0.62	
$B \to \pi$	0.61	-0.61	$f^+_{B\to\pi}(0)_{th} = 0.4 \pm 0.1$ [32]

Tab. 2a: Comparison with some different theoretical models

GROUP	$f^+_{D\to K}(0)_{calc}$	MODEL
Gibilisco - Preparata	0.61	ACD Theory
Dominguez - Paver	0.75 ± 0.05	QCD Sum Rules [19]
Aliev - Eletskii - Kogan	0.60 ± 0.10	QCD Sum Rules [19]
Crisafulli - Martinelli - Sachrajda	0.74 ± 0.17	Lattice Calculation [20]
Lubicz - Martinelli - Sachrajda	0.63 ± 0.08	Lattice Calculation [20]
Bauer - Stech - Wirbel	$0.75 \div 0.82$	Constituent Quark Model [33]

Tab. 2b: Comparison with some different theoretical models

GROUP	$f^+_{D\to\pi}(0)_{calc}$	MODEL
Gibilisco - Preparata	0.44	ACD Theory
Dominguez - Paver	0.75 ± 0.05	QCD Sum Rules [19]
Crisafulli - Martinelli - Sachrajda	0.70 ± 0.20	Lattice Calculation [20]
Lubicz - Martinelli - Sachrajda	0.58 ± 0.09	Lattice Calculation [20]

Tab. 3: Vectorial and axial form factors for the D meson

CHANNEL	FORM FACTOR	GIBILISCO-PREPARATA	DIFFERENT MODELS
			ON LATTICE:
$D \to \bar{K}^{**}$	V_0	1.07	0.86 ± 0.10
	A_1	0.65	0.53 ± 0.03
	A_2	0.57	0.19 ± 0.21
			(Lubicz-Martinelli [20])
$D \to \omega$	V_0	0.91	
	A_1	0.44	
	A_2	0.47	
$D \to \rho$	V_0	0.91	0.78 ± 0.10
	A_1	0.43	0.45 ± 0.04
	A_2	0.45	0.02 ± 0.26
			(Lubicz-Martinelli [20])

Tab. 4: Vectorial and axial form factors for the D_s meson

CHANNEL	FORM FACTOR	OUR MODEL
$D_s \to \phi$	V_0	1.13
	A_1	0.71
	A_2	0.43
$D_s \to K^{*+}$	V_0	0.95
	A_1	0.63
	A_2	0.42

Tab. 5: Vectorial and axial form factors for the B meson

CHANNEL	FORM FACTOR	OUR MODEL
$B \to D^{*+}$	V_0	0.38
	A_1	0.31
	A_2	0.05
$B \to \omega$	V_0	0.12
	A_1	0.20
$B \to \rho$	A_2	0.02

Tab. 6a: Form factors for $D \to K^*$.

MODEL	$A_1(0)$	$A_2(0)$	$V(0)$	$R_2 = A_2(0)/A_1(0)$	$R_V = V(0)/A_1(0)$
OUR	0.65	0.57	1.07	0.88	1.65
I-S [34]	0.8	0.8	1.1	1.0	1.4
B-W [35]	0.9	1.2	1.3	1.3	1.4
G-S [36]	0.8	0.6	1.5	0.75	1.9
K-S [37]	1.0	1.0	1.0	1.0	1.0
L-M [20]	0.53	0.19	0.86		

Tab. 6b: E691 Experimental form factors for $D \to K^*$ (see [21]).

Form factors	E691 Data [21]
$A_1(0)$	$0.46 \pm 0.55 \pm 0.05$
$A_2(0)$	$0.0 \pm 0.2 \pm 0.1$
$V(0)$	$0.9 \pm 0.3 \pm 0.1$
$R_2 = A_2(0)/A_1(0)$	$0.0 \pm 0.5 \pm 0.2$
$R_V = V(0)/A_1(0)$	$2.0 \pm 0.6 \pm 0.3$

Tab. 6c: Ratios Γ_L/Γ_T and Γ_+/Γ_- for the decay $D^+ \to \bar{K}^{*0} e^+ \nu_e$

MODEL	Γ_L/Γ_T	Γ_+/Γ_-
OUR	1.26	0.22
I-S [34]	1.1	
B-W [35]	0.9	
G-S [36]	1.2	
K-S [37]	1.2	
L-M [20]	1.51 ± 0.27	
E691 exp [21]	$1.8^{+0.6}_{-0.4} \pm 0.3$	$0.15^{+0.07}_{-0.05} \pm 0.03$
WA82 exp [22]	$0.6 \pm 0.3^{+0.3}_{-0.1}$	
MARKIII exp [23]	$0.5^{+1.0}_{-0.1}{}^{+0.1}_{-0.2}$	

Tab. 7: Results: $D^+ \to PP$

CHANNEL	$\Gamma_{calc.}$ (10^{11} sec^{-1})	$\Gamma_{exp.}$ (10^{11} sec^{-1})	GROUP
$D^+ \to \bar{K}^0 \pi^+$	0.34	0.26 ± 0.04 0.23 ± 0.04 ± 0.06	PDG E691
$D^+ \to \pi^+ \pi^0$	0.003	< 0.050	PDG
$D^+ \to K^+ \bar{K}^0$	0.046	0.064 ± 0.019 ± 0.012	E691
$D^+ \to \eta \pi^+$	0.009	< 0.085	PDG

Tab. 8: Results: $D^0 \to PP$

CHANNEL	$\Gamma_{calc.}$ (10^{11} sec^{-1})	$\Gamma_{exp.}$ (10^{11} sec^{-1})	GROUP
$D^0 \to \bar{K}^0 K^0$	0.028	$0.031^{+0.017}_{-0.012}$ ± 0.005	CLEO
$D^0 \to \eta \bar{K}^0$	0.12	0.20 ± 0.03 ± 0.04	CLEO
$D^0 \to \pi^0 \pi^0$	0.013	0.021 ± 0.005 ± 0.005	CLEO

Tab. 9: Results: $D^+ \to PV$

CHANNEL	$\Gamma_{calc.}$ (10^{11} sec^{-1})	$\Gamma_{exp.}$ (10^{11} sec^{-1})	GROUP
$D^+ \to \bar{K}^{*0} \pi^+$	0.21	0.55 ± 0.18 ± 0.23	MARKIII
$D^+ \to \bar{K}^{*0} K^+$	0.021	0.041 ± 0.018 ± 0.010	MARKIII
$D^+ \to \rho^+ \bar{K}^0$	0.56	0.64 ± 0.07 ± 0.21	MARKIII
$D^+ \to K^{*+} \bar{K}^0$	0.034		
$D^+ \to \omega \pi^+$	0.001	< 0.056	PDG
$D^+ \to \phi \pi^+$	0.008	0.064 ± 0.006 ± 0.011 0.054 ± 0.010	E691 PDG
$D^+ \to \rho^0 \pi^+$	0.002	0.007 ± 0.005 ± 0.001	E691

Tab. 10: Results: $D^0 \to PV$

CHANNEL	$\Gamma_{calc.}$ $(10^{11}\ sec^{-1})$	$\Gamma_{exp.}$ $(10^{11}\ sec^{-1})$	GROUP
$D^0 \to \rho^+ K^-$	1.52	1.85 ± 0.26	PDG
$D^0 \to \omega \bar{K}^0$	0.31	0.88 ± 0.36 0.81 ± 0.21 ± 0.24	PDG CLEO
$D^0 \to \bar{K}^{*0} \pi^0$	0.31	0.47 ± 0.14	PDG
$D^0 \to \rho^0 \bar{K}^0$	0.33	0.40 ± 0.14	PDG
$D^0 \to \phi \bar{K}^0$	0.17	0.19 ± 0.04	PDG
$D^0 \to \eta \bar{K}^{*0}$	0.075	< 0.33	E691
$D^0 \to K^{*+} K^-$	≤ 0.10 *	$(0.16^{+0.08}_{-0.06} \pm 0.04)$	E691

* In order to have $|\cos\alpha| < 1$

Tab. 11: Results: $D_s^+ \to PV$

CHANNEL	$\Gamma_{calc.}(10^{11}\ sec^{-1})$	$\Gamma_{exp.}(10^{11}\ sec^{-1})$	GROUP
$D_s^+ \to \phi \pi^+$	0.68 (BR = 3.0%)	0.78 ± 0.27 0.74 ± 0.36 ± 0.09 0.72 ± 0.16 ± 0.11 > 0.76 0.74 ± 0.22 0.79 ± 0.18 0.78 ± 0.22	CLEO TASSO ARGUS MARKIII HRS E691 Mean Value

We calculate also:
$\Gamma(D_s^+ \to K^{*+} \bar{K}^0)/\Gamma(D_s^+ \to \phi\pi^+) = 0.98$
Ratio (PDG) = 0.92 ± 0.32
Ratio (CLEO) = 1.2 ± 0.25

Tab. 12: Results: $D_s^+ \to PP$

CHANNEL	$\Gamma_{calc.}/\Gamma(D_s^+ \to \phi \pi^+)$	$\Gamma_{exp.}/\Gamma(D_s^+ \to \phi \pi^+)_{CLEO}$	GROUP
$D_s^+ \to \eta \pi^+$	0.65	0.52 ± 0.13 ± 0.08	PDG
$D_s^+ \to \bar{K}^0 \pi^+$ (*)	0.045	≤ 0.17	PDG

(*) Here we have the disconnected contribution only.

Tab. 13: Results: $B \to PP$

CHANNEL	$\Gamma_{calc.}$ (sec^{-1})	$\Gamma_{exp.}$ (sec^{-1})	GROUP
$B^- \to D^0 D_s^-$	0.066×10^{11}	$(0.21 \pm 0.10) \times 10^{11}$ $(0.18 \pm 0.09 \pm 0.03) \times 10^{11}$	CLEO ARGUS *
$B^- \to D^0 \pi^-$	0.020×10^{11}	$(0.021 \pm 0.010) \times 10^{11}$ $(0.015 \pm 0.006 \pm 0.004) \times 10^{11}$	PDG ARGUS
$B^- \to \pi^- \pi^0$	2.31×10^7	$< 0.17 \times 10^{10}$	PDG
$\bar{B}^0 \to D^+ D_s^-$	0.063×10^{11}	$(0.084 \pm 0.049) \times 10^{11}$ $(0.120 \pm 0.091 \pm 0.042) \times 10^{11}$	CLEO ARGUS *
$\bar{B}^0 \to D^+ \pi^-$	0.042×10^{11}	$(0.026 \pm 0.010) \times 10^{11}$ $(0.034 \pm 0.007 \pm 0.007) \times 10^{11}$ $(0.036 \pm 0.019 \pm 0.010) \times 10^{11}$	PDG ARGUS CLEO
$\bar{B}^0 \to D_s^+ \pi^-$	0.17×10^9	$< 0.91 \times 10^9$	PDG
$\bar{B}^0 \to \pi^+ \pi^-$	0.92×10^8	$< 0.63 \times 10^8$	PDG
$\bar{B}^0 \to K^- \pi^+$	0.68×10^7	$< 0.63 \times 10^8$ $< 0.91 \times 10^8$	PDG ARGUS
$\bar{B}^0 \to \pi^0 \pi^0$	0.26×10^7	$< 0.32 \times 10^7$	Crystal Ball *

* Preliminary data

Tab. 14: Results: $B^- \to PV$

CHANNEL	$\Gamma_{calc.}$ (sec^{-1})	$\Gamma_{exp.}$ (sec^{-1})	GROUP
$B^- \to D^0 \rho^-$	0.36×10^{10}	$(0.96 \pm 0.30 \pm 0.30) \times 10^{10}$	ARGUS
$B^- \to D^0 D_s^{*-}$	0.75×10^{10}	$(1.2 \pm 0.9 \pm 0.2) \times 10^{10}$	ARGUS *
$B^- \to D_s^- D^{*0}$	0.23×10^{10}	$(1.0 \pm 0.7 \pm 0.1) \times 10^{10}$	ARGUS *
$B^- \to D^{*0} \pi^-$	0.35×10^8	$(0.29 \pm 0.10 \pm 0.09) \times 10^{10}$ $(0.22 \pm 0.30) \times 10^{10}$	ARGUS PDG
$B^- \to J/\Psi K^-$	0.083×10^{10}	$(0.38^{+0.24}_{-0.17} \pm 0.08) \times 10^{10}$ $(0.05 \pm 0.02 \pm 0.01) \times 10^{10}$	LEP * ARGUS
$B^- \to \psi(2S) K^-$	0.021×10^{10}	$(0.13 \pm 0.06 \pm 0.03) \times 10^{10}$ $(0.16 \pm 0.12) \times 10^{10}$	ARGUS PDG
$B^- \to \rho^0 \pi^-$	0.28×10^8	$< 1.11 \times 10^8$	PDG
$B^- \to K^- \rho^0$	0.20×10^7	$< 0.52 \times 10^8$	PDG

* Preliminary data

We use $\tau_{B^\pm} = 13.5 \times 10^{-13}\ sec$, $\tau_{B^0} = 14.2 \times 10^{-13}\ sec$ (ALEPH Data [29])

Tab. 15: Results: $\bar{B}^0 \to PV$

CHANNEL	$\Gamma_{calc.}$ (sec^{-1})	$\Gamma_{exp.}$ (sec^{-1})	GROUP
$\bar{B}^0 \to D^+ D_s^{*-}$	0.08×10^{11}	$(0.19 \pm 0.12 \pm 0.03) \times 10^{11}$	ARGUS *
$\bar{B}^0 \to D^+ \rho^-$	0.10×10^{11}	$(0.15 \pm 0.10) \times 10^{11}$ $(0.06 \pm 0.03 \pm 0.02) \times 10^{11}$	PDG ARGUS
$\bar{B}^0 \to D_s^- D^{*+}$	0.023×10^{11}	$(0.099 \pm 0.070 \pm 0.021) \times 10^{11}$	ARGUS *
$\bar{B}^0 \to D^{*+} \pi^-$	0.019×10^{11}	$(0.023^{+0.008}_{-0.007}) \times 10^{11}$ $(0.020 \pm 0.006 \pm 0.004) \times 10^{11}$ $(0.020 \pm 0.010 \pm 0.006) \times 10^{11}$	PDG ARGUS CLEO
$\bar{B}^0 \to J/\Psi \bar{K}^0$	0.009×10^{11}	$< 0.035 \times 10^{11}$	PDG
$\bar{B}^0 \to D^0 \rho^0$	0.002×10^{11}	$< 0.02 \times 10^{11}$ $< 0.004 \times 10^{11}$	PDG CLEO
$\bar{B}^0 \to \rho^+ \pi^-$	0.012×10^{10}	$< 0.043 \times 10^{11}$	PDG
$\bar{B}^0 \to \rho^- \pi^+$	0.018×10^{10}	$< 0.043 \times 10^{11}$	PDG
$\bar{B}^0 \to K^{*-} \pi^+$	0.11×10^{8}	$< 0.31 \times 10^{9}$	PDG
$\bar{B}^0 \to \bar{K}^0 \rho^0$	0.33×10^{6}	$< 0.41 \times 10^{9}$	PDG

* Preliminary data

Tab. 16: Results: Semileptonic Decays $D, B \to P\, e\, \nu_e$

CHANNEL	$\Gamma_{calc.}$ ($10^{11}\ sec^{-1}$)	$\Gamma_{exp.}$ ($10^{11}\ sec^{-1}$)	GROUP
$D^+ \to \bar{K}^0 e^+ \nu_e$	0.54	$(0.56 \pm 0.08 \pm 0.15)$ $(0.61^{+0.15}_{-0.10} \pm 0.07)$ $(0.56 \pm 0.09 \pm 0.12)$	E691 MARKIII E653
$D^0 \to K^- e^+ \nu_e$	0.54	(0.81 ± 0.09) $(0.81 \pm 0.12 \pm 0.09)$ $(0.88 \pm 0.07 \pm 0.14)$ $(0.91 \pm 0.07 \pm 0.17)$	PDG MARKIII CLEO E691
$D^0 \to \pi^- e^+ \nu_e$	0.02	$(0.09^{+0.05}_{-0.03} \pm 0.01)$	MARKIII
$D_s \to K^0 e^+ \nu_e^-$	0.029	$(0.045)^*$	
$D_s \to \eta\, e^+ \nu_e$	0.39	$(0.46)^*$	
$\bar{B}^0 \to D^+ e^- \bar{\nu}_e$	0.17	0.15 ± 0.07 $0.17 \pm 0.03 \pm 0.02$ $0.15 \pm 0.05 \pm 0.04$	PDG CLEO ARGUS
$B^- \to D^0 e^- \bar{\nu}_e$	0.17	$0.135 \pm 0.05^{+0.08}_{-0.05}$	CLEO
$B^- \to \pi^0 e^- \bar{\nu}_e$	0.004	< 0.016	Crystal Ball

* Estimates of the expected rates (see Ref. 39).

Common Problems and Trends of Modern Physics, pp. 125–135
ed. T. Bressani, S. Marcello and A. Zenoni

©1992 World Scientific Publishing Company

MESONS IN QCD: DECAYS

L. Gamberale
Dipartimento di Fisica, Università diMilano, via G. Celoria 16, 20133 Milano, Italy
and
INFN Sezione di Milano, via G. Celoria 16, 20133 Milano, Italy

ABSTRACT

Starting from the knowledge of the meson eigenstates at rest of the QCD hamiltonian derived in the framework of ACD it is possible to calculate the 3-meson vertex adopting a perturbative strategy. Results are in good agreement with experiment.

1. Introduction

QCD is universally considered to be the correct theory for the description of strong interactions. Nevertheless the way in which the theory is usually developed is far from being satisfactory and, above all, does not explain one of the most important characteristics of quark dynamics: confinement. However there exists an alternative way of solving the theory, developed by G.Preparata during the last 15 years, which is able to include and more importantly to explain confinement in a natural manner. This work follows this second line of research and is concerned with the calculation of the transition amplitudes between meson states. In section 2 I will discuss the theoretical motivations of the validity of this line of research and in section 3 I will briefly describe the effective hamiltonian of the theory. In sections 4 and 5 the explicit calculation of the 3-meson vertex and numerical results are presented.

2. The vacuum structure of QCD and the resulting hamiltonian.

The generally accepted theory of QCD is based on the notion of *Asymptotic Freedom* (AF) which is considered the theoretical basis for the applicability of the perturbative expansion at high energy scales. This idea is based on the assumption that the perturbative vacuum, i.e. the quantum state on which we construct the perturbation theory, is a good approximation of the true vacuum, i.e. the quantum state of minimum energy.

The study of the minimum of the QCD energy density performed with variational techniques, initiated by J.K.Savvidy in 1977 [1], continued by Nielsen and Olesen [2] and later by G.Preparata [3–5], has shown that this assumption is completely wrong. The outcome of this work is that there exist quantum states of the gauge field (Savvidy states) that have an energy density lower than that of the perturbative vacuum by a quantity diverging like Λ^4 (ultraviolet cut-off). This means that at any scale the perturbative vacuum is not a good approximation of the true vacuum implying that the standard perturbative procedure is theoretically unfounded.

The probable structure of the true QCD vacuum stems from the study of the magnetic condensation of the Savvidy states and is supplied by the *chromomagnetic liquid* (CML). The CML is a collection of needle-shaped domains uniformly distributed in all points and all directions of the space in which the chromomagnetic field assumes huge values (proportional to Λ^2) and in which the color charges are confined. In this way the color charges placed in a given needle follow an independent dynamics with respect to those placed in different needles, and as a result in each needle the dynamics is 1+1-dimensional (like in QCD_2, which is a naturally confined theory). The reason why it is not possible to observe the existence of the needles is that they rotate: we can only observe an average on directions.

In this scenario it has been possible to construct an effective theory of QCD (Anisotropic Chromo Dynamics, ACD) [6–8] in an "enlarged" space-time in which to each point of the usual space-time is associated a Lorentz-invariant pseudosphere in order to include the "notion" of the needle-shaped domains. In the next section I describe briefly the structure of the ACD hamiltonian which can be considered the effective QCD hamiltonian on the true vacuum.

3. The hamiltonian structure: confining and pair-creation terms.

In ACD the elementary fields are the gauge and matter fields defined on the needle. The matter field on the needle is defined by:

$$q_{\vec{u}}(\vec{x},0) = \int \frac{d^3\vec{k}}{(2\pi)^{\frac{3}{2}}} \sqrt{\frac{m}{E_{\vec{k}\cdot\vec{u}\vec{u}}}} \left[u_r(\vec{k}\cdot\vec{u}\vec{u}) b_{\vec{u}r}(\vec{k}) e^{i\vec{k}\cdot\vec{x}} + v_s(\vec{k}\cdot\vec{u}\vec{u}) d^{\dagger}_{\vec{u}s}(\vec{k}) e^{-i\vec{k}\cdot\vec{x}}\right] . \quad [3.1]$$

On the other hand the observable fields do not depend on the $\vec{u}$ variable. Thus we need to define a "collective" field

$$q(\vec{x},0)=\int\frac{d^3\vec{k}}{(2\pi)^{\frac{3}{2}}}\sqrt{\frac{m}{E_{\vec{k}}}}\left[u_r(\vec{k})b_r(\vec{k})e^{i\vec{k}\cdot\vec{x}}+v_s(\vec{k})d_s^{\dagger}(\vec{k})e^{-i\vec{k}\cdot\vec{x}}\right] \qquad [3.2]$$

which is the usual matter field in the 4-dimensional space-time and is related to the elementary fields $q_{\vec{u}}(\vec{x},0)$ by a Bogoliubov transformation. The ACD hamiltonian has the following structure:

$$H=H_{\mathrm{Kin}}+H_{\mathrm{int}}, \qquad [3.3]$$

where

$$H_{\mathrm{Kin}}=\int d^3\vec{x}:\bar{q}(\vec{x},0)[-i\vec{\gamma}\cdot\vec{\nabla}+m]q(\vec{x},0): \qquad [3.4]$$

is the usual kinetic hamiltonian defined on the collective fields and

$$H_{\mathrm{int}}=\sum_{\vec{u}}\int d^3\vec{x}d^3\vec{y}\;\; j_{\vec{u}}^{\mu a}(\vec{x},0)\Delta_{\vec{u}\;\mu\nu}^{ab}(\vec{x}-\vec{y})j_{\vec{u}}^{\nu b}(\vec{y},0) \qquad [3.5]$$

is the interaction hamiltonian where

$$j_{\vec{u}}^{\mu a}(\vec{x},0)=:\bar{q}_{\vec{u}}(\vec{x},0)\gamma^{\mu}\frac{\lambda^a}{2}q_{\vec{u}}(\vec{x},0): \qquad [3.6]$$

are the current operators defined on the needle $\vec{u}$ and

$$\Delta_{\vec{u}\;\mu\nu}^{ab}(\vec{x})=\lim_{\lambda\to 0}\int\frac{d^3\vec{k}}{(2\pi)^3}\frac{e^{i\vec{k}\cdot\vec{x}}}{(\vec{k}\cdot\vec{u})^2+\lambda^2}(g_{\mu\nu}+n_{\mu}n_{\nu})\delta^{ab} \qquad n=(0,\vec{u}) \qquad [3.7]$$

is the "longitudinal" gluon propagator on the needle $\vec{u}$ which carries an instantaneous interaction.

Let us describe briefly H_{int}. By decomposing the color current in two parts

$$j_{\vec{u}}^{\mu a}(\vec{x},0)=j_{\vec{u}}^{\mu a}(\vec{x},0)^{(0)}+j_{\vec{u}}^{\mu a}(\vec{x},0)^{(1)} \qquad [3.8]$$

in such a way that $j_{\vec{u}}^{\mu a}(\vec{x},0)^{(0)}$ contains $bb^{\dagger}$ and $dd^{\dagger}$ terms only, we can write the interaction hamiltonian in the following form:

$$H_{\mathrm{int}}=H^{(0)}+H'$$

where $H^{(0)}$ contains only $j^{(0)}j^{(0)}$ terms and H' contains $j^{(0)}j^{(1)}, j^{(1)}j^{(0)}$ and $j^{(1)}j^{(1)}$ terms. With this decomposition the $H^{(0)}$ term has the following characteristics:

- it preserves only the number of quarks, antiquarks and gluons separately
- it is the only infra-red singular term of the hamiltonian. It contains the confining term.

On the other hand the H' term:

- preserves the difference of the number of quarks and antiquarks; it is a pair-creation term and is responsible of the transitions between hadronic states
- is regular in the coordinate space
- can be treated as a perturbation.

The structure of $H^{(0)}$ and H' allows us to adopt a perturbative strategy which consists in solving the eigenvalue problem for the operator $H_{\text{Kin}} + H^{(0)}$ then making a perturbation with H'. Note that $H_{\text{Kin}} + H^{(0)}$ commutes with the number operators of quarks, antiquarks and gluons and it is then possible to write a Shroedinger equation of the form:

$$\left[H_{\text{Kin}} + H^{(0)}\right] |n\rangle = E_n |n\rangle \qquad [3.9]$$

where $|n\rangle$ belongs to the Fock subspace with a given number of particles. In this work I will concentrate on the $N_q = 1, N_{\bar{q}} = 1, N_g = 0$ sector that describes mesons. The quantum state that describes a meson at rest has the following structure:

$$|n\rangle = \frac{\delta^{cc'}}{\sqrt{3}} F^{ff'} S^{ss'} \int d^3\vec{k} \phi_n(|\vec{k}|) Y_{lm}(\hat{k}) b^{\dagger}_{cfs}(\vec{k}) d^{\dagger}_{c'f's'}(-\vec{k}) |0\rangle \qquad [3.10]$$

where $b^{\dagger}_{cfs}$ and $d^{\dagger}_{c'f's'}$ represent the creation operators of the "collective" fields. Matrices F and S are the flavour and spin wave functions while c and c' are color indexes. By substituting this expression in eq. [3.9] we obtain an equation for ϕ_n which determines successfully the meson spectrum and the meson wave functions ϕ_n in terms of six parameters: 5 quark masses and the string tension [9].

4. Boosting the wave functions. Pion electromagnetic form factor.

The diagonalization of $H_{\text{Kin}} + H^{(0)}$ can be carried out only for mesons at rest since the calculation of the interaction potential in the "collective" space is a very complicated problem when the meson is in motion.

On the other hand if we want to calculate transition amplitudes and other hadronic parameters it is necessary to know the wave functions at any $\vec{p}$. Hence we need to determine a boost operator that links the wave functions in a given frame with the wave functions in any frame of reference. Writing the meson state in the generic frame as follows:

$$|M,\vec{p},J^{PC}\rangle = \int d^3\vec{k}\Phi(\vec{p},\vec{k})_{cc',ff',ss'} b^{\dagger}_{cfs}(\frac{\vec{p}}{2}+\vec{k}) d^{\dagger}_{c'f's'}(\frac{\vec{p}}{2}-\vec{k})\,|0\rangle \qquad [4.1]$$

we need to know the transformation

$$\Phi(\vec{0},\vec{k})_{cc',ff',ss'} \xrightarrow{\text{Boost}} \Phi(\vec{p},\vec{k})_{cc',ff',ss'}. \qquad [4.2]$$

This problem has been solved approximately by G. Preparata [10] and the meson wave function with momentum $\vec{p}$ (chosen in the 3-direction) can be written as

$$\Phi(\vec{p},\vec{k})_{cc',ff',ss'} = \frac{\delta^{cc'}}{\sqrt{3}} F^{ff'} S^{ss'} \sqrt{\frac{\sqrt{\vec{p}^2+M_n^2}}{M_n}\frac{\partial w_3}{\partial k_3}(\vec{p},\vec{k})}\phi_n(|\vec{w}(\vec{p},\vec{k})|) Y_{lm}(\hat{w}(\vec{p},\vec{k})) \quad [4.3]$$

where

$$\vec{w}(\vec{p},\vec{k}) = (k_1,k_2,w_3(\vec{p},\vec{k})) \qquad [4.4]$$

and $w_3(\vec{p},\vec{k})$ is the solution of the differential equation

$$\frac{\partial w_3}{\partial k_3}(\vec{p},\vec{k}) = \frac{\sqrt{(\frac{1}{2}\vec{p}+\vec{k})^2+m_q^2}+\sqrt{(\frac{1}{2}\vec{p}-\vec{k})^2+m_{\bar{q}}^2}-\sqrt{\vec{p}^2+M_n^2}}{\sqrt{\vec{w}^2+m_q^2}+\sqrt{\vec{w}^2+m_{\bar{q}}^2}-M_n}. \qquad [4.5]$$

This method has been applied to the calculation of the pion electromagnetic form factor in the $q^2 < 0$ region. Using the Bethe-Salpeter formalism it can be shown that in the Breit-frame ($\vec{p}' = -\vec{p}$) the electromagnetic current operator has the form (elastic case)

$$\langle M,-\vec{p},J^{PC}|\,J_\mu(0)\,|M,\vec{p},J^{PC}\rangle = F(-4p^2)2\sqrt{\vec{p}^2+M_n^2}\delta_{\mu 0} \qquad [4.6]$$

and

$$F(-4p^2) = \frac{1}{2\sqrt{\vec{p}^2+M_n^2}}\int\frac{d^3\vec{k}}{(2\pi)^3}\Phi^*(\vec{p},\vec{k}+\frac{\vec{p}}{2})_{s's}\Phi(-\vec{p},\vec{k}-\frac{\vec{p}}{2})_{ss''}$$

$$[\frac{e_q m_q}{\sqrt{E^q_{\vec{k}+\vec{p}}E^q_{\vec{k}-\vec{p}}}}\bar{u}_{s''}(\vec{k}-\vec{p})\gamma^0 u_{s'}(\vec{k}+\vec{p}) + \frac{e_{\bar{q}} m_{\bar{q}}}{\sqrt{E^{\bar{q}}_{\vec{k}+\vec{p}}E^{\bar{q}}_{\vec{k}-\vec{p}}}}\bar{v}_{s''}(\vec{k}-\vec{p})\gamma^0 v_{s'}(\vec{k}-\vec{p})] \qquad [4.7]$$

where the wave function are normalized according to

$$\int\frac{d^3\vec{k}}{(2\pi)^3}|\Phi(\vec{p},\vec{k})|^2 = 2\sqrt{\vec{p}^2+M_n^2}. \qquad [4.8]$$

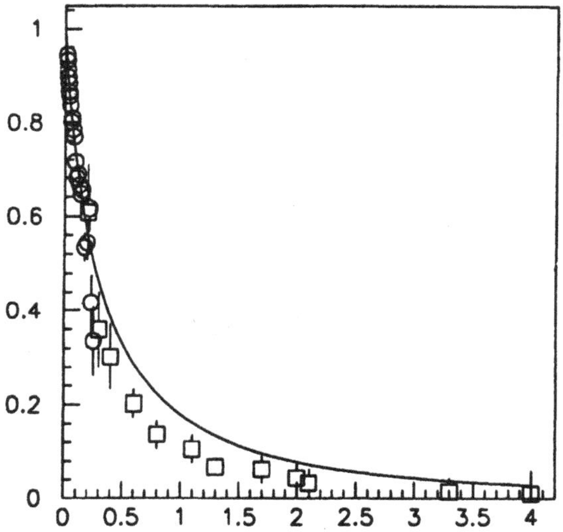

Fig. 1 **Computed electromagnetic form factor of the pion compared whit the experimental data of NA7 collaboration and C.J.Bebek et al..**

Figure 1 shows the computed form factor for the pion together with the experimental data (NA7 experiment n and C.J.Bebek et al. [11-14])

5. The 3-meson vertex

In this section I discuss the method I have adopted in the calculation of transition amplitudes between hadronic states. To focus our ideas let us consider the matrix element

$$\langle M_2,\vec{p}_2,M_3,\vec{p}_3|\,S\,|M_1,\vec{p}_1\rangle \qquad [5.1]$$

which represents the transition amplitude of meson 1 with momentum $\vec{p}_1$ into mesons 2 and 3 with momentum $\vec{p}_2$ and $\vec{p}_3$ respectively. To lowest order in perturbation expansion of the S matrix the amplitude has the form:

$$\mathcal{M} = \langle M_2,\vec{p}_2,M_3,\vec{p}_3|\int d^3\vec{x}\sum_{\vec{u}} j_{\vec{u}}^{\mu a}(\frac{\vec{x}}{2},0)\Delta_{\vec{u}\;\mu\nu}^{ab}(\vec{x}-\vec{y})j_{\vec{u}}^{\nu b}(-\frac{\vec{x}}{2},0)\,|M_1,\vec{p}_1\rangle \qquad [5.2]$$

By substituting the explicit expression of eq. [4.1] for the meson states in eq. [5.2] we are left with

$$\mathcal{M} = \int \frac{d^3\vec{k}_1}{(2\pi)^3}\frac{d^3\vec{k}_2}{(2\pi)^3}\frac{d^3\vec{k}_3}{(2\pi)^3}\Phi_1(\vec{p}_1,\vec{k}_1)\Phi_2^*(\vec{p}_2,\vec{k}_2)\Phi_3^*(\vec{p}_3,\vec{k}_3)K_{\vec{p}_1,\vec{p}_2,\vec{p}_3}(\vec{k}_1,\vec{k}_2,\vec{k}_3) \qquad [5.3]$$

where

$$K_{\vec{p}_1,\vec{p}_2,\vec{p}_3}(\vec{k}_1,\vec{k}_2,\vec{k}_3) = (2\pi)^9\langle 0|\, b(\frac{\vec{p}_2}{2}+\vec{k}_2)d(\frac{\vec{p}_2}{2}-\vec{k}_2)b(\frac{\vec{p}_3}{2}+\vec{k}_3)d(\frac{\vec{p}_3}{2}-\vec{k}_3)\cdot$$

$$\cdot H_{\rm int}b^\dagger(\frac{\vec{p}_1}{2}+\vec{k}_1)d^\dagger(\frac{\vec{p}_1}{2}-\vec{k}_1)\,|0\rangle \qquad [5.4]$$

where I have omitted color, flavour and spin indexes for simplicity. The kernel $K_{\vec{p}_1,\vec{p}_2,\vec{p}_3}(\vec{k}_1,\vec{k}_2,\vec{k}_3)$ can be decomposed into a sum of two contributions and takes the form

$$K_{\vec{p}_1,\vec{p}_2,\vec{p}_3}(\vec{k}_1,\vec{k}_2,\vec{k}_3) = (2\pi)^6\langle 0|\, b(\frac{\vec{p}_2}{2}+\vec{k}_2)b(\frac{\vec{p}_3}{2}+\vec{k}_3)d(\frac{\vec{p}_3}{2}-\vec{k}_3)\cdot$$

$$\cdot H_{\rm int}b^\dagger(\frac{\vec{p}_1}{2}+\vec{k}_1)\,|0\rangle(2\pi)^3\delta^3(\frac{\vec{p}_2}{2}-\vec{k}_2-(\frac{\vec{p}_1}{2}-\vec{k}_1))+$$

$$(2\pi)^6\langle 0|\, b(\frac{\vec{p}_2}{2}+\vec{k}_2)d(\frac{\vec{p}_2}{2}-\vec{k}_2)d(\frac{\vec{p}_3}{2}-\vec{k}_3)\cdot$$

$$\cdot H_{\rm int}d^\dagger(\frac{\vec{p}_1}{2}-\vec{k}_1)\,|0\rangle(2\pi)^3\delta^3(\frac{\vec{p}_3}{2}+\vec{k}_3-(\frac{\vec{p}_1}{2}+\vec{k}_1)). \qquad [5.5]$$

The explicit calculation of the creation potential presents the same difficulties of the confining potential [8,9]. Nevertheless by choosing the frame of reference in which the meson connected with the gluon line is at rest (see Fig. 2) it is possible to carry out the sum over the $\vec{u}$ directions in a simple way.

Thus the expression of the first diagram in that particular frame is given by

$$\mathcal{M}^{(1)} = \int d^3\vec{x}d^3\vec{x}'\left[\int\frac{d^3\vec{k}}{(2\pi)^3}\Phi_2^*(\vec{p},\vec{k}-\frac{\vec{p}}{2})\Phi_1(\vec{p},\vec{k}-\frac{\vec{p}}{2})e^{i\vec{k}\cdot\vec{x}}\right]\cdot$$

$$\cdot\left[\int\frac{d^3\vec{q}}{(2\pi)^3}\Phi_3^*(\vec{0},\vec{q})e^{-i\vec{q}\cdot\vec{x}'}\right]V(\vec{x},\vec{x}')$$

where Φ_1,Φ_2,Φ_3 are the wave functions normalized according to eq. [4.8] and where $\vec{p}$ is directed along the 3-axis and whose modulus satisfies the kinematical condition

$$\sqrt{p^2+M_1^2} = \sqrt{p^2+M_2^2}+M_3. \qquad [5.6]$$

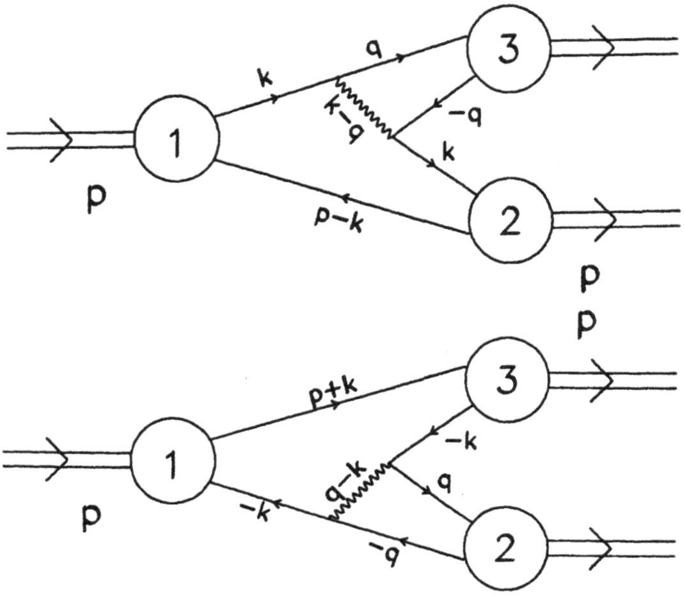

Fig. 2 **Graphical representation of the two contributions to the decay amplitude each one calculated in the rest frame of the meson connected with the gluon line.**

$V(\vec{x},\vec{x}')$ is the creation potential in the coordinate space and in the case of equal quark masses reads

$$V(\vec{x},\vec{x}') = \frac{\mu^2}{2\pi^2\sqrt{3}}\delta^2(\hat{x}-\hat{x}')\frac{v(mr,mr')}{rr'}\left[-i\mathbb{1}^{(1)}\vec{\sigma}^{(2)}\mathbb{1}^{(3)} + \vec{\sigma}^{(1)}\times\vec{\sigma}^{(2)}\mathbb{1}^{(3)}\right]\cdot\hat{x} \qquad [5.7]$$

where m is the quark mass and

$$v(\rho,\rho') = \int_0^{+\infty} dt[K_0(\rho-t)K_0(\rho'-t) - K_0(\rho+t)K_0(\rho'+t)]. \qquad [5.8]$$

Figure 3 shows the behaviour of $v(\rho,\rho')$. It can be noted that $v(\rho,\rho')$ has a quasi-local structure and the potential can be approximated for $mr > 1$ by

$$V(\vec{x},\vec{x}') = \frac{\mu^2}{2\pi^2\sqrt{3}}\frac{4.9}{m_q}\delta^3(\vec{x}-\vec{x}'). \qquad [5.9]$$

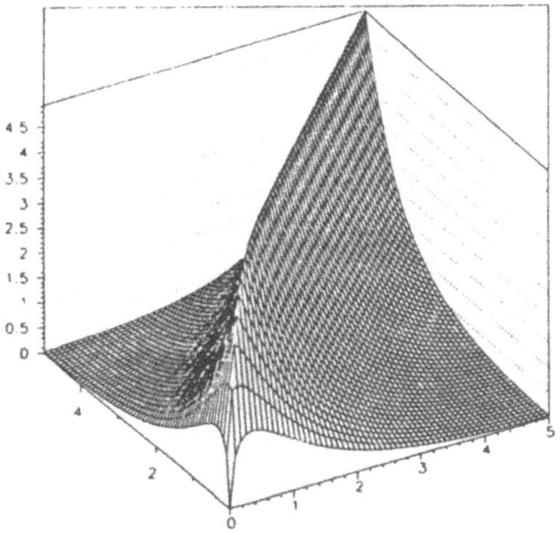

Fig. 3 Behaviour of the function v of eq. [5.8].

Introducing the explicit expression (eq. [4.3]) of the boosted wave functions for the $L = 0$ case we obtain

$$\mathcal{M}^{(1)} = \frac{\mu^2}{2\pi^2\sqrt{3}} \frac{1}{(2\pi)^3\pi\sqrt{4\pi}} < \text{Flavour} >< \text{Spin} > \int_0^{+\infty} dr dr' \; u(r)v(r,r')F(r') \qquad [5.10]$$

where

$$u(r) = \int_0^{+\infty} dq q^2 j_0(qr)\phi_3(\vec{0}, |\vec{q}|)r \qquad [5.11]$$

is the reduced wave function of meson 3 at rest and

$$F(r') = \int_0^{+\infty} dk k^2 j_1(kr')r' \int_{-1}^{1} d\cos\theta \phi_1(\vec{p}, \vec{k} - \frac{\vec{p}}{2})\phi_2^*(\vec{p}, \vec{k} - \frac{\vec{p}}{2}) \cos\theta. \qquad [5.12]$$

The integral in eq. [5.10] can be carried out numerically and the calculation of the second contribution can be accomplished in a similar way by choosing the frame of reference in which meson 2 is at rest. The calculation of the transition amplitude between mesons is in principle completely solved and has been done explicitly for the $\rho^0 \to \pi^+\pi^-$ decay and the computed transition amplitude is

$$\mathcal{M}_{\rho^0 \to \pi^+\pi^-} = 4.1 \, \text{GeV}. \qquad [5.13]$$

Since the Lorentz invariant transition amplitude of a vector meson into two pseudoscalar mesons is given by

$$\mathcal{M}^{\lambda} = \epsilon_{\mu}^{\lambda}(p_2 - p_3)^{\mu} g_{V\to PP} \qquad [5.14]$$

where p_2 and p_3 are the four-momentum vectors of the outgoing mesons and ϵ_{μ}^{λ} is the polarization vector of the decaying meson with helicity λ, it follows that $g_{V\to PP}$ is given by the relation

$$g_{V\to PP} = \mathcal{M}_{\lambda=3}/2\tilde{p} \qquad [5.15]$$

where $\tilde{p}$ is the momentum of the decay products in the rest frame of the decaying particle. Exploiting these relations we obtain

$$g_{\rho^0\to\pi^+\pi^-} = 5.72$$

to be compared with the experimental value

$$g^{\exp}_{\rho^0\to\pi^+\pi^-} = 6.01$$

a very encouraging result!

6. Conclusions.

The success of the $\rho^0 \to \pi^+\pi^-$ calculation is another piece of evidence of the goodness of the theory since now developed. The calculation will be soon applied to the calculation of $K^* \to K\pi$, $\phi \to K^+K^-$, $\phi \to K^0\overline{K^0}$. It is also possible to calculate Vector Meson Dominance contributions that, in conjunction with the electromagnetic matrix element shown in section 4, will allow us to compute $V \to P\gamma$ processes.

Acknowledgements

I am grateful to doc. M.Scorletti and doc. M.Verpelli for helpful discussions and in particular to prof. G.Preparata for his help. I also thank L.Pirovano for his collaboration in the theoretical development of the calculation.

REFERENCES

1. J.K.Savvidy *Phys. Lett.* **B71** (1977) 71

2. N.K.Nielsen and P.Olesen *Nucl. Phys.* **B134** (1978) 376
3. M.Consoli and G.Preparata *Phys. Lett.* **B154** (1985) 411
4. G.Preparata *Nuovo Cim.* **A96** (1986) 366
5. L.Gamberale,G.Preparata and S.S.Xue *Nuovo Cim.* **A105** (1992) 309
6. J.L.Basdevant and G.Preparata *Nuovo Cim.* **A67** (1982) 19
7. J.L.Basdevant P.Colangelo and G.Preparata *Nuovo Cim.* **A71** (1982) 445
8. see G.Preparata's talk at this school
9. see M.Scorletti's talk at this school
10. L.Cosmai, M.Pellicoro and G.Preparata *Nucl. Phys.* **B228** (1983) 31
11. NA7 Collaboration *Phys. Lett.* **B138** (1984) 454
12. C.J.Bebek et al. *Phys. Rev.* **D17** (1978) 1693
13. C.J.Bebek et al. *Phys. Rev.* **D9** (1974) 1229
14. C.J.Bebek et al. *Phys. Rev.* **D13** (1976) 25

Common Problems and Trends of Modern Physics, pp. 137–146
ed. T. Bressani, S. Marcello and A. Zenoni
©1992 World Scientific Publishing Company

RESONANCE INTERFERENCE AND SPIN ASYMMETRIES

R. Barni
Dipartimento di Fisica, Università di Milano, via G. Celoria 16, 20133 Milano, Italy

G. Preparata
Dipartimento di Fisica, Università di Milano
and
INFN, Sezione di Milano, via G. Celoria 16, 20133 Milano, Italy

P.G. Ratcliffe
INFN, Sezione di Milano, via G. Celoria 16, 20133 Milano, Italy

ABSTRACT

We present a model for the spin asymmetries in inclusive reactions at high energy. In this model, the asymmetries arise from the interference between the various channels through which the detected particle is produced: in particular we take care, in the calculation of the reaction amplitudes, of the resonances that are coupled to the particle. Using a Triple-Regge parameterization and the SU(6) coupling constant we are able to fit the unpolarized cross-sections in the fragmentation region and, subsequently, to predict the asymmetries. We have applied our model to two outstanding problems: the polarization of the Λ^0 produced in reactions pN and the single-spin asymmetry of the π produced in $p^\uparrow p$ reactions.

1. INTRODUCTION

During the last fifteen years a great quantity of data has been collected from inclusive reactions at high energy. One of the most surprising result is that there are large spin asymmetries in many reactions. For example there are large polarizations in $pN \rightarrow$ hyperons $(\Lambda^0, \Sigma^+, \Sigma^-, \Sigma^0, \Xi^0$ and $\Xi^-)$ and also in $\pi^\pm p, K^\pm p \rightarrow \Lambda^0$ (but small asymmetries in $p^\uparrow p \rightarrow \Lambda^0$ or Σ^0), large asymmetries in $p^\uparrow p \rightarrow$ mesons $(\pi^\pm, \pi^0, K^\pm, K^0)$, while there is not polarization in the reactions $pp \rightarrow p$ and $pp \rightarrow \bar{\Lambda}^0$ [1,2].

The experimental data show a common pattern: the asymmetries seem to be quite independent from the energy of the reaction (for $pp \rightarrow \Lambda^0$ there are data with p_{LAB} from 10 to 2000 GeV) and from the nuclear composition of the target (for the same reactions there are data for N = H,Be,Cu,Pb) and also the kinematical behaviour is quite similar. The asymmetries increase almost linearly throughout the fragmentation region ($x_F > 0.2$) and increase with p_T until they reach a plateau at $p_T \sim 0.5 - 1$ GeV/c and remain constant also in the high p_T domain (there are data until $p_T \sim 3 - 4$ GeV/c). There are also data on asymmetries in the central region ($x_F \sim 0$) at high p_T. The magnitude of these asymmetries is relevant (a few tens percent) while the sign depends on the particle produced.

From a theoretical point of view, for an inclusive reaction $a + b \rightarrow c + X$, single-spin observables are related, via the Mueller optical theorem, to the 3-body elastic amplitudes [3,4] :

$$
\begin{aligned}
E_c \frac{d^3\sigma}{d^3p_c}^{incl} &= \sum_{h_a h_b h_c} Disc_{M^2}\ A_{h_a h_b h_c, h_a h_b h_c} \\
P_c \cdot E_c \frac{d^3\sigma}{d^3p_c}^{incl} &= 2\ \mathrm{Im} \sum_{h_a h_b} Disc_{M^2}\ A_{h_a h_b +, h_a h_b -} \qquad [1.1] \\
A_a \cdot E_c \frac{d^3\sigma}{d^3p_c}^{incl} &= 2\ \mathrm{Im} \sum_{h_b h_c} Disc_{M^2}\ A_{+h_b h_c, -h_b h_c} \ .
\end{aligned}
$$

As one can see, in order to have large asymmetries, we need large imaginary parts and large spin-flip amplitudes in the high energy and high p_T region, which was a rather unexpected thing in the standard point of view [5].

2. THE MODEL

Following the idea of G. Preparata, [6] we have developed a model for such phenomena in which the asymmetries arise from the interference of indirect production channels. In particular we have considered, beside the direct channel, production through a resonant state belonging to the $\frac{1}{2}^+$ octet or $\frac{3}{2}^+$ decuplet.

$$
\begin{aligned}
&a + b \rightarrow c + X \\
&a + b \rightarrow R_1 + X \rightarrow c + d + X \qquad \text{where } R_1 \in \frac{1}{2}^+ \qquad [2.1] \\
&a + b \rightarrow R_2 + X \rightarrow c + d + X \qquad \text{where } R_2 \in \frac{3}{2}^+ \ .
\end{aligned}
$$

Now, we need a parameterization of these production channels throughout the fragmentation region and we have chosen the Triple-Regge approximation, which has already been applied successfully to high energy phenomenology [4] and the parameter of which (Regge residues and trajectories) are well known from low energy reactions [7]. For the indirect channels we have used the Born approximation which separates the production part, which has been evaluated via the Triple-Regge approximation as the direct one, from the decay part. The decay amplitudes have been evaluated from the decay widths with the SU(6) coupling constants, using the standard phenomenological Lagrangians [8] . In order to simplify the model, we have considered only the leading Regge trajectories (the upper ones), the first resonant states (belonging to the baryon octet and decuplet) and an unbroken SU(6) symmetry. The Regge trajectory has been interpolated between the linear form (in $t > 0$ region) and a constant (for $t \rightarrow -\infty$), as suggested by quark-parton model [9] and by experimental data on the unpolarized cross-sections.

Before analyzing the applications of the model, we can do some observations. The first one is that, since the non-fragmentation vertex is dominated by Pomeron exchange, the amplitudes become essentially $\sqrt{s}$ independent, so our calculations are applicable only in the scaling region at high energy and the results are almost energy independent. The second one is that, since both the Regge model and the decay Lagrangians have a rich spin structure, we have no suppression of spin-flip amplitudes respect to spin non-flip ones. The third one is that the indirect production channels interfere and, since the amplitude for the octet channel is almost completely real (being the resonance off-shell) and that for the decuplet channel is imaginary (being the resonance on-shell), the interference contribution is almost completely imaginary. So, in our model, we have met the two pre-conditions for having large spin asymmetries. Since, the asymmetries arise mainly from the interference of the indirect production channels, our model will give effectively large spin asymmetries only if these channels contribute considerably to the total cross-section in the fragmentation region.

The only parameters, which have been let free in our model, are the absolute normalization of the amplitudes and the Triple-Regge vertex, that has been parameterized to reproduce the exponential p_T dependence of the unpolarized cross-sections data. All the parameters have been fixed in the fit of the unpolarized cross-sections, so our calculation of spin asymmetries is parameter-free. In the next sections we present the application of the model to two interesting reactions.

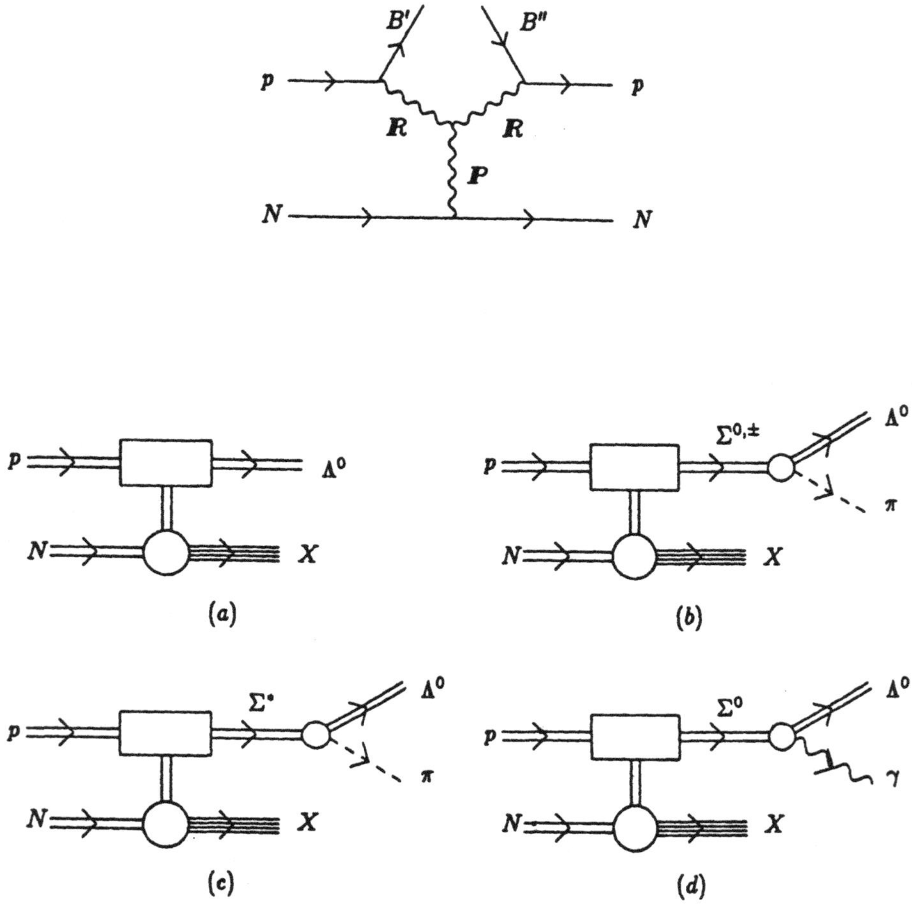

Fig. 1 Triple-Regge approximation of the Mueller optical theorem and the production channels for the reaction $p + p \to \Lambda^{\circ} + X$.

3. POLARIZATION IN $p+p \to \Lambda^0 + X$

As regards the production of Λ^0, the relevant states, involved in the indirect channels considered in our model, are the $\Sigma(1189)$ and the $\Sigma^*(1385)$. They are coupled to the Λ^0 through the decay into $\Lambda^0 + \pi$, so this implies that the Σ are off-shell. Besides the direct production channel, there is another important contribution to the production of Λ^0, the one arising from the electromagnetic decay of the $S0$. Summarizing, we have considered dominant the four production channels:

$$\begin{aligned} p+p &\to \Lambda^0 + X \quad \text{(direct)} \\ p+p &\to \Sigma^0 + X \quad \to \quad \Lambda^0 + \gamma + X \\ p+p &\to \Sigma^*(1385) + X \quad \to \quad \Lambda^0 + \pi + X \\ p+p &\to \Sigma + X \quad \to \quad \Lambda^0 + \pi + X \quad (\Sigma \text{ is off}-\text{shell})\ . \end{aligned} \qquad [3.1]$$

Here, the relative importance of the indirect production channels respect to the direct one is known to be considerable, since the ratio between the cross sections for production of Σ^0 and Λ^0 was measured to be about 0.3 in the fragmentation region [10] and that of Σ^{*+} and Λ^0 is about 0.2 in the same kinematical region [11] . So we expect a large polarization, arising from the interference of the last two channels and also from the decay of the Σ^0 (if, by means of a similar mechanism, also the Σ^0 are produced polarized, as the experimental data seem to confirm [2], then the Λ^0 coming from the decay retain part of that polarization: $P_{\Lambda^0} = -\frac{1}{3} P_{\Sigma^0}$ from simple quark-model and SU(6) considerations). We want to point out that, since a large fraction of the Λ^0 comes from the indirect channels, any self-consistent model for the polarization must include their contribution and so one is naturally lead to our model. The last observation is that, for the production channels mentioned above, the dominant Regge exchanges are $\mathcal{R} = K^*$ and K^{**}, which are exchange degenerate, thus suppressing any contribution to polarization arising from the Regge model (as suggested in [3]).

For our analysis, we have used the data collected at FNAL with protons of 400 GeV/c in reactions on hydrogen or beryllium targets [12,13]. We have tested the goodness of our parameterization in the fit of the unpolarized cross-section in the fragmentation region. Then, we have calculated the Λ^0 polarization predicted by our model and the results are very encouraging, since we obtain, not only the correct sign and magnitude, but also the right kinematical behaviour as shown in Fig. 2 [14].

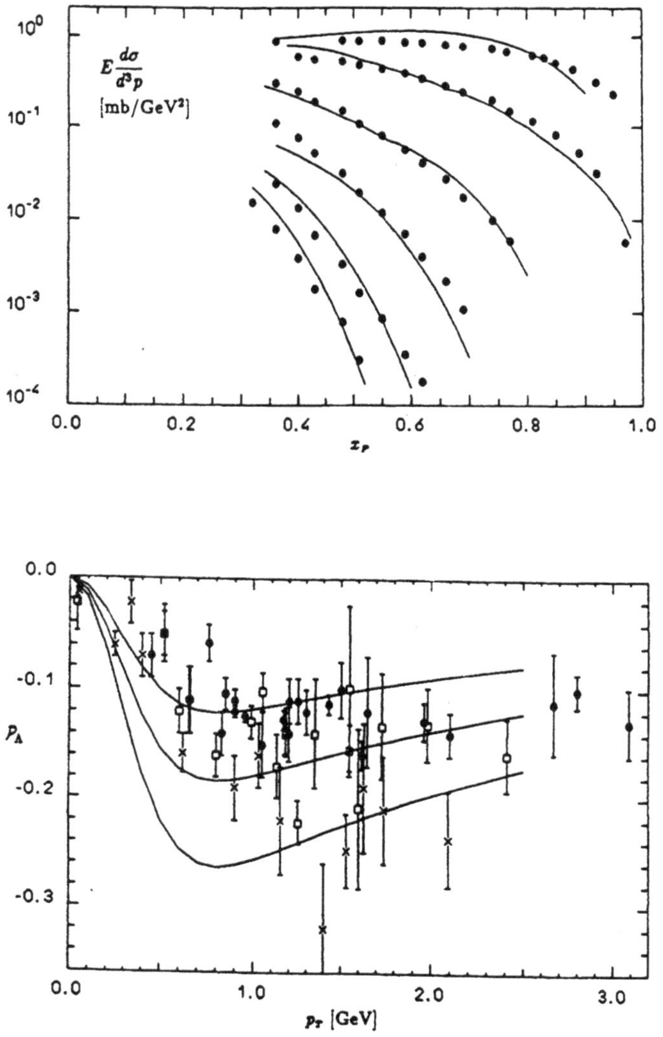

Fig. 2 Invariant cross-section and Λ° polarization for the reaction $p + p \to \Lambda^\circ + X$.

A similar analysis can be performed on the polarization of the Σhyperons [15] but here the dominant indirect contribution comes not only from the decuplet resonances (the coupling of Σ^* with the Σ s is suppressed respect to that with the Λ^0, as we know from the decay widths of the Σ^*), but also from $\Lambda(1405)$. As a result, our outcomes are not so good as for the Λ^0 [14] .

4. ASYMMETRY IN $p^\uparrow + p \rightarrow \pi + X$

The analysis of these reactions does not differ substantially from that of Λ^0 polarization. Here the relevant indirect production states are $N(938)$ (which necessarily will be off-shell) and $\Delta(1232)$, which decay into $N\,\pi$ and give the necessary interference. The dominant production channels are:

$$\begin{aligned} p+p &\rightarrow \pi + X \quad \text{(direct)} \\ p+p &\rightarrow \Delta(1232) + X \quad \rightarrow \quad N + \pi + X \\ p+p &\rightarrow N + X \quad \rightarrow \quad N + \pi + X \quad (\text{N is off}-\text{shell})\,. \end{aligned} \qquad [4.1]$$

The dominant Regge exchanges for the last two channels are $\mathcal{R} = \rho$ and A_2 and so we can do the three following observations: the first one is that other indirect production channels via meson resonances (for example ρ or ω) are suppressed because the Regge exchanges involved are the baryon ones ($\mathcal{R} = \Delta, N$), which have Regge trajectory that lies lower than the meson ones, and so we can safely restrict ourselves to the three channels mentioned above, the second one is that also the direct production channel is suppressed for the same reason and so the resulting asymmetry is larger than the corresponding Λ^0 polarization, especially for $x_F \rightarrow 1$ and at large p_T, the third one is that also here the dominant exchanges are degenerate and so the asymmetry arises only from the interference of the two indirect channels.

We have used the data collected at CERN ISR at $\sqrt{s} = 45 GeV/c$ [16] for the unpolarized cross-section and the data collected at FNAL with 200 GeV/c polarized protons for the asymmetry of π^+, π^-, π^0 in the fragmentation region [17] . The results for the asymmetries are shown in Fig. 3 [18].

5. CONCLUSIONS

Our model can be adapted straightforwardly to all the other inclusive reactions interesting as regards to the spin asymmetries, in particular we can explain why there

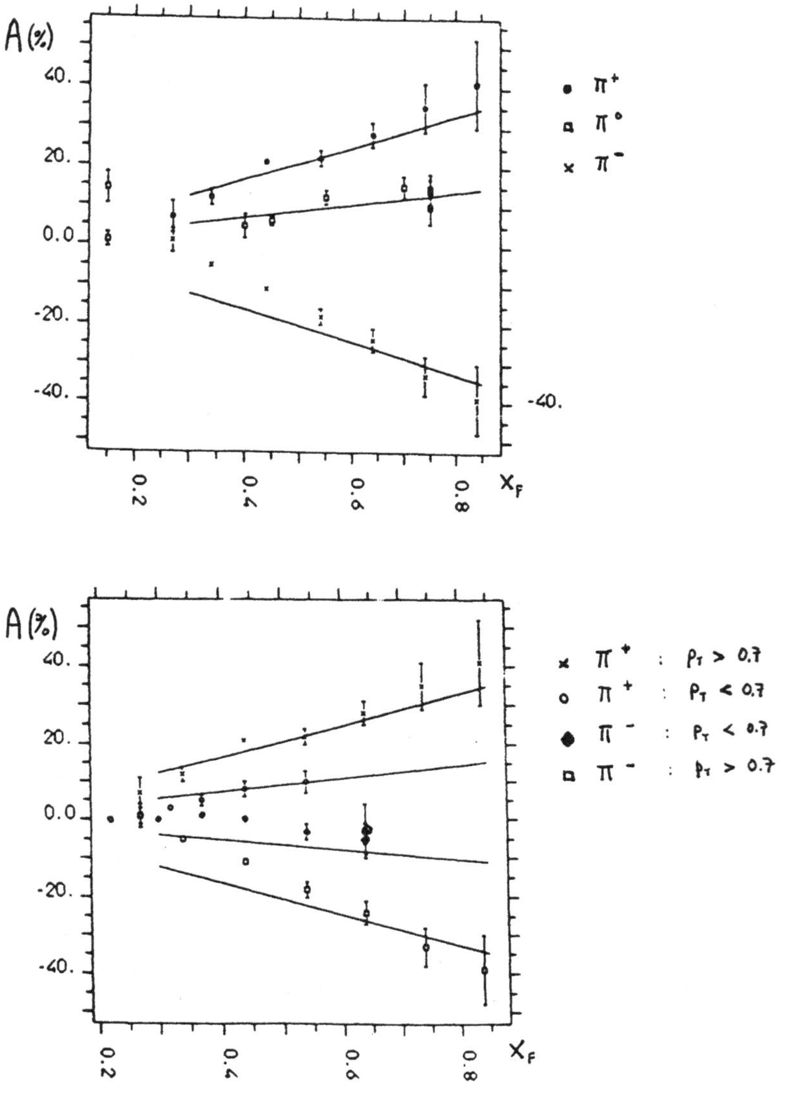

Fig. 3 Kinematical behaviour of the asymmetry in p_T and x_F for the reaction $p^{\uparrow} + p \to \pi + X$.

is not polarization in the reactions $p+p \rightarrow p+X$ [19] and $p+p \rightarrow \bar{\Lambda}^0+X$ [20]. Regarding the first reaction, one can observe that the direct channel is dominated by the Pomeron exchange, while the indirect ones are dominated by the Regge exchanges $\mathcal{R}=\rho, A_2$, which have lower trajectories and thus are suppressed respect to the other channel: so the polarization is completely diluted by the predominant direct channel, which becomes more and more preeminent as $x_F \rightarrow 1$ and at large p_T. For the second reaction, we observe that the production mechanism is not a fragmentation one, but rather similar to that involved in the production of the Λ^0 in the central region (as one can see from the ratio of the production cross-sections of $\bar{\Lambda}^0$ and Λ^0 throughout the full kinematical range [12]), where there is no polarization.

We have applied our model successfully to the analysis of two apparently different phenomena: Λ^0 polarization and π asymmetry, and this strongly suggests that there is only an underlying mechanism in the large phenomenology of spin asymmetries in inclusive reactions. The description is given entirely in terms of familiar aspects of hadron dynamics and without recourse to ad hoc assumptions as to magnitudes or kinematical dependence of the dynamics [21]. In conclusion, we believe this mechanism to be the interference of the different production channels of the particle.

REFERENCES

1. K.J Heller, Proc. VIII Int. Symp. on High-Energy Spin Physics, ed. K.J. Heller, AIP conf. proc. 187 (1988).
2. B.E. Bonner et al., *Phys. Rev. Lett.* **61** (1988) 1918.
3. C. Bourrelly, E. Leader and J. Soffer ,*Phys. Rep.* **59** (1980) 96.
4. S.N. Ganguli and D.P. Roy , *Phys. Rep.* **67** (1980) 201.
5. G.L. Kane, J. Pumplin and W. Repko, *Phys. Rev. Lett.* **41** (1978) 1689.
6. G. Preparata, presented atHigh energy physics with polarized beams and polarized targets , Ed. J.Soffer (Basel 1981),120.
7. A.C. Irving and R.P. Sworden , *Phys. Rep.* **34** (1977) 120.
8. G. Hohler et al., *Nucl. Phys.* **95B** (1975) 210.
9. R. Blackenbecler and S.J. Brodsky , *Phys. Rev.* **10D** (1974) 2973.
10. M. Sullivan *et al.*, *Phys. Lett.* **B142** (1984) 451.
11. H. Kinichi, *Phys. Lett.* **B72** (1978) 411.
12. L.G. Pondrom , *Phys. Rep.* **122** (1985) 60.
13. B. Lundberg et al., *Phys. Rev.* **40D** (1989) 3557.
14. R. Barni, G. Preparata and P.G. Ratcliffe, *A Simple Explanation of Hyperon Polarization at High p_T*, Milano preprint, MITH 91/21, submitted to *Phys. Rev. Lett.*

15. L. Deck *et al.*, *Phys. Rev.* **D31** (1983) 1.
16. J. Singh et al., *Nucl. Phys.* **140B** (1978) 189.
17. D.L. Adams *et al.*, *Phys. Lett.* **B261** (1991) 201; **B264** (1991) 462; **B276** (1992) 531.
18. R. Barni, G. Preparata and P.G. Ratcliffe, *Initial-State Spin Asymmetry and Final-State Interference in Large-Angle Pion Production*, Milano preprint, MITH 92/5, submitted to *Phys. Rev. Lett.*
19. R.O. Polvado *et al.*, *Phys. Rev. Lett.* **42** (1979) 1325.
20. K.J. Heller *et al.*, *Phys. Rev. Lett.* **41** (1978) 607.
21. T.A. Degrand and H.I. Miettenen, *Phys. Rev.* **D24** (1981) 2419;
B. Andersson, G. Gustafson and G. Ingelman, *Phys. Lett.* **B85** (1979) 417;
J. Qiu and G. Sterman, *Phys. Rev. Lett.* **67** (1991) 2264;
D. Sivers, *Phys. Rev.* **D41** (1990) 83; **D43** (1991) 261.

Common Problems and Trends of Modern Physics, pp. 147–157
ed. T. Bressani, S. Marcello and A. Zenoni
©1992 World Scientific Publishing Company

THE FINAL STATES IN e^+e^- EVENTS

M. Verpelli
Dipartimento di Fisica, Università di Milano, via G. Celoria 16, 20133 Milano, Italy

ABSTRACT

The high energy limit of ACD allows us to formulate a model of e^+e^- annihilation events that explains the final state structure. Implemented in the EPOS montecarlo programme, the model reproduces the experimental data avoiding an excess of parameters.

1. Introduction

The physics of e^+e^- has always been an unceasing source of news in studying the hadronic world: the evidence of color degree of freedom provided by $R^{e^+e^-}$ (the ratio of the hadronic cross section to the pointlike μ pair creation), the discovery of the jets, gave a fundamental contribution to impose the QCD as the theory of strong interactions. A complete description of the process, from the initial annihilation to the complex structure of final states, must face aspects at first sight contradictory of quark physics: the free field partonic behaviour at high energies and the absence of gluons and quarks as final states, that is the confinement: the theory must be dynamically realized in such a way as to avoid the appearance of quark degrees of freedom in its physical states. The standard approach [1] utilizes a high energy perturbative expansion around the non interacting free quark fields: nothing can be said about the confinement that is supposed to emerge from a phase transition in the long distances regime where the coupling constant diverges. Thus the world of quarks and gluons appears to be decoupled with that of hadrons: the real world. The theory dwells only in the first, unphysical world: the construction of final states consists, essentially, in phenomenological parametrizations. The study of dynamical stability of the free vacuum, started with Savvidy [2] and brought to completion by Preparata, [3] has revealed that things go differently due to the self interaction of the gauge fields: the perturbative vacuum is not stable but collapses to a gluon condensate which is locally strongly anisotropic and in which the color flux flows in only one direction giving rise, by the Gauss theorem, to a linear confining potential; the particle excitations out of this vacuum are treated perturbatively via pair creation.

Whilst PQCD consist first in a perturbative development of a shower of partons (quarks and gluons) and after in a non perturbative and not understood rearrangement in the physical states, ACD (the realization of QCD on the true vacuum) evolves the initial state down to the hadrons in a set of perturbative steps involving only physical confined states [†]. We report now the three fundamental results of ACD, the basis of our construction of e^+e^- final states:

- The physical states are made of a minimum numbers of quark ($q\bar{q}$ or qqq).
- Their evolution is a perturbative pair creation at every scale (this is strongly supported by the validity of Zweig rule also for the decay of ϕ in KK).
- The final hadrons come from the decay of initial confined $q\bar{q}$ or qqq states.

2. The high energy limit

A calculable model of an e^+e^- event is obtainable in a simplified version of ACD which utilizes $q\bar{q}$ wave functions in the high mass limit; we started from the work reported in [4] to make it as close as possible to ACD. The wave function describing the q ($\bar{q}$) component of a meson state in its center of mass frame, has the following features:

- Is confined:

$$\Psi = 0 \quad \text{for} \quad |x| \geq R_s(M), \quad |t| \geq R_t(M) \quad \text{and} \quad R_{s,t} \to R^2 M. \qquad [2.1]$$

- Is free:

in the allowed region obeys to the free Dirac equation.

The propagator obtained with these requirements is:

$$G(p^2) = \begin{cases} \dfrac{\sin(R^2(p^2 - m_q^2))}{(p^2 - m_q^2)} & p^2 \simeq m_q^2 \\ \dfrac{1}{p^2} & |p^2| \gg m_q^2 \end{cases} \qquad [2.2]$$

[†] see the talks presented by M. Scorletti and L. Gamberale at this School and references therein

This expression shows the absence of poles on mass shell (confinement) and the free field behaviour at short distances. The following step is to obtain, starting from the above expression, an asymptotic mass spectrum in terms of a set of Regge trajectories as in fig. 1:

$$M^2_{n,l} = \pi R^{-2}(2n + l) \qquad [2.3]$$

where l is the angular momentum, n the radial excitation and R represents the rate of increase of the hadronic radius with mass.

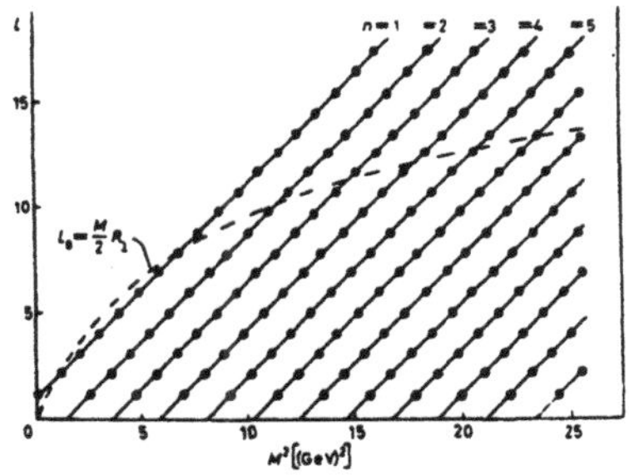

Fig. 1 The Regge trajectories

Due to the finite range of strong interactions, there is a maximum value of the angular momenta coupled in a physical process that is given by:

$$l_o = R_T M/2 \qquad \text{with} \quad R_T \simeq 0.5 \text{ fm}. \qquad [2.4]$$

The crucial observation is that at fixed but large M there is a set of degenerate states in l and one can describe a general $q\bar{q}$ state of mass M as a linear coherent superposition of states with a definite l ranging from zero to l_o. So we are led to consider the following state:

$$\Psi(P, k, \Omega) = \sum_{l=0}^{l=l_o} \sum_{m=-l}^{m=l} \psi_{nl}(p, k) Y_l^m(\Omega) \qquad [2.5]$$

where $\psi_{nl}(p, k)$ is the meson state wave function, $p^2 = M^2$, k is the momentum of the quarks in the center of mass, Y_l^m are the spherical harmonics, Ω is an arbitrary spatial direction. The important feature of this state is that, being a coherent superposition of physical states is also a Final State. We immediately see that this is a promising

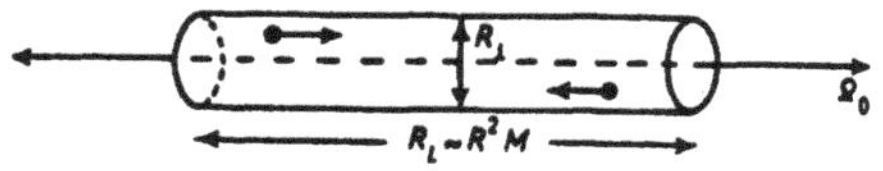

Fig. 2 The spatial structure of a FS

way, if we look at the structure of a FS showed in fig. 2: a tubelike system in which the quarks move on shell as plane waves along the direction Ω.

In this way one obtains a partonic behaviour starting with $q\bar{q}$ permanently bound states: the Fire String. The longitudinal size of this confining domain increases with the mass as R^2M, while the transverse radius R_T is fixed, thus reproducing the jet structure of the events. Following the outlined strategy, one must now calculate the perturbative evolution of this state:. the decaying of a FS in two FS through a pair creation caused by a quantum fluctuation in the field of a gluon that in ACD has a propagator that remembers the direction of the color flux between the two quarks:

$$D_{\mu\nu} = \frac{-(g_{\mu\nu} + n_\mu n_\nu)}{(q \cdot n)^2} \qquad [2.6]$$

The polarization vector n_μ is obtained by requiring that it is normalized to minus one, orthogonal to q to have a pole in the propagator and that its spatial part is aligned with the center of mass direction of quark motion. The process is described by diagrams of the type:

Fig. 3 Left: diagram contributing to FS decay. Right: reduced partonic diagram

in which is considered the decaying of a meson state with fixed angular momentum, the total FS decay width is recovered after, performing the sum over l. Using the quark propagator [2.6] and the diffractive behaviour of $q\bar{q}$ mediated by a Pomeron trajectory, the diagram is reduced to one in which the cut is on quark legs, implying that the FS reproduces the partonic behaviour. Now, we have seen that the gluon is essentially on shell; it travels a considerable distance inside the FS. In this way it is possible to separate the calculation in two stages:

1) The gluon emission by a quark leg.

2) The gluon decay in a $q\bar{q}$.

In the first step one considers the bubble representing the meson state which contains: a Dirac matrix describing the spin structure (γ_5 for a pseudoscalar and γ_μ for a vector) and an angular wave function. The latter, according to ACD, is given by Bessel special function of $|\vec{p}|$, the modulus of the spatial part of the quark momentum in the C.M., multiplied by R, the spatial extent of the excited system than can be evaluated form the Heisemberg principle:

$$j_l(R|\vec{p}|)_{R|\vec{p}|\to 0} \simeq (R|\vec{p}|)^l \quad \text{and} \quad (R/2)^2 \simeq 1/(p)^2. \qquad [2.7]$$

Thus the angular factor, normalized to one when the quark is on shell ($|\vec{p}| = M/2$), is:

$$(f)^l = \left(\left(\frac{R|\vec{p}|}{1+R|\vec{p}|}\frac{1+RM/2}{RM/2}\right)^l . \qquad [2.8]$$

The physical implications are very interesting: when the quark is excited near the mass shell, R is maximum and the excitation of colour flux (the gluon) travels along the FS seeing all the partial waves in the sum, producing strong anisotropies: the decaying is collinear with the FS direction giving the two jet structure. But when the quark is off shell, R is minimum, and only the low (isotropic) waves are non zero in this region, so it is possible to obtain the three jets events As consequence there is no "quark jet" nor "gluon jet" but all the jets in an event have the same dynamical origin.

Returning to the calculation, attention must be paid to current conservation. Due to the fact that the meson vertex is not local, a piece must added to in order to restoring the Ward identity. We are left with a process composed by three amplitudes: fig. 4

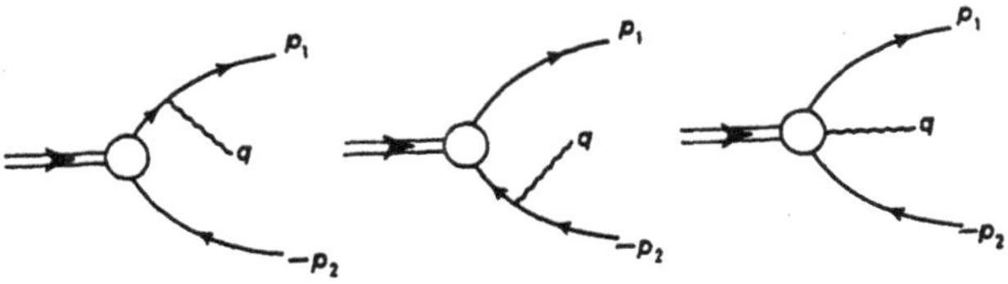

Fig. 4 The three bremsstrahlung-like amplitudes

Squaring to obtain the matrix element, the gluon can not be treated as free but it propagates through a pair creation in the FS, so the sum over the polarization states is substituted with $D_{\mu\nu}$, the gluon propagator, to obtain the final result:

$$\sum_{l=0}^{l=l_o} \frac{qp_1}{qp_2}\Big(f(|\vec{p}|)\Big)^{2l}(2l+1)P_l(\cos\theta) + \Big(p_1 \Longleftrightarrow p_2\Big) \qquad [2.9]$$

Turning to the gluon decay, it is the well know amplitude for a vector particle that goes in two spin 1/2 particles:

$$\Gamma \div (1+\cos^2\theta). \qquad [2.10]$$

Neglecting the quark masses (the heavy quark pair creation is suppressed) the presencė of four delta at the vertex ($\delta(q-k-l), \delta(q^2), \delta(l^2), \delta(k^2)$), imposes the following kinematical constraints:

$$k = \lambda q \quad l = (1-\lambda)q \qquad [2.11]$$

that is k, l, q are collinear. In this way $\cos\theta$ can be reexpressed as a function of λ and the masses of the two FS are:

$$M_{FS_1} = (p_1+k)^2 \simeq \lambda p_1 \cdot q \quad \text{and} \quad M_{FS_2} = (p_2+k)^2 \simeq (1-\lambda)p_2 \cdot q. \qquad [2.12]$$

The treelike process of sequential FS decay ends when the mass is so low that, according to [2.4], the sum over l involves only a few partial waves thus projecting the FS in a lowest lying meson state. This "fragmentation process" requires only one free parameter: R_T; there is but another parameter introduced by the decay, described elsewhere [4], of a FS in two barions. It is worth noticing that in the tree process we do not have as parameters as ratio of vector to pseudoscalar mesons creation, or the suppression of strange quark creation. We also stress that in the other more popular models, the jet structure and the scaling behaviour are imposed *ab externo* in the fragmentation functions by the statement of the form of the integral equation and of its kernel. Also the p_T, the transverse momentum, is obtained adopting a gaussian distribution around a p_T mean. In this way the theoretical value of models is obscured by a great deal of phenomenological parametrisation. [5]

To better explain the end steps of a FS decay we see that for a $u\bar{u}$ system:

$$l_o \simeq 1 \quad \text{for} \quad M_{FS} \simeq 0.9\ \text{GeV} \quad \text{and} \quad l_o \simeq 2 \quad \text{for} \quad M_{FS} \simeq 1.6\ \text{GeV} \qquad [2.13]$$

when the tree decay produces a FS in the mass range below 0.9 GeV, this state decays into the lowest lying mesons (ρ, ω, a, η) weighted according $SU_f(3)$ coefficients; if the range is between 0.9 to 1.6 GeV, it is let to decay as an incoherent superposition of a π and a previous meson.

3. The EPOS montecarlo results

In fig 6/8 are reported the simulation results of differential distributions at the Z^o and the energy dependence of mean quantities from 14 to 150 GeV. We notice that EPOS reproduces the quantities usually adopted to determine α_S, as jet-rates and asymmetry in energy-energy correlation. In our vision in fact the hard physics of very short distances (the physics of α_S) is not involved in the FS decay which is a soft perturbative process; what is really determined by α_S are the correction to the cross section that involves the pointlike coupling with the electroweak current, so we are led to consider $R^{e^+e^-}$ as the quantity most naturally related to α_S. More than hundred experimental points are now at our disposal and their simple analysis carried in [6] and showed in fig 5, strongly favours the QCD dynamics realized by ACD. Let us now get to the quantities used to describe an event. Defining:

$$M_{i,j} = 1/n \sum_{k=1}^{n} p_i(k)p_j(k) \quad \text{and} \quad \Lambda_1 < \Lambda_2 < \Lambda_3, \quad \Lambda_1 + \Lambda_2 + \Lambda_3 = 1, \qquad [3.1]$$

where $\Lambda_{i=1,2,3}$ are the eigenvalues of M and n_i the eigenvectors. The sphericity S and the aplanarity A are:

$$S = \frac{3}{2}(1 - \Lambda_3), \quad A = \frac{3}{2}\Lambda_1 \qquad [3.2]$$

Collinear events give $S = 0$, isotropic $S = 1$; for coplanar events $A = 0$. The mean value of the transverse momenta (with respect to n_3) are:

$$p_T^{out} = M_{11} = 1/n \sum_{k=1}^{k=n} (p(k) \cdot n_1)^2, \quad p_T^{in} = M_{22} = 1/n \sum_{k=1}^{k=n} (p(k) \cdot n_2)^2 \qquad [3.3]$$

One also defines the thrust and the minor:

$$T = 2\frac{\sum_{k+} p(k) \cdot n}{\sum_k |p(k)|}, \quad M = 2\frac{\sum_{k+} p(k) \cdot n_3}{\sum_k |p(k)|} \qquad [3.4]$$

where n is chosen to maximize T. The angular correlation of energy flow is obtained with the energy-energy correlation:

$$EEC(\theta) = \frac{1}{\Delta\theta N_{events}} \sum_{N_{events}} \int_{\theta-\Delta\theta/2}^{\theta+\Delta\theta/2} d\theta^o \delta(\theta^o - \theta_{i,j}) E_i E_j / E_{vis}^2 \qquad [3.5]$$

where E_i is the energy of the particle i, E_{vis} is the energy observed in the event, $\Delta\theta$ is the bin width. Two jet events contributes near $\theta = 0$ and $\theta = 180$. To emphasize the three jet events the $AEEC$ is introduced:

$$AEEC(\theta) = EEC(180 - \theta) - EEC(\theta). \qquad [3.6]$$

Finally, the string effect is related to three jet events in which the two less energetic jets are symmetric with respect to the most energetic; choosing those events in which a hard lepton is present in the first jet and in one of the remaining, a depletion of particles is evident between this two jets even if from the kinematics one aspect symmetry. For the PQCD-inspired models, the two jets with the lepton are quark jet while the third is a gluon jet. For the Fire String model there is no difference between jets, and the effect is related to a mass effect due to the presence of energetic leptons.

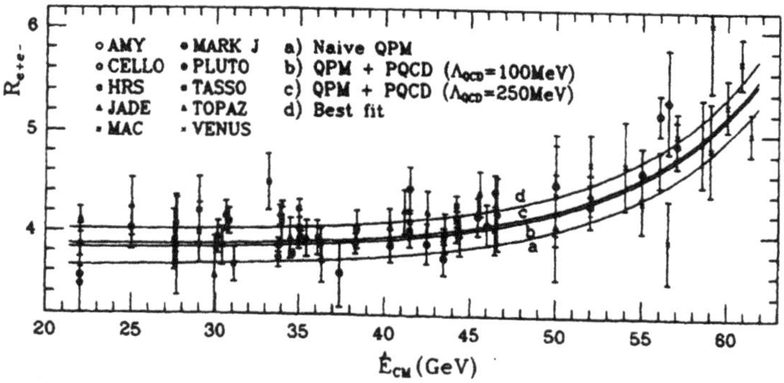

Fig. 5 $R^{e^+e^-}$ data, compared to (a) QPM prediction; the PQCD correction for (b) $\Lambda_{QCD} = 100$ MeV, (c) $\Lambda_{QCD} = 100$ MeV and (d) the best fit

REFERENCES

1. G. Altarelli *Phys. Rep.* **97** (1983) 32.
2. Savvidy *Phys. Lett.* **B71** (1977) 133
3. G. Prepapata *Nuovo Cim.* **103A** (1990) 1073
4. L. Angelini, L. Nitti, M. Pellicoro, G. Preparata *Riv. Nuovo Cim.* **6** (1983) 1
5. M. Bengtsson, T. Sjöstrand *Nucl. Phys.* **B289** (1987) 810
6. G. Preparata, P. G. Ratcliffe, M. Verpelli *Mod. Phys. Lett.* **A6** (1991) 605

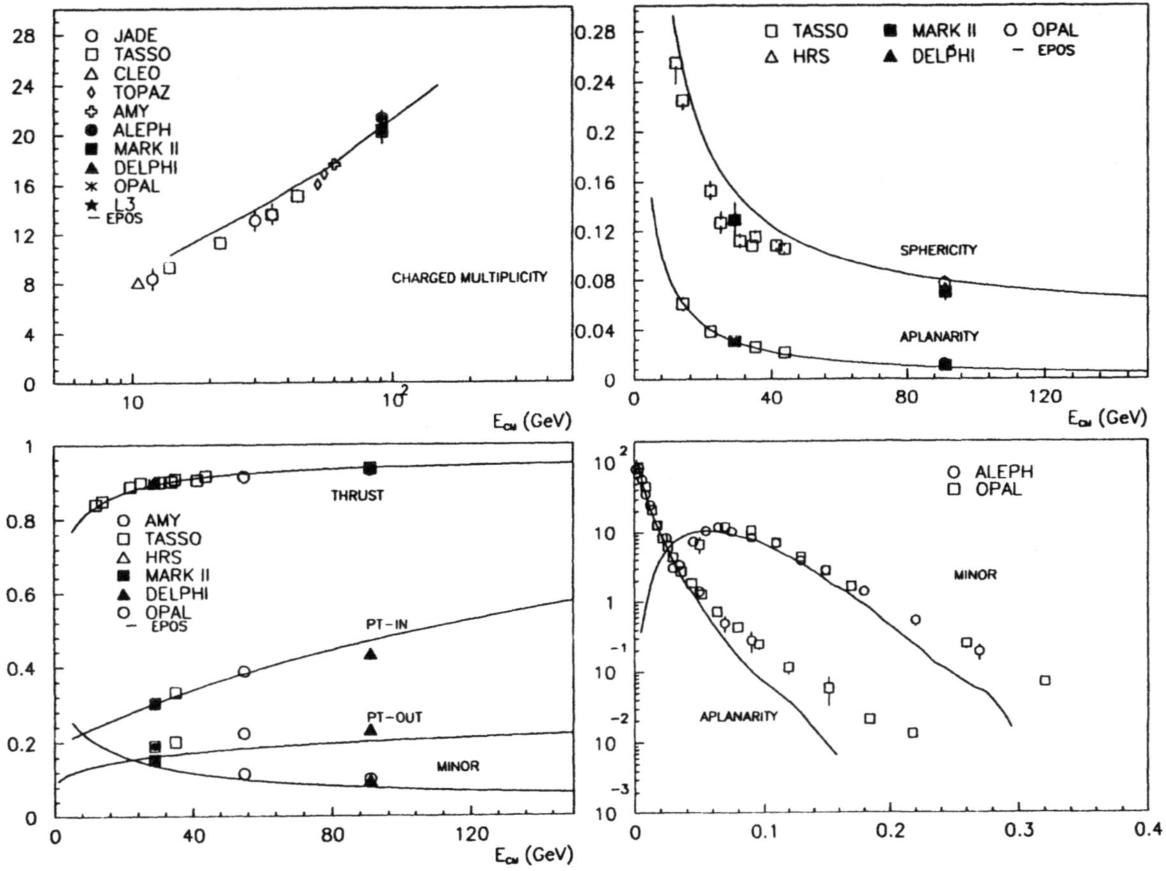

Fig. 6 The mean quantities energy dependence

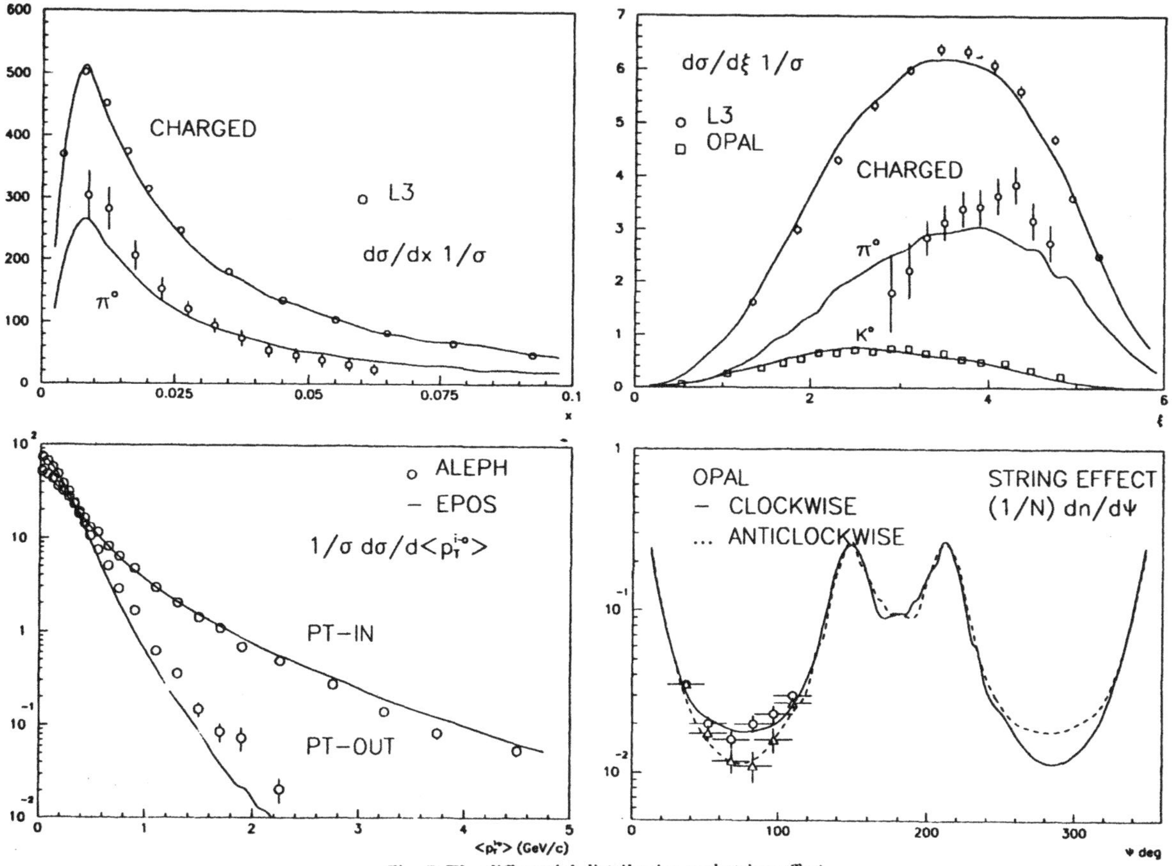

Fig. 7 The differential distribution and string effect

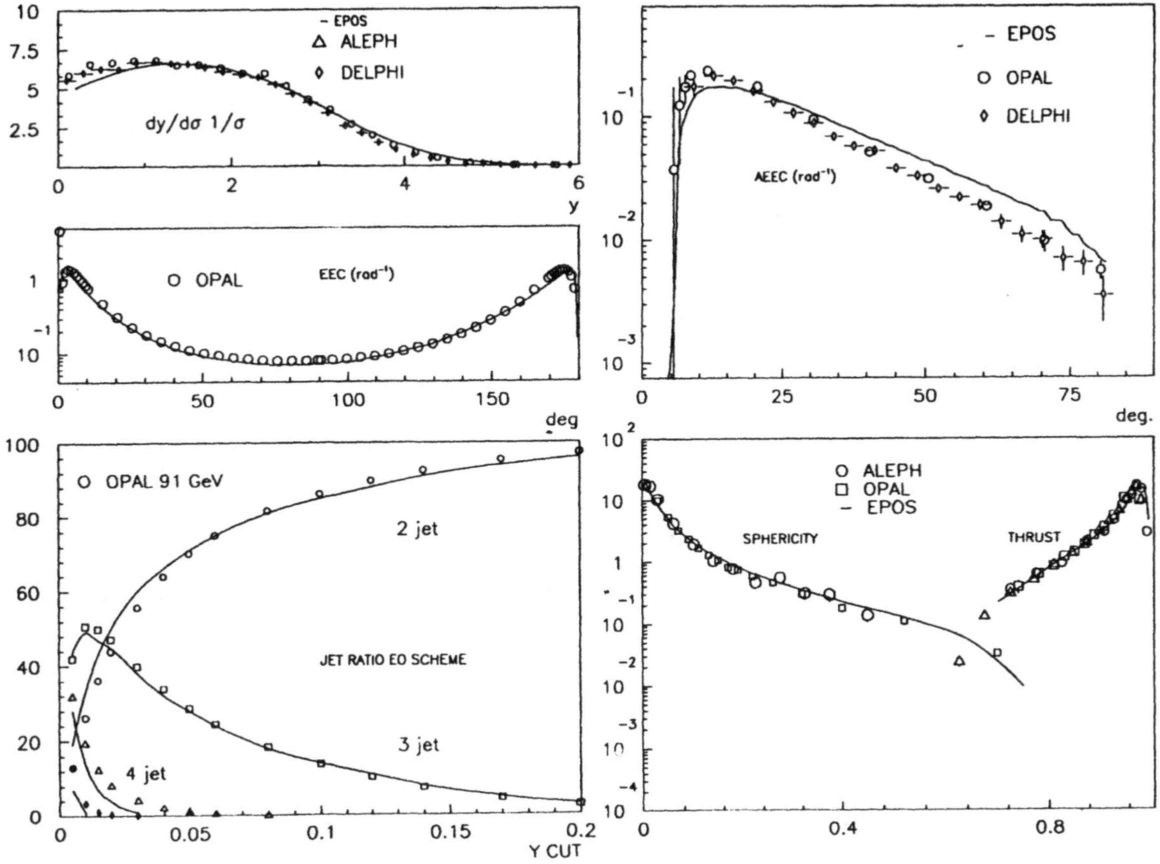

Fig. 8 The quantities related to α_S

Common Problems and Trends of Modern Physics, pp. 159–166
ed. T. Bressani, S. Marcello and A. Zenoni

©1992 World Scientific Publishing Company

MESONS IN QCD: SPECTROSCOPY

M. Scorletti
Dipartimento di Fisica, Università di Milano, via Celoria 16, 20133 Milano, Italy

ABSTRACT

An alternative strategy for solving QCD leads to an effective lagrangian on the highly non-trivial vacuum of the theory, which supports confinement of coloured states and spontaneous chiral symmetry breaking. We are able to obtain the relation between "current" and "constituent" quark masses and the Schroedinger-like equation for bound states. Our overall fit of the mesonic spectrum is in satisfactory agreement with experimental data. We also compute the current-particle matrix elements for pseudoscalar and vector states.

1. Bound states in QCD

We address in this contribution a crucial problem of hadronic physics, namely how the long-distance observed hadronic behaviour, whose major features are confinement and chiral symmetry breaking, could be related to the usual QCD lagrangian, which is universally believed to describe hadronic dynamics. The starting point [1] of our approach is a non-perturbative analysis of the structure of the QCD ground-state, both at short and at large distances. The results of a five-year long investigation are condensed in ref. 2 where, based on these findings, a general strategy has been developed to completely solve QCD. We recall here the most important conclusions of that analysis:

(a) the QCD vacuum has (most likely) the structure of a liquid of needle-shaped magnetic domains, of length of the order of one fermi. The quantum colour fluctuations (quarks and gluons) inside such domains are confined, and their "residual" effective interactions are dominated by a constant force (linear potential) at long distances and by a Coulomb-like (one gluon-exchange) potential at short distances.

(b) The (effective) Hamiltonian of the system on this peculiar ground state can be written

$$H = H^{(0)} + H', \tag{1.1}$$

where $H^{(0)}$ contains the infrared-singular (long-distance) part of the colour interaction, and commutes with $\hat{N}_q$, $\hat{N}_{\bar{q}}$ and $\hat{N}_g$ — the quark, antiquark and number operators —, while H' (which is infrared regular) describes pair creation and annihilation, and we assume that it can be treated perturbatively. The diagonalization of $H^{(0)}$, carried out independently in each $(N_q, N_{\bar{q}}, N_g)$ sector compatible with global colour neutrality, describes a world of stable, asymptotic, hadrons (the "Primitive World") that interacts perturbatively via H'.

As a first step in the strategy just described to completely solve QCD, we perform the calculation of the meson spectrum, i.e. we carry out the diagonalization of $H^{(0)}$ in the $(N_q = 1,\ N_{\bar{q}} = 1,\ N_g = 0)$ sector:

$$H^{(0)}\,|M\rangle = M\,|M\rangle. \tag{1.2}$$

The mesonic state $|M\rangle$ is simply (we are in the rest frame)

$$|M\rangle = \frac{\delta^{\alpha\beta}}{\sqrt{3}} Q^{ij} \chi^{rs} \int d^3\mathbf{p}\,\psi_{ij}^{rs}(\mathbf{p}) b^\dagger(\mathbf{p}, i, r, \alpha) d^\dagger(-\mathbf{p}, j, s, \beta)\,|\Omega\rangle \tag{1.3}$$

where δ, Q and χ are colour, flavour and spin wave functions. We refer to ref.3 for the explicit expressions of hamiltonian and effective potential, the discussion about the choice of the gluon propagator and all the details of the calculation.

Here we only recall that, introducing the reduced wave function $u_{ij,l}^{rs}(p)$, we can write

$$\psi_{ij}^{rs}(\mathbf{p}) = Y_{l,m}(\hat{\mathbf{p}}) p^{-1} u_{ij,l}^{rs}(p); \tag{1.4}$$

then the eigenvalue equation in momentum space reads

$$\left(\sqrt{p^2 + m_q^{(0)^2}} + \sqrt{p^2 + m_{\bar{q}}^{(0)^2}} - M_J\right) u_J(p) + \int_0^\infty dp' V_J(p, p') u_J(p') = 0 \tag{1.5}$$

where J collects all the quantum numbers of the mesonic state, and $V_J(p, p')$ is the fourier transform of the non-local potential $V_J(r, r')$. For the equal mass case, and with the quotation marks recalling the usual spin-dependent term of the one-gluon-exchange vector potential, we have

$$\begin{aligned} V_J(r,r') =& \mu^2\delta(r-r')r - \frac{\text{“}4}{3}\frac{\alpha_s\text{”}}{r} - \frac{3\mu^2}{\pi^2}K_0[m^{(0)}(r-r')]+ \\ &+\frac{\mu^2}{2\pi^2}(1+S_{q\bar{q}})\int_{-\infty}^{\infty}\frac{dt}{Ch^2(t)}K_0[m^{(0)}\sqrt{r^2+r'^2+2rr'Ch(2t)}]. \end{aligned} \qquad [1.6]$$

We would like to stress here a few points concerning our result:

(a) $V_J(r,r')$ looks like one of the many existing variants of the successful linear + Coulomb potential. We emphasize once again that [1.6] is not a phenomenological, QCD-inspired expression, with the linear term inserted by hand to incorporate confinement: $V_J(r,r')$ is *the* QCD potential, calculated without any assumption from the basic QCD lagrangian.

(b) The non-local terms containing the K_0 Bessel function depend drastically on the quark masses, and cause [1.5] to have a logarithmic singularity for vanishing masses; this is a good signal for chiral symmetry breaking.

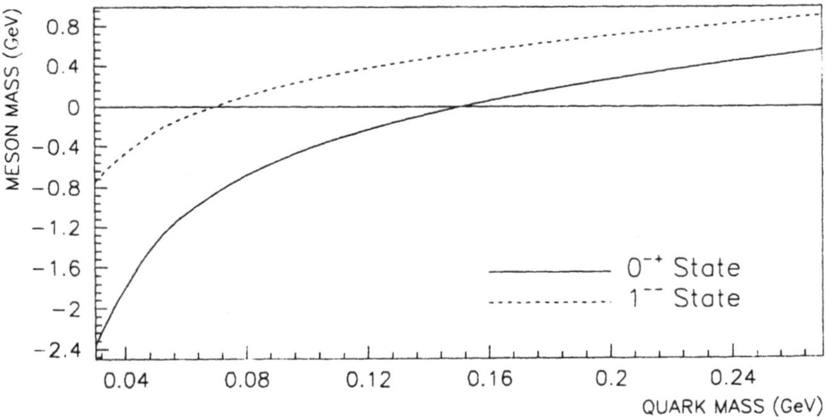

Fig. 1 Quarkonium vs. quark mass. The critical mass is $m^* \simeq 151\,\text{MeV}$.

2. What Happens in the chiral limit?

In fact, when we solve [1.5] for small quark masses, we obtain the remarkable situation depicted in fig. 1: for $m^{(0)} < m^*$ the pseudoscalar state acquires negative energy.

Thus $|\Omega\rangle$ cannot be the true ground state of the theory, and we must quantize it in terms of a new ground state $|0\rangle$ and new creation and annihilation operators, which are now relative to a new mass parameter, the constituent mass m_c. This can be achieved [4,5] via a Bogoliubov transformation , and the simple result for the hamiltonian is:

$$H^{(0)}(m^{(0)}) = H^{(0)}(m_c) + H^p \qquad [2.1]$$

$$H^p = (m^{(0)} - m_c) \int d^3\mathbf{x} : \bar{q}(\mathbf{x})q(\mathbf{x}) : . \qquad [2.2]$$

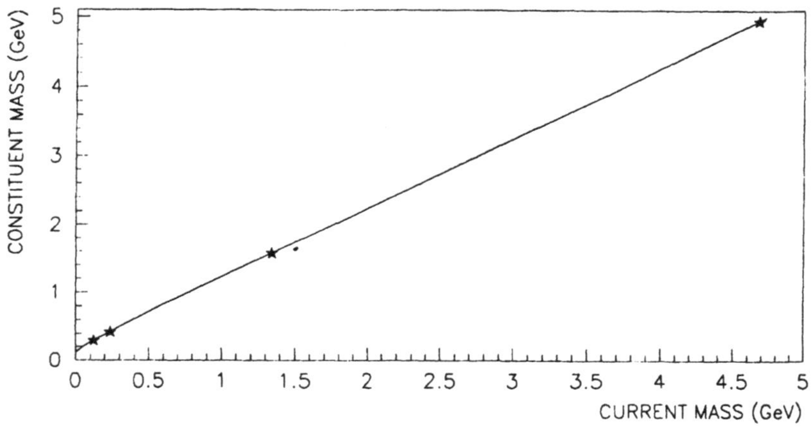

Fig. 2 Constituent vs. current quark mass. Values of the masses used in our calculations are marked by stars.

Let us consider now, according to

$$\langle 0| A^\mu(0) |P\rangle = i\sqrt{2} f_P \qquad [2.3]$$

and to

$$i\partial_\mu A^\mu(0) = -2m^{(0)}\bar{q}(0)\gamma_5 q(0), \qquad [2.4]$$

that, in the rest frame,

$$\frac{\langle 0| i\partial_\mu A^\mu(0) |P\rangle}{\langle 0| A^\mu(0) |P\rangle} = -2m^{(0)}\frac{\langle 0| \bar{q}\gamma_5 q |P\rangle}{\langle 0| \bar{q}\gamma_0\gamma_5 q |P\rangle} = \frac{i f_P M_P^2}{i f_P M_P} = M_P. \qquad [2.5]$$

Recalling that M_P is nothing but the expectation value of the hamiltonian on the pseudoscalar state $|P\rangle$, we obtain:

$$\langle P| H^{(0)}(m_c) + H^p |P\rangle + 2m^{(0)}\frac{\langle 0| \bar{q}\gamma_5 q |P\rangle}{\langle 0| \bar{q}\gamma_0\gamma_5 q |P\rangle} = 0, \qquad [2.6]$$

a relation between constituent and current mass, whose solution is showed in fig. 2. We are also able to calculate [4]

$$\lim_{m^{(0)}\to 0} \langle 0|\, \bar{q}_\alpha(0) q^\alpha(0)\, |0\rangle = -6 \int \frac{d^3\mathbf{p}}{(2\pi)^3} \frac{m_c^*}{\sqrt{p^2 + m_c^{*2}}} \propto -m_c^* \Lambda^2 \qquad [2.7]$$

where Λ is the ultraviolet cut-off and $m_c^* = \lim_{m^{(0)}\to 0} m_c(m^{(0)}) = 110\,\text{MeV}$.

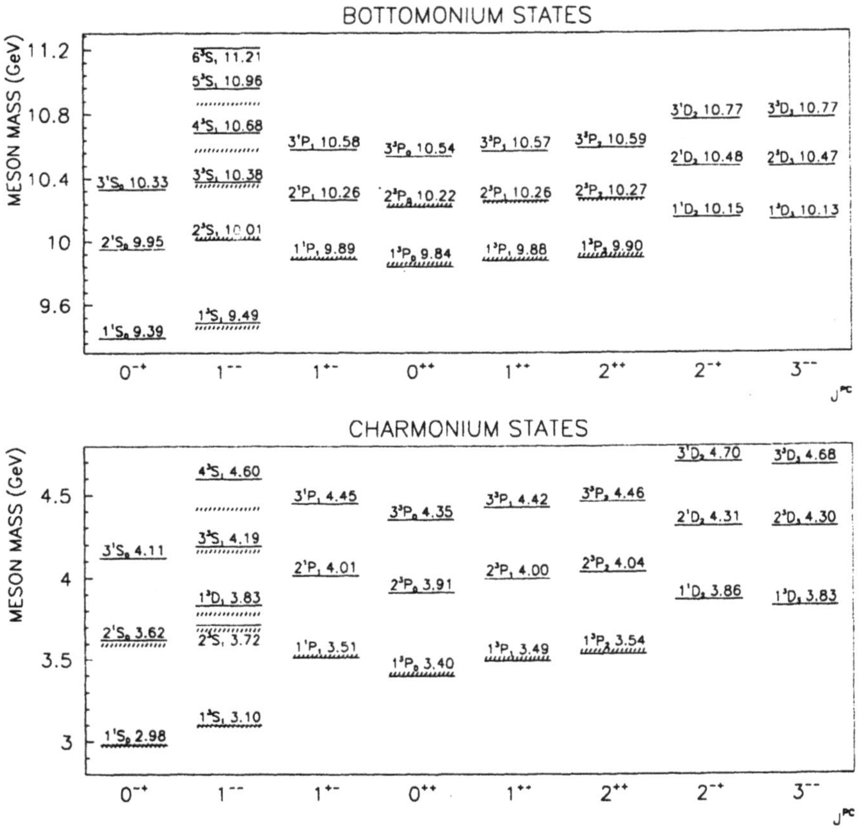

Fig. 3 Comparison between our theoretical calculation (solid lines) and the available experimental data (dashed areas) for the heavy quarkonia. We set $m_c^{(0)} = 1.345$, $m_b^{(0)} = 4.685$, $\mu = 0.49$, $\alpha_s = 0.24$. Units are in GeV.

This is a standard way to show that the vacuum $|0\rangle$ breaks chirality and this fact, together with the conservation of the axial current in the chiral limit meets the hypotheses of the Goldstone theorem [6]: the vacuum rearrangement we have just described is due to spontaneous chiral symmetry breaking, and the π appears to be the Nambu-Goldstone boson of the chiral symmetry.

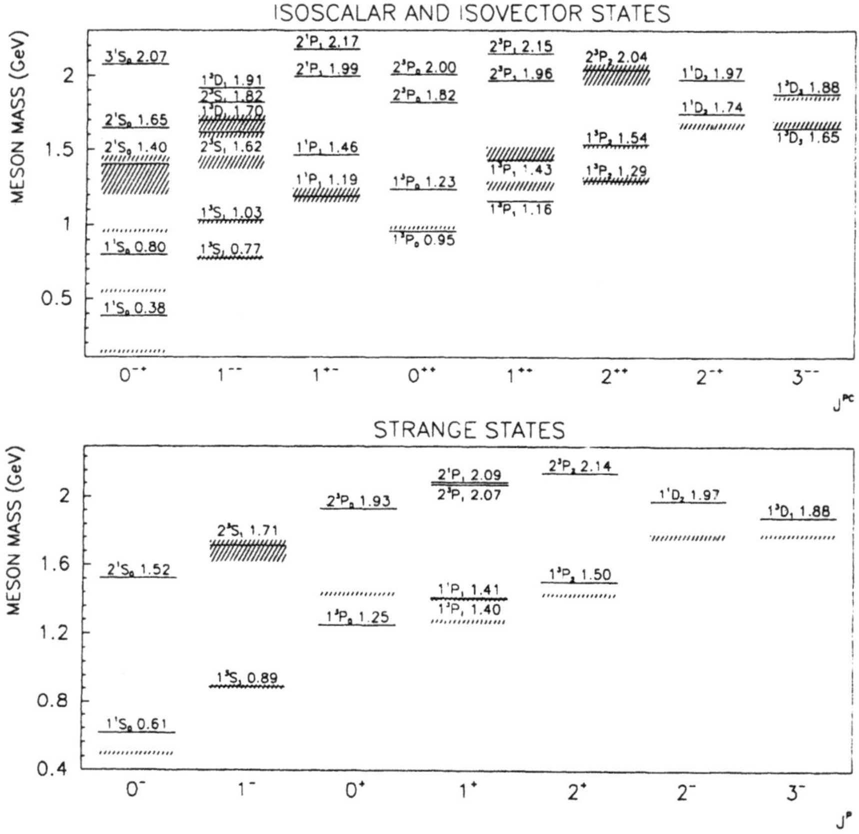

Fig. 4 Comparison between our theoretical calculation (solid lines) and the available experimental data (dashed areas) for the light mesons. We set $m_u^{(0)} = 0.125$, $m_s^{(0)} = 0.237$, $\mu = 0.49$, $\alpha_s = 0.24$. Units are in GeV.

The eigenvalue equation for mesons on the new vacuum looks like [1.5], with $m^{(0)}$ replaced by m_c and with a new diagonal term due to H^p. Figs. 3,4 shows our results for the spectrum: they substantially agree with available data, with the important exception of light pseudoscalar states: this is due to the fact that, at this stage of the calculation, we do not take into account either the Q^2 dependence of α_s or flavour mixing in the $SU(3)_f$ sector (which is produced by H'), so that isoscalar and isovector states are degenerate in our spectrum.

3. Decay constants

Let us define the current-particle matrix elements

$$\begin{aligned} \langle 0|\, A^{\mu}_{ij}(0)\, |P(q)\rangle &= iQ_{ij}\sqrt{2} f_P q^{\mu} \\ \langle 0|\, V^{\mu}_{ij}(0)\, |V(q,\epsilon)\rangle &= Q_{ij} f_V \epsilon^{\mu} \end{aligned} \qquad [3.1]$$

where

$$\begin{aligned} A^{\mu}_{ij}(x) &= \bar{q}_i(x)\gamma^{\mu}\gamma_5 q_j(x) \\ V^{\mu}_{ij}(x) &= \bar{q}_i(x)\gamma^{\mu} q_j(x) \end{aligned} \qquad [3.2]$$

Table I. Theoretical pseudoscalar and vector decay constants. The experimental values, where available, are shown for comparison.

Pseudoscalar state	f_P (MeV)	Vector state	f_V (GeV^2)
π	135 (93)	ρ, ω	0.24 (0.17)
K	118	K^*	0.28
η, η'	133	Φ	0.30 (0.24)
D	134 (< 200)	D^*	0.58
D_s	154	D^*_s	0.59
B	113	B^*	1.07
η_c	200	J/Ψ	1.11 (1.19)
		$\Psi(2S)$	0.97 (1.04)
η_b	229	Υ	3.12 (6.76)
		$\Upsilon(2S)$	2.59 (4.88)

are the usual axial and vector current. $|P(q)\rangle$ and $|V(q,\epsilon)\rangle$ denote pseudoscalar and vector mesons with fixed momentum and polarization (see [1.3]); this results in (see also ref.7):

$$f_P = \sqrt{\frac{3}{2}\frac{1}{2\pi M_P}} \int_0^{\infty} dpp \sqrt{\frac{(E_i+m_i)(E_j+m_j)}{E_i E_j}} \left(1 - \frac{E_j - m_j}{E_i + m_i}\right) u_P(p) \qquad [3.3]$$

$$f_V = \frac{\sqrt{3}}{2\pi}\int_0^\infty dpp\sqrt{\frac{(E_i+m_i)(E_j+m_j)}{E_iE_j}}\left(1+\frac{1}{3}\frac{E_j-m_j}{E_i+m_i}\right)u_V(p), \qquad [3.4]$$

where the wave functions are normalized according to

$$\int_0^\infty dp|u_J(p)|^2 = 2M_J. \qquad [3.5]$$

We note that the wave functions that we used in [3.3] and [3.4] do not contain the contribution of Coulomb interaction, for it causes the wave functions to have bad asymptotic momentum behaviour and the matrix elements to diverge. It is clear that such divergences are renormalized by taking into account the quark wave-function renormalization, however we have not yet performed this, which is why the decay constants that we obtain — shown in table I— are too large for the light states and too small for the heavy ones.

REFERENCES

1. G. Preparata, *Nuovo Cim.* **96A** (1986) 366.
2. G. Preparata, *Nuovo Cim.* **103A** (1990) 1073.
3. R. Garattini, S. Mattina and G. Preparata, A new QCD Approach to the Physics of Quarkonia, Milano Univ. preprint MITH 91/12 (1991).
4. P. Castorina, P. Cea, P. Colangelo, G. Nardulli and G. Preparata, *Phys. Lett.* **115A** (1982) 487.
5. N.N. Bogoliubov, Sov.Phys.JETP **34** (1958) 41.
6. J. Goldstone, *Nuovo Cim.* **19** (1961) 154.
7. P. Cea, P. Colangelo, G. Nardulli and G. Preparata, *Phys. Lett.* **115B** (1982) 310.

Common Problems and Trends of Modern Physics, pp. 167–179
ed. T. Bressani, S. Marcello and A. Zenoni
©1992 World Scientific Publishing Company

MASS GENERATION IN THE STANDARD MODEL

She-Sheng Xue[†]
INFN, National Laboratory at Frascati, Rome Italy

ABSTRACT

A proposal of mass generation in the Standard Model (SM) is presented, based on the idea that the violent fluctuations of quantum gravity at the Planck scale plays a nature regulator for the SM with fundamental length $a_p \sim 10^{-33}$cm, thus gauge invariant four-interactions are necessary due to the "no-go" theorem of Nielsen and Ninomiya, forbidding a sensible formulation of the usual electroweak action on a such "Planck lattice". We present that for certain values of the new four-fermion couplings a spontaneous violation of SM chiral symmetry emerges which (i) avoids the "no-go" theorem, (ii) produces a $\bar{t}t$ condensate model without appearance of the low-energy Higgs boson. The mass generation of other fermions and intermediate gauge bosons, are briefly discussed.

1. Introduction

In this lecture, I shall outline and discuss that the gauge principle realized in the SM prevents from massive fermions and gauge bosons, and the quantum gravity acts as a nature regulator called the "Planck lattice"(PL) which breaks gauge symmetry and thus gauge invariant four-fermion interactions are necessary by the "no-go" theorem of Nielsen and Ninomiya. In this framework of the SM in the PL, gauge principle is maintained and however all fermions as well as intermediate gauge bosons can acquire their masses. The structure of the lecture will thus be as follows. In Section 2 the possibility of quantum gravity may endow space-time with "foam" structure and the usual Higgs mechanism for mass generation in the SM are briefly discussed. Section 3 will provide an brief outline of the "no-go" theorem of Nielsen

† E-mail: xue@milano.infn.it and mailing address: INFN-section of Milan, Via Celoria 16, Milan Italy

and Ninomiya and the necessity of the gauge invariant four-fermion interactions. The resulting $\bar{t}t$-condensate without the low-energy Higgs boson will be presented in Section 4. The discussions of other fermions and intermediate gauge bosons acquiring their masses are briefly followed in Section 5 and Section 6. And finally I shall close in Section 7 with a few comments and conclusions.

2. The Planck Lattice

The very-small-scale structure of space-time, the arena of physical reality, has recently attracted a great deal attention. The realization of the important role that quantum gravity plays in determining such a structure, coupled with the grave difficulties that this theory faces in the usual, perturbative formulation at distance smaller than the Planck length $a_p \sim 10^{-33}$cm (the Planck mass $\Lambda_p \sim 10^{19}$GeV), is the basic motivation of a number of theoretical proposals to overcome such difficulties and provide a solution of this fascinating problem. As well known, the very popular superstring theory belongs to this class of proposals, overcoming the problems of the conventional quantum gravity by postulating that the fundamental "constituents" of space-time, the physical events, at distance smaller than a_p exhibit the rather complex structure of a space-time "string" instead of a simple space-time point. In this way at distance smaller than a_p a whole brave new world emerges, described by the rich spectrum of excitations of the string, whose phenomenology, however, seen from our vantage point appears, to say the least, remote.

At the opposite end we may conceive that precisely due to the violent quantum fluctuations that the gravitational field must exhibit at a_p, space-time somehow "end" there. Either by the creation of a "foam" [1], or by some other mechanism which we need not discuss here, one may conceive that as a result the physical space-time gets endowed with a fundamental length, a_p, and thus the basic arena of physical reality becomes a lattice with lattice constant a_p.

It is this possibility that we wish to explore in a fundamental problem of the SM, the problem of fermion and intermediate gauge boson masses. As is well known the gauge-symmetry principle as realized in the electroweak sector $SU_L(2) \otimes U_Y(1)$ demands that, at lagrangian level, all fermions and gauge bosons must be massless. In order to avoid an obvious theoretical disaster and save the gauge principle, the Higgs mechanism [2] had to be grafted upon the beautiful gauge lagrangian, with a completely *ad hoc* negative squared mass in order to secure a spontaneous symmetry breaking mechanism for the generation of fermion masses as well as gauge boson masses. It is for this reason that most people regard the Higgs mechanism at best as a simple, rough approximation to real situation, and deny such field any fundamental reality. And this is perfectly in line with what one learns from superconductivity, where Higgs mechanism was introduced as a simple, phenomenological means to describe the more complicated dynamics of Cooper pairs.

3. The Necessity of Four-fermion Interactions

Suppose now that, following the considerations reported in Section 2, we wish to write the SM without Higgs field on the PL, a profound result obtained more than ten years ago [3] in the form of a "no-go" theorem, tells us that there is no consistent way to transpose straightforwardly on a lattice the lagrangian of the continuum theory. While this is not the place for a detailed discussion of this important result, we limit ourselves to remarking that the origin of the "no-go" theorem is quite simple, stems from the peculiar dispersion relation of fermion on a lattice. According to it, in the long wave-length limit more than one species appear in the spectrum and, for instance of a left-handed Weyl field ψ_L with chiral charge, say $\chi = -1$, there appear four right-handed and four left-handed species. A situation which is clearly at variance with experimental observation.

Wilson [4] has shown how can one modify the lagrangian by adding a simple bilinear term, so as to remove the unwanted "replicas" from the long wave-length regime. However, this can only be done by sacrificing chiral invariance, and "no-go" theorem shows that no bilinear modification can be made to obey the chiral gauge-principle. Thus if we are to go ahead in our program of putting the SM on the PL without sacrificing the chiral gauge-principle, the "no-go" theorem tells us that the simple transposition of the continuum lagrangian must be supplemented by extra-terms that are least quadrilinear in fundamental fermion fields. Familiarity with the original Nambu-Jona-Lasinio model [5] [6] immediately reminds us that such quadrilinear terms can be made to obey the chiral gauge-principle. Thus an $SU_c(3)\otimes SU_L(2)\otimes U_Y(1)$ chiral-gauge invariant lagrangian which evades *in principle* the "no-go" theorem can be written on the PL as [7]

$$S_{PL} = S_G + \sum_F (S_D^F + S_{NJL}^{F1} + S_{NJL}^{F2}), \tag{3.1}$$

where S_G is the usual Wilson gauge-action, S_D the usual Dirac action, which will both be analyzed in a forthcoming article [8], F=l (F=q) denotes its lepton (quark) sector. The simplest new quadrilinear NJL-terms are as follows [9]:

$$S^{F1}_{NJL} = G_1 \sum_x \{\bar{\psi}^{Fi}_L(x) \cdot \psi^{Fj}_R(x) \bar{\psi}^{Fj}_R(x) \cdot \psi^{Fi}_L(x)\} \qquad [3.2]$$

and

$$S^{F2}_{NJL} = \frac{G_2}{2} \sum_{\pm\mu,x} \left[\bar{\psi}^{Fi}_L(x) G^L_\mu(x) U^c_\mu(x) \cdot \psi^{Fj}_R(x + a_\mu) \bar{\psi}^{Fj}_R(x) G^R_\mu(x) U^c_\mu(x) \cdot \psi^{Fi}_L(x + a_\mu)\right], \qquad [3.3]$$

where the indices i, j denote fermion families; the Dirac indices are denoted by scalar product ".". The gauge link $U^c_\mu(x)$ connects left-and right-handed quark fields in neighboring points so as to have the $SU_c(3)$ gauge symmetry. The chiral gauge links $G^L_\mu(x)$ ($G^R_\mu(x)$) connect left-handed (right-handed) fermion fields to enforce $SU_L(2) \otimes U_Y(1)$ chiral gauge symmetry. $G_{1,2}$ are two, yet unspecified, Fermi-type $O(a_p^2)$ coupling constants which are assumed universal for both the lepton and quark sectors. Thus action [3.1] is invariant under chiral gauge symmetries $SU^c(3) \otimes SU_L(2) \otimes U_Y(1)$ and a global $U_L(3) \otimes U_R(3)$ in generation space. In addition it evades, at least in principle, the "no-go theorem". The question now is: does it yield a long wave-length spectrum in agreement with observation (free from the unwanted doublers)?

4. $\bar{t}t$ Condensate and Composite Particles

In order for the action [3.1] to get rid of unwanted doubler and avoid *in practice* the "no-go" theorem, it is necessary that the quadrilinear terms S^{F1}_{NJL} and S^{F2}_{NJL} develop a dynamical chiral symmetry breaking through the following non-zero vacuum expectation values [7] (V_4 is the 4-dimensional volume)

$$\begin{aligned} m^{ij}_F &= -\frac{G_1}{2V_4} \sum_x \langle \bar{\psi}^F_i(x) \psi^F_j(x) \rangle; \\ \bar{r}^{ij}_F &= \frac{G_2}{4V_4} \sum_{\mu,x} \{\langle \bar{\psi}^F_{Li}(x) U^c_\mu(x) \psi^F_{Rj}(x+a_\mu) \rangle + \text{h.c.}\}. \end{aligned} \qquad [4.1]$$

Indeed, should this happen, one would obtain the following effective lattice action of the Wilson type [4]:

$$\begin{aligned} S^{eff}_{PL} &= S_G + S_D \\ &+ \sum_{xF} \left\{ \bar{\psi}^F(x) M_F \psi^F(x) - \frac{1}{2} \sum_\mu \left(\bar{\psi}^F(x) G^L_\mu(x) \bar{r}_F G^R_\mu U^c_\mu(x) \psi^F(x + a_\mu) + \text{h.c.} \right) \right\}, \end{aligned} \qquad [4.2]$$

where $M_F = m_F + 4\bar{r}_F$. Thus through dynamical symmetry breaking the obligatory (in order to get rid of the doublers) Wilson term ($r_F = a_p\bar{r}_F$) gets produced together with a mass term (m_F) which, according to our action, must necessarily come with it. In this way the evasion of the "no-go theorem" entails an extra bonus: the generation of a fermion mass term. The $SU_L(2) \otimes U_Y(1)$ and $U_L(3) \otimes U_R(3)$ symmetries are clearly broken and the only surviving gauge symmetries $SU_c(3)$ and $U_{em}(1)$ [9]. Given the quadrilinear NJL-terms [3.2] and [3.3], we construct an effective potential in terms of m_F, $\bar{r}_F$ [8]. A non-trivial dynamical symmetry breaking may emerge only if the matrices m_F, $\bar{r}_F$ obey a set of coupled, self-consistent equations obtained from variation of the effective potential. We call these equations "gap equations", that turn out to have the following approximate matrix-form in the weak-isospin space [9]

$$\begin{aligned} m_F a_p &= 2g_1 \int_l \frac{m_F a_p}{\mathrm{Den}^F(l)}, \\ r_F &= g_1 \int_l \frac{r_F \sin^2 \frac{l_\mu}{2}}{\mathrm{Den}^F(l)}, \\ r_F &= -g_2 \int_l (\cos) \frac{m_F a_p + 2r_F \sin^2 \frac{l_\mu}{2}}{\mathrm{Den}^F(l)} \end{aligned} \qquad [4.3]$$

where $g_{1,2}a_p^2 = N_c G_{1,2}$; $r_F = a_p\bar{r}_F$; $\int_l = \int_{-\pi}^{\pi} \frac{d^4 l}{(2\pi)^4}$; $(\cos) = \sum_\mu \cos l_\mu$ and

$$\mathrm{Den}^F(l) = \sin^2 l_\mu + (m_F a_p + 2r_F \sin^2 \frac{l_\mu}{2})^2. \qquad [4.4]$$

We see that for any $g_2 \neq 0$, one gets $r > 0$, which removes the doublers through a Wilson-type mechanism. By using now the gap equation [4.3] for $m_F a_p \neq 0$ and the chain approximation, for the Goldstone modes $\langle\bar{\psi}_i(x)\gamma_5\psi_i(x)\bar{\psi}_i(0)\gamma_5\psi_i(0)\rangle$ and scalar modes $\langle\bar{\psi}_i(x)\psi_i(x)\bar{\psi}_i(0)\psi_i(0)\rangle$, we find the inverse propagators

$$\Gamma_{Fp}^{-1}(q) = B^2(q) I_p^F(q); \quad I_p^F(q) = -\frac{N_c}{2} \int_l \frac{\sum_\mu \left(\cos^2 l_\mu + r_F^2 \sin^2 l_\mu\right)}{\mathrm{den}^F\left(l + \frac{q}{2}\right) \mathrm{den}^F\left(l - \frac{q}{2}\right)}; \qquad [4.5]$$

and

$$\Gamma_{Fs}^{-1}(q) = 2\left(B^2(q) I_s^F(q) + 4M_F^2\right); \; I_s^F(q) = \frac{N_c}{4} \int_l \frac{\sum_\mu (\cos^2 l_\mu)}{\mathrm{den}^F\left(l + \frac{q}{2}\right) \mathrm{den}^F\left(l - \frac{q}{2}\right)}, \qquad [4.6]$$

where $B^2(q) = \sum_\mu \left(\frac{4}{a_p^2} \sin^2 \frac{q_\mu a_p}{2}\right)$ and

$$4M_F^2 = 4 \int_l \frac{\left[m_F + \frac{2r_F}{a_p} \sin^2 \frac{l_\mu}{2}\right]^2}{\mathrm{den}^F\left(l + \frac{q}{2}\right) \mathrm{den}^F\left(l - \frac{q}{2}\right)}. \qquad [4.7]$$

For $q_\mu a_p \ll 1$, we can calculate numerically the position of the pole of the scalar mode, which turns out to be of the order of the Planck mass instead of $4m_F^2$. Indeed one gets

$$4M_F^2 = 4m_F^2 + 0.8r\frac{m_F}{a_p} + 0.9\frac{r^2}{a_p^2}, \qquad [4.8]$$

which for $r = 1$ pushes this pole at the Planck mass, making it thus disappear from the observable, low energy spectrum. Analogously, we find charged Goldstone modes appearing in the flavored channels corresponding to the quantum numbers of the $W^\pm$ bosons.

We shall now analyse the solutions of [4.3]. If we look for "physical" solutions, for which the eigenvalues of the mass matrix are such that $m_F a_p \ll 1$ and the doublers are removed by $r_F \neq 0$, then equation [4.3] for r approximately decouples and becomes

$$1 = -g_2 \int_l (\cos) \frac{2\sin^2\frac{l_\mu}{2}}{\sin^2 l_\mu + 4r^2(\sin^2\frac{l_\mu}{2})^2}. \qquad [4.9]$$

Notice that, assuming a universal coupling constant G_2 in [10], the Wilson parameters for quarks and leptons are different because of the color number N_c. The solution $r = r_F(g_2)$ is reported in fig. 1.

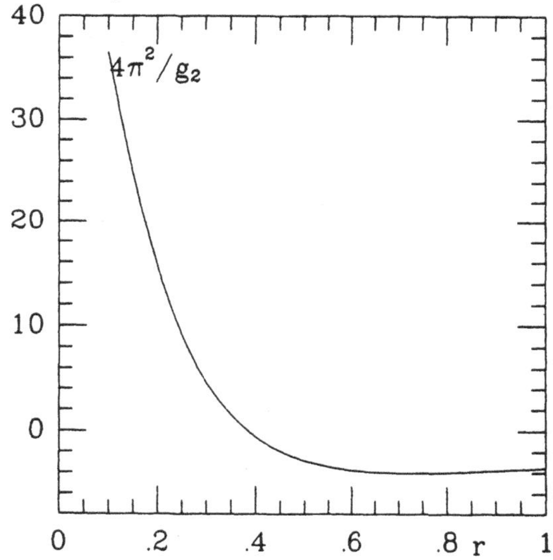

Fig. 1 $\frac{4\pi^2}{g_2}$ as a function of r

In order to understand the implications of obtaining physical solutions $m_F a_p \ll 1$, we use the gap-equations [4.3] to draw the phase diagram (fig. 2), where the phases ($m < 0$) are separated from the phase ($m > 0$) by critical lines, on which $m = 0$.

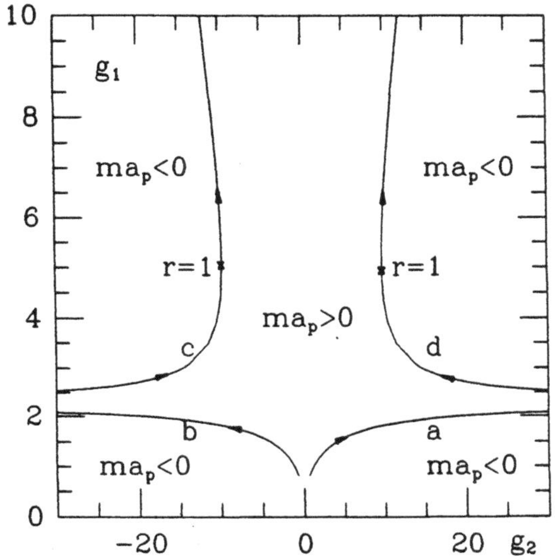

Fig. 2 phase diagram in terms of g_1 and $g_2, a, b(r = 0 \to 0.385)$ and $c, d(r = 0.385 \to 1 \to \infty)$.

This phase diagram is in agreement with the continuum NJL-model that possesses only the coupling constant G_1. The requirement of obtaining physical solutions for quarks $N_c = 3$ with $\bar{r}_q a_p = r_q \sim O(1)$ and $m_F a_p \ll 1$ can clearly be met by a coupling constant $g_2 \gg 1$, in this case one has $r \to 0.385$. On the other hand the coupling g_1 is required to lie in the vicinity of the critical value g_c^2. This holds for quarks, for it is easy to see that for leptons ($N_c = 1$) equation [4.3] does not allow for massive solutions. Thus leptons at this stage remain massless. In order to have $\bar{t}t$ condensate and get rid of doublers, we chose the mass matrix, which only has one non-zero eigenvalue, and $r = rI$ (I is identity matrix) for every fermions, which are the solution to gap equation [4.3].

5. Fermion Masses without Goldstone Bosons

In this section we wish to outline our strategy to proceed further in our computation of fermion masses, by considering the contributions arising from the photon

γ, $W^{\pm}$, and Z° interaction that we have so far neglected. The problem we must now solve is the evaluation of the self-energy function of the fermions with the SM action and supplemented by the interactions [3.2] and [3.3] on the Planck lattice. In order to render this problem manageable we shall work in the"rainbow" approximation of the Dyson equation, which neglects all vertex and gauge propagator corrections. Thus on the Planck lattice we must solve the Dyson-Schwinger equation which we give in diagrammatic form fig. 3.

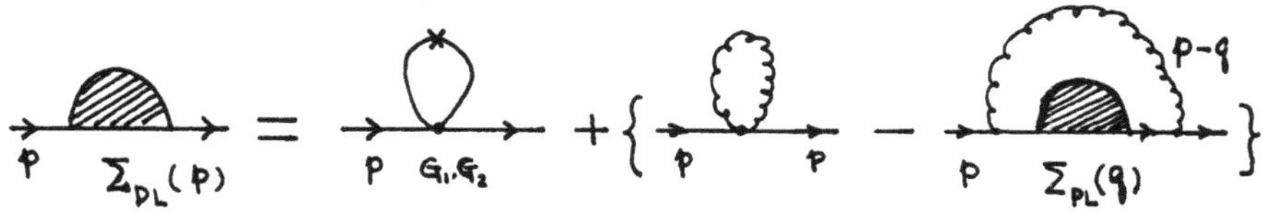

Fig. 3 The diagrammatic form of Dyson-Schwinger equation

The first diagram in the left-hand side represents the result of the first stage calculation of ref. [9], while the second is a contact interaction typical of gauge theories on a lattice. Please note that in the Dyson equation we have not included colored gluon exchanges for they are related to color confinement, producing an infinite mass-shift for quarks and all colored states. We shall return on this point in a future publication; for, as a matter of fact, in this paper our sole intention is to study the structure of the solutions of equation (fig. 3) on a Planck lattice and compare it with the rather large amount of extant works on the continuum version of the Dyson equation. A final remark on our approximation: the smallness of the coupling constants associated with electroweak exchanges makes it quite plausible.

Let's now turn to equation (fig. 3), where, for illustration purposes, we keep only the exchange of a massive vector gauge-field of mass m_v coupled to a vector current. We can write it as:

$$\Sigma_{PL}(p) = W_r(p) + \bar{g}^2 \int_{-\frac{\pi}{a_p}}^{\frac{\pi}{a_p}} \frac{d^4q}{(2\pi)^4} \frac{1}{S(p-q)^2 + m_v^2}$$
$$\left(\delta_{\mu\nu} - \xi \frac{S_\mu(p-q)S_\nu(p-q)}{S(p-q)^2 + m_v^2}\right) \mathrm{tr}\left[V^{(2)}_{\mu\nu}(p,p) - V^{(1)}_\mu(p,q)\frac{1}{\gamma_\rho D_\rho(q) + \Sigma_{PL}(q)} V^{(1)}_\nu(p,q)\right], \quad [5.1]$$

where $S_\mu(l)=\frac{2}{a_p}\sin\frac{l_\mu a_p}{2}$, $D_\mu(l)=\frac{1}{a_p}\sin(l_\mu a_p)$ and $\bar{g}^2=e^2(\frac{g^2(N^2-1)}{2N})$ for $U_{em}(1)(SU(N))$ gauge group. The $W_r(p)$ is the self-energy of fermion generated through the Nambu-Jona Lasinio mechanism [5]. As discussed in ref. 9 for one-loop contribution, $W_r(p) = M_t + \frac{2r}{a_p}\sum_\mu \sin^2\frac{p_\mu a_p}{2}$; M_t is non-vanishing for the top-quark only and, however r is different from zero for all fermions. The vertices [10] are ($k_\mu = \frac{(p_\mu+q_\mu)a_p}{2}$)

$$V^{(1)}_\mu(p,q) = (\gamma_\mu \cos k_\mu + r\sin k_\mu)\,; \quad V^{(2)}_{\mu\nu}(p,q) = a_p\,(-\gamma_\mu \sin k_\mu + r\cos k_\mu)\,\delta_{\mu\nu}. \quad [5.2]$$

In this paper, the only property of $W_r(p)$; that we need specify, is its being $O(a_p)$ for $pa_p \ll 1$.

We limit our attention to fermions other than the top quark and to values of the external momentum p such that $pa_p \ll 1$, and in this kinematical region the self energy function $\Sigma_{PL}(p)$ should not and does not differ from its continuum limit version $\Sigma_c(p)$. One of the main novelties of [5.1] is the non trivial interplay between the continuum-limit region, i.e., for momenta ($qa_p \ll 1$), and the truly discrete region, which is probed for momenta $qa_p \simeq 1$. In order to study such interplay it is important to introduce a "dividing scale" ϵ, such that $pa_p \ll \epsilon \ll \pi$. Separating the integration region in [5.1] into two regions $[0,\epsilon]^4$ and $[\epsilon,\pi]^4$, we may write our integral equations as

$$\Sigma_c(p) = \frac{\lambda}{4\pi^2}\int_{\Lambda=\epsilon\Lambda_p} d^4q \frac{1}{(p-q)^2+m_v^2}\frac{\Sigma_c(q)}{q^2+m^2} + \delta_{PL}(r,\epsilon), \quad [5.3]$$

where $\lambda = \frac{3\bar{g}^2}{4\pi^2}$, and the continuum-limit integral equation, following the common practice, has been linearized, i.e., in the denominator one sets $\Sigma_c(q) \simeq m$, the "physical" mass of the fermion, and the Landau gauge $\xi = 1$, in which the self-energy function for $m_v = 0$ contains a mass renormalization only, is chosen for simplicity. As for $\delta_{PL}(r,\epsilon)$, the contribution to the integral equation from the discrete lattice region, we may write it as ($l_\mu = q_\mu a_p$):

$$\delta_{PL}(r,\epsilon) \simeq \frac{\lambda r}{a_p}\int_{[\epsilon,\pi]^4} \frac{d^4l}{12\pi^2}\frac{1}{(4\sin^2(\frac{l_\mu}{2}))}\left[\frac{1}{2} - \frac{\frac{1}{r}G_{PL}(l)(-\cos^2(\frac{l_\mu}{2}) + r^2\sin^2(\frac{l_\mu}{2})) + \sin^2(l_\mu)}{G_{PL}(l)^2+\sin^2(l_\mu)}\right]$$
$$-\lambda\int_{[\epsilon,\pi]^4}\frac{d^4l}{4\pi^2}\frac{\Sigma_c(l)}{(4\sin^2(\frac{l_\mu}{2}))}\left[\frac{-\cos^2(\frac{l_\mu}{2}) + r^2\sin^2(\frac{l_\mu}{2})}{G_{PL}(l)^2+\sin^2(l_\mu)}\right], \quad [5.4]$$

where we have decomposed $\Sigma_{PL}(l)$ as $a_p\Sigma_{PL}(l) = a_p\Sigma_c(l) + G_{PL}(l)$, and $G_{PL}(l) = W_r(l)$ is non zero only in the discrete lattice region, i.e. for $l \in [\epsilon,\pi]^4$. The dependence on the external momentum pa_p is omitted in $\delta_{PL}(\epsilon,r)$ because of $pa_p \ll l \in [\epsilon,\pi]^4$.

Note that (i) there is a perfect cancellation of gauge-dependent term $a_p^{-1}\xi$ between the contact and the "rainbow" diagrams, that is guaranteed by Ward's identities; (ii) $\delta_{PL}(\epsilon, r)$ cannot vanish if r, as in our case, does not vanish. Thus, on the Planck lattice, the self energy integral equation [5.3] acquires an inhomogeneous term even for $M_t = 0$. This implies the very important consequence that [5.3] admits only massive solutions provided $\lambda > 0$ [11]. The most appealing aspect of this result is that on the Planck lattice mass gets generated for all fermions, but the top-quark, without the appearance of Goldstone bosons: the reason for this fact, extremely important phenomenologically, is the connection of the inhomogeneous term $\delta_{PL}(\epsilon, r)$ with a non zero r-value that, as emphasized, already breaks the chiral symmetry of the four-fermion interaction.

At a moment we are not able to show the cancellation of a_p^{-1} divergence by intrinsic requirement of hidden symmetries in the theory, this linear divergence can be cancelled by simply adding counterterm. We can now write the finite part of $\delta_{PL}(r, \epsilon)$ is

$$\delta^f_{PL}(r,\epsilon) = -\int_{[\epsilon,\pi]^4} \frac{d^4l}{4\pi^2} \frac{\Sigma_c(l)}{(4\sin^2(\frac{l_\mu}{2}) + m_v^2)} \left[\frac{-\cos^2(\frac{l_\mu}{2}) + r^2\sin^2(\frac{l_\mu}{2})}{G_{PL}(l)^2 + \sin^2(l_\mu)}\right]. \qquad [5.5]$$

Let's now address the important question of the ϵ-independence of our result. The introduction of this "dividing scale" in [5.3] is, apart from the requirement $pa_p \ll \epsilon \ll \pi$, rather arbitrary thus no dependence on ϵ should appear in our final results. In order for such independence to occur, as it must occur, it is clear that the ϵ-dependence from the continuum integral in eq. [5.3], that is clearly logarithmic, must be compensated by an analogous logarithmic term arising in the calculation of $\delta_{PL}(r, \epsilon)$. Thus segregating the $\ell n\, \epsilon$-term in $\delta^f_{PL}(r, \epsilon)$, we may write

$$\delta^f_{PL}(r,\epsilon) = \lambda\delta^1_{PL}(r)\,\ell n\,\epsilon + \lambda\delta^0_{PL}(r), \qquad [5.6]$$

where $\delta^o_{PL}(r)$ and $\delta^0_{PL}(r)$ is independent of ϵ, [5.3] becomes

$$\Sigma_c(p) = \frac{\lambda}{4\pi^2}\int_{\Lambda_p} d^4q \frac{1}{(p-q)^2 + m_v^2} \frac{\Sigma_c(q)}{q^2 + m^2} + \lambda\delta^o_{PL}(r). \qquad [5.7]$$

This is an inhomogeneous equation admitting a finite solution for small λ. We will report this result soon [12].

6. Neutrino and Intermediate Gauge Boson Masses

The D.S. equation we adopted in above contains only vector-like coupling vertices and massless gauge bosons. We are going to focus on the D.S. equation involving V-A couplings and massive gauge bosons. Neutrinos, which belong to a special category in fermion family, have only weak interaction with massive intermediate gauge bosons. In the lattice with spacing a_p, we not only have left-handed neutrinos but also right-handed neutrinos since again there exists a Wilson term which is essentially an interacting term between left- and right- handed species. It is expected that they would acquire small masses by weak interaction with heavy intermediate gauge bosons. Since weak interactions change fermion flavours, the fermion mass matrix we will calculate should not be diagonal in the basis of weak interaction eigenvectors. Thus we expect that the different eigenvalues (fermion masses) would be obtained if the fermion mass matrix is diagonalized and the transfer matrices would be related to the Cabibbo-Kobayashi-Maskawa matrix.

Let us turn to intermediate gauge boson masses. The photon and gluon (without considering their self-interaction) are massless since a vector-like gauge symmetry can be maintained in an appropriate regularization (via a cutoff a_p). This is guaranteed by the so-called Elitzur's theorem [13], which states: a spontaneous breaking of local symmetry for a symmetrical gauge theory (with cutoff and vector-like gauge coupling) without gauge fixing is impossible. On the other hand, the intermediate gauge bosons are endowed with masses because they couple to fermions in a chiral fashion and chiral-gauge symmetry cannot be maintained in any regularization. We have been trying to prove that it is impossible to maintain gauge symmetry in a regularized gauge theory where the gauge coupling is in chiral fashion (parity violating) [14]. Thus, instead of adding local gauge-variant counterterms to lagrangian for maintaining gauge symmetry order by order, as one usually does for perturbative QED and QCD, we should keep the local gauge-variant terms from the calculation of the two-point function, which turn out to be mass terms of gauge bosons. The gauge boson masses obtained in this way are related to the sum of all fermion masses [15], which are finite once the spontaneous symmetry breakdown happens and fine tuning is performed (as described in section 4). In this way, we will see the counterterm (divergence) structure of a gauge theory is not affected by the occurrence of spontaneous symmetry breakdown. In another words, we do not need new counterterms after fermions and intermediate boson become massive. It is then expected to be renormalizable and all effects from the short wave-length regime ($\Lambda = \epsilon\Lambda_p$) would be renormalized away so as to have meaningful results defined on the physical mass-shell in the long wave-length regime. In addition, the triangle-anomalies from the calculations of the three-point function are cancelled naturally by the fermion contents of the theory, thus conservation of matter source current and Ward identities are preserved and the theory is unitary. This is also important for renormalizability.

7. Conclusions and Remarks

In conclusion, we have shown that on the Planck lattice, whose "raison dêtre" may well reside in the violent quantum fluctuations of the metric field at the Planck scale a_p, a consistent SM requires the addition of extra terms, quadrilinear in the Dirac fields, whose coupling constants can be determined to induce the emergence of the $\bar{t}t$-condensate model with $m_t a_p \ll 1$ and $0 < r \leq 1$: m_t gives rise to the scale of electroweak breakdown and $r \neq 0$ endows the composite scalar and the mirror fermions with masses at the Planck scale, making them disappear from the low-energy spectrum. We should also stress that, differently from the continuum $\bar{t}t$-condensate model, its Planck lattice version disposes in a nice way of the scalar composite, thus leading to the disappearance from the spectrum of a particle that would resemble the Higgs boson. The implication of this conclusion for the present and future phenomenology is too obvious to need further comments. In this framework, we are proceeding the calculations of fermion and gauge boson masses, and other problems we mentioned in this paper.

Acknowledgments

We wish to thank Prof. E. Iarocci, Director of the Laboratori Nazionali di Frascati of INFN, for the kind hospitality extended to us at his institution, where we have been proceeding this work.

REFERENCES

1. C.W. Misner, K.S. Thorne and J.A. Wheeler, *Gravitation* (Freeman, San Fransisco, 1973);
   J. Greensite, *Phys. Lett.* **B255** (1991) 375.

2. P.W. Higgs, *Phys. Rev. Lett.* **12** (1964) 132; **13** (1964) 508.
3. H.B. Nielsen and M. Ninomiya, *Nucl. Phys.* **B185** (1981) 20; **B193** (1981) 173; *Phys. Lett.* **B105** (1981) 219.
4. K. Wilson, in *New phenomena in subnuclear physics* (Erice, 1975) ed. A. Zichichi (Plenum, New York, 1977).
5. Y. Nambu and G. Jona-Lasinio, *Phys. Rev.* **122** (1961) 345.
   Y. Nambu, in New theories in physics, Proc. XI Int. Symp. on elementary particle physics, eds. Z. Ajduk, S. Pokorski and A. Trautman (World Scientific, Singapore, 1989).
6. W.A. Bardeen, C.T. Hill and M. Linder, *Phys. Rev.* **D41** (1990) 1647;
   V.A. Miranski, M. Tanabashi and K. Yamawaki, *Mod. Phys. Lett.* **A4** (1989) 1043;*Phys. Lett.* **B221** (19(1989)) 117;
   W.J. Marciano, *Phys. Rev. Lett.* **62** (1989) 2793; *Phys. Rev.* **D41** (1990) 219.
7. G. Preparata and S.-S. Xue, *Phys. Lett.* **B264** (1991) 35; MITH 92/1 submitted to Phys. Rev. Lett.; Nucl. Phys.B (Proc. Suppl. for Lattice '91) (1992).
8. G. Preparata and S.-S. Xue, in preparation.
9. G. Preparata and S.-S. Xue, MITH 92/1 submitted to Phys. Rev. Lett..
10. L.H. Karsten and J. Smit, *Nucl. Phys.* **B183** (1981) 103.
11. T. Maskawa and H. Nakajima, *Prog. Theo. Phys.* **52** (1974) 1326; **54** (1975) 860.
12. G. Preparata and S.-S. Xue, in preparation.
13. S. Elitzur, *Phys. Rev.* **D12** (1975) 3978;
    G.F. De Angelis, D. de Falco and F. Guerra, *Phys. Rev.* **D17** (1978) 1624.
14. S.-S. Xue, Nuovo Cimento (Note Brevi) A105(1992)131; MITH 91, submitted to Phys. Lett. B.
15. S.-S. Xue, MITH 90/15, MITH 90/14, to be published in Nuovo Cimento A(1992).

II. SUBNUCLEAR PHYSICS

Common Problems and Trends of Modern Physics, pp. 183–200
ed. T. Bressani, S. Marcello and A. Zenoni
©1992 World Scientific Publishing Company

ANALYSIS OF MESON RESONANCES IN LOW ENERGY $N\bar{N}$ ANNIHILATION

Alberto ROTONDI

Dipartimento di Fisica Nucleare e Teorica, Via Bassi 6, 27100 Pavia, Italy
and
INFN Sezione di Pavia, Via Bassi 6, 27100 Pavia, Italy

Abstract

The methods of analysis of resonances produced in the low energy nucleon-antinucleon annihilation are reviewed. Some uncertain or critical points are put in evidence, and their practical consequences on the final results are discussed. The approach is pedagogical.

1. Introduction

The discovery of meson resonances in the sixties and seventies led to a beautiful confirmation of the existence of quark-antiquark ($q\bar{q}$) states, as predicted by the non relativistic quark model. As it is explained in any textbook on particle physics, many of these resonances could be interpreted as well distinct Breit-Wigner (BW) peaks over a smooth phase space background, so that data analysis was rather easy in these cases. This explains why hadron spectroscopy is considered among non specialists merely as bump-hunting.

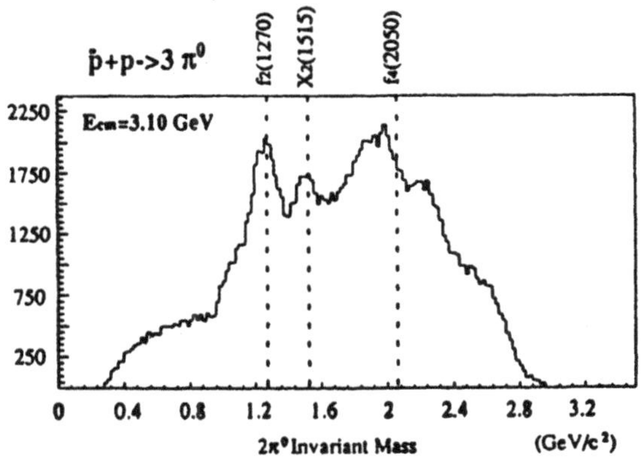

Fig.1: $p\bar{p} \to 3\pi^0$ from the E760 experiment: projection of the $3\pi^0$ Dalitz plot to the $\pi^0\pi^0$ mass axis. In addition to the three shown resonances, probably many other states are present. So far, a satisfactory analysis of this spectrum is still missing (From ref.1).

However, in many experiments a different picture emerges: often many broad final states with similar quantum numbers overlap (see fig. 1). In this context, the reliability of the methods of resonance analysis is a point of crucial importance.

In this paper we present an overview of these methods applied to the $N\bar{N}$ annihilation at low energy, a channel which is at present object of intensive experimental studies at LEAR (CERN), since the possibility to reveal non $q\bar{q}$ (exotic) mesons is envisaged.

The reader interested not only in the methods of analysis, but also in the physics of the $p\bar{p}$ interaction, can find exhaustive and recent reviews on this topic in refs.2,3.

2. The $p\bar{p}$ interaction: some general aspects

Let us consider the reaction $p\bar{p} \dashrightarrow (\pi\pi)\pi... \dashrightarrow n\pi$, where the parenthesis indicates possible resonant intermediate states and n is an integer; the following selection rules for the parity P, the charge conjugation C and the G-parity hold[4,10]:

$p\bar{p}$ initial state	boson-antiboson intermediate state	n π final state
$P = (-)^{L+1}$	$P = (-)^{L}$	wave function parity
$C = (-)^{L+S}$	$C = (-)^{L+S}$	$C = +1$
$G = (-)^{L+S+I}$	$G = (-)^{L+S+I}$	$G = (-)^{n}$

where L,S and I are the orbital angular momentum, spin and isospin quantum numbers, respectively.

According to the current PDG rules[5], for the initial or intermediate final states we use the notation $^{2S+1}L_J$, $I^G(J^{PC})$, where J is the total angular momentum quantum number.

The number of independent kinematical variables for a n meson final state is given by[6] ($n \geq 3$):

$$(3 \text{ x number of mesons}) - (\text{total 4-momentum constraints}) - \text{isotropy} = 3n - 4 - 3 \qquad (1)$$

For example, in the reaction $p\bar{p} \dashrightarrow \pi^+\pi^-\pi^0$, we have two final independent kinematical variables; a good choice is $X=M^2(\pi^+\pi^0)$ and $Y=M^2(\pi^-\pi^0)$, where M^2 is the square of the invariant mass. In this case the Dalitz theorem, which states that in the absence of resonances the X-Y plot shows uniform density[6], is very useful. In order to avoid kinematical reflections[6], it is essential, in fitting theory to data, to use spectra involving a number of kinematical variables ≥ the number given by eq.1 . In the three pion final state a fit to the Dalitz plot satisfies this requirement. For further details on data reduction techniques for fitting resonances, see ref.7.

3. The final state interaction (FSI) model

The prevalent model used in the analysis of $p\bar{p}$ meson resonances is the old final state interaction (FSI) model of Watson[9]. We explain this model for the simple case of a quasi two body (QTB) $p\bar{p} \dashrightarrow R\pi \dashrightarrow (\pi\pi)\pi$ reaction, with three pions in the final state and an intermediate resonant state R decaying into two pions (see fig. 2).

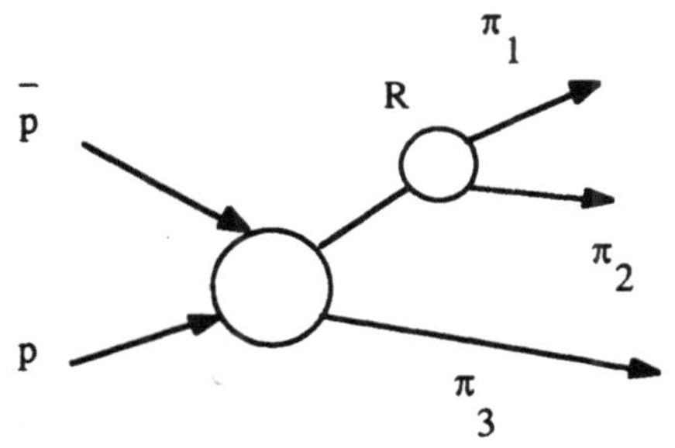

Fig. 2: quasi two body (QTB) decay of the $p\bar{p}$ initial state in a pion and a resonance R

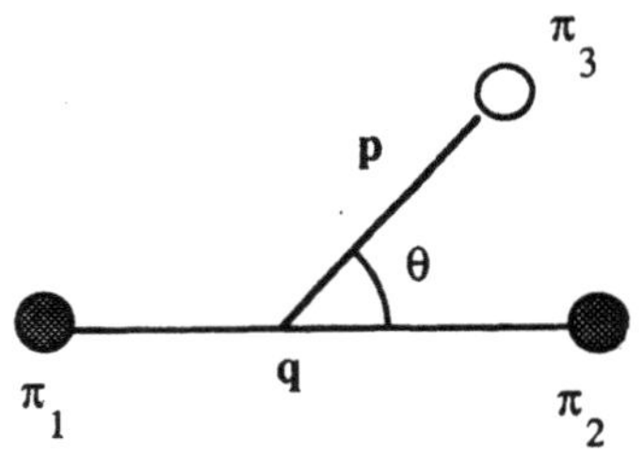

Fig. 3: **q** is the 3-momentum difference of the two resonant (shaded) pions in their c.m.frame; **p** is the spectator 3-momentum in the general c.m. frame

The three particle spectrum (Dalitz plot) in this case is given by:

$$D[p\bar{p} \to (\pi_1\pi_2)\pi_3] = [\sum_{J^{PC}} |T_{J^{PC}}|^2]\ \mathrm{LIPS}(p\bar{p} \to \pi_1\pi_2\pi_3) \qquad (2)$$

where LIPS is the Lorentz Invariant Phase Space density[6], T is the transition matrix and the sum runs over all the different (distinguishable) states J^{PC} of the $p\bar{p}$ initial state. The transition matrix is usually written in the following form:

$$T_{J^{PC}} = \sum_i a_i\, Z_i(J^{PC},R,\pi_3)\, BW(m_i) \tag{3}$$

where BW is the "usual" relativistic Breit-Wigner function

$$BW(m) = \left(\frac{m}{q}\right)^{1/2} \frac{\Gamma(m)^{1/2}}{m_o^2 \; - \; m^2 - i m_o \Gamma(m)} \tag{4}$$

with m_o the nominal resonance mass, m the invariant mass of the $\pi_1\pi_2$ system, and q the 3-momentum difference of the two pion momenta in their c.m. frame ($q=(m^2-4m^2_\pi)^{1/2}$). The term $Z(J^{PC},R,\pi_3)$ is the transition matrix for the intermediate state $p\bar{p} \to R\pi_3$, and a_i is an isospin weight (if any). The sum indicates a possible symmetrization over all the identical particles.

For a derivation of all the above formulae we refer to the book of Pilkuhn [10] and to the review article of Jackson[11] . Here we want only to emphasize some points of importance for the following discussion.

Firstly, the term $\Gamma(m)$ in eq.4 represents the internal decay dynamics of the resonance R. The more usual form is[10]

$$\Gamma(m) = \int |T(R \to \pi_1\pi_2)|^2 \, d(LIPS(R \to \pi_1\pi_2)) \equiv \left(\frac{q}{q_o}\right)^{2l+1} \Gamma \, \frac{m_o}{m} \tag{5}$$

where Γ is the nominal resonance width, l is the relative angular momentum of the two decaying pions and q_o is q calculated at $m = m_o$. Eq.5 takes into account the centrifugal barrier effects through the power to 2l+1 and the unknown decay dynamics through the empirical factor Γ. Many other parametrizations of $\Gamma(m)$ are found in the literature, so that the form of this term shows cyclical variations with time[10-16] . Fortunately, all these different forms give BW functions that differ in shape only along the tail of the resonance, where the BW form itself is questionable. Presumably, these differences do not affect significantly the data analysis.

Secondly, the transition matrix $Z(J^{PC},R,\pi_3)$ assumes, in the final state interaction approximation, a form depending only on angular momentum effects. In meson spectroscopy, when the final particles have zero spin and usually there is no polarization, this matrix is simply a Zemach tensor[17] (this explains our notation). How to build-up in practice these tensors for a specific reaction is explained elsewhere in these proceedings[7,8].

The above discussion concerns some mathematical techniques; what about the physical limitations of the FSI model? We want to discuss here three important points:

a) the initial state effects taken into account by the model are the P,C,G selection rules, the 4-momentum conservation and the angular momentum conservation only;

b) the BW parametrization assumes complete independence between the production and the decay modes of the resonance. This assumption is very well founded in compound nucleus theory, where the BW form was for the first time introduced[18], but it appears questionable in $p\bar{p}$ annihilation: a compound nucleus can have 10^6 internal oscillations before decaying, whereas a meson resonance with a 100 MeV width and dimension of 1 fm has probably a number of internal cycles of about 10 (ref.19);

c) the model also assumes the absence of secondary final state interactions between the resonance and the other spectator mesons. This last assumption seems to be justified *a posteriori* by the data. Let us consider, for example, the reaction $p\bar{p} \to \rho\pi \to \pi^+\pi^-\pi^0$: if a secondary final state interaction took place, the $\pi^\pm$ flying near the π^0 in the ρ rest frame could give measurable effects on the $\cos\theta$ decay distribution of the ρ (see fig.3 for the definition of this angle). These effects have never been observed[20].

In conclusion, we can affirm that the success by the analysts of the FSI model described above is probably due to its simplicity and to its success to describe data.

4. How to perform practical calculations

We consider, as an example, the reaction $p\bar{p} \to \pi^+\pi^-\pi^0$ from the $^3S_1$, $0^-(1^{--})$ $p\bar{p}$ initial state. According to the QTB model (and to the experimental data), this channel is dominated by the $\rho(770)$ production only. By means of the selection rules previously discussed, one obtains the following scheme:

$I^G(J^{PC})$ initial state	final state	l, L	(T;T3) dipion	Clebsch-Gordan (CG) isospin coefficients
$0^-(1^{--})$	$(\pi\pi)_1\pi_L$	1, 1	(1; 1, 0, -1)	$-\frac{1}{\sqrt{6}}(1, 1, 1)$

where l and L are the relative orbital angular momenta of the dipion and of the recoiling pion, respectively, and the isospin combinations T=1, T_3=1, 0, -1 correspond to the channels $\rho^+\pi^-$, $\rho^-\pi^+$, $\rho^0\pi^0$. Since the initial state is isoscalar, the isospin CG coefficients are given by (the example is for the $\rho^0\pi^0$ case):

$$<\pi\pi^+,\pi\pi^-|\rho\rho^\circ><\rho\rho^\circ,\pi\pi^\circ|00>=<11,1-1|10><10,10|00>=-\frac{1}{\sqrt{6}}$$

The non relativistic Zemach tensor, for a l=1 dipion which recoils against an L=1 pion, is given by **qxp**, where the two momenta are defined in fig. 3. Hence, eq.2 becomes, with obvious notation:

$$\begin{aligned} D(p\bar{p} \to \pi^+\pi^-\pi^\circ) = \; & |T(\mathbf{q},\mathbf{p})|^2 \;\; LIPS(p\bar{p} \to \pi^+\pi^-\pi^\circ) \;\; = \\ & N\,|BW(\rho^+)\,\mathbf{q}_+ x\mathbf{p}_- + BW(\rho^-)\,\mathbf{q}_- x\mathbf{p}_+ + BW(\rho^\circ)\,\mathbf{q}_o x\mathbf{p}_o\,|^2 \cdot \qquad (6) \\ & LIPS(p\bar{p} \to \pi^+\pi^-\pi^\circ) \end{aligned}$$

where N is a normalization constant.

Note that the indistinguishable $J^{PC}=1^{--}$ intermediate states $(\pi^+\pi^-)\pi^0$, $(\pi^+\pi^0)\pi^-$, $(\pi^-\pi^0)\pi^+$ are summed in interference. Physically this is analogous to the two (or more) slit problem of Quantum Mechanics: the π^0 assumes at the same time the role of a decay product of a ρ^+, ρ^- resonating state and that of a recoiling particle against a ρ^0. These different configurations are weighted differently according to the BW functions and the Zemach tensors. Note also that BW are complex functions and the Zemach tensors are real, so that the amplitude is a complex vector **V**: its squared modulus is $\mathbf{V}\cdot\mathbf{V}^*$.

If, instead of the ρ alone, we had another resonance with different combinations of isospin and orbital l,L quantum numbers, the amplitude T of eq.6 would become:

$$\begin{aligned} & |\, x_1\, T_1(\mathbf{p},\,\mathbf{q}) \; + \; x_2\, e^{i\phi}\, T_2(\mathbf{p},\,\mathbf{q})|^2 \;\; = \\ & x_1^2\, |T_1|^2 + x_2^2\, |T_2|^2 + 2x_1x_2\cos\phi\, Re(T_1T_2^*) \; + \; 2x_1x_2\sin\phi\, Im(T_1T_2^*) \qquad (7) \end{aligned}$$

with x_1, x_2, ϕ real parameters.

For the complete analysis of the 3 pion decay, with the full listing of the states, see refs.14-16, 24; see also ref.7 for the explanation of some computational aspects of eqs.6-7.

In practice, one has to deal with an experimental Dalitz plot $D_{exp}(p\bar{p} \rightarrow \pi^+\pi^-\pi^o)$ to be fitted with the theoretical prediction of eqs.2-6. Usually these equations are calculated by sampling a phase space event (for example with the GENBOD routine of the CERN library) and then by weighting this event with the square modulus of the transition amplitude. In this way it is possible to have, for each experimental spectrum or histogram, the corresponding theoretical one, and to perform the fit. The free parameters are the branching ratios and phases between resonances (the parameters x_1, x_2, ϕ of eq.7) and the masses and widths of the resonance candidates (the known resonance masses are usually kept fixed!).

In general, the analysis of resonances must give their mass M, width Γ and quantum numbers J^{PC}. M and Γ are contained in the BW functions, which determine mainly mass invariant spectra, whereas the spin-parity J^{PC} is determined mainly by the Zemach angular momentum tensors (constructed under a precise spin-parity hypothesis), that determine the decay angular distributions. For the three meson case (two independent variables), a fit to the Dalitz plot already contains the fit to every invariant mass and angular distribution. When more mesons are present in the final state, a combined fit to mass spectra and angular distributions (suitably chosen) permits, if the statistics is enough, to measure unambiguously the resonance.

For a clear explanation of these techniques see refs.14-16: these are recommended readings for the beginner.

5. The acceptance effects

We consider again, as an example, a Dalitz plot D(X,Y). For a perfect apparatus, the theoretical plot is given by an eq.2-like formula:

$$D(X,Y) = |T(X,Y)|^2 \ LIPS(X,Y) \tag{8}$$

This is nearly the case of bubble chamber experiments, where one has 4π acceptance and only small scanning and measurement distortion effects. For the new generation of non visualizing experiments, instead, the apparatus acceptance introduces very often dramatic effects, see fig.4.

In this case eq.8 transforms in a folding integral:

$$D(X,Y) = \int_{\text{APPARATUS}} |T(X',Y')|^2 \, LIPS(X',Y') \; \varepsilon(X',Y',X,Y) \, dX'dY' \qquad (9)$$

where the *true* values X', Y' are smeared out by the acceptance function ε to give the *observed* values X, Y. This integral should be calculated at each step of the fitting procedure, over the whole apparatus! Note that with the word 'apparatus' one should intend not only hardware, but also the reconstruction software , because both have an efficiency depending on the event topology. This is a formidable and impossible task, so that approximations are needed.

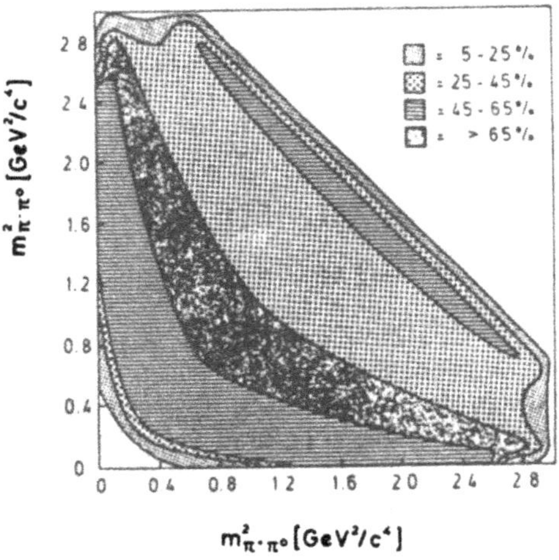

Fig.4: highly non uniform Monte Carlo acceptance plot resulting from an uniform Dalitz plot injected into the apparatus. (From ref.15).

A widely used procedure consists in the application to eq.9 of the mean value theorem, so that one obtains:

$$D(X,Y) \cong \Sigma_\Delta \; \langle |T(X',Y')|^2 \rangle_\Delta \int_\Delta LIPS(X',Y') \, \varepsilon(X',Y',X,Y) \, dX'dY'$$

$$\cong \Sigma_\Delta \; \langle |T(X',Y')|^2 \rangle_\Delta \; LIPS_{MC}(X,Y)_\Delta \qquad (10)$$

where Δ indicates a small cell in the X,Y space, and $LIPS_{MC}$ is phase space spectrum folded into the apparatus. In practice, one samples a lot of events from phase space, simulates by Monte Carlo the detector response and writes the Raw Data Tape (RDT file). Then, this simulated RDT is read by the chain of reconstruction programs, which produces the Data Summary Tape (DST file). This file contains the reconstructed 4-momenta of the particles at the interaction vertex (vertices). This simulated DST, having as input the pure phase space, is exactly the term $LIPS_{MC}$ of eq.10. Then, the simulated DST events are weighted with the "physics", that is with the square modulus of the transition amplitude, $|T(X,Y)|^2$. Note that it is essential that the simulated DST file contains, event by event, the true momenta X', Y' corresponding to the reconstructed ones X,Y, because both quantities enter in eq.10. Finally, the histograms are booked with these simulated events, and the fit is performed to the experimental histograms.

6. Fitting procedures

In the case of many interfering amplitudes, the transition matrix T of eq.10 has a very complex structure, outlined by eq.7 in the case of two amplitudes only. In this simple case, one has four terms, weighted with the coefficients x_1, x_2, cosφ, sinφ. In general, with N interfering amplitudes one has N^2 terms. In the following, we will call these terms 'amplitudes' or 'histograms', with the following meaning: $Re(A_1 A^*_2)$ is an 'amplitude', [$Re(A_1 A^*_2)$ LIPS(X,Y)] is an 'histogram'. The masses and widths of the resonances are contained in these 'amplitudes' or 'histograms'. Hence, if one varies these parameters during the fit, at each minimization step 'amplitudes' or 'histograms' for each simulated event have to be recalculated. Since this is a cumbersome and time consuming procedure, often one prefers to fix masses and widths of resonances, and to store at the beginning 'amplitudes' or 'histograms' in a file; then, the fit is performed by varying the parameters x_1, x_2, cosφ, sinφ,...... only. The fit is repeated many times over a grid of different mass and width values, and the absolute minimum of the chi-square or of the likelihood function is found 'manually' by means of a parabolic interpolation among the relative minimum values of this grid. The procedure is sketched in fig.5.

7. Does the phase space exist?

Contrarily to a widespread belief, the phase space is not a physical (measurable) entity, but merely a mathematical one. Indeed, it appears as the density of states in the fundamental eq.2 and in all its subsequent derivations, as we have explained above.

However, very often phase space is assumed as a *physical* background underlying the resonance peaks. Is this picture true? Are the non interacting pions

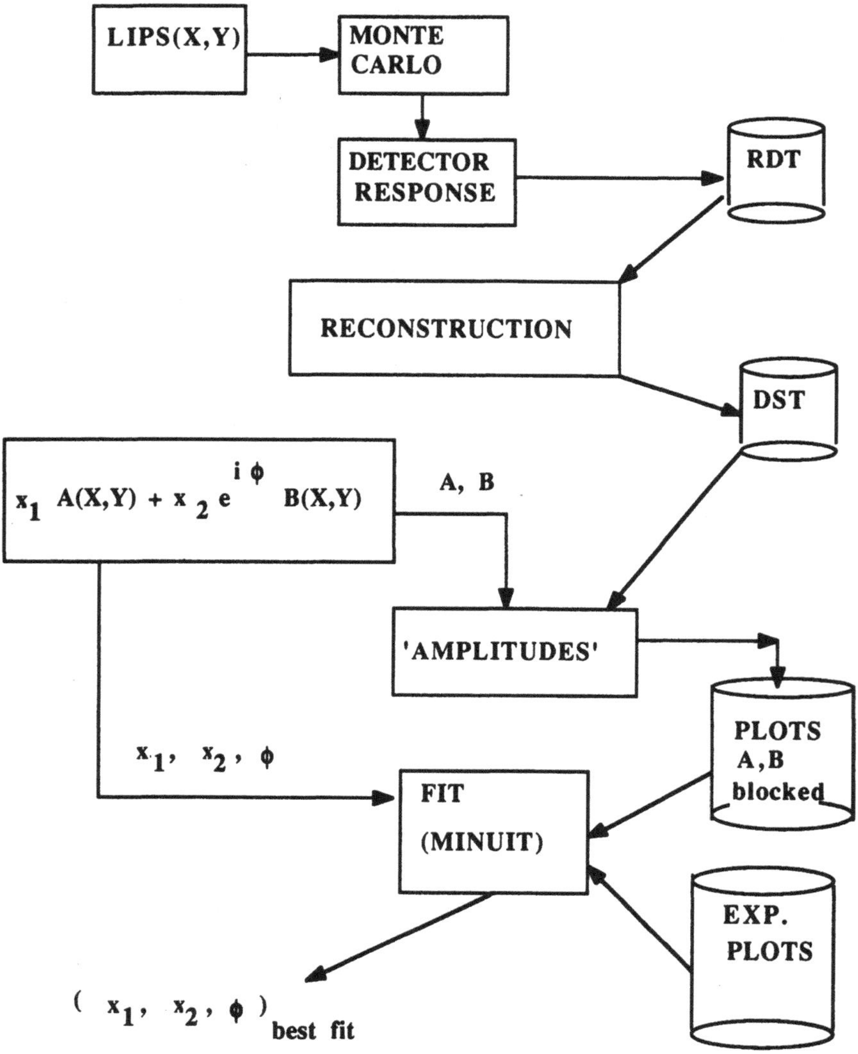

Fig. 5: a sketch of the procedure for the fit of the calculated spectra to the experimental ones

flying into space correlated only by 4-momentum conservation? We can show that it is not so by means of the following example. Let consider the $\bar{p}n \rightarrow \pi^- \pi^- \pi^+$ annihilation and suppose that no resonance is produced: in this case it is reasonable to suppose that the pions are emitted incoherently from the $N\bar{N}$ annihilation 'fireball' and that their momenta will obey to the phase space distributions, since in eq.2 the square modulus of the transition matrix should be equal to one. However, we have two identical boson in the final state, so that we should expect that symmetry correlations will modify this simple picture. In fact, this effect was observed and explained many years ago by Goldhaber et al.[21], and it is known as the GGPL effect. (see fig 6). It is important to note that the treatment of Bose-Einstein correlations in the annihilation final state assumes that the pions are emitted *incoherently* from the source [22].

However, we know that in the annihilation many resonances are produced, and that the production of a resonance in the intermediate state introduces *coherence* among the decaying pions. Hence, we have two conflicting pictures of the annihilation process:

a) in the absence of resonances the pions are produced incoherently, and the phase space density must be multiplied by Bose-Einstein (GGPL) correlation factors;

b) when resonances are produced, there is (partial?) coherence among the pions; the Bose-Einstein effects can be taken into account by symmetrizing properly the transition matrix elements of eq.2 (see also eq.3 and the discussion below).

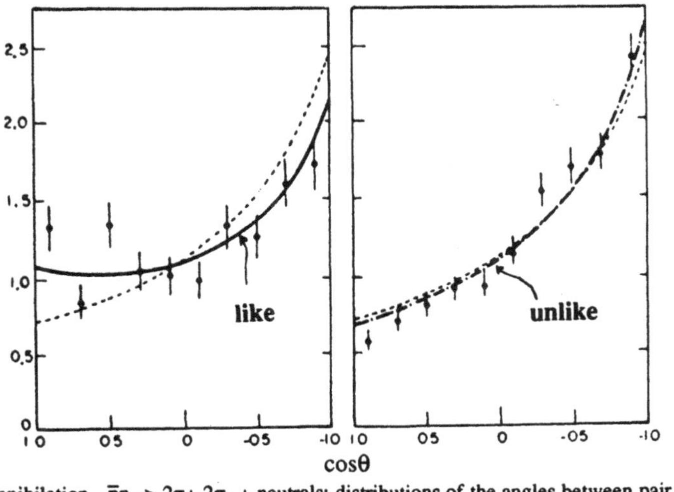

Fig. 6: annihilation $\bar{p}p \rightarrow 2\pi^+ 2\pi^-$ + neutrals: distributions of the angles between pair of like and unlike pions in the c.m. The dotted line is the phase space prediction, the solid line is obtained with the inclusion in the phase space integral of factors taking into account symmetrization. (From ref.21)

So far it is not clear how to merge these two different pictures: the common trend is to assume the picture b), that is a total coherence in the presence of resonances, and the inclusion of the symmetrization effects in the transition matrix elements only. This is probably correct, because it seems that all the annihilation channels are saturated by the resonance production, so that the assumption of the picture b) leads usually to a rather good description of the data. However, if a non resonant background with identical particles is added to the decay amplitude, it should be corrected with Bose-Einstein correlation functions[23]. Unfortunately, these functions require the radius of the source and the coherence strength as additional parameters. Often one prefers to avoid this complication by assuming for the background the dipion sigma interaction, which is discussed in the next section.

8. The I=0 S-wave dipion interaction

In the low energy region (0.5-1.5 GeV) the $\pi\pi$ interaction has been widely studied through reaction like pp --> $pp\pi^+\pi^-$ (see fig. 8) or πp --> $\pi\pi p$. The phase shift analysis of these reactions[24] revealed the existence of a very interesting and complicated S-wave interaction, called the coherent, non-resonant (i.e not BW shaped) ($\pi\pi$) or σ dipion interaction. Probably, this interaction is due to the coherent overlap of at least three $0^+(0^{++})$ resonances, decaying in $\pi\pi$ or $K\bar{K}$[24]. In fig.7 the behaviour of the phase shift is shown. For a discussion, see also the PDG review[5].

Hence, it seems more realistic to put as a background in $N\bar{N}$ annihilation spectra this interaction instead of 'pure' phase space, because, whatever happens in this channel, it is automatically taken into account when the phase shifts obtained from the parametrization of real data are used.

The transition matrix elements of eq.3 are written in this case by substituting the BW term with the usual partial wave expression of the transition amplitude:

$$T_{J^{PC}} = \sum_i a_i \, Z_i(J^{PC},R,\pi_3) \, \frac{1}{2i}\left(\eta(m_i)\, e^{2i\delta(m_i)} - 1\right) \qquad (11)$$

where the inelasticity η and the phase shift δ are taken from the fit of experimental data, of which many parametrizations exist [24].

We note that the use of this background relies on the implicit assumption that the pion production mechanism is essentially the same both in $N\bar{N}$ annihilation and in

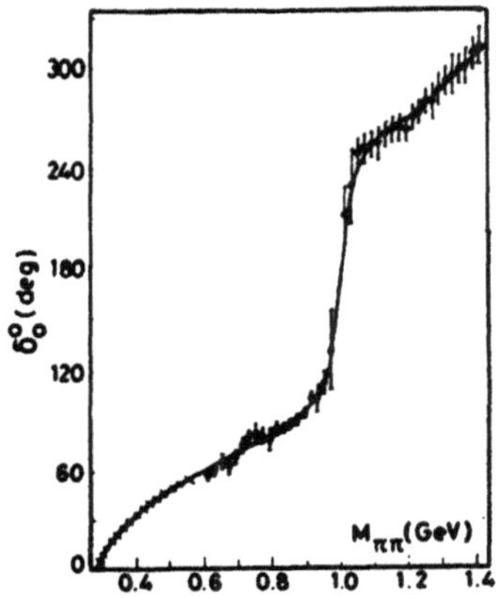

Fig.7: I=0 S-wave phase shift for $\pi\pi$ scattering. Note the resonant behaviour at the $K\bar{K}$ threshold (from ref.24).

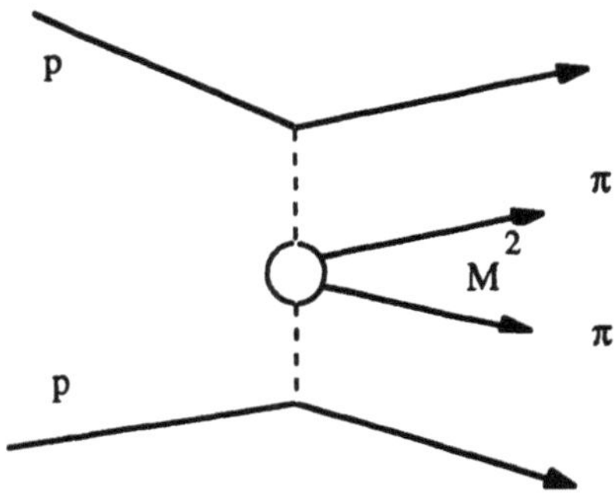

Fig.8: the double (Pomeron) exchange graph controlling central dimeson production in pp-->pp($\pi\pi$).

πp scattering and pp central collisions (see fig. 8), an hypothesis which is not based on safe theoretical grounds. Moreover, nothing is known about a possible 3π or 4π interaction, but the assumption that the corresponding diagrams should be unfavoured compared to the binary ones.

In order to show that these assumptions are not harmless for what concerns data analysis, we report here some examples.

In ref.16 the reaction $p\bar{p}$ --> $\pi^+\pi^-\pi^0$ has been analysed and a new resonance AX has been found, with M= 1565±20 MeV and Γ = 170±40 MeV. Pure phase space was used as a background. In ref.25 the reaction $p\bar{p}$ --> $\pi^0\pi^0\pi^0$ was analysed with the dipion σ interaction as a coherent background, and the parameters of this resonance were found to be M= 1515±10 MeV and Γ = 120±10 MeV. Although compatible, these values show noticeable differences: are they due to the different analysis approaches?

Much more striking is the case of the reaction $\bar{p}\,n \to X\,\pi^- \to (\pi^+\pi^-\pi^+\pi^-)\,\pi^-$ studied in deuterium. An analysis assuming an X-->ρρ decay and usual LIPS in eq.2 led to the proposal of a X(1410) (0^{++}) resonance candidate[26]. Another analysis assuming again the X-->ρρ decay and introducing into LIPS the GGPL corrections led to a X(1477) (2^{++}) resonance candidate[27]. More recently, a reanalysis of this channel under the hypothesis of X-->ρρ and X-->σσ (the dipion interaction) decays and usual LIPS gave X(1386) (0^{++}) as a result[28]. After 26 years of studies we still do not know what exactly happens in this channel!

9. The combinatorial background

We explain this point again in the simplest case of the 3 pion annihilation. Let us consider the $\bar{p}n \rightarrow (\pi^+\pi^-)\pi^-$ reaction, with $(\pi^+\pi^-)$ resonant intermediate states. Which of the two negative pions (let's say π_1^-, π_2^-) participate to the resonance? The answer of Quantum Mechanics is known to any undergraduate student: both ones. This effect is correctly taken into account by the FSI model when eq.2 it is properly symmetrized in the presence of identical particles. We note that the model deals correctly with two different types of interference:

a) 'π-interference' (symmetrization): *two or more* identical particles participate to the *same* resonance;

b) 'BW-interference': the *same* particle participates to *two or more* resonances (see eq.6 and the discussion below).

When the $(\pi^+\pi^-)$ experimental invariant mass spectrum has to be plotted, in order to consider both negative pions one is forced to consider 'two entries per event', that is to plot on the same spectrum the $(\pi^+\pi_1^-)$ and $(\pi^+\pi_2^-)$ invariant masses. This fact, in the presence of many identical particles, give rise to a large structurless combinatorial background, due to the particles that have participated 'weakly' to the resonances (for instance, a 1.5 GeV pion has a low probability to participate to the $\rho(770)$ resonance). We note that in the FSI model this effect is also correctly taken into account by the Zemach tensors and the BW functions that act as weights in eqs.2,6.

When fitting the theory to data, the combinatorial background is not a problem, because the theoretical plots are constructed with exactly the same entries of the experimental ones, so that all the effects previously mentioned are correctly taken into account.

However, often it is necessary to examine the spectra before fitting, in order to have the feeling of the situation, that is to know which resonances are involved in the spectra, how large is the phase space contribution, to decide even whether the spin-parity analysis and the fit are feasible or not, etc. In these cases the presence of the combinatorial background can mask the effective contribution of some resonance, making difficult to draw reliable conclusions. For this reason experimentalists usually subtract the combinatorial background and begin to work with the so-called difference spectra; since the exact meaning of these spectra is often object of discussion, we examine this method in some detail. In the reaction here considered, the method consists in the following procedure:

a) to consider the two entry invariant mass spectrum $(\pi^+\pi_1^-)$ and $(\pi^+\pi_2^-)$;

b) to generate the one entry invariant mass spectrum $(\pi_1^-\pi_2^-)$;

c) to subtract spectrum b) from spectrum a), obtaining the one entry difference spectrum for the $(\pi^+\pi^-)$ invariant mass.

It is possible to show[29] that if one neglects the π-interference (it is equivalent to say that only one negative pion comes from the resonance), the charge conjugation invariance assures that in the difference spectrum the non resonant (combinatorial) background is eliminated. In fact, if one assumes that a) the π_1^- only participates to the resonance, b) the other negative pion is distinguishable, c) the phase space for the non resonant $(\pi^+\pi_2^-)\pi_1^-$ and $(\pi_1^-\pi_2^-)\pi^+$ configurations is the same by C invariance, the eq.2 gives, for the difference spectrum:

$$
\begin{aligned}
D[p\bar{p} \rightarrow (\pi^+\pi^-)\pi^-] &= |BW\,(\pi^+\pi_1^-)|^2\, LIPS\,[(\pi^+\pi_1^-)\pi_2^-] + \\
&\quad LIPS\,[(\pi^+\pi_2^-)\pi_1^-] - LIPS\,[(\pi_1^-\pi_2^-)\pi^+] \\
&= |BW\,(\pi^+\pi_1^-)|^2\, LIPS\,[(\pi^+\pi_1^-)\pi_2^-] \qquad (12)
\end{aligned}
$$

The reliability of the method is expected to improve with increasing the resonance mass, decreasing its width and increasing the pion multiplicity[29].

This difference spectrum technique seems to be effective in enhancing the presence of possible resonances[27,29], but it should be used with some grain of salt, as it is shown in fig.9. For instance, it is clear from this figure that the determination of the ρ mass, width and production rate (branching ratio) as the maximum, FWHM and area under the ρ peak is incorrect and channel dependent. A better procedure is the fit of the reaction Dalitz plot and the determination of the best fit parameters (M, Γ and the coefficients x_1, x_2, ϕ of eq.7) of all the resonances involved in the channel. The branching ratios are then defined as the square of the x_1, x_2,... coefficients[14-16]. The resonance masses and widths determined in this way by the fit are independent of the measured channel (within the limits of validity of the FSI model), because both π and BW-interferences are taken correctly into account. On the contrary, the branching ratios are still channel dependent, due to the presence of the BW-interference. However, we recall that the concept itself of branching ratio for a single resonance tends to be meaningless in the presence of a coherent interaction with other resonances[2,3,30].

In conclusion, the right choice of the spectra for the fit is that discussed in sect.2: the spectra must have globally the right number of independent variables and must be a combination of invariant mass and angular distributions (for the three particle case, the Dalitz plot is the best choice). In this context, the results of the fit are insensitive to the use of difference rather than combinatorial spectra. However, the possible presence of negative values in the difference spectra, due to the π-interference effects, can make problematic the fit of these histograms.

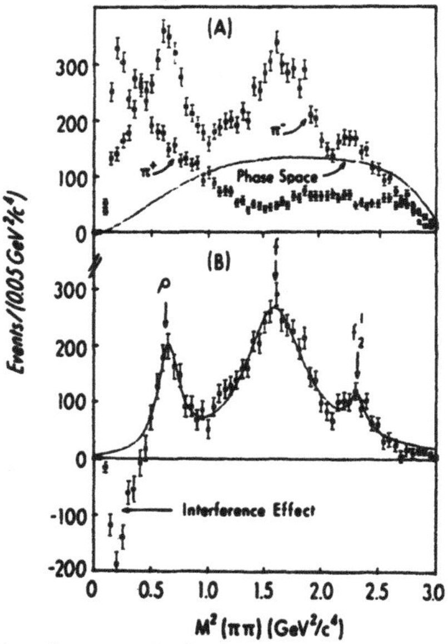

Fig.9: the $\pi\pi$ invariant mass distributions recoiling against the $\pi^+$and the π^- in the $2\pi^-\pi^+$ channel (A) and their difference spectrum (B). The presence of the well known ρ and f resonances emerges clearly, together with a new candidate f'_2 (now the $f_2(1520)$). Note the large interference effects to the ρ left side (from ref.29).

10. When do two resonances interfere?

According with Quantum Mechanics, two states do not interfere when they can, *even in principle*, be experimentally distinguished. In our case, this happens when two resonances are produced in different $p\bar{p}$ J^{PC} initial states (see sect .3) *or* their decay length is greater than the effective annihilation range (that is, the decay mesons are emitted far from the resonance production region). Since in our case this range is about 1 fm (ref.3), it follows that all the resonances with Γ<20 MeV ($\hbar c / \Gamma > 10$ fm) should be summed incoherently in the transition amplitude. These resonances, for the low energy $p\bar{p}$ annihilation, are listed below:

resonance	Γ (MeV)	decay length = $\hbar c$ / Γ (fm)
$\eta(547)$	0.00119	$1.65\ 10^5$
$\omega(783)$	8.43	23
$\eta'(958)$	0.198	996
$\phi(1020)$	4,43	44

In alternative to the incoherent summation with other resonances, a more practical and probably better procedure is the separate extraction of these resonances from the data, with the background subtraction or kinematics fitting techniques.

11. Conclusions

$N\bar{N}$ annihilation is a powerful source of light mesons. These resonances are numerous and broad in the 1-3 GeV mass region, so that their unambiguous identification is a challenging task. In these notes we made a short overview of the current analysis methods of this kind of reactions, emphasizing some still critical and uncertain points. Among these, we recall the limits of the FSI model, the background parametrization and the approximations of the fitting procedures.

The analyst has to decide in each specific case, by checking the stability of the solution, to what extent these uncertainties affect the final results of her/his analysis.

References

1) M.A. Kunze, in *Physics at SuperLear*, eds. C Amsler and D. Urner, Instiute of Physics Conference Series **124** (Bristol, 1992)

2) C. Amsler and F. Myhrer, *Ann. Rev. of Nucl. and Part. Science*, **41**(1991)219; U.Gastaldi, *Nucl Phys. B (Proc.Suppl.)* **23B**(1991)224

3) C. B. Dover et al., BNL report n.47212, (Brookhaven, 1992), to appear in *Prog. in Particle and Nuclear Physics*

4) R. Armenteros and B. French, *High Energy Physics*, **4** (Academic Press, 1969), p.237

5) Particle Data Group, *Phys. Rev. D* **45**(1992)1
6) E. Byckling and K. Kajantie, *Particles Kinematics* (Wiley, London,1973)
7) V. Filippini, *this volume*
8) N. Semprini-Cesari, *this volume*
9) K.M. Watson, *Phys. Rev.* **88**(1952)1163
10) H. Pilkuhn, *The interactions of Hadrons*, (Wiley, New York, 1967)
11) J.D. Jackson, *Nuovo Cim.* **34**(1964)1644
12) J. Pisut and M. Roos, *Nucl. Phys.* B **6**(1968)325
13) C. Amsler et al., Crystal Barrel collaboration CERN Report, *Phys. Lett.*, in press
14) M. Foster et al., *Nucl. Phys.* B **6**(1968)107
15) B. May et al., *Z. Physik* C **46**(1989)191
16) B. May et al., *Z. Physik* C **46**(1989)203
17) C. Zemach, *Phys. Rev.* B **133**(1964)1201
18) G. Breit and E. Wigner, *Phys. Rev.* **49**(1936)519
19) D. C. Peaslee, in *LEAP '90*, eds. P. Carlson, A. Kerek and S. Szilagyi, (World Scientific, Singapore, 1991) p. 319
20) E. Klempt, in ref 19, p. 273
21) G. Goldhaber et al., Phys. Rev. C **120**(19960)300
22) M. Gyulassy et al., *Phys. Rev.* C **20**(1979)2267
23) H.Q. Song et al, *Z. Phys. A*, **342**(1992)439
24) L. Au et al., *Phys. Rev.* D **35** (1987)1633
25) E. Acker et al., *Phys. Lett.* B **260**(1991)249
26) A. Bettini et al., *Nuovo Cim.* **42A**(1966)695
27) D. Bridges et al., *Phys. Rev. Lett.* **57**(1986)1534
28) M. Gaspero, in Nucleon Antinucleon Interaction, Moscow Conference, 1991, in press
29) D. Bridges et al, *Phys. Rev. Lett.* **56**(1986)215
30) J. Vandermeulen, *Z. Phys. A* **342**(1992)329

Common Problems and Trends of Modern Physics, pp. 201–211
ed. T. Bressani, S. Marcello and A. Zenoni

©1992 World Scientific Publishing Company

ANTIPROTON CAPTURE IN DEUTERIUM

ALBERTO MASONI

INFN - Sezione di Cagliari, via Ada Negri 18
Cagliari, I-09127, Italy

ABSTRACT

The problems in the determination of the initial angular momentum states from which $\bar{p}$-d annihilation occurs are discussed. An overview of the experimental situation is performed. The lack of coherence and the limited number of available experimental data let the problem still open even if some very recent experimental and theoretical results provide convincing evidence against a strong P wave contribution both in liquid and in gas.

1. Introduction

The measurement of the initial angular momentum states in the $\bar{p}$-N annihilation is of great importance for the comprehension of the annihilation dynamics[1]. The matter has been object of controversial discussion since the first measurements performed by bubble chamber experiments in the early sixties. The poor quality of the antiproton beams represented a severe limitation to any measurement performed with low energy antiprotons. With the startup of the Low Energy Antiproton Ring (LEAR) at CERN the situation changed completely. A pure, intense and monochromatic $\bar{p}$ beam is available. Therefore quite a comprehensive set of measurements, both in liquid and in gas, is available for Hydrogen. On the contrary Deuterium data are still scarce and controversial.

2. The antiproton - Deuterium system

The $\bar{p}$ Deuterium atom wave function can be written as

$$\psi_D(\rho,\delta) = \psi_A(\rho)\psi_N(\delta) \tag{1}$$

where $\psi_A(\rho)$ is the atomic wave function for the $\bar{p}$ bound to the Deuterium and $\psi_N(\delta)$ is the nuclear wave function describing the nucleon in the Deuterium, ρ is the distance between the $\bar{p}$ and the two nucleon center of mass while δ is the distance between the two nucleons (see fig. 1). To describe the $\bar{p}$ interacting with one nucleon another coordinate system is appropriate and we introduce the $\bar{p}$ nucleon distance (r) and the distance between the spectator nucleon and the center of mass of the system constituted by the $\bar{p}$ and the nucleon on which it annihilates (d) (see fig. 1). Therefore we have two angular momentum states according to the chosen coordinate system. In one case we consider the relative angular momentum states between the antiproton and the Deuterium nucleus. In the other the angular momentum state between the antiproton and the nucleon on which the annihilation occurs. In the first case we will call it orbital, in the second one wave. Because of the nucleon Fermi motion they can be different and, starting from an S-orbital we can have S, P and higher waves of the $\bar{p}$ - N system. In practice, for experiment at rest only S and P states contribute.

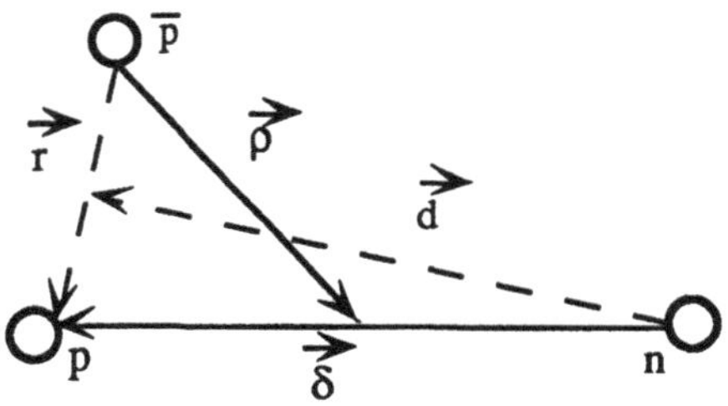

Figure 1. The $\bar{p}$ - deuterium system.

Following reference[2] one can compute, from the wave function Fourier transformed, the momentum distribution of the spectator nucleon. Figure 2 (a) and b)) shows the S and P wave spectator nucleon momentum distribution for annihilation from S and P orbitals respectively. In case of annihilation from P-orbitals there is small difference between the S and P wave momentum distribution. In both cases the shape is that of the familiar Hulthén distribution[2]. The situations changes completely in the case of annihilation from S-orbitals. The P-wave spectrum is completely different from the S-wave one. In the case of S-wave annihilation the spectator nucleon momentum distribution has again the classical Hulthén shape. On the contrary the P-wave spectrum shows a characteristic high energy tail (see figure 2). This fact has a considerable experimental relevance for two reasons. First an accurate measurement of the spectator nucleon momentum distribution allows to get information concerning the $\bar{p}$ - N and the $\bar{p}$ - nucleus angular momentum states. Second, selection cuts on the final states in the data analysis, can induce selection into the P-wave fraction of the annihilation as noted by Bizzarri et al.[2] and Reifenröther and Klempt [3].

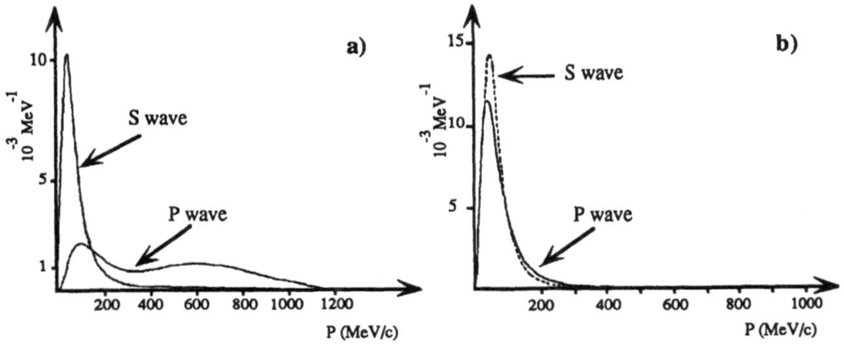

Figure 2. Momentum distribution of the spectator nucleon for S and P wave annihilations starting from S orbitals (a) and from P orbitals (b).

3. Branching ratios and angular momentum states

The absolute and relative branching ratios for π-π and K-K final states provide information on the angular momentum states. Measurements can be performed selecting the initial angular momentum state. Several methods are available. It is possible to choose annihilation channels which can proceed only through a particular state because of parity, total angular momentum, or other conservation rules. Another method is the tagging with X-rays in coincidence. The target density can also be employed.

In Hydrogen a comprehensive set of measurements exists (see table 1). The situation is definitely different in Deuterium (see table 2) where the data are scarcer and controversial. Indicating with $f(m_1 m_2\text{-}N_S)$ the branching ratio for a given two mesons plus spectator nucleon final state:

$$f\,(m_1m_2\text{-}N_S) = \frac{N(m_1m_2\text{-}N_S)}{N(\text{all-}N_S)} \tag{2}$$

(where $N(m_1m_2\text{-}N_S)$ is the number of events in a given final state with two mesons m_1, m_2 and a spectator nucleon N_S and $N(\text{all-}N_S)$ is the total number of annihilation with a spectator nucleon.

Table 1
Branching ratios in 10^{-3} for $N - \bar{N}$ annihilations into π-π and $K - \bar{K}$ in gaseous and liquid H_2. B_0 and B_1 are the branching ratios into $\pi^+\pi^-$ in pure S and P wave respectively. C_0 and C_1 are the corresponding S and P wave branching ratios for K^+K^-.

Liquid H_2					
$f\,(\pi^+\pi^-)$	3.33 ± 0.17	21	$f\,(K^+K^-)$	1.01 ± 0.05	10
$f\,(\pi^o\pi^o)$	0.48 ± 0.10	17	$f\,(K^o{}_sK^o{}_s)$	0.004 ± 0.003	10
	0.14 ± 0.03	18	$f\,(K^o{}_sK^o{}_l)$	0.76 ± 0.04	10
	0.206 ± 0.014	19			
	0.25 ± 0.03	20			
Gaseous H_2	(NTP)				
$f\,(\pi^+\pi^-)$	4.30 ± 0.15	21	$f\,(K^+K^-)$	0.69 ± 0.040	21
			$f\,(K^o{}_sK^o{}_l)$	0.36 ± 0.06	10
$\frac{f\,(\pi^+\pi^-)}{f\,(K^+K^-)}$	0.163±0.011	21			
B_0	3.19 ± 0.20	21	C_0	1.08 ± 0.05	21
B_1	4.81 ± 0.49	21	C_1	0.287 ± 0.051	21
Gaseous H_2	(15 atm.)				
$\frac{f\,(\pi^+\pi^-)}{f\,(K^+K^-)}$		22			

Following table 2 we see that several measurements are available for the ratio

$$r = \frac{N(\pi^-\pi^o\text{-}p_S)}{N(\pi^+\pi^-\text{-}n_S)} \tag{3}$$

but they are in clear disagreement.

Table 2
Branching ratios in 10^{-3} for $N - \bar{N}$ annihilations into π-π and $K - \bar{K}$ in gaseous and liquid D_2.

Liquid D_2					
$f(\pi^+\pi^- n)$	4.2 ± 0.12	6	$f(K^\circ K^- p)$	1.47 ± 0.21	9
$\frac{f(\pi^-\pi^\circ p)}{f(\pi^+\pi^- n)}$	0.68 ± 0.07	6	$f(K^\circ_s K^\circ_s n)$	0.036 ± 0.010	9
	0.70 ± 0.05	8	$f(K^\circ_s K^\circ_l n)$	0.36 ± 0.04	9
	0.55 ± 0.05	8			
	2.07 ± 0.05	7			
Gaseous D_2	(NTP)				
$f(\pi^+\pi^- n)$	4.30 ± 0.150	14	$f(K^+K^- n)$	0.69 ± 0.040	14
$\frac{f(K^+K^- n)}{f(\pi^+\pi^- n)}$	0.36 ± 0.08	14			
	0.27 ± 0.02	14			

The measurement of r is important because it allows to determine the P-wave contribution for $\bar{p}$-D_2 annihilation into $\pi\pi$. Indeed the branching ratio $f(\pi^+\pi^-\text{-}n_S)$ can be split into an S-wave $S(\pi\pi)$ and a P-wave $P(\pi\pi)$ part. $f(\pi^-\pi^\circ\text{-}p_S)$ has only S-wave because $\bar{p}n \to \pi^-\pi^\circ$ P-wave annihilation is forbidden for G-parity conservation. Charge independence requires [4]:

$$f(\pi^+\pi^-\text{-}n_S) = \frac{1}{2} f(\pi^-\pi^\circ\text{-}p_S) + 2 f(\pi^\circ\pi^\circ\text{-}n_S) \qquad (4)$$

Similarly $\bar{p}p \to \pi^\circ\pi^\circ$ proceeds only by P-wave because of parity and angular momentum conservation. Therefore in the hypothesis of no P-wave contribution must be r = 2. Under the hypotheses discussed above the P-wave fraction will be:

$$P(\pi\pi) = \frac{3f(\pi^\circ\pi^\circ\text{-}n_S)}{f(\pi^+\pi^-\text{-}n_S) + f(\pi^\circ\pi^\circ\text{-}n_S)} \qquad (5)$$

So, from the values of r reported in table 2 the relative $\pi^\circ\pi^\circ$ fraction can be obtained and the P-wave contribution for $\bar{p}$ annihilation in the $\pi\pi$ channel can be inferred. The values quoted by the experiments are listed in table 3. A total disagreement is evident.

These results were obtained without taking into account the initial state interaction of the $\bar{p}$d atom. If we consider the possibility of a different angular momentum state of the $\bar{p}$d atom in respect of the $\bar{p}$N angular momentum state, following ref. [3] we obtain:

$$\frac{P(\pi\pi)}{S(\pi\pi)} = \frac{f(\pi^+\pi^- \text{-} n_S) - \frac{1}{2} f(\pi^-\pi^\circ \text{-} p_S)}{\frac{1}{2} f(\pi^-\pi^\circ \text{-} p_S)} \tag{6}$$

Indicating by R_{pn} the ratio between annihilation on protons and annihilation on neutrons(R_{pn} = 1.33±0.03 [5]) we will have:

$$r = \frac{N(\pi^-\pi^\circ \text{-} p_S)}{N(\pi^+\pi^- \text{-} n_S)} = \frac{f(\pi^-\pi^\circ \text{-} p_S)}{f(\pi^+\pi^- \text{-} n_S)} \frac{1}{R_{pn}} \tag{7}$$

Combining 6) and 7) one obtain:

$$P(\pi\pi) = 1 - \frac{rR_{pn}}{2} \tag{8}$$

Which leads to a value for $P(\pi\pi)$ of 0.55 ± 0.05 from $r = 0.68$ [6]. To be compared with $P(\pi\pi) = 0.75 \pm 0.08$ quoted by the experiment [6].

Table 3
P-wave contribution for $\bar{p}$ annihilation in the $\pi\pi$ channel in liquid D_2 ($P_{LD}(\pi\pi)$) as obtained from the ratio $\frac{f(\pi^-\pi^\circ p)}{f(\pi^+\pi^- n)}$.

$P_{LD}(\pi\pi)$	0.75 ± 0.08	6
	< 0.06 (95% CL)	7

Following the procedure of ref.[3] the total P-wave fraction in liquid deuterium (P_{LD}) can also be deduced. Actually the branching ratio for the ($\pi^+\pi^-$-n_S) and ($\pi^-\pi^o$-p_S) channels can be expressed as

$$f(\pi^+\pi^- \text{-} n_S) = (1-P_{LD})B_0 + P_{LD}B_1 \tag{9a}$$

$$f(\pi^-\pi^o \text{-} p_S) = 2(1-P_{LD})B_0 \tag{9b}$$

Where B_0 and B_1 are the π-π branching ratios, in pure S and P wave respectively, measured in gaseous H_2. Their values are listed in table 1. Using eq. 6,7 and 9 we can write:

$$P_{LD} = \frac{2 - rR_{pn}}{2 + \frac{rR_{pn}(B_1 - B_0)}{B_0}} \tag{10}$$

From the value r = 0.68 [6] the following values for P_{LD} can be calculated:

$$P_{LD} = 0.45 \pm 0.05 \quad (R_{pn}=1.33) \quad P_{LD} = 0.56 \pm 0.04 \quad (R_{pn}=1) \tag{11}$$

From the value r = 2.07 [7] the following values for P_{LD} can be calculated:

$$P_{LD} = -0.20 \pm 0.02 \quad (R_{pn}=1.3) \quad P_{LD} = -0.02 \pm 0.02 \quad (R_{pn}=1) \tag{12}$$

The hypothesis R_{pn}=1 means that the initial state interaction is not taken into account. For r = 2.07 only R_{pn}=1 gives an acceptable result.

The results 11) and the 12) are in complete disagreement. Let us try to examine which arguments are in favor and which are against each one of them.

The result 11) is supported by the following considerations.

1) Another experiment [8] measured r obtaining similar values (0.70 ± 0.05 and 0.55 ± 0.05 according to the chosen normalization)

2) The spectator nucleon momentum distribution both in the ($\pi^+\pi^-$-n_S) [6] and in ($K^o{}_S K^o{}_S$ n) [9] final state shows an high momentum tail characteristic of a strong P-wave fraction (see fig. 2).

3) From the measurement of $\bar{p}d \to K^0 \bar{K}^0$ branching ratios obtained by Bizzarri et al. [9] the P-wave fraction results:

$$P_{LD} = 0.19 \pm 0.12 \quad (13a)$$

This result is obtained without doing any hypothesis on the origin of the P wave annihilation. Therefore they suggest that, under the hypothesis of annihilation mainly from S orbitals (as apparent from the spectator neutron momentum shape quoted above) using the absorption length estimated by Bryan and Phillips their calculations [2] lead to an even higher P- wave contribution:

$$P_{LD} \approx 0.28 \quad (13b)$$

Combining the branching ratio measurements of $(K - \bar{K} n)$ by Bizzarri et al. [9] and $(K - \bar{K})$ on Hydrogen by Doser et al. [10], assuming that the $\bar{p}p$ annihilation frequencies are not affected by the presence of the spectator neutron and using the procedure of ref. [3] the following value for P_{LD} has been inferred:

$$P_{LD} = 0.55 \pm 0.09 \quad (14)$$

in agreement with the 11).

4) The $f_2(1270)$ production in $\bar{p}p$ annihilation has been measured to be small from S states [11] and big from P states [12]. Since in $\bar{p}d$ annihilation a strong peak at the $f_2(1270)$ mass is observed [13] a strong P wave contribution is expected.

On the other side P wave contribution in deuterium gas was measured to be small by two experiments:

$$P_{GD} = -0.05 \pm 0.20 \quad (15)$$ [3,14]

$$P_{GD} = 0.18 \pm 0.07 \quad (16)$$ [15]

These two values are obtained using the $K^+K^-/\pi^+\pi^-$ ratio from $\bar{p}$ annihilations in gaseous D_2. The values are listed in table 2. They are substantially in agreement despite the large error which affects the first one. They provide also a strong evidence against a sizeable P wave contribution in liquid D_2. Indeed P wave contribution must be larger in gas than in liquid. The argument is that in liquid or in a dense gas the collisions with the neighboring atoms cause the $\bar{p}$ to cascade into an S state by Stark effect. On the contrary in a low density gas the majority of the $\bar{p}$ will annihilate from a P state. This has been noted by Day Snow and Sucher [16] and confirmed by several experimental evidences in liquid and gaseous H_2. Despite the first ($\pi^o\pi^o$) branching ratio measurements [17] P wave has been measured to be suppressed in liquid [18,19,20]. In gaseous H_2 targets P wave contribution is much greater (~50 %) [21], but decreases with higher target density [22]. Other confirmations of the validity of the arguments based on the Stark mixing can be found in ref. [5].

Summarizing the S and P wave fraction has been calculated, from branching ratio measurements, both for inclusive and exclusive final state production. The two available measurements in gas [14,15] report a P wave contribution < 25 % at one standard deviation. This result is definitely against a strong P wave contribution in liquid for the reasons explained above. In liquid we have a set of measurements [6,8,9] which give a P wave fraction of about 50 % and one measurement [7] which indicates total S wave dominance, consistent with the results in gas.

3. The neutron spectator momentum distribution

As reported in section 2 the neutron spectator distribution, in the case of P wave annihilation from S orbitals, shows a relevant high momentum tail. Selection cuts, at the trigger level or in the data analysis, might suppress the P wave contribution. Since the LEAR experiments [7,14] introduced collinearity cuts to suppress the background, Reifenröther and Klempt [3] proposed the following argument to explain why the experiments [7,14] measured a weak P wave contribution. The collinearity cuts are, actually, cuts on the neutron spectator momentum. Therefore a good fraction of the P wave contribution coming from S orbitals is removed. Moreover, the disagreement suggests that most of the annihilations occur from S orbitals of the $\bar{p}$d atom.

To test this hypothesis different criteria were adopted in the analysis of Obelix data[15]. Much looser collinearity cuts ($\Delta\phi < 25°$, $\Delta\theta < 25°$) were applied, corresponding to a cut on the spectator momentum at about 300 MeV/c. Moreover a study of the spectator neutron momentum distribution was performed. It did not show the high energy tail characteristic of a strong P wave fraction. The results were in agreement with that obtained by the branching ratio measurement. One peculiar characteristic of the data analysis performed by Obelix was the particle identification provided by the time-of-flight. A remarkable result was that, without the particle identification, the kinematical fit with a missing particle can lead very misleading results especially in the reconstruction of the spectator momentum distribution .

4. Conclusions

From the experimental side the most recent data are coherently all against a considerable P wave contribution both in liquid [7] and in gas [14,15]. On the contrary all the data coming from bubble chamber experiments [5,6,8] report a strong P wave contribution in liquid D_2 in complete disagreement with the results [7,14,15]. The result [15] weakened considerably the tentative of Reifenröther and Klempt 3 to explain the disagreement in terms of different selection cuts employed in the data analysis. Also the result [8] disagrees with the hypothesis of P-wave removing by cuts on spectator momentum because in ref. [8] collinearity cuts were employed, but a large P-wave contribution is reported. One must note also that in the theoretical work by Sand et al. [23] total S wave dominance in liquid D_2 is expected and the agreement with the result [7] is reported. In another paper by Locher and Zou [24] the P wave contribution is calculated to be very small in liquid, again in agreement with [7], and of the order of 20 % in gas. The consistency with the result [14] is reported. The results [15] were at that time not available but they are in remarkable agreement.

In summary the situation cannot be considered completely clear even if the most recent results are coherent in indicating that P wave contribution appear to be suppressed in $\bar{p}d$ annihilation both in liquid and in gas.

References

1 see e.g. U. Gastaldi in 'Meson Spectroscopy at LEAR', *Nucl. Phys. B (Proc. Suppl.)* **B23** (1991) 224-262.

2 R. Bizzarri et al., *Nucl. Phys.* **B69** (1974) 298.

3 G. Reifenröther and E. Klempt, *Phys. Lett.* **B245** (1990) 129.

4 H. Lipkin and M. Peshkin, *Phys. Rev. Lett.* **28** (1972) 862.

5 R. Bizzarri et al., *Nuovo Cimento* **A53** (1968) 956.

6 L. Gray et al., *Phys. Rev. Lett.* **30** (1973) 1091.

7 A. Angelopulos et al., *Phys. Lett.* **B212** (1988) 129.

8 D. Bridges et al., *Phys. Lett.* **B180** (1986) 313.

9 R. Bizzarri et al., *Nucl. Phys.* **B69** (1974) 307.

10 M. Doser et al., *Phys. Lett.* **B215** (1988) 792.

11 M. Foster et al., *Nucl. Phys.* **B6** (1968) 107.

12 B. May et al., *Phys. Lett.* **B225** (1989) 450.

13 T. E. Kalogeropoulos et al., *Phys. Rev.* **D24** (1981) 1739.

14 J. Riedelberger et al., *Phys. Rev.* **C40** (1989) 2717.

15 A. Adamo et al., accepted by *Phys. Lett.* **B**

16 T.B. Day, G.A. Snow and J. Sucher, *Phys. Rev.* **3** (1960) 864.

17 S. Devons et al., *Phys. Rev. Lett.* **27** (1971) 1614.

18 G. Bassompierre et al., *Proc. IV European Antiproton Symposyum (Barr),* **Vol. I** (1978) 139.

19 L. Adiels et al., *Z. Phys.* **C35** (1987) 15.

20 M. Chiba et al., *Phys. Rev.* **D38** (1988) 2021.

21 M. Doser et al., *Nucl. Phys.* **A486** (1988) 493.

22 R. Adler et al., *Phys. Lett.* **B267** (1991) 154.

23 P. Sand, A.S. Jensen, A. Miranda and G.C. Oades, *Nucl. Phys.* **A523** (1991) 517.

24 M.P. Locher and B.S. Zou, *PSI Preprint* **PSI-PR-91-35** (1991)

Common Problems and Trends of Modern Physics, pp. 213–226
ed. T. Bressani, S. Marcello and A. Zenoni
©1992 World Scientific Publishing Company

$\overline{N}N$ ANNIHILATION IN $\vec{T} = 1$ STATES

Alessandro Feliciello
Istituto Nazionale di Fisica Nucleare - Sezione di Torino
I-10125 Torino, Italy

and

Luca Venturelli
Dipartimento di Elettronica per l'Automazione, Università di Brescia and
Istituto Nazionale di Fisica Nucleare - Sezione di Torino
I-25060 Brescia, Italy

ABSTRACT

First results on $\bar{n}p$ annihilations at 100 ÷ 290 MeV/c into three and five pions, studied by the OBELIX collaboration at LEAR, are reported. The pure $\vec{T}= 1$ initial state, without the complication of a spectator proton (as in $\bar{p}d$ annihilation) gives access to a new window in the investigation of $\overline{N}N$ interactions. The $(\pi^+\pi^-)$ and $(\pi^+\pi^+\pi^-\pi^-)$ invariant mass spectra in the 3- and 5- charged pion final states show some interesting structures when considered together with previous experimental results on $\bar{p}d$ and $\bar{p}p$ annihilations: the first one is accounted for by the well known $\rho^0(770)$ and $f_2(1270)$ resonances and a further enhancement around 1540 MeV; the best fit for the second one allows to identify the already reported structure $\zeta(1480)$ and indicates the presence of a second object around 1650 MeV.

1. Introduction

The study of the antinucleon-nucleon ($\overline{N}N$) interaction at low energies (i.e. below 2 GeV) has received a great boost with the advent of the Low Energy Antiproton Ring (LEAR) at CERN.

The main goal of the detectors on the floor at LEAR (Crystal Barrel, Jetset, OBELIX) is the discovering of exotic hadrons, whose existence is predicted by quantumchromodynamics (QCD) which is our present description of strong interaction[1]. It not only describes how quarks and antiquarks interact, but also predicts that the gluons, the quanta of the field, can bind to form mesons. Such bound states are called glueballs when they are composed entirely of valence gluons (gg or ggg), while hybrids when they consist of a mixture of valence quarks, antiquarks and gluons ($q\overline{q}g$). In addition multiquarks mesons (e.g. $\overline{q}^2q^2$) are also foreseen. The properties of this new form of matter are predicted only in a gross way due to the lack of a detailed understanding of color confinement in QCD. Several models have attempted to calculate the glueball spectrum as well as their properties such as widths, decay modes and branching ratios. No consensus exists on the their expected properties while there is agreement in assigning them quite low masses: some glueballs should have masses in the 0.5 to 2 GeV region. If to the above point we add that the $\overline{N}N$ annihilation involves the annihilation of quark pairs and the emission of gluons, we can infer how the study of low energy $\overline{N}N$ interactions could be useful to discover gluonium states. However the 0.5 to 2 GeV region has been intensively explored by hadron spectroscopy using various probes during the last 30 years and no structure has unambiguously been identified as gluonium or hybrid. In order to explain these negative results, alternatively to the drastic hypothesis of non existence of glueballs, it is possible to bring forward several reasons: glueballs are so broad that they are hidden by background contributions; glueballs have been observed but misidentified as quarkonium states; glueballs are only produced in "gluon rich" processes. The identification of a new hadronic structures as a glueball is not immediate. The finding of a new meson having J^{PC} quantum numbers not allowed for $q\overline{q}$ states would be the best signature for the existence of gluonium matter.

The discovering of new hadronic states (particularly if broad objects) can be helped by the capability of selecting the reaction initial quantum numbers whose conservation restricts the possible final-state configurations. So it is possible to isolate resonances which otherwise may be overwhelmed by large background contributions. To this purpose new techniques have been proposed and developed in the last years[2]. For example $\overline{p}p$ annihilation at rest occurs in S or P wave: in liquid hydrogen the S wave contribution is dominant, while in very low density hydrogen there is a substantial P-wave annihilation. At normal density, where S- and P-wave contributions are comparable, the ASTERIX

Collaboration has collected a large statistical sample of data. By detecting low-energy X-rays the P-wave annihilations have been tagged; in this way ASTERIX has found a new state $f_2(1565)$[3].

The study of $\bar{p}n$ or $\bar{n}p$ annihilations offers the advantage, respect to $\bar{p}p$ annihilation, of having a pure isospin state ($\vec{T} = 1$). Although $\bar{p}n$ and $\bar{n}p$ channel studies are equivalent, the former are obtained from $\bar{p}d$ reactions and have worse momentum resolution because of the uncertainty in the neutron momentum due to Fermi motion in the deuteron. One of the peculiarities of the OBELIX facility is the possibility of using $\bar{n}$ as a probe for meson spectroscopy. Such a projectile has been never used extensively before: in fact, to our knowledge, the only measurements so far reported are due to Banerjee et al.[4].

2. The OBELIX spectrometer

OBELIX is a large acceptance (~ 3π) magnetic spectrometer, capable of detecting all the products following the annihilation of antinucleons on nucleon or nuclei. It consists of four sub-detectors arranged between and around the poles of the Open Axial Field Magnet (OAFM)[5], which provides a field on an unobstructed volume of ~ $3m^3$, with a maximum intensity of 0.5 T along its axis. The detector complex is characterized by a cylindrical geometry. Moving out radially from the center of the apparatus, at first one finds the target (gaseous H_2, D_2, liquid H_2, D_2) depending on the measurements one wishes to perform. In the case of gaseous target its aluminized mylar wall acts as the cathode of the central detector, a Spiral Projection Chamber (SPC)[6], which provides the interaction vertex determination and pattern recognition of the annihilation prongs. It also allows the detection of X-rays following the deexcitation of the $\bar{p}$-p atom and the identification of K^0_S for triggering purposes. Then a first layer of 30 thin (1 cm) scintillators (tofino), arranged as the staves of a barrel, again for triggering purposes and time of flight measurements. The outer Jet Drift Chamber (JDC)[7] is devoted to the tracking of particles. It allows the measurement of the particle momenta and of their $\Delta E/\Delta x$ and contributes to the determination of the interaction vertex and of the topology of the events. It is surrounded by a second barrel of 90 scintillators (tofone), 4 cm thick; thanks to the tofino and tofone segmentation the time of flight system (TOF) provides the first level trigger on the multiplicities of the charged annihilation products, on some selected topologies and on particle discrimination (in particular between K and π)[8]. Finally, the High Angular

Resolution electromagnetic calorimeter (HARGD)[9], structured in four independent modules consisting of planes of Limited Streamer Tubes, interleaved with Lead sheets (~ 10 radiation length in total), is addressed to the detection of the γ (mainly from the π^0 decay) and can provide a first level trigger on neutrals.

Fig. 1 shows schematically how the spectrometer has been set up to perform the meson spectroscopy program with $\bar{n}$.

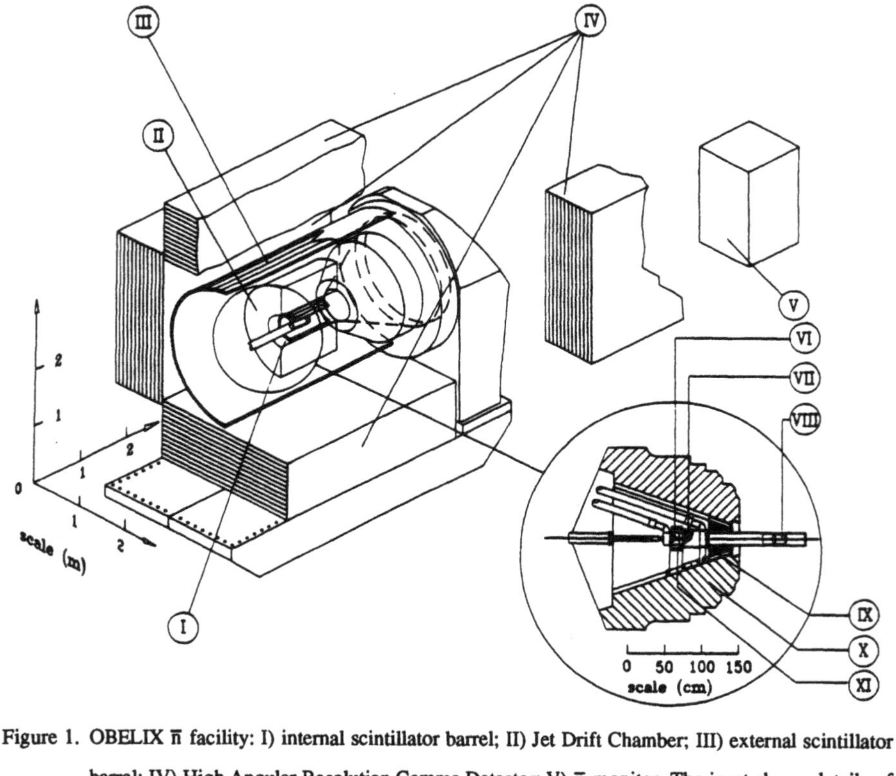

Figure 1. OBELIX $\bar{n}$ facility: I) internal scintillator barrel; II) Jet Drift Chamber; III) external scintillator barrel; IV) High Angular Resolution Gamma Detector; V) $\bar{n}$ monitor. The inset shows details of the target region: VI) LH_2 $\bar{n}$ production target; VII) Lead collimator; VIII) LH_2 $\bar{n}$ reaction target; IX) Lead shield; X) upstream magnet pole; XI) veto counter.

The $\bar{n}$ were produced via the charge exchange reaction (CEX) $\bar{p}p \rightarrow \bar{n}n$ on a 15 cm long LH_2 target, having a diameter of 6 cm and located in the upstream hollow pole of the

OAFM. A scintillator veto box, put in anticoincidence, enclosed the target and was used to trigger on neutrals coming from the interaction of ~ 300 MeV/c LEAR $\bar{p}$, with a flux of ~ 10^6/s, brought to rest in LH_2. A Lead foil, 5 mm thick, surrounded the production target, in order to convert the γ, from π^0 produced in $\bar{p}p \rightarrow n\pi^0$, into e.m. showers. The $\bar{p}$ produced, while slowing down to 98 MeV/c (the threshold for CEX reaction), a continuous spectrum of $\bar{n}$. A suitably shaped Lead collimator, 80 cm long, defines a "beam" of $\bar{n}$[10] produced at ~ 0°, which impinges on a second LH_2 target, 25 cm long and with a diameter of ~ 15 cm, replacing the SPC detector in the center of the apparatus. The $\bar{n}$ beam flux was about 30 $\bar{n}/10^6$ $\bar{p}$ and ~ 30% of them annihilated in the target: the spot on the interaction target had a diameter of ~ 8 cm. The parameters of the beam of the non-interacting $\bar{n}$ (size, intensity, spectrum, etc.) were continuously monitored by a $\bar{n}$ detector (ANTIN)[11], located downstream the spectrometer. Exploiting the good timing properties of the TOF device, the momentum of each $\bar{n}$ annihilating was reconstructed with an error of less than 10 MeV/c, by means of the information coming from tofino[12] (see Fig. 2). The start signal was provided by a 0.2 mm thick scintillator, placed upstream the production target.

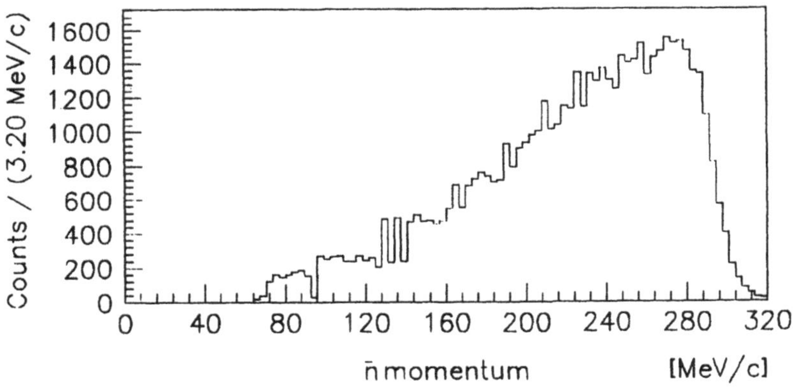

Figure 2. $\bar{n}$ momentum spectrum reconstructed by the OBELIX spectrometer.

3. Meson spectroscopy with $\bar{n}$

With the apparatus in the above described configuration ~ 6.7×10^5 events with different first level triggers have been collected in total: ~ 2.1×10^5 requiring a prong multiplicity of at least 1 on both tofino and tofone detectors, ~ 1.9×10^5 requiring a multiplicity of at least 2 on tofino and of at least 1 on tofone and, finally, ~ 2.7×10^5 requiring a multiplicity of at least 2 again on both barrels.

In such a way the obtained samples of data, besides containing some interesting annihilation channels (i.e. $\bar{n}p \rightarrow \pi^+\pi^+\pi^-$ and $\bar{n}p \rightarrow \pi^+\pi^+\pi^+\pi^-\pi^-$), allowed the evaluation of possible biases introduced by the trigger. The events were analyzed by means of the OBELIX general reconstruction program (OXPR) and first of all we looked for 3-body and 5-body exclusive final states from $\bar{n}p$ annihilations: they represent, in fact, the simpler and cleaner processes that may be selected for meson spectroscopy studies.

From the original data sample we first selected events having a reconstructed vertex in the fiducial volume, a cylinder just a little smaller than the interaction target. In this way the events under study were reduced to ~ 2×10^5. The second cut was concerning the event topology: we in fact selected events having 3 or 5 prongs reconstructed in the JDC and belonging to the reconstructed vertex, with the correct charge distributions. Finally we submitted the remaining events (~ 6×10^4 $\pi^+\pi^+\pi^-$ and ~ 1.5×10^4 $\pi^+\pi^+\pi^+\pi^-\pi^-$) to a 4-C kinematical fit, in order to check the two above exclusive hypothesis with a probability larger than 10%. Fig. 3 shows the squared missing mass distribution for the ($\pi^+\pi^+\pi^-$) and ($\pi^+\pi^+\pi^+\pi^-\pi^-$) events, before and after the kinematical fit cut.

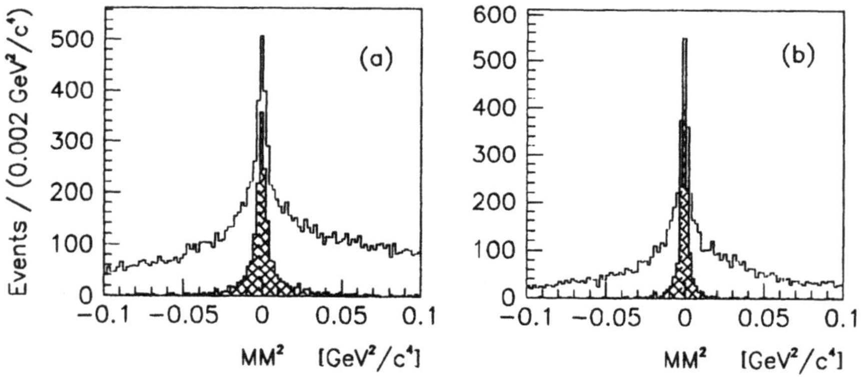

Figure 3. a) Squared missing mass distribution from 3-prong $\bar{n}p$ annihilations. The hatched area corresponds to ($\pi^+\pi^+\pi^-$) exclusive final states as selected by the kinematical fit. b) Analogous distribution for 5-prong $\bar{n}p$ annihilations.

It is worth to notice that a very clean peak, centered at $|MM|^2 = 0$ with a FWHM of 0.006 GeV^2/c^4 was obtained; since this width is a factor two lower than the squared π^0 mass value, we were able to exclude any background in the final sub-samples, consisting of 1837 and 1334 events, respectively.

The invariant mass distribution of the $(\pi^+\pi^-)$ pairs extracted from the final sample of the 3π events is presented in Fig. 4a and it is compared with the most natural reference, namely the results from bubble chamber experiments[13] ($\bar{p}n$ on deuteron). One can easily identify the peaks corresponding to the known $\rho^0(770)$ and $f_2(1270)$ resonances. In addition it is evident the presence of a third structure here indicated as X^0, having a mass of ~ 1540 MeV/c^2.

The overall efficiency of apparatus (geometrical acceptance + reconstruction program) for detecting objects at different masses, decaying into $(\pi^+\pi^-)$ pairs and recoiling against a π^+, was evaluated by means of the OBELIX general simulation programs (OBEX & OXDG) based on GEANT3 package[14]. The generated events, with the simple hypothesis of an isotropically decay of the resonance, flowed through the same reconstruction chain of the real events: we found that the efficiency is uniform. More in detail it is of $(44 \pm 2)\%$, $(45 \pm 2)\%$ and $(46 \pm 2)\%$ for objects with mass of 770, 1270 and 1540 MeV/c^2, respectively.

The invariant mass spectrum of Fig. 4a contains two entries per event, corresponding to the two different ways of combining the three final state π to form a $(\pi^+\pi^-)$ pair. As suggested by Bridges et al.[15] the effect of the combinatorial background can be removed to first order (neglecting possible dynamic and interference effects) by subtracting from $(\pi^+\pi^-)$ invariant mass spectrum the one of the $(\pi^+\pi^+)$ pairs, which is structureless and can be easily fitted to a smooth polynomial function (see Fig. 4b).

The difference spectrum obtained (Fig. 4d) shows the remarkable feature that it can be fully accounted for by only the three above mentioned resonances, without phase space contribution. In other words the annihilation of a pure $\vec{T} = 1$ initial state into 3π exclusive final state seems to proceed only through three two-body intermediate states. The negative values in low mass region may be interpreted, following again Bridges et al.[15], as due to interference effects.

Before going on with the analysis it was checked whether the response of the spectrometer was symmetric to π^+ and π^- over the full momentum range, by studying $(\pi^+\pi^-)$ and $(\pi^+\pi^+\pi^-\pi^-)$ inclusive events following annihilation of $\bar{p}$ at rest in H_2: no differences have been found in the inclusive spectra within 2σ.

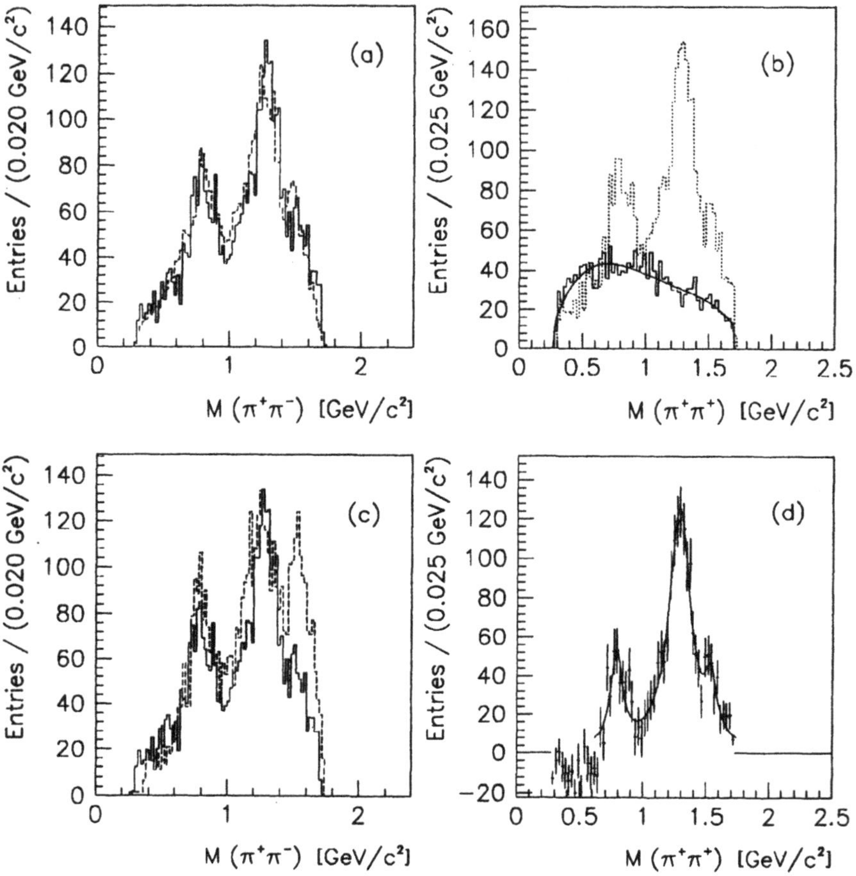

Figure 4. a) $(\pi^+\pi^-)$ invariant mass distribution from $(\pi^+\pi^+\pi^-)$ exclusive events (solid line) compared to the analogous spectrum of Ref. 15 (dashed line). b) $(\pi^+\pi^-)$ (dotted line) and $(\pi^+\pi^+)$ (solid line) invariant mass distributions with superimposed the polynomial fit. c) $(\pi^+\pi^-)$ invariant mass distributions for $\bar{n}p \rightarrow \pi^+\pi^+\pi^-$ (solid line) and $\bar{p}p \rightarrow \pi^+\pi^+\pi^0$ (dashed line) (Ref. 25). d) Difference spectrum as explained in the text; a fit to three Breit-Wigner functions has been superimposed to the data .

A fit of the difference spectra to three non interfering Breit-Wigner functions yielded the values for masses and widths reported in Tab.1, together with the results from Ref.15 and the relative production rates. Even if these results were very good, also a fit to two Breit-Wigner curves has been attempted taking into account only the ρ^0 and the f_2 resonances. However these last results were much less satisfactory and furthermore the resulting value for the $f_2(1270)$ mass was shifted upwards to 1293 MeV/c^2.

Table 1. Results of fits made to difference spectra for 3π exclusive channel.

Final state	π^+X	M_X [MeV/c^2] (this exp.)	Γ_X [MeV/c^2] (this exp.)	M_X [MeV/c^2] (Ref. 15)	Γ_X [MeV/c^2] (Ref. 15)	Relative B.R. (%) (this exp.)
$2\pi^+\pi^-$	$\pi^+\rho^0$ $\hookrightarrow\pi^+\pi^-$	785 ± 9	112 ± 25	806 ± 6	140 ± 12	16.9 ± 2.3
	π^+f_2 $\hookrightarrow\pi^+\pi^-$	1281 ± 6	206 ± 19	1258 ± 3	262 ± 8	71.8 ± 6.2
	π^+X^0 $\hookrightarrow\pi^+\pi^-$	1540 ± 15	132 ± 37	1522 ± 7	59 ± 12	11.3 ± 2.4
						100.0 ± 10.9

From the comparison between our data and those of Ref.15 a systematic shift of the resonances masses of ~ 20 MeV/c^2 towards higher values and a remarkable difference in their widths arose. Anyway, present results agree better with the best fit values for the $\rho^0(770)$ and $f_2(1270)$ resonances[16]. This fact can be explained by the absence of rescattering effects in the present case of $\bar{n}p$ annihilation, while final state interactions of the produced mesons with the spectator proton are expected in $\bar{p}n$ annihilations[17]. Furthermore it has been shown that, due to this rescattering the apparent mass of the states depends on the momentum of the proton in $\bar{p}d$ annihilation[18]: more precisely a higher spectator momentum is associated with the state of lower mass and larger width. The difference between the mass of $\rho^0(770)$ and $f_2(1270)$ and the PDG ones[16], could be attributed to a possible systematic error on the measurement of final state pion momenta and, for the low mass region, to the negative value in the difference spectrum. The peak observed at ~ 1540 MeV/c^2 lies in the same mass region of the known AX(1565)[19] and $f_2(1515)$[20] resonances, reported by ASTERIX and Crystal Barrel collaborations, respectively.

The same analysis strategy was applied to the 5π exclusive event sub-sample. Fig. 5a shows the invariant mass distribution of the ($\pi^+\pi^+\pi^-\pi^-$) combinations, again compared with analogous results obtained in $\bar{p}n$ annihilation[15]. In this case the spectrum contains three entries for each event. The histogram shown in Fig. 5b represents the invariant mass distribution of the ($\pi^+\pi^+\pi^+\pi^-$) combinations (two for each event) whose fit to a polynomial function has been subtracted from the ($\pi^+\pi^+\pi^-\pi^-$) one. The obtained difference spectrum, to be once more compared with that of Ref. 15, is reported in Fig. 5c. This comparison shows a reasonable agreement, although not so good as that noticed for the 3π events. In fact, the present difference spectrum can be described in terms of a broad structure centered at ~ 1500 MeV/c^2, here indicated as $\overline{X}^0$, and of a second narrow structure ($\tilde{X}^0$) with a mass around 1640 MeV/c^2. Results of a fit to two non interfering Breit-Wigner functions are reported in Tab. 2, together with that ones of Ref. 15. Again a prudential fit to a single Breit-Wigner function has be done, obtaining very unsatisfactory results.

Table 2. Results of fits made to difference spectra for 5π exclusive channel.

Final state	π^+X	M_X [MeV/c^2] (this exp.)	Γ_X [MeV/c^2] (this exp.)	M_X [MeV/c^2] (Ref. 15)	Γ_X [MeV/c^2] (Ref. 15)	Relative B.R. (%) (this exp.)
$3\pi^+2\pi^-$	$\pi^+\overline{X}^0$ ↳ $\pi^+\pi^-\pi^+\pi^-$	1490 ± 9	172 ± 15	1477 ± 5	116 ± 9	86.1 ± 6.9
	$\pi^+\tilde{X}^0$ ↳ $\pi^+\pi^-\pi^+\pi^-$	1647 ± 7	58 ± 20	-	-	13.9 ± 2.3
						100.0 ± 9.2

As concerns the main peak it is conceivable to identify it with the structure already observed at 1480 MeV/c^2 in $\bar{p}n$ annihilations, named ζ(1480)[21]. Again the shift towards larger mass values can be attributed to the absence of rescattering of final state pions on the

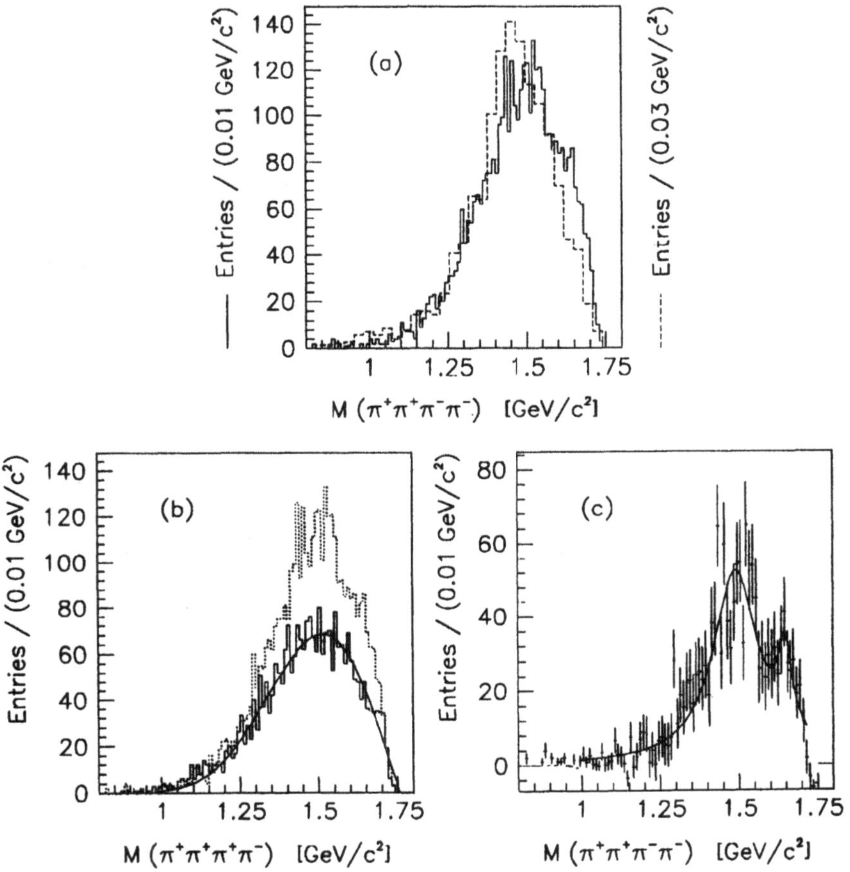

Figure 5. a) $(\pi^+\pi^+\pi^-\pi^-)$ invariant mass distribution from $(\pi^+\pi^+\pi^+\pi^-\pi^-)$ exclusive events (solid line) compared to the analogous spectrum of Ref. 15 (dashed line). b) $(\pi^+\pi^+\pi^-\pi^-)$ (dotted line) and $(\pi^+\pi^+\pi^+\pi^-)$ (solid line) invariant mass distributions with superimposed the polynomial fit. c) Difference spectrum as explained in the text; a fit to two Breit-Wigner functions has been superimposed to the data.

spectator proton[17]. More surprising is instead the presence of the second structure that could represent a relevant result. It should be noted in fact, that a resonance at (1643 ± 7) MeV/c^2 with $\Gamma < 70$ MeV/c^2 was reported by the GAMS collaboration as a tensor meson[22].

At the present analysis state, due to the limited statistics it was not attempted a spin-parity assignment for the discussed resonances. Nevertheless, as suggested in Ref. 15, the state at ~ 1540 MeV/c^2, seen in the 2π channel, and that one at ~ 1490 MeV/c^2, seen in the 4π channel, may have the same $I^G(J^{PC}) = 0^+(2^{++})$. This means, in other words, that they can be considered, in spite of the present observed mass difference, the same object. Following this hypothesis we found for the ratio of the branching ratio into 4π- to 2π-decay a value of (7.6 ± 1.6). A correction factor of 1.53, taking into account the different acceptance of the OBELIX spectrometer for 3π and 5π exclusive final states, was included in the evaluation of the above ratio. The obtained value is considerably lower than that inferred from the bubble chamber experiments[15,21], namely 28.5. On the other hand Dover, in a very detailed analysis of $\overline{N}N$ quasi nuclear bound states[23], noticed that this value was probably overestimated and predicted a value of ~ 2.4 for the $\rho\rho(4\pi)/2\pi$ ratio. The OBELIX result, though different from this theoretical expectation, seems to support the hypothesis that the structure observed in 3π and 5π exclusive channels is a quasi-nuclear state: several independent models[23,24] predict, in fact, a dominance of the 4π decay mode over the 2π one for $\overline{N}N$ quasi-nuclear bound states.

At this point one can try to make some considerations about the $(\pi^+\pi^-)$ invariant mass spectrum, taking into account results from other experiments. The signal for the enhancement around 1540 MeV/c^2 is rather weak, in accordance with the previous $\bar{p}n$ measurements[15]. However, looking at the spectrum obtained with the same apparatus, but with $\bar{p}$ at rest (see Fig. 3d), one can observe a rather strong signal[25] in the same mass region, in full agreement with what was found by the ASTERIX experiment[3,19]. Several checks have been performed in such a way to exclude the possibility that this striking difference can be attributed to instrumental bias. The most plausible explanation is based on the fact that in the $\bar{p}p$ channel there is a noticeable contribution of phase space; furthermore interference effects between ρ^+ and ρ^- (produced in the $^1P_1$ wave) falling in the same mass region of the AX, may considerably enhance the "peak". Such an effect is completely absent in the case of $\bar{n}p$ annihilation into three charged pions, since the charged ρ are not produced.

This situation seems to be quite clear and satisfactory. Nevertheless the question about the interpretation of the results coming from Crystal Barrel collaboration[20] is still open. In fact, studying the $\bar{p}p \rightarrow \pi^0\pi^0\pi^0$ reaction, a large peak at ~ 1515 MeV/c^2 has been found in the $(\pi^0\pi^0)$ invariant mass spectrum. A qualitative comparison between the two spectra is not conceivable, since in the $\bar{n}p$ annihilation ρ^0 is significatively produced. So the mistery on the nature of this unexpected peak remains: in fact the initial state of the two experiment are exactly the same ($^1S_0$, $^3P_1$, $^3P_2$), with a similar P-wave percentage; furthermore in both cases a pure state on isotopic spin is selected, in the initial state for OBELIX, in the final one for Crystal Barrel.

4. Conclusions

Much work remains to do in order to clarify the meson spectroscopy scenario. The nature of the different observed objects can be assessed by the exact determination of their quantum numbers, which requires much more data to be treated with the partial-wave analysis method[26,27,28].

Undoubtedly the choice of using $\bar{n}$ for meson spectroscopy has demonstrated its full validity, providing a sample of clean and of good quality data that have shown rather interesting features and can allow, with higher statistics, a major understanding of this intriguing field.

5. Acknowledgements

It is a pleasure to thank Dr. E. Rossetto for her contribution to the data analysis and for the help in preparation of this document.

6. References

1. See, for example, F.E. Close, *Rep. Prog. Phys.* **51** (1988) 833.
2. U. Gastaldi, *Nucl. Phys.* **B23** (Proc. Suppl.) (1991) 224.
3. B. May et al., *Phys. Lett.* **B225** (1989) 450.
4. S. Banerjee et al., *Z. Phys.* **C32** (1986) 163.
5. H. Gordon et al., *Nucl. Instr. and Meth.* **196** (1982) 303.
6. A. Zenoni, *Common Problems and Ideas of Modern Physics*, Eds. T. Bressani, B. Minetti and A. Zenoni (World Scientific, Singapore, 1992) 289.
7. O. Botner et al., *Nucl. Instr. and Meth.* **176** (1982) 315.

8. G.C. Bonazzola et al., *Nucl. Instr. and Meth.* **A306** (1991) 305.
9. A.Adamo et al., Proc. of Workshop on *Nuncleon Anti Nucleon Interactions*, Moscow (USSR), July 8 - 11 (1991) to be published.
10. M. Agnello et al., to be published on *IEEE Trans. Nucl. Sci.*.
11. T. Bressani et al., *Nucl. Instr. and Meth.* **A292** (1990) 563.
12. F. Iazzi, *Intense Hadron Facilities and Antiproton Physics*, Eds. T. Bressani, F. Iazzi and G. Pauli (SIF, Bologna 1990) 171.
13. L.Gray et al., *Phys. Rev.* **D27** (1983) 307.
14. R. Brun et al., *GEANT3* (CERN Data Handling Division, DD/EE/84-1, 1987) 1.
15. D. Bridges et al., *Phys. Rev. Lett.* **56** (1986) 215.
16. Particle Data Group, *Phys. Lett.* **B239** (1990) 1.
17. V.M. Kolybasov et al., *Phys. Lett.* **B222** (1989) 135.
18. S. Ahmad et al., *Physics at LEAR with Low Energy Antiprotons*, Eds. C. Amsler et al., (Harwood Academic, Chur, 1988) 447.
19. B. May et al., *Z. Phys.* **C46** (1990) 203.
20. E. Aker et al., *Phys. Lett.* **B260** (1991) 249.
21. D. Bridges et al., *Phys. Rev. Lett.* **57** (1986) 1534.
22. F.G. Binon et al., *Nucl. Phys.* **B8** (Proc. Suppl) (1984) 24.
23. C.B. Dover et al., *Phys. Rev.* **C43** (1991) 379.
24. I.S. Shapiro, *Phys. Rep.* **C35** (1978) 129.
25. A. Adamo et al., Proc. 4th Conf. on *Intersection between Particle and Nuclear Physics*, Tucson (AZ) USA (1991) to be published.
26. A. Rotondi, *this volume*.
27. V. Filippini, *this volume*.
28. N. Semprini Cesari, *this volume*.

Common Problems and Trends of Modern Physics, pp. 227–238
ed. T. Bressani, S. Marcello and A. Zenoni
©1992 World Scientific Publishing Company

Nucleon Electromagnetic Form Factors in the time-like region: first results from the FENICE experiment

M.Bertani
University of Ferrara and INFN Sezione di Ferrara, Italy
A.Antonelli
INFN Laboratori Nazionali di Frascati, Italy
C.Bini
University of Rome "La Sapienza" and INFN Sezione di Roma 1, Italy
E.Pasqualucci
University of Rome "Tor Vergata" and INFN Sezione di Roma 2, Italy

(Presented by C.Bini)

ABSTRACT

The nucleon electromagnetic form factors have been extensively studied in the space-like region, but only some measurements for the proton have been done in the time-like region. The FENICE experiment aims to perform the first measurement of the neutron electromagnetic form factor in the time-like region by detecting the process $e^+e^- \to n\bar{n}$ at the ADONE storage ring in Frascati. The results of a test-run at the J/ψ energy are presented here. The $J/\psi \to n\bar{n}$ branching ratio is given together with the $J/\psi \to e^+e^-$ width Γ_{ee}, and the $J/\psi \to p\bar{p}$ branching ratio. An estimate of the neutron form factor at 9.6 GeV^2 is also discussed.

1. Introduction.

The first experimental evidence of the electromagnetic structure of the nucleons, came from the measurements of the anomalous magnetic momenta of proton and neutron during the '40s [1]. Such measurements showed that the proton is not a Dirac elementary particle and that the neutron interacts electromagnetically in spite of its zero net charge. During the '50s a big effort was devoted to the systematic study of the nucleon electromagnetic structure, expecially at the Stanford linear accelerator by Hofstadter and co-workers [2]. In these experiments an electron beam was used as an electromagnetic probe on hydrogen and deuterium targets. Results were given in terms of functions of the square transfer momentum $F_1(q^2)$ and $F_2(q^2)$ called Dirac and Pauli form factors (FF) respectively [3]. In these experiments $q^2<0$ hence the F_1 and F_2 in this region are referred to as space-like FF.

In the '70s the first experimental investigation on time-like FF ($q^2>0$ region) was carried out at both e^+e^- storage ring experiments ($e^+e^- \to$ nucleon-antinucleon processes) and at antiproton annihilation on fixed target experiments (antiproton-proton $\to e^+e^-$). All the measurements of this kind done until now concern the proton time-like FF and no measurement exists on neutron time-like FF. A comparison between neutron and proton FF is required as a possible experimental check of many theoretical models of the nucleon structure. Furthermore a renewed interest in this field came after the surprising result of the EMC collaboration on the proton spin [4] (the so-called proton-spin crisis). Hence the measurement of the time-like neutron FF for different values of q^2 is a very important tool to improve our knowledge of the nucleon structure.

In the next Section, a summary of both experimental and theoretical status of the art is briefly presented with special emphasis on recent and somehow surprising experimental results on the proton time-like form factor. In Sect.3 the main features of the FENICE experimental apparatus are described and in Sect.4 some preliminary results from a first test-run are presented and discussed.

2. The Nucleon Electromagnetic Form Factors: a brief review.

In order to compare the experimental results with different theoretical approaches, two combinations of $F_1(q^2)$ and $F_2(q^2)$ are introduced, called Electric (G_e) and Magnetic (G_m) Form Factors. In a special reference frame (the Breit frame), they represent respectively the non spin-flip (electric) and spin-flip (magnetic) currents.

In Fig.1 the experimental space-like values of G_e and G_m as a function of $t=-q^2<0$ are shown for the proton [5]. The solid line represents the so-called dipole fit, namely a behaviour like

$$G \sim \frac{1}{(q^2 - m_0{}^2)^2} \tag{1}$$

where $m_0{}^2$ is a phenomenological parameter related to the proton spatial dimensions. Such dipole behaviour can be interpreted in two different ways. In a non-relativistic approach G_e and G_m are the Fourier transform of the charge and magnetic moment space distributions so that the function Eq.1 indicates an exponential behaviour for these space distributions. Asymptotically, in a field theory approach, a simple dimensional argument [6] provides a connection between the G functions and the number k of elementary constituents by means of the relation

$$G \sim \left(\frac{1}{q^2}\right)^{k-1} \tag{2}$$

By comparing Eq.2 and Eq.1 it can be argued that in the limit $q^2 >> m_0{}^2$ the prediction Eq.2 is satisfied for k=3.

Similar clues come from the data on neutron space-like FF [5].

In the time-like region, the processes $e^+e^- \to N\bar{N}$ or $N\bar{N} \to e^+e^-$ are looked for. The differential cross-section for the $e^+e^- \to N\bar{N}$ process is given by

$$\frac{d\sigma}{d\Omega} = \frac{\alpha^2\beta}{4s}\left[|G_m|^2 (1+\cos^2\theta) + \frac{4M^2}{q^2} |G_e|^2 \sin^2\theta\right] \tag{3}$$

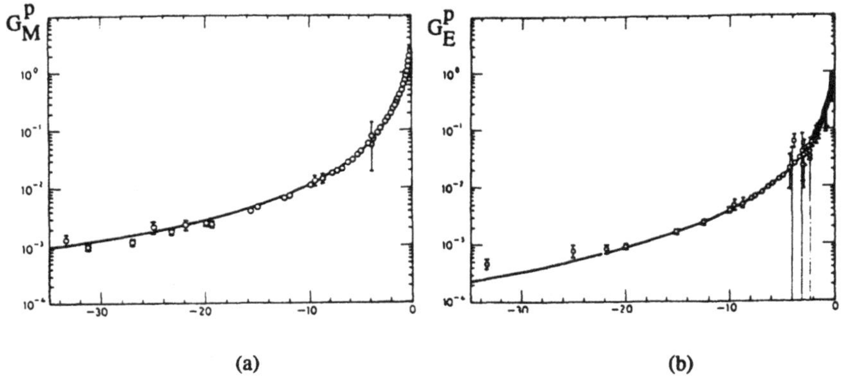

Fig.1 Experimental space-like Magnetic (a) and Electric (b) proton Form Factors as a function of q^2. The superimposed curves are dipole fits (see Eq.1).

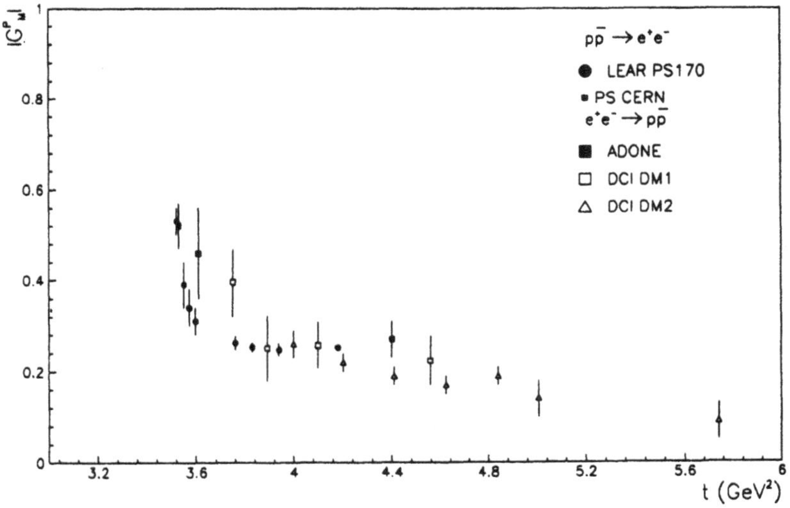

Fig.2 Experimental data on proton time-like form factors.

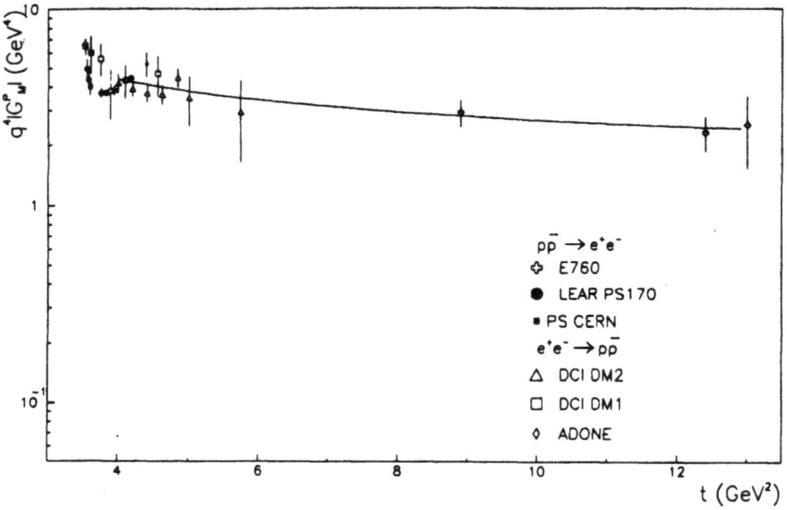

Fig.3 Experimental proton time-like Form Factor (multiplied for q^4) as a function of q^2. The dipole fit is shown for energies in excess of 4 GeV.

where θ is the angle between the electron beam and the nucleon direction, β and M the nucleon speed and mass respectively and s the square of the energy in the center of mass. According to CPT invariance, the FF G_e and G_m must be considered the analytical continuation of the corresponding space-like FF. It can be seen from Eq.3 that a measurement of $|G_e|$ and $|G_m|$ can be done by fitting the angular distribution of the nucleon-antinucleon pairs produced. At high q^2 the contribution of the term involving $|G_e|$ becomes very small and a very accurate angular measurement would be required in order to extract $|G_e|$ from the data. Near the nucleon production threshold the angular distribution is nearly isotropic as expected and $|G_e|=|G_m|$. Such an assumption is made in all the available measurements.

In Fig.2 data on low-q^2 proton time-like FF are shown coming from both kinds of experiments [7]. It can be noticed that the recent low-q^2 data from the experiment LEAR PS170 are in agreement with previous results but, due to smaller errors, they show a completely unexpected low-energy proton FF behaviour. In Fig.3 more recent high-q^2 data [8] are included. A dipole fit (see Eq.1) is found to describe well the data for q^2 values exceeding 4 GeV^2.

Theoretical models [9] give estimates of the expected ratio between neutron and proton time-like FF. A variable r can be introduced according to

$$r = \frac{\sigma(e^+e^- \to n\bar{n})}{\sigma(e^+e^- \to p\bar{p})} \qquad (4)$$

QCD predictions on FF are available for high-q^2 only. In this regime a nucleon should be considered as a diquark-quark system, the leading quark being an u-quark for the proton and a d-quark for the neutron. Thus it should be $r \sim (q_d/q_u)^2 = 0.25$ where q_d and q_u are the electric charge of the d-quark and of the u-quark respectively [10].

For low-q^2 values, the Vector Meson Dominance (VMD) approach is usually assumed. Estimates for the value of r from different VMD models [11] are compared in Tab.1. The disagreement between the quoted VMD predictions is mainly due to the lack of informations on the vector meson spectroscopy in the GeV region.

Tab.1 Low-q^2 r estimates from VMD models

VMD predictions		r
Cabibbo-Gatto	(1961)	~ 14
Korner-Kuroda	(1977)	~ 2
Voci et al.	(1982)	~100
Dubnicka	(1988)	~ 25

Thus an experimental investigation of both low-q^2 and high-q^2 r values is required. This is the main goal of the FENICE experiment.

3. The FENICE experiment

ADÓNE [12] is an e^+e^- storage ring operating at a center of mass energy range $1.5 < E_{c.m.} < 3.1$ GeV. At the FENICE interaction point it is $\beta_x = 4.5$ m and $\beta_y = 1.0$ m. A mean luminosity of $\sim 10^{29}$ cm^{-2}s^{-1} at $E_{c.m.} = 2$ GeV is provided. The beam lifetime is about 6 hours. Assuming r=1 (see eq.6) an $e^+e^- \to n\bar{n}$ rate of ~ 5 evts/day is expected at this energy. It should be noticed that the inverse process $n\bar{n} \to e^+e^-$ (antineutron beam on deuterium) using conventional antineutron sources, would provide a rate lower by one order of magnitude at least.

Antineutrons can be detected looking at the products of the $\bar{n}$ annihilation on nuclei. The time of flight (TOF) technique [13] allows a measurement of the antineutron speed. Thick scintillator counters can be used to detect low energy neutrons. A neutron detection efficiency of about 1% per cm of scintillator is expected [14].

The FENICE detector [15] is a non-magnetic $(3/4) \times 4\pi$ apparatus, expecially designed for the identification of antineutron annihilation. In Fig.4a an overall view of the detector is shown. A first layer of 5 mm thick scintillation counters sorrounds the 2 mm thick Fe beam pipe. The central tracking detector consists of 8 planes of 1 cm^2 cross section limited streamers tubes (LST) [16]. The outer detector consists of eight octants made of alternative layers of 2 m long plastic scintillator slabs (NE110 viewed

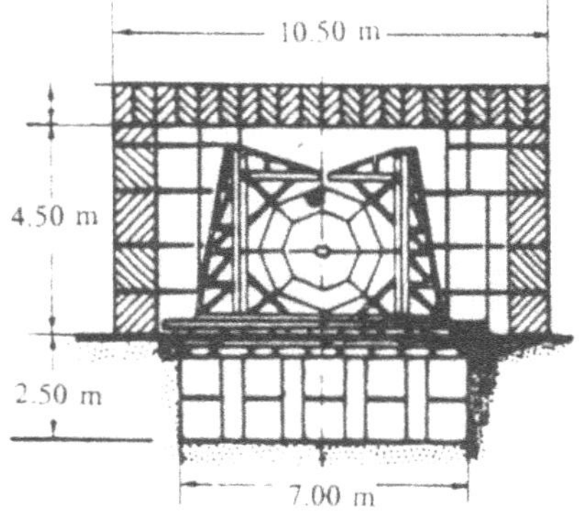

(a)

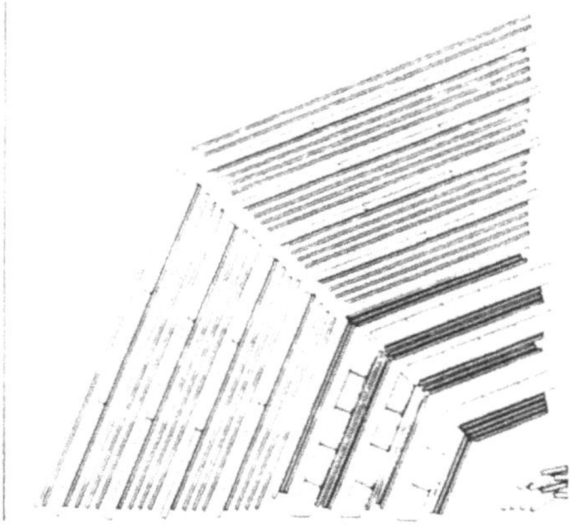

(b)

Fig.4 The FENICE experimental layout: (a) overall view and (b) stratigraphy of two octants.

at both ends by photomultipliers EMI 9813KB), LST and iron absorbers. In Fig.4b the stratigraphy of one octant is shown. In the inner lighter part, called neutron detector, a basic structure consisting of a 5 cm thick scintillator plane and three LST layers is repeated three times giving a 15 cm overall scintillator thickness. In the outer heavier part, called antineutron detector, a basic structure consisting of four 5 mm thick iron absorbers, four LST layers and one 2 cm thick scintillator plane is repeated four times. After the last structure two more LST planes are added. The overall thickness corresponds to about one interaction length and eight radiation lengths. The whole detector is covered by a 1 m thick concrete roof and by a double layer of resistive plate counters (RPC) [17] used as veto counters for charged cosmic rays eventually simulating an interaction in the apparatus [18]. The trigger signal is provided by scintillators and RPC counters. Many logical triggers are implemented for different final state topologies. The trigger logic for antineutron annihilation does not include a back to back scintillator neutron signal. (The neutron signal can be used off-line to select a $n\bar{n}$ events sub-sample). An overall neutron detection efficiency ~ 20% is expected depending on the total energy. Scintillators are also used as TOF detectors with a single scintillator time resolution ranging from 0.5 to 1 ns.

Finally a fast luminosity monitor based on the detection of single bremsstrahlung photons [19] is operating, providing a real-time luminosity measurement.

4. Preliminary results and discussions.

The FENICE experiment started the data taking on April 1991. At the J/ψ resonance energy (3.1 GeV) about 50 nb^{-1} integrated luminosity was collected in one week at a trigger rate of about 10 Hz. After a first energy scan around the J/ψ peak energy, the machine energy was fixed at the peak value.

The off-line data analysis procedure consists of four steps. In the first step an off-line computer program rejects cosmic events with straight tracks escaping the veto system and machine-background events due to the interaction of off-momentum beam particles with the beam pipe; the overall rejection is about 97%. In the second step e^+e^-, $\mu^+\mu^-$, $\gamma\gamma$ and multi-hadrons events are selected providing TOF calibration samples and luminosity measument. In the third step a further selection is performed both for $n\bar{n}$ and for $p\bar{p}$. The $n\bar{n}$ selection is essentially based on TOF measurements and provides a further rejection power ~ 90%. The $p\bar{p}$ selection is based on a search for two collinear tracks in the central detector and provides a rejection power ~ 99.5%. About 10^4 $n\bar{n}$ and some 10^2 $p\bar{p}$ are left at this stage. Finally in the last step of the analysis the two samples of $n\bar{n}$ and $p\bar{p}$ candidates are visually inspected at a graphic terminal using the FENICE Event Display program.

In the following the results on $J/\psi \to e^+e^-$ partial width Γee and on $p\bar{p}$ and $n\bar{n}$ branching ratios are presented. Finally an estimate of the neutron time-like FF at the J/ψ energy is discussed.

4.1. J/ψ-> e+e- partial width Γee.

The multi-hadronic cross section as a function of the center of mass energy is shown in Fig.5. A gaussian fit gives a value of 1.5 MeV for the energy resolution of the ADONE storage ring in agreement with the Accelerator Division measurement. The $J/\psi \to e^+e^-$ partial width can be obtained from the formula [20]

$$\Gamma_{ee} = \frac{M^2}{6\pi^2}\frac{\Gamma_{tot}}{\Gamma_{mh}} \int \sigma(e^+e^- \to m.h.)\, dE \qquad (5)$$

where M is the J/ψ mass and Γ_{tot} and Γ_{mh} are the total and multi-hadronic widths respectively. The integral is derived from our experimental data and then is corrected taking into account the radiative corrections [21]. A value of $\Gamma_{ee} = 5.0 \pm 0.5$ KeV is found in good agreement with previous results [22]. The error is mainly due to systematic uncertaintes on absolute normalization and efficiency. To reduce these uncertaintes $p\bar{p}$ and $n\bar{n}$ branching ratios are evaluated respect to the multi-hadronic events.

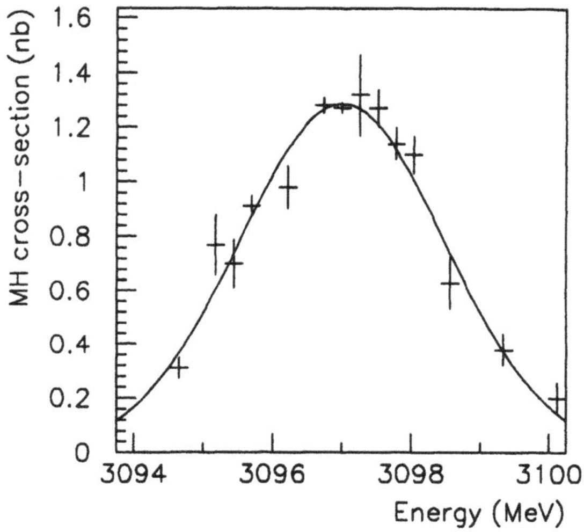

Fig.5 Multi-hadronic cross-section as a function of the C.of M. energy. The gaussian fit is also shown.

4.2. J/ψ->p$\bar{p}$ branching ratio.

A sample of 51 p$\bar{p}$ candidates out of 33.2 nb^{-1} was found. They exhibit two collinear tracks in the central detector. One of them (the proton) goes through the detector, the other one (the antiproton) interacts giving an annihilation star. The overall efficiency taking into account trigger, filter and selection, is 0.33. A $\beta=v/c$ measurement both for the proton track and for the antiproton annihilation vertex is provided. In Fig.6 the $1/\bar{\beta}$ distribution for the 51 candidates is shown, $\bar{\beta}$ being the mean proton-antiproton β value. The mean of this distribution corresponds to a β value of 0.77 $\pm$0.04 in good agreement with the expected value of 0.79. A background contribution of 3.6 events in the same $1/\bar{\beta}$ range coming from events J/ψ -->p$\bar{p}$+X or J/ψ -->$\pi^+\pi^-$, K^+K^-, $\pi^+\pi^-\pi^0$ is expected.The branching ratio turns out to be $(2.02\pm0.32)\times10^{-3}$ in agreement with the world average [24] $(2.16\pm0.11)\times10^{-3}$.

4.3. J/ψ->n$\bar{n}$ branching ratio.

n$\bar{n}$ candidates are selected looking for an annihilation vertex in a fiducial volume of the detector defined in order to increase the signal/background ratio. Once the vertex is found, events with further tracks or showers pointing to the interaction region are rejected.A sample of 50 $n\bar{n}$ candidates out of 33.2 nb^{-1} integrated luminosity at the J/ψ peak energy is selected. The overall efficiency taking into account trigger, off-line computer "filters" and visual inspection is 0.12. For each event the velocity β is given by the annihilation vertex position neglecting the source length and

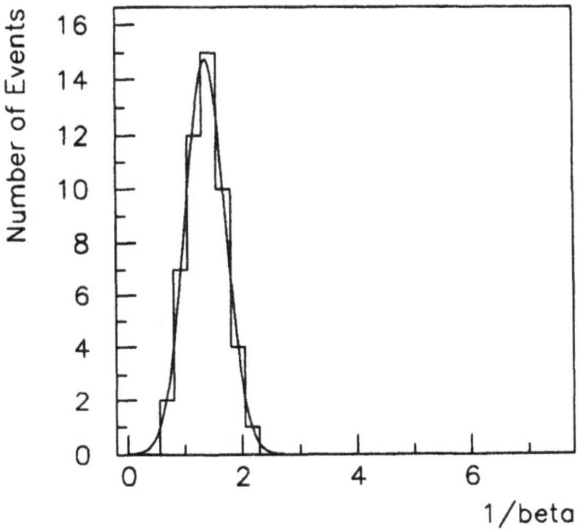

Fig.6 $1/\bar{\beta}$ distribution for 51 J/ψ -->$p\bar{p}$ candidates.

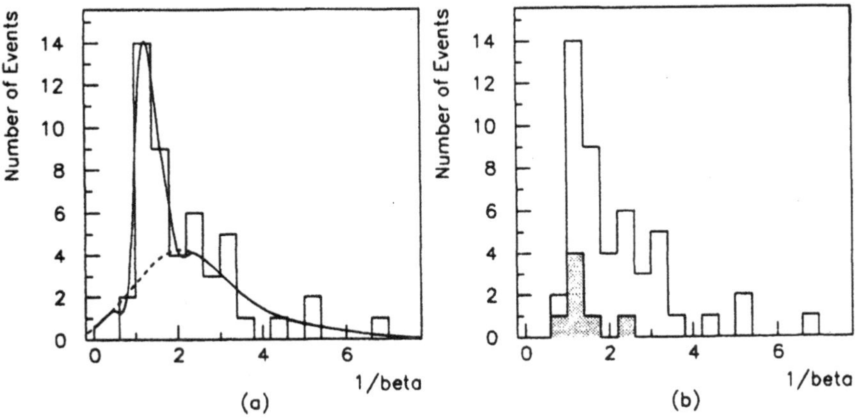

Fig.7 (a) $1/\beta$ distribution for 50 J/ψ -->$n\bar{n}$ candidates. The three-component fit described in the text is also shown. (b) The $1/\beta$ distribution is compared with the same distribution for events with a neutron signal (shaded area).

by the $\bar{n}$ annihilation time.

In Fig.7a the $1/\beta$ distribution for the 50 candidates is shown. The superimposed curve is obtained by fitting the data with the sum of the expected gaussian distribution of the $n\bar{n}$ events plus a two-component background. The two background contributions (dashed curve) come from cosmic rays annihilation and from events J/ψ -->$n\bar{n}$+X with X escaping the detector respectively. The $1/\beta$ distribution of the latter agrees with that of a special sample of selected events with further showers pointing to the e^+e^- interaction point. 23 events with an estimated background of 7 are in the signal region. A branching ratio of $(1.9\pm0.6)\times10^{-3}$ is found, in agreement with the only one previously published [23] $(1.8 \pm 0.9)\times10^{-3}$. 7 events out of the 50 candidates exhibit a neutron signal from a scintillator back to back with respect to the vertex position. In Fig.7b the $1/\beta$ distribution of these events is compared with that of the 50 candidates. It can be argued that the use of the neutron signals makes the background negligible at the expense of a lower efficiency.

It should be noticed that the errors on both $n\bar{n}$ and $p\bar{p}$ branching ratios are mainly statistical.

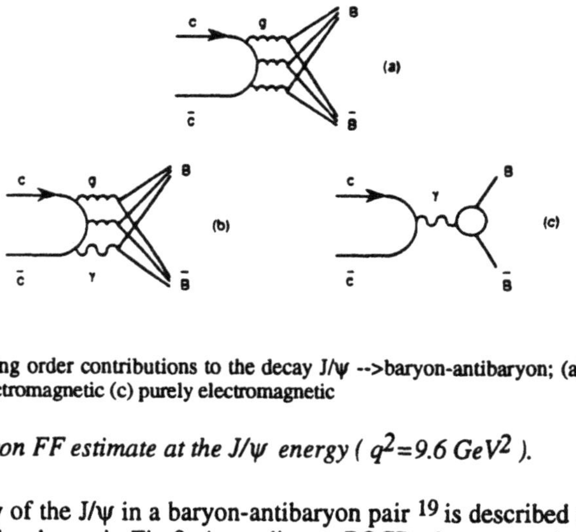

Fig.8 Leading order contributions to the decay J/ψ -->baryon-antibaryon; (a) purely strong decay, (b) strong-electromagnetic (c) purely electromagnetic

4.4. Neutron FF estimate at the J/ψ energy (q^2=9.6 GeV^2).

The decay of the J/ψ in a baryon-antibaryon pair [19] is described at the lowest order by the graphs shown in Fig.8. According to PQCD, the strong-electromagnetic amplitude (Fig.8b) is zero for the decay in $n\bar{n}$ and is small (-3% of the purely strong amplitude (Fig.8a)) for the decay in $p\bar{p}$[10,25]. Perturbative QCD requires the amplitude of the purely strong contribute to be real. Furthermore dispersion relations require asymptotyc symmetry at high q^2 between space-like and time-like FF. Therefore both purely electromagnetic and purely strong amplitudes can be assumed to be real numbers [25]. Let A be the purely strong isospin-conserving amplitude (Fig.8a), E_p

and E_n the purely electromagnetic amplitudes (proportional to the electromagnetic form factors at q^2=9.6 GeV^2 G_p and G_n respectively) for decays in $p\bar{p}$ and in $n\bar{n}$ respectively (Fig.8c); the ratio

$$r = \frac{BR(J/\psi \to n\bar{n})}{BR(J/\psi \to p\bar{p})} \tag{6}$$

can be written in the form

$$r = \left(\frac{A+E_n}{A-0.03A+E_p}\right)^2 \tag{7}$$

An estimate of E_n based on Eq.6 can be obtained using the FENICE experimental value for r and the extrapolated value from the data of Fig.3 for E_p. We estimate a value of $G_n(q^2=9.6\ \mathrm{GeV}^2)=0.012 \pm 0.026$. Such value turns out to be positive and more than a standard deviation away from the G_n value expected using dispersion relations arguments. It is worth noting that a more accurate measurement of this quantity could be a very crucial test for the perturbative QCD assumption of reality of the amplitude A. Measurements of the strong amplitude phase with respect to the electromagnetic one for J/ψ decays in meson pairs [20], give indeed a value close to 90 degrees, suggesting a purely immaginary strong decay amplitude.

5. Conclusions

The experimental investigation on nucleon electromagnetic FF is still far from being complete. The FENICE experiment is now operating at the Frascati National Laboratory of I.N.F.N. to study the neutron FF in the time-like region. Such measurement has never been done before. A preliminary study of the $J/\psi \to n\bar{n}$ decays has been done and the result does not fit very well the PQCD expectations. Data at energies around the nucleon-antinucleon production threshold are now being collected and analyzed.

References

1. W.W.Havens Jr,I.I.Rabi and L.J.Rainwater, *Phys.Rev.* **72**(1947)634; E.Fermi and L.Marshall, *Phys.Rev.* **72**(1947)1139.
2. A complete review of these experiments can be found in R.Hofstadter, *Electron scattering and nuclear and nucleon structure* (W.A.Benjamin,New York 1963).
3. In many text-books the theory of electromagnetic form factors is presented. One of these is F.Halzen and A.D.Martin, *Quarks and Leptons* (Wiley 1984).
4. EMC collaboration *Phys. Lett.* **B206** (1988) 364
5. A complete review on these results can be found in M.Gourdin, *Phys. Rep.* **11**(1974)29.
6. S.J.Brodsky,G.R.Farrar,*Phys.Rev.* **D11**(1975)1309.

7. M.Castellano et al. *Nuovo Cimento* **A14**, 1 (1973); B.Delcourt et al. *Phys.Lett.* **B86**, 395 (1979); D.Bisello et al. *Nucl.Phys.* **B224**, 379 (1983); G.Bassompierre et al. *Nuovo Cimento* **A73**, 347 (1983); G.Bardin et al. *Phys.Lett.* **B255** (1991).
8. P.DalPiaz, private comunication.
9. A review of theoretical models can be found in 5) or in R.Baldini Ferroli and M.E.Biagini, LNF-89/090(P) (1989).
10. V.L.Chernyak,I.R.Zhitnitsky, *Nucl.Phys.* **B246** (1984)52
11. N.Cabibbo and R.Gatto, *Phys.Rev.* **124** (1961) 1577; T.Massam and A.Zichichi, *Nuovo Cimento* **43** (1966) 1137; J.G.Korner and M.Kuroda, *Phys.Rev.* **D16** (1977) 2165; P.Cesselli, M.Nigro and C.Voci, Proc. of Workshop on Phys. at LEAR Erice (1982); S.Dubnicka, *Nuovo Cimento* **A100** (1988) 1
12. F.Amman et al. Proceed. Int. Accel. Conf. Geneva 1971
13. T.Bressani et al., *Nucl.Instr. and Meth.* **A292** (1990) 563.
14. P.Macciotta et al., *IEEE Trans. on Nucl. Sci.* **33** (1985)
15. A.Antonelli et al. LNF 87-18(R) (1987); A.Antonelli et al. submitted to *IEEE Trans on Nucl. Sci.* (1992).
16. G.Battistoni et al. *Nucl.Instr. and Meth.* **164** (1979) 57; M.Spinetti, in *Hadronic Physics at Interm. Energies,* North-Holland (1987) p.409.
17. R.Santonico and R.Cardarelli, *Nucl.Instr. and Meth.* **187** (1981) 377.
18. M.Tavani and D.Zanello, Rome I University Int.Rep. 805 (1983).
19. H.C.Dehne et al., *Nucl. Instr. and Meth.* **116** (1974) 345
20. A complete review on J/y decays can be found in L.Kopke and N.Wermes, *Phys.Rep.* **174** (1979) 67.
21. J.D.Jackson LBL Intern. Note JDJ/74-3 2-12-1974
22. See for example J.P.Alexander et al. *Phys. Rev.* **D37** (1988) 56 and references therein.
23. H.J.Besch et al., *Phys. Lett.* **B78** (1978) 347.
24. I.Peruzzi et al., *Phys. Rev.* **D17** (1978) 2901; R.Brandelik et al., *Z. Phys.* **C1** (1979) 233; M.W.Eaton et al., *Phys. Rev.* **D29** (1984) 804, D.Pallin et al., *Nucl. Phys.* **B292** (1987) 653.
25. M.Claudson,S.L.Glashow and M.B.Wise, *Phys. Rev.* **25** (1982) 1345.

III. NUCLEAR PHYSICS

Common Problems and Trends of Modern Physics, pp. 241–251
ed. T. Bressani, S. Marcello and A. Zenoni
©1992 World Scientific Publishing Company

DECAY OF HYPERNUCLEI

TULLIO BRESSANI

Dipartimento di Fisica Sperimentale dell'Universita', Torino (Italy) and
Istituto Nazionale di Fisica Nucleare, Sezione di Torino (Italy)

ABSTRACT

The phenomenology of Λ-hypernuclei decay is shortly reviewed. Some emphasis is given to the possibility of a violation of the $\Delta T=1/2$ rule in non-mesonic decay of Λ-hypernuclei, as well to the possibility of π^+ mesonic decay, related to the H-particle existence. Some ideas on the realization of dedicated measurements at the Φ-factory DAΦNE are put forward.

1. Introduction

Hypernuclear Physics is an exciting area of nuclear science whose importance is increasing with time. Support to this assumption is the fact that two of the major projects of machines for the year 2000, KAON at Vancouver [1] and PILAC at Los Alamos [2], place Hypernuclear Physics at top priority in their scientific programs. At three previous courses of this School, mostly devoted to Intermediate Energy Hadronic Physics, I gave lectures devoted to various aspects of Hypernuclear Physics [3,4,5]. In this lecture I will present an aspect that was only marginally discussed in the past, that of the Λ-hypernuclei decay. The presentation will be at very elementary level, with hints of possible developements at the DAΦNE complex [6], the recently approved ($e^+ e^-$) collider with high luminosity at Frascati. The reader interested to a deeper insight into the field can find a complete and up-to-date review in the paper of Cohen [7].

2. Decay of Hypernuclei

A free Λ-hyperon decays into a nucleon and a pion through the weak non-leptonic decay:

$$\Lambda \left\langle \begin{array}{ll} \nearrow\; p + \pi^- & (\sim 64\%) \\ \searrow\; n + \pi^0 & (\sim 36\%) \end{array} \right. \tag{1}$$

The lifetime τ_Λ is 2.63±0.02 $*10^{-10}$ s, the energy released is Q = 37 MeV and the corresponding momentum of the nucleon is about 100 MeV/c, or approximately 5 MeV of kinetic energy.

The decay (1) may occur with an isospin change (ΔT) of 1/2 or 3/2. An s-quark converts to a u- (or a d-) quark through W boson exchange. The V-A Hamiltonian describing this decay can be written as:

$$\mathcal{H}_{V\text{-}A} = \frac{G_F}{\sqrt{2}} \sin\theta_c \cos\theta_c \bar{u}\gamma_\mu (1-\gamma_5)\, s\, \bar{d}\gamma^\mu (1-\gamma_5)\, u + c.c. \qquad (2)$$

θ_c being the Cabibbo angle. Such a model assumes equal contributions for $\Delta T = 1/2$ and $\Delta T = 3/2$ transitions. However, the experimental branching ratio (1), as well as the branching ratios of other strange particles decays, implies that the $\Delta T = 1/2$ amplitude is enhanced by at least one order of magnitude. It is immediate to see in fact that the experimental ratio between the two dominating decays (~ 1.9) is given approximately by the square of the Clebsch-Gordan coefficients in the $\Delta T = 1/2$ amplitude:

$$|1/2, -1/2\rangle = \sqrt{2/3}\ |p\,\pi^-\rangle - \sqrt{1/3}\ |n\,\pi^0\rangle \qquad (3)$$

The $\Delta T = 1/2$ amplitude is, more precisely, ~ 20 times larger than the $\Delta T = 3/2$ amplitude.

Λ-hypernuclei in their ground state, or in low-lying excited states with a γ-decay lifetime larger than the free Λ lifetime, decay also via weak-interaction mechanisms. However, the weak decay of the free Λ-hyperon changes totally when it is embedded in nuclear matter. The binding energy B_Λ increases linearly with A, reaching the saturation value of 23 MeV for heavy hypernuclei, and the phase space for the mesonic decay is greatly reduced. The final-state nucleon produced in the decay has a very low momentum (< 100 MeV/c), and the process is consequently Pauli-blocked. I recall that the nucleon Fermi momentum in a nucleus is ~ 270 MeV/c.

It has long time been recognized that, because the mesonic decay mode would be strongly inhibited in all but the lightest hypernuclei [8)], the primary decay channel would be the non-leptonic weak reaction $\Lambda N \rightarrow NN$, referred to as non-mesonic decay. The energy release is approximately 176 MeV, corresponding to a momentum of each of the final nucleons of ~ 417 MeV/c. This decay mode has a much larger phase space, compared to that of the mesonic mode, and the final nucleons are not Pauli-blocked. A measurement of the relative branching ratio for the non-mesonic channels:

$$\Lambda + p \rightarrow n + p \qquad (4)$$

$$\Lambda + n \rightarrow n + n \qquad (4')$$

could provide information about the structure of the weak Hamiltonian, in particular on the relative importance of the $\Delta T = 1/2$ and $\Delta T = 3/2$ amplitudes.

However, particular care must be devoted to many different effects related to the nuclear medium (initial and final state interactions, nuclear structure,...) that can strongly distort the true weak interaction mechanism. A very detailed discussion of all these effects can be found in Ref. 7.

3. Experimental data

In spite of the great physics interest for the decay of hypernuclei, the data up-to-now available are scarce and not precise. This is due to the tremendous difficulty in producing abundantly hypernuclei in their ground states and detecting the products of their decay.

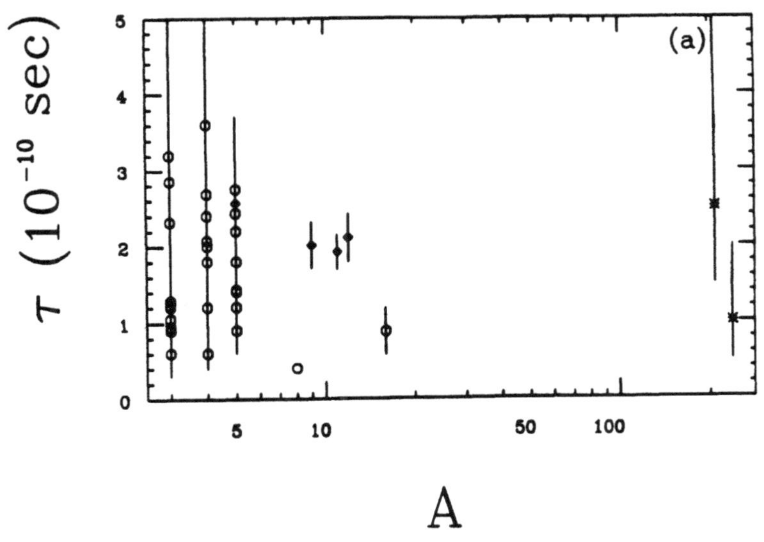

Fig. 1 Present experimental values of the lifetime τ of the Λ-hyperon in hypernuclei as a function of A (from Ref. 7)

Extremely hard is to study the non-mesonic decay (4') for which two fast neutrons have to be detected in coincidence and related to the hypernuclear ground states. Fig. 1, taken from Ref. 7, shows the values of the lifetime τ_Λ of the Λ-hyperon in various hypernuclei, as a function of the mass number A. One may observe the big errors affecting the measurements and the fact that, substantially, it is not possible with the present data to affirm that τ_Λ in nuclei is very different from the value for a free Λ ($2.63 * 10^{-10}$ s). It is interesting to note that the values of τ_Λ reported in Fig. 1 were obtained by quite different techniques. The values for the light hypernuclei ($A \leq 5$) are from old bubble chamber and emulsion experiments, a part a new recent measurement for $^5He_\Lambda$ from a counter experiment at the Brookhaven AGS [9]. The values for $9 \leq A \leq 12$ were also given by this last experiment. The lifetime for $^{16}O_\Lambda$, which appears to be the most different from that of free Λ, was performed by an unusual technique, by bombarding with a ^{16}O beam a polyethylene target and was affected by a large background [10]. Finally, τ_Λ for $^{209}Bi_\Lambda$ and $^{238}U_\Lambda$ was measured at CERN LEAR by studying the delayed fission following $\bar{p}$

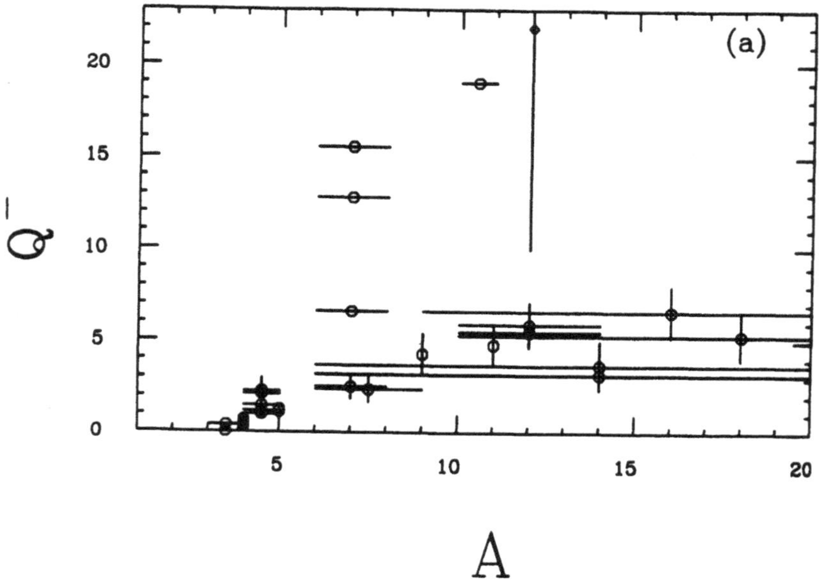

Fig. 2 Ratio of non-mesonic to π^- mesonic decay rates as a function of A (from Ref. 7).

annihilation at rest [11]. The most recent analysis [12] of this experiment has provided for the lifetime of heavy hypernuclei a value of $(1.25 \pm 0.15) * 10^{-10}$ s. In view of the small error this experiment can be considered as the first one providing strong evidence for a decrease of τ_Λ in nuclear matter.

Fig. 2 shows the ratio Q^- of non-mesonic to π^- mesonic decay as a function of A. Again one may notice the large errors affecting the measurements, relevant also to the mass number of the decaying hypernucleus. However, a genereal trend showing the dominance of non mesonic decays over the mesonic ones for $A \geq 3$ is evident. It is expected that the ratio between the π^- and π^0 mesonic modes could be modified with respect to the free Λ ratio (1) in light nuclei due to the Pauli blocking, more effective for a proton than a neutron and vice-versa. A systematic study of the mesonic decay for the p-shell hypernuclei was performed by Itonaga [13], and a measurement of the π^0 decay of $^{12}C_\Lambda$, recently performed at KEK [14], though affected by a large error, was in agreement with the calculation.

Fig. 3 shows the ratio n of the $\Lambda + n \rightarrow n + n$ non-mesonic decay to the total

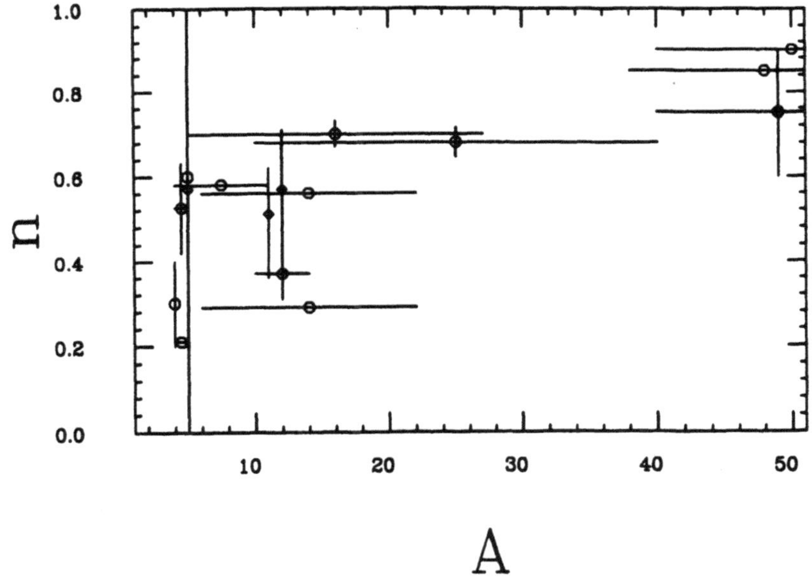

Fig. 3 Ratio of neutron stimulated non-mesonic decay rate to the total non-mesonic decay rate as a function of A (from Ref. 7).

non-mesonic decay as a function of A. The errors are very large, since we are dealing with the decay experimentally most difficult to detect. An increase of n with A is apparent.

A very recent analysis of the neutron stimulated non-mesonic decay for light hypernuclei ($^4H_\Lambda$, $^4He_\Lambda$ and $^5He_\Lambda$), performed by Schumacher [15], showed an apparent violation of the $\Delta T = 1/2$ rule. In light hypernuclei, $^1S_0$ and $^3S_1$ initial ΛN spin configurations with isospin 1/2 can decay non-mesonically to six NN final state spin configurations with isospin of either 1 or 0. For any reaction mechanism leading to T = 1 final states, the ratio of neutron-neutron decay (R_n) to proton- neutron decay (R_p) is given by:

$$\frac{R_n}{R_p} = \frac{\lambda^2 + 4 - 4\lambda}{2\lambda^2 + 2 + 4\lambda} \tag{5}$$

where λ is the ratio of the isospin 3/2 decay amplitude to the isospin 1/2 decay amplitude. Defining furthemore $R_{N\,S}$ the rate for the $\Lambda N \to NN$ decay from spin state S, it follows that for the pure $\Delta T = 1/2$ interactions we have $R_{n0} = 2R_{p0}$ (T = 1 final state only) and $R_{n1} \le 2R_{p1}$ (both T = 0 and 1 final states). Hence the quantity which is obtained by Schumacher is the ratio $r = R_{n0}/R_{p0}$, which for pure $\Delta T = 1/2$ interactions must be 2, and for pure $\Delta T = 3/2$ interactions would be 1/2. He finds from the data $r = 0.20 \pm 0.14$, suggesting a rather large violation in the case of light hypernuclei. However, the above error was obtained treating the errors on the experimental values as independent and uncorrelated. A more careful analysis of the possible correlation between the errors leads to the conclusion that the pure $\Delta T = 1/2$ hypothesis is excluded at about the 2σ level.

This remark strongly supports the interest in performing accurate measurements on the decay modes of hypernuclei.

4. There is a π^+ decay mode of hypernuclei ?

The H particle has ben the subject of theoretical and experimental studies following the Jaffe proposal [16] based upon the MIT bag model. A recent careful review on the H particle was performed by Guaraldo [17]. Although there has been no evidence of the existence of the H, no experiment can exclude the possibility of strongly bound H particle. Recent lattice QCD calculations [18], performed with the largest lattice size even used for the H mass, shows that the mass could be quite light, even lighter than that of two nucleons. In that case the H becomes stable not only against strong decays but also weak decays. On the other hand, if the H particle is lighter than two nucleons, a nucleus becomes unstable against $\Delta S = -2$ double β decay producing the H as final product. The most stringent lower limit on the H mass, as given by double β decay is 1800 MeV.

Kishimoto [19] has pointed out that there is a window in the H mass (1870-1880 MeV), for which the existence of the H could manifest itself in a new hadronic weak decay mode, the decay with a π^+ in the final state. The elementary reactions would be:

$$\Lambda + p \rightarrow H + \pi^+ \tag{6}$$

$$\Lambda + n \rightarrow H + \pi^0 \tag{6'}$$

The resulting π meson would have, assuming for example a mass of 1850 MeV for the H, ~ 64 MeV for the decay (6), ~ 70 MeV for the decay (6'). The important point in the Kishimoto's observation is that the presence of monochromatic π^+ following the decay of

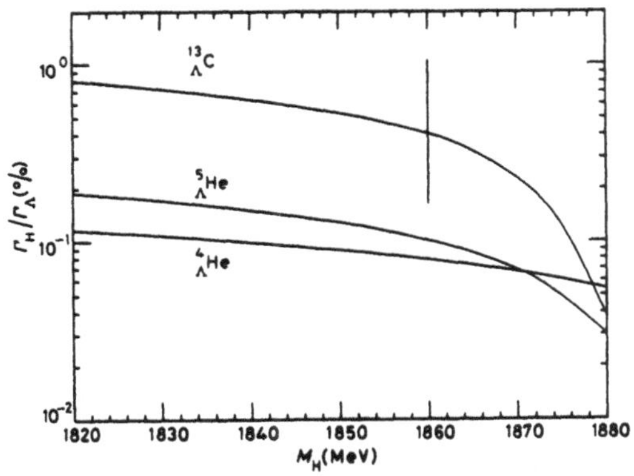

Fig. 4 Decay rate of π^+ mesonic decay as a function of the H mass, for the hypernuclei $^{13}C_\Lambda$, $^5He_\Lambda$ and $^4He_\Lambda$. The ordinate scale is in (%) of the total decay rate (from Ref. 19).

hypernuclei would be an unique signature for the existence of the H and would at the same time provide a measurement of its mass. The decay (6'), with a π^0 in the final state, is difficult , if not impossible, to utilize to this purpose, since conventional mesonic decay modes, though strongly suppressed as discussed in Sec. 2 and 3, are competing.

Kishimoto evaluated the decay rates for the decay (6) and the results are shown by Fig. 4, as a function of the H mass, for $^4He_\Lambda$, $^5He_\Lambda$ and $^{13}C_\Lambda$. The obtained π^+ decay rate is ~ 10^{-2} - 10^{-3} of that of free Λ decay when the H mass is below 1870 MeV, a value within the reach of spectrometers and machines presentely available.

5. Measurement of Hypernuclear Decays with FINUDA

At the previous course of this School I presented [5] a proposal for a detector (FINUDA) dedicated to experiments on Hypernuclei production with K^- at rest, to be installed at the Frascati Φ-factory DAΦNE. The detector is particularly well suited for the detection of the decay products from Hypernuclei in their ground state. The reason stems in its wide angular acceptance ($\sim 2\pi$) and on the fact that the source of particles is at the center of the apparatus.

Let me briefly recall the features of FINUDA [20, 21], whose scheme is shown by Fig. 5. It consists of a solenoid with a radius of ~ 1 m and a lenght of ~ 2 m, with a maximum field intensity of 1.5 T at the center. The interaction region of the (e^+, e^-)

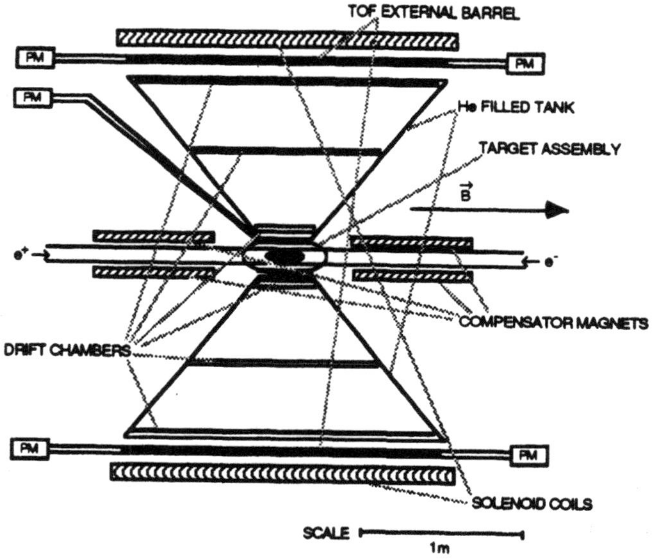

Fig. 5 Layout of the proposed magnetic spectrometer FINUDA.

colliding beams, at the center of the apparatus, is surrounded by a cylindrical nuclear target in which the 15 MeV K^- from the Φ-meson decay are stopped. The target is sandwiched between layers of thin detectors: a scintillator barrel for first level trigger purposes and microstrip chambers that provide the first point on the charged particle's trajectories (fast π^- from the hypernuclei formation reaction, π^- and protons from the hypernuclei decay).

Three layers of thin-walled cylindrical or poligonal drift chambers allow the tracking of the particles, with an estimated momentum resolution $\Delta p/p$ of ~ 0.2% fwhm. The outermost detector is a cylindrical barrel of scintillators (~ 80 elements, 200*80*10 cm^3 each) viewed at both ends by fast photomultipliers. It will allow Time-of-Flight measurements, provide the first level trigger and detect neutrons from non-mesonic decays.

With this apparatus I expect a production rate of 50 ev/hour for a single hypernuclear final state produced at the rate of 10^{-3} stopped K^-, without tagging the event with the K^+, for a luminosity L of the machine of 10^{32} cm^{-2} s^{-1}. The event rate drops to 20 ev/hour if the tagging with the K^+ is required. At the design luminosity value $L = 10^{33}$ cm^{-2} s^{-1} the event rate will obviously increase by an order of magnitude.

The apparatus was optimized in order to achieve the best momentum resolution on the π^- from the formation reaction, but is also the best one for studying the hypernuclei decays. The direct measurement of the hypernuclear life-time could be performed by means of the internal barrel of thin scintillators (8 or 16), surrounding the target and of small dimensions (10-16 cm). A time resolution of less than 100 ps fwhm is within the limits of the present technology, and the simultaneous measurement of the relative time differences among the tagging K^+, the π^- from the formation reaction (prompt) and the π^- or proton from the hypernuclear decay (delayed), with good statistics, will allow a determination of the lifetime τ_Λ with an error of less than 5%.

As said before, the momenta of the π^- from mesonic decays and of the proton from non-mesonic decays can be easily determined by means of the magnetic spectrometer. The neutrons will be detected by the outer barrel of scintillators, with an efficiency of ~ 10%. By adding two scintillator's walls, of the same thickness (10 cm), orthogonal to the axis of the solenoid and enclosing it, a solid angle of ~ 4π for neutron detection will be reached.

For the measurement of the hypernuclear decay the tagging with the K^+, useful in order to reach the best momentum resolution, will be probably not necessary. In fact it will be enough to have a momentum resolution sufficient to isolate the ground state. I expect then, at $L = 10^{32}$ cm^{-2} s^{-1}, 3 ev/hour for the $\Lambda + p \rightarrow n + p$ non mesonic weak interaction and 1 ev/hour for the $\Lambda + n \rightarrow n + n$ decay mode. I recall that the counter experiment that provided the only measurement on non-mesonic decays obtained less than 10 pairs of neutrons in coincidence in total, for several weeks of data taking. FINUDA will provide the same number in 10 hours at $L = 10^{32}$ cm^{-2} s^{-1} in 1 hour at $L = 10^{33}$ cm^{-2} s^{-1}. This enormous difference is not only due to the hypernuclei production rate, but to a great extent to the large solid angle of detection and to the fact that the same apparatus is used both for the spectroscopy of the π^- from the hypernuclei formation reaction and for the detection of the decay products. In the BNL apparatus these tasks were performed by different detectors, subtending small solid angles.

Concerning finally the possible detection of π^+, related to H-particle production, the crucial point is the evaluation of the background of π^+ following K^- capture in nuclei and due to the decay of quasi-free Σ^+ produced in the reaction $K^- + p \rightarrow \Sigma^+ + \pi^-$ on protons of the target. By extrapolating down to 100 MeV/c the experimental π^+ spectra reported by Hayano et al.[22] for a ^{4}He target, we may estimate that the background contribution will be the same than a signal due to a π^+ production rate, via H decay, of 10^{-5} /stopped K^-. The experiment seems feasible!

6. Conclusions

I have shown that a dedicated experimental study of the decay of Λ-hypernuclei may give access to fundamental problems of particle physics. In particular, a study of the neutron stimulated non-mesonic decay rate compared to the total non-mesonic decay rate may shed light on the nature of the empirical $\Delta T = 1/2$ rule for weak interactions. Present data suggest, at a 2σ level, a violation of such a rule and a strong research program on several nuclear targets is necessary in order to confirm this result. Furthermore, a new hadronic decay of Λ in nuclei, the π^+ mesonic one, could be the manifestation of the existence of the H particle, if its mass is in the range (1800-1870) MeV.

The recently approved Φ-Factory at Frascati and the FINUDA detector, under design, are the best machine and apparatus, at present, for such a research program.

7. References

1. KAON Factory Proposal, edited by TRIUMF (Vancouver, Canada), 1985.
2. PILAC User Group Report on the Physics with PILAC, Rep.LA-UR-95-150, Los Alamos, U.S.A., 1992.
3. T. Bressani, in *Hadronic Physics at Intermediate Energy*, eds. T. Bressani and R. A. Ricci (North Holland, Amsterdam, 1986), p. 223.

4. T. Bressani, in *Hadronic Physics*, eds. R. Cherubini, P. Dalpiaz and B. Minetti (North Holland, Amsterdam, 1989), p. 151.
5. T. Bressani, in *Common Problems and Ideas of Modern Physics* , eds. T. Bressani, B. Minetti and A. Zenoni (World Scientific, Singapore, 1992), p. 211.
6. M. A. Preger, in *Common Problems and Ideas of Modern Physics* , eds. T. Bressani, B. Minetti and A. Zenoni (World Scientific, Singapore, 1992), p. 277.
7. J. Cohen, in *Progress in Particle and Nuclear Physics*, **Vol. 25** (Pergamon Press, 1990), 139.
8. W. Cheston and H. Primakoff, *Phys. Rev.* **92** (1953), 1537;
H. Primakoff, *Nuovo Cimento* **3** (1956), 1394;
M. Rudermann and R. Karplus, *Phys. Rev.* **102** (1956), 247.
9. J. J. Szymansky et al. *Phys. Rev.* C **43** (1991), 849.
10. K. J. Nield et al. *Phys. Rev.* C **13** (1976), 1263.
11. J. P. Bocquet et al. *Phys. Lett. B* **182** (1986), 146;
J. P. Bocquet et al. *Phys. Lett. B* **192** (1987), 312.
12. T. A. Armstrong et al. CERN-PPE/92-106 (1992).
13. K. Itonaga, *Nuovo Cimento* **102A** (1989), 501.
14. M. Sakaguchi et al., *Nuovo Cimento* **102A** (1989), 511.
15. R. Schumacher, in *Proc. 1991 20th INS Symposium on Hypernuclear and Strange particle Physics, Shimoda (Japan), December 9-12, 1991*, to be published in *Nucl. Phys. A* (Abstract C-8).
16. R. L. Jaffe, *Phys. Rev. Lett.* **38** (1977), 195, 1617 (E).
17. C. Guaraldo, in *Hadronic Physics at Intermediate Energy II* , eds. T. Bressani B. Minetti and G. Pauli (North Holland, Amsterdam,1987), p. 139.
18. Y. Iwasaki, T. Yoshie and Y. Tsuboi, *Phys. Rev. Lett.* **60** (1988), 1371.
19. T. Kishimoto, *Nuovo Cimento* **102A** (1989), 553.
20. M. Agnello et al., in *Proc. 1991 20th INS Symposium on Hypernuclear and Strange particle Physics, Shimoda (Japan), December 9-12, 1991*, to be published in *Nucl. Phys. A* (Abstract C-8).
21. P. Gianotti and E. Rossetto, this volume.
22. R. S. Hayano et al., *Nuovo Cimento* **102A** (1989), 437.

Common Problems and Trends of Modern Physics, pp. 253–267
ed. T. Bressani, S. Marcello and A. Zenoni
©1992 World Scientific Publishing Company

η PRODUCTION BY PROTON INTERACTION NEAR THRESHOLD

E. CHIAVASSA and E. SCOMPARIN
Dipartimento di Fisica Sperimentale Universita' di Torino - INFN Torino

ABSTRACT

η production by nucleon-nucleon and proton-nucleus interaction near the threshold is reviewed. A short summary of the subthreshold p-A production measurements done at SATURNE with PINOT spectrometer is presented. Data are discussed in the frame of the folding model.

1. Introduction

The η meson has a mass of 548.8 MeV, isospin I=0, lifetime of $0.75 \cdot 10^{-18}$ s, contains $s\bar{s}$ pair in addition of $u\bar{u}$ and $d\bar{d}$ pairs and its interaction with nucleons is weaker than π-N interaction.

Its lifetime, even if too short for doing η beams, is long enough to allow a path-length greater than the nucleus radius and therefore its production in nuclei is a tool for studying η-N interaction [1].

In particular it has been realized [2,3] that near the threshold the ηN system is dominated by the S_{11} $N^*(1535)$ resonance that decays in 50% of the cases via η emission. Unfortunately the parameters of this resonance are not very well known [4], but using its main characteristics an optical η-nucleus potential has been evaluated and an η-nucleus bound state has been conjectured [5].

Furthermore the η production in nucleus-nucleus interaction has been proposed as a way to obtain information about high density nuclear matter [6].

Experiments with π and protons [1,7,8] have been reported, with results proving the interest in studying the η production.

In this context a program for measuring η production near the proton threshold has been developed at SATURNE using the two γ rays spectrometer PINOT[9]. In this lecture we will expose the motivation, the experimental method and the results of this investigation.

2. p-N and p-A η production near the threshold

2.1. η production on nuclei

The production of η in nuclei by means of proton interaction has been studied at SATURNE in light nuclei detecting the residual charged nucleus generated in the reaction. In this way the excitation function for the pd$\to^3$Heη reaction has been measured in CM forward and backward regions [7] for proton kinetic energies from 0.9 to 2.6 GeV in lab. frame. With the same technique the dp$\to^3$Heη has been studied near the threshold[8]. Between other interesting results the last experiment proved that near the threshold the amplitudes of the dp→^{3}Heη and dp→^{3}Heπ^0 are similar despite of the greater momentum transfer involved in the η production. Different theoretical explanation of this behaviour have been proposed [2,3]; in ref.3 a dominance of 3 body mechanisms in the reaction has been considered.

Concerning η production in heavier nuclei, data come from the PINOT experimental program that will be discussed in sect. 3; in the meantime two theoretical calculations have been developed [10,11].

In the first one Cassing et al. [10] evaluated the η production in nuclei by means of the folding model used to fit data on the K$^+$ production. In this model the differential cross sections for η production is given by :

$$E_\eta d^3\sigma_\eta/d^3k = N_1(A)\cdot[(E_\eta d^3\sigma_\eta/d^3k)_{prim} + (E_\eta d^3\sigma_\eta/d^3k)_{sec}\cdot C(A)]$$

$$\cdot exp(-r_0 A^{1/3}/\lambda_\eta(k))$$

in which two processes are distinguished. The primary process refers to the η production in pN→pNη reaction while the secondary production is due to the πN→ηN reaction induced by pions originated in a NN primary interaction. $N_1(A)$ denotes the number of first chance NN collisions and is determined in the Glauber approximation;

$$C(A) = \frac{(R^2 - \lambda^2)/4}{R^2}$$

accounts for the probability that a pion (produced in the first collision) rescatters in the finite volume of the nucleus;

$$\lambda = (\rho_0\sigma_{pN})^{-1} + (\rho_0\sigma_{\pi N})^{-1}$$

where ρ_0 is the nuclear density. Finally

$$\lambda(\eta) = [\sigma_{\eta N}^{abs}(k)\rho_0]^{-1}$$

In the calculation of the two contributions the elementary NN processes are considered taking into account the nucleon momenta distribution given by an Hartree-Fock calculation. All the necessary elementary cross sections are obtained parametrizing the existing experimental data, that are in the η case very poor. The η reabsorption is evaluated considering a dominance of the N*(1535) production in the ηN process and using the detailed balance to extract the $\sigma_{\eta N}^{abs}$ from the $\sigma_{\pi N\to\eta N}$ measured value. The elastic scattering of the η is not taken in consideration.

The second theoretical evaluation of η production is due to a russian group and is based on an intranuclear cascade model [11]. In this model the η production originates from a series of successive quasifree collisions of primary and secondary particles with the nucleons of the nucleus. The nucleus is considered as a neutron and proton Fermi gas enclosed in a spherical potential well with diffusive boundary. The cross sections for the nucleon-nucleon interaction are considered to be the same as in vacuum; the

Pauli principle is applied to block the interaction in the nuclear matter. All the elementary cross sections are obtained fitting the existing experimental data.

Before discussing the experimental results with respect to the theoretical predictions, let us examinate the present knowledge of the elementary NN→NNη reaction.

2.2. The pp→ppη reaction

A quick overlook to experimental data shows that there is no information from the threshold (T_p=1.26 GeV) to T_p=2 GeV. De Paoli et al.[12] suggest a parametrization of the experimental data via a fit to the functional form

$$\frac{A(\sqrt{s}-\sqrt{s_0})}{B+(\sqrt{s}-\sqrt{s_0})^2}$$

This fit (see fig. 1) is rather satisfactory for the high energy points $T_p \geq 3$ GeV) but its behaviour close to threshold is clearly an extrapolation; the same functional form fits rather well the data for the reaction pp→ppπ^0 also close to threshold.

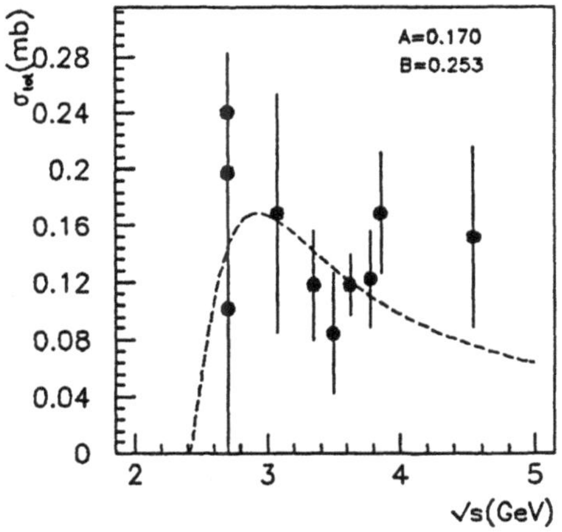

Fig. 1. Experimental data on pp→ppη reaction, and parametrization by De Paoli et al.

The raising interest for recent experiments of η production in nuclei has now lead to the formulation of theoretical microscopic calculations for the pp→ppη reaction; three models have been presented till now[13,14,15]. All of them are based on a perturbative approach and assume the dominance of N*(1535) in the production mechanism; i.e. the nucleon resonance is excited via one-boson-exchange, and the decay of the resonance produces the η. Nevertheless they reach different conclusions with regard to magnitude and energy dependence of the calculated cross sections.

The model of Germond and Wilkin[13] studies the reaction pp→ppη near threshold, in the energy region where the emitted protons are in a relative $^1S_0$ state, which they call d*; the η is supposed to be in s-wave with respect to d*. The exchanged meson can be ρ, π, η (the ω is not supposed to contribute because it does not seem to couple appreciably with the N*(1535)); the ρ exchange term dominates. The results of the calculation depend strongly on the choose of the wave function for the d* state. Using a soft-core potential for d*, with high energy distortion and Coulomb repulsion in the d* wave functions, these authors calculate the cross section plotted in fig. 2 in the case of pure ρ exchange (other mesons exchange contributes no more than 20% to the total cross sections). The results indicate a very fast increase of cross section values with energy. The energy range plotted corresponds to $1.258 \leq T_p \leq 1.271$ GeV, i.e. the model provides us with cross section values very near threshold.

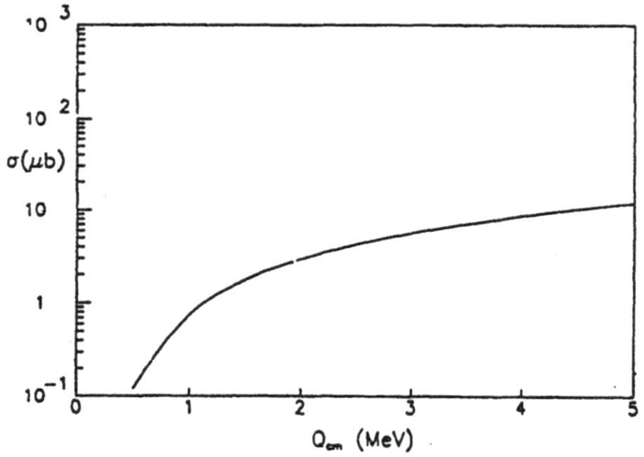

Fig. 2. Cross section for pp→ppη reaction, calculated by Germond and Wilkin

The other two models investigate the cross section behaviour also at higher energies. The model proposed by Vetter et al.[14] describes the nucleon-nucleon interaction by means of a lagrangian which contains the interactions vertices NNπ, NNσ, NNρ, NNω, NNη, NN*π, NN*σ, NN*ρ, NN*ω and NN*η. The coupling constants in the sector NNx are calculated fitting the experimental data on nucleon-nucleon scattering; the NN*π and NN*η coupling constants are obtained fitting the partial decay widths of the N*(1535). The NN*ρ vertex is calculated taking into account the 5% B.R. of N*(1535) going to Nρ , while NN*ω coupling constant is obtained supposing that

$$g_{NN^*\pi}/g_{NN\pi} = g_{NN^*\omega}/g_{NN\omega}$$

The analogous coupling constant for σ turns out to be negligible. The processes taken into account are sketched in fig. 3. The calculated cross section (fig. 3) rises very fast with energy near threshold as in Germond and Wilkin model, but now the contribution of the ρ, even if important, is not so dominant. π and ω exchange contributes with nearly equal importance. The different weight of various mesons contribution originates because Germond and Wilkin do not include ω contribution while Vetter et al. do, and because Germond and Wilkin deduce the ρ coupling in the vector dominance hypothesis, while Vetter et al. rely on the B.R. value for $N^*(1535)\to N\rho$.

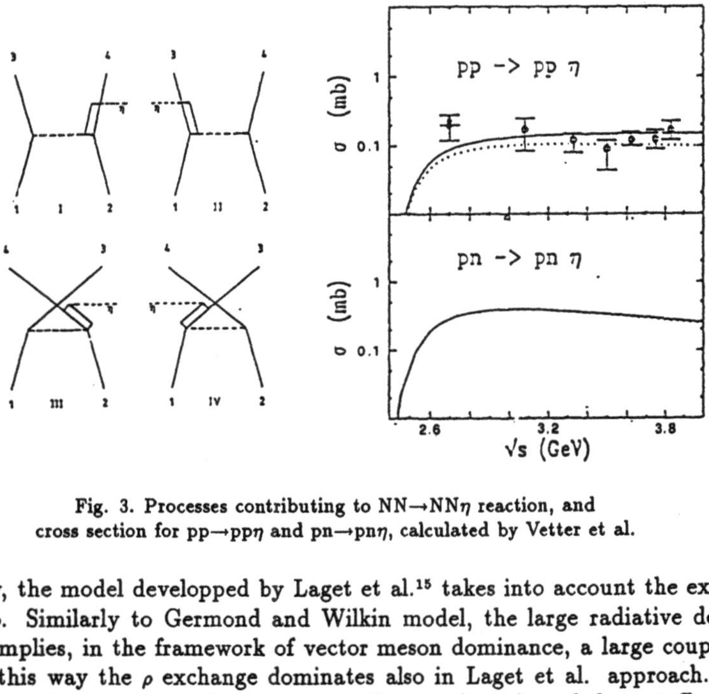

Fig. 3. Processes contributing to $NN\to NN\eta$ reaction, and cross section for $pp\to pp\eta$ and $pn\to pn\eta$, calculated by Vetter et al.

Finally, the model developped by Laget et al.[15] takes into account the exchange of π and ρ. Similarly to Germond and Wilkin model, the large radiative decay of $N^*(1535)$ implies, in the framework of vector meson dominance, a large coupling to the ρ. In this way the ρ exchange dominates also in Laget et al. approach. Form factors are used at meson-nucleon vertices. Tuning the value of the cutoff mass in the ρ dipole form factor to the few experimental values, Laget obtains a cross section (fig. 4) which rises faster than Vetter one. Its value , near treshold , is about a factor of 3 higher than Vetter's.

2.3. The $np\to\eta X$ reaction

Two neutron induced reaction can produce η: $np\to np\eta$ and $np\to d\eta$. For the former no experimental data exist, while for the latter an analysis of old $np\to dX$ data[16] has allowed to calculate the cross section very near threshold, obtaining $\sigma_{tot}= 100\ \mu b$.

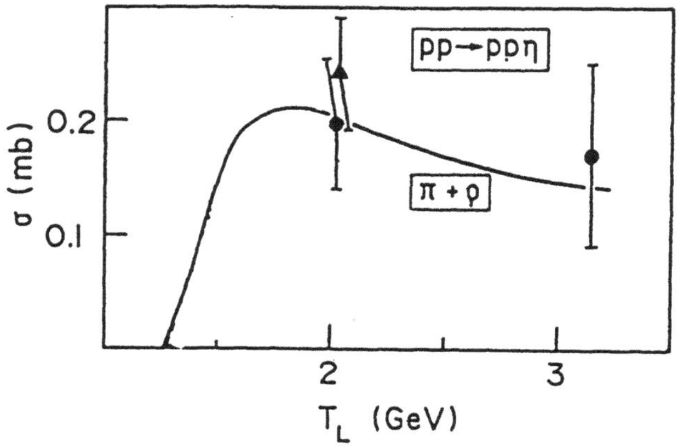

Fig.4. Cross section for pp→ppη reaction, calculated by Laget et al.

The theoretical models described before give also predictions for neutron induced η production. All of them agree on the fact that, mainly for isospin reasons, this reaction should have higher cross sections than the proton induced one. Germond and Wilkin calculate values for the np→dη that are in good agreement with the only experimental data point and predict values one order of magnitude less for pp→ppη. Laget et al., calculating np→dη under the same assumptions of the proton induced reaction, agree with Germond and Wilkin about the expected ratio between np→dη and pp→ppη. The calculus by Vetter et al. is different, because they calculate cross sections for np→npη reaction. Anyway they agree with the two other models on the fact that the difference in magnitude for the cross sections of the two reactions is mainly due to isospin, but the ratio np→npη over pp→ppη varies from a value of about 4 near threshold decreasing towards high energy (fig. 3).

In conclusion, the overlook on the actual knowledge of proton and neutron induced η reaction on nucleons presents a lack of experimental data that has to be compared with three very accurate calculations. The actual situation is such that there is a great incertitude about the pp→ppη and pn→pnη cross sections, but in any case it seems that near the threshold considerable differences between proton and neutron production exist. Experimental measurements of these reactions are needed and are in program at Saclay [17].

3. Measurements with PINOT spectrometer

The 2γ rays spectrometer PINOT has been built in Turin and installed at SAT-

URNE in Saclay to perform a program of measurements on proton η production. The spectrometer profites of the fact that in 39% of the cases the η meson decays producing two γ rays. The two identical arms of the spectrometer look at a target located in a proton beam where the η mesons may possibly be produced; the energies E_1 and E_2 of the two γs are measured and their relative angle ψ determined. The two photon invariant mass, defined as $M_{\gamma\gamma} = 2sin(\psi/2)\sqrt{E_1E_2}$ is then reconstructed and the η recognized; then the meson kinetic energy T is calculated using the expression

$$T = \frac{M_0}{sin(\psi/2)\sqrt{1-X^2}} - M_0$$

where M_0 is the η mass, $X=(E_1-E_2)/(E_1+E_2)$ is the asimmetry parameter.

To realize this goal the two arm of PINOT are made of a couple of thin active converters followed by thin plastic scintillators, lucite Cerenkov counters and a calorimeter of thick blocks. One arm is sketched in fig. 5.

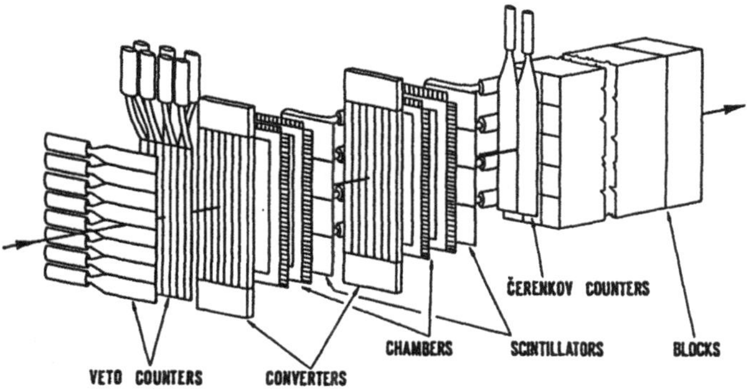

Fig. 5. Pictorial view of one arm of PINOT spectrometer

The converter consists in seven bars of SCG1-C glass scintillator of $5{\cdot}70{\cdot}4.5$ cm^3; its thickness corresponds to one radiation length. The shower that develops in one of the two converters stops in the calorimeter made of 15 blocks $15{\cdot}15{\cdot}60$ cm^3. The scintillators after each converter assure a good time coincidence between the two arms. The Cerenkov before the blocks identify without ambiguity the showers. The most relevant characteristics for η detection of the instrument are reported in table 1 when the two arms are symmetrically positioned at a distance of 1.4 m from the target and at an angle $\xi=66^0$ from the beam axis.

Table 1. PINOT main characteristics.

γ ENERGY RESOLUTION	$\sigma_\gamma = 0.8\sqrt{E}$ (E in MeV)
η INVARIANT MASS RESOLUTION	$\sigma = 20$ MeV
η ENERGY INTERVAL MEASURED	0 - 300 MeV
η ANGULAR RESOLUTION (energy dependent)	$\sigma_{\theta_\eta} = 2^0$ - 8^0
η KINETIC ENERGY RESOLUTION (energy and X dependent)	$\sigma_{T_\eta} = 2$ - 10 MeV for $\lvert X\rvert \leq 1$
EFF. ACCEPTED SOLID ANGLE (energy and X dependent)	2 - 10 msr for $\lvert X\rvert \leq 0.3$
η DETECTION EFFICIENCY	20 %
η ENERGY INTERVAL ACCEPTED	$\pm$ 30 MeV for $\lvert X\rvert \leq 0.3$

More detailed description of the spectrometer may be found elsewhere [18]. The spectrometer has been installed at the line 5 of SATURNE that produces a proton beam with a maximum energy of 1 GeV. In this line therefore only the subthreshold production in nuclei can be studied. Presently the spectrometer is operating on a line where a maximum proton energy of 2.5 GeV is available allowing to study the near threshold η production on nucleons and nuclei.

Subthreshold measurements have been done on Li, C, Al, Cu and Au with target thicknesses ranging from 0.6 to 4 g/cm^2. The proton beam intensity was about 10^9 p/s. In fig. 6 we present a $M_{\gamma\gamma}$ invariant mass spectrum where is clearly visible the peak due to η mesons. The low energy peak in the spectrum is due to π^0 while the large continuum contribution between the 2 peaks is generated by γ from multiple π^0 production.

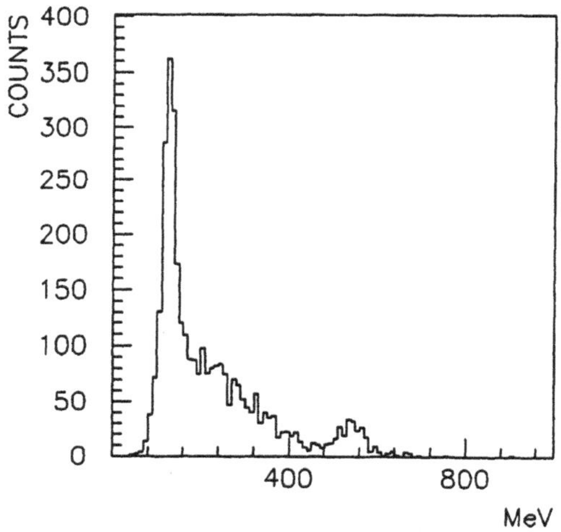

Fig. 6. $\gamma\gamma$ **invariant mass spectrum.**
The process is p+Al$\rightarrow\gamma\gamma$+X , with T_p = 1 GeV.

In table 2 we give a summary of the measurements until now done at SATURNE using nuclear targets.

Table 2. Measurements done at SATURNE with PINOT spectrometer.

Reaction $p+A \to X+\eta$	Proton energy(MeV)	ηdetected energies(MeV)	Angular range	PINOT geometry
Li	900	40 – 80	$0^0 - 40^0$	$\xi = 66^0$
Li	1000	40 – 80	$0^0 - 40^0$	$\xi = 66^0$
C	800	40 – 80	$0^0 - 40^0$	$\xi = 66^0$
C	900	40 – 80	$0^0 - 40^0$	$\xi = 66^0$
C	1000	40 – 80	$0^0 - 40^0$	$\xi = 66^0$
Al	1000	40 – 80	$0^0 - 40^0$	$\xi = 66^0$
Cu	900	40 – 80	$0^0 - 40^0$	$\xi = 66^0$
Cu	1000	40 – 80	$0^0 - 40^0$	$\xi = 66^0$
Au	1000	40 – 80	$0^0 - 40^0$	$\xi = 66^0$
B	1000	20 – 250	$0^0 - 20^0$	$45^0 \le \xi \le 80^0$

From these measurements we were able to extract the following information:

1) The energy dependence of the inclusive double differential cross section on ^{12}C at θ_η=30^0 and T_η=60 MeV.

2)The A dependence of the inclusive double differential cross section for proton energies of 900 and 1000 MeV and θ_η=30^0 and T_η=60 MeV for nuclei ranging from Li to Au.

The η detection probability for measurements 1) and 2) is reported in fig. 7.

3)The energy spectrum of η mesons produced in B by 1000 MeV protons in the angular range between 0^0 and 20^0. To perform this measurement the opening angle of the two arms of the spectrometer was moved from ξ=45^0 to ξ=80^0.

We try to compare these data with the theoretical previsions of the folding model using the code kindly made available to us by W.Cassing [10]. In fig. 8 we present our data compared with the previsions of the model.

It clearly appears that all our experimental data are underestimated. In particular the behaviour with A seems not to be well reproduced. This fact has been considered by the Torino group due to the too crude assumptions done in the folding model to treat the η reabsorption. A model has been developed therefore to calculate the η generation point in the target nucleus using Glauber theory and assuming a Saxon-Wood nuclear density distribution for A$\geq$16 and an armonic oscillator density for lighter nuclei [19]. In figure 9 the predictions of the folding model corrected for this η reabsorption calculation are reported. It is evident the better agreement with the data, in particular about the behaviour of the cross sections versus A.

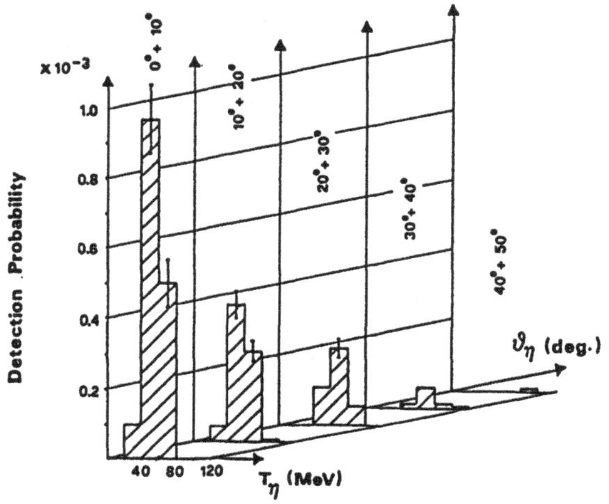

Fig. 7. η detection probability when the two arms are placed symmetrically with respect to beam axis, at $\xi=66^0$.

Finally, it is clear that the absolute magnitude of the cross sections is not well reproduced. This quantity depends mainly on the values elementary cross sections for the reactions pp→ppη and np→npη. The absolute magnitude of our measured cross sections can be reproduced fairly well if, in the frame of the folding model, one assumes for pp→ppη reaction the behaviour proposed in Laget calculation, and for np→npη the same energy behaviour and a factor 4 in the absolute magnitude (see fig. 10). This factor is predicted by Vetter et al. in his np→npη calculation near threshold.

The agreement between the data and the theoretical calculation is quite good in the case of A dependence of doubly differential cross section at T_p=1000 MeV and T_p=900 MeV. The trend of the excitation curve on ^{12}C is reproduced and this fact could indicate the fact that the Fermi motion of nucleons, which becomes an essential feature at very subthreshold incident proton energies, is well treated. On the contrary the energy spectrum of η, measured on B, is not well reproduced. We must however remark that the authors of the folding model[10] underline the difficulties in interpretating differential η spectra. In their opinion, this fact is principally due to the lack of experimental data about the elementary reactions pp→ppη and np→npη reaction close to threshold.

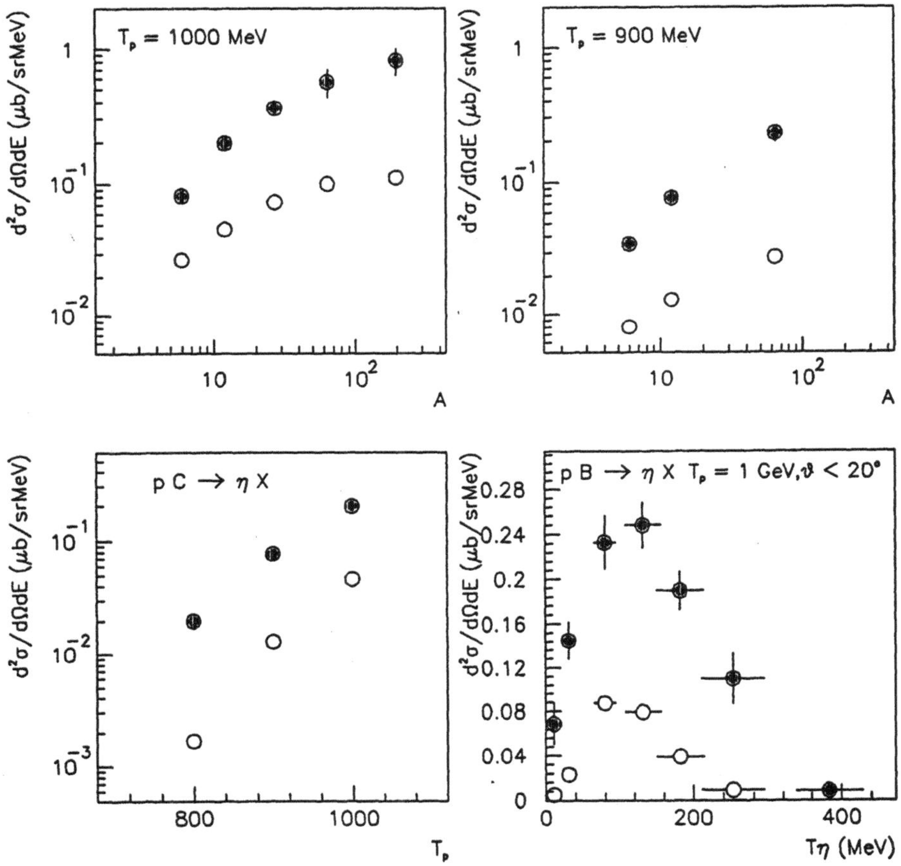

Fig. 8. PINOT subthreshold η production data.
Close symbols are experimental points,
open symbols are the predictions of the model by Cassing et al.

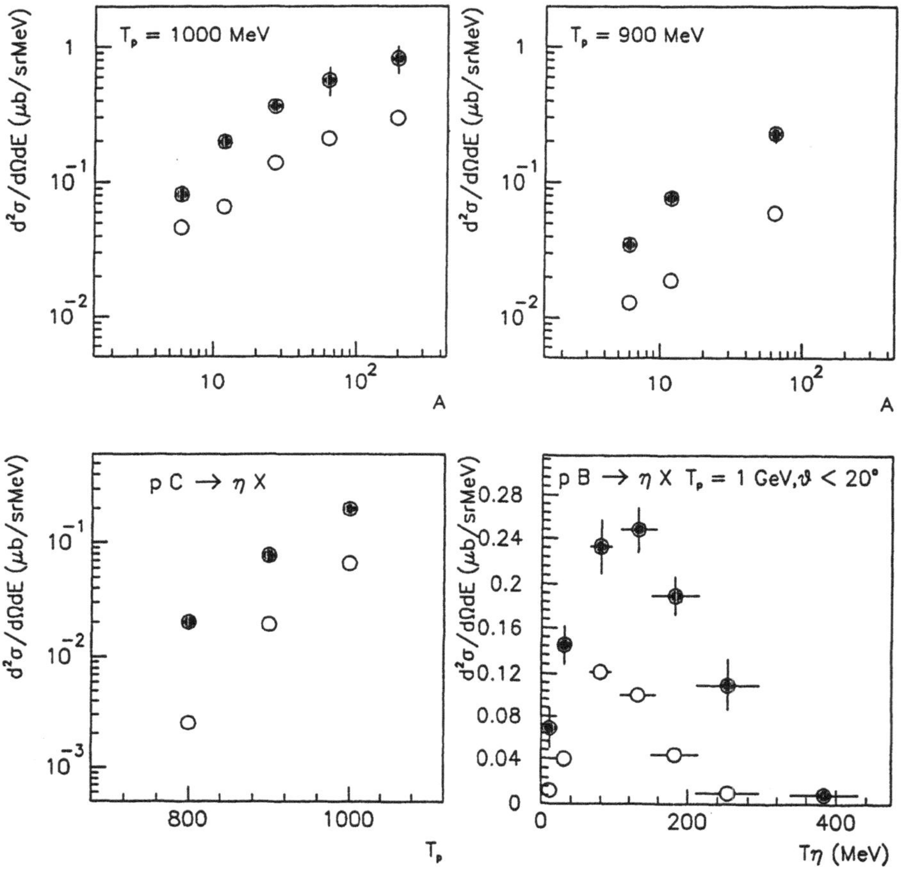

Fig. 9. PINOT subthreshold η production data.
Close symbols are experimental points,
open symbols are the predictions of the model by Cassing et al.,
with more realistic calculation of η absorption

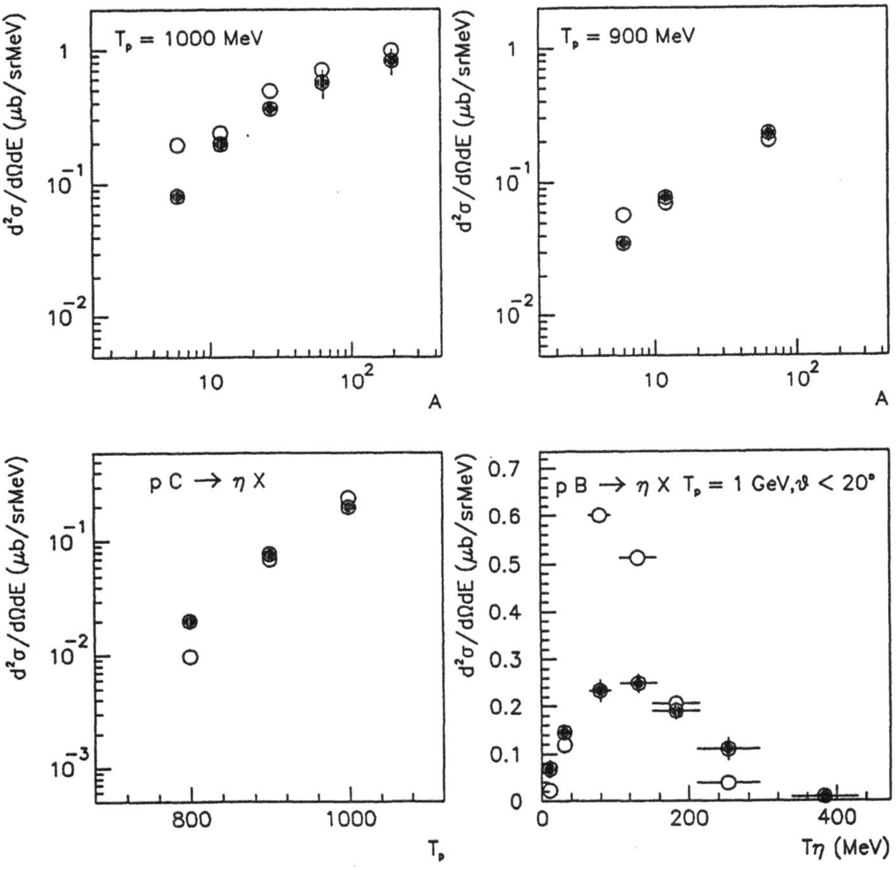

Fig. 10. PINOT subthreshold η production data.
Close symbols are experimental points,
open symbols are the predictions of the model by Cassing et al.,
with more realistic calculation of η absorption,elementary cross section
as calculated by Laget et al., and assuming $\sigma(np \rightarrow np\eta) = 4\sigma(pp \rightarrow pp\eta)$

4. Conclusions

In this lecture we have presented a short review of the studies until now performed with the PINOT spectrometer, together with a description of the physics motivation of such studies and a presentation of theoretical efforts made in the very last years to study the features of η production on nucleon and nucleus. Some features of these processes are now better understood, e.g. the important role played by η absorption in the reaction mechanism in the case of nuclear production. Forthcoming experiments at SATURNE with the PINOT spectrometer, studying the elementary cross section for the $pp \rightarrow pp\eta$ reaction, will hopely help to tune the theoretical models in their basic ingredients, i.e. the differential and total cross section for the η production in NN interaction. In this way it is hoped that the characteristics of η production and interaction and the role played by $N^*(1535)$ in the reaction mechanism, will be understood with much greater detail.

REFERENCES

1) J.C. Peng, *Proc. of the Workshop on Production and Decay of Light Mesons,* Paris 1988, ed. by P. Fleury, World Scientific.
2) J.F. Germond and C. Wilkin, *Journ. of Phys.*G15(1988), 437.
3) J.M. Laget and J.F. Lecolley, *Phys. Rev. Lett.*61(1988), 2069.
4) Particle Data Group, J.J. Hernandez et al., Review of Particle properties, *Phys. Lett.*B239(1990), 1.
5) E. Oset et al., *Phys. Rev.*C44(1991), 738.
6) W. Cassing et al., *Phys. Rep.*188(1990), 365.
7) P. Berthet et al., *Nucl. Phys.*A443(1985), 589.
8) J. Berger et al., *Phys. Rev. Lett.*61(1988), 919.
9) E. Chiavassa et al., *Nucl. Phys.*A538(1992), 121c.
10) W. Cassing et al., *Z. Phys.*A340(1991), 51.
11) A.S. Iljinov, private communication.
12) A. DePaoli et al., *Phys. Lett.*B219(1989) 194.
13) J.F. Germond and C. Wilkin, *Nucl. Phys.*A518(1990), 308.
14) T.Vetter et al., *Phys. Lett.*B263(1991), 153.
15) J.M. Laget et al., *Phys. Lett.*B257(1991), 254.
16) F. Plouin et al., *Phys. Rev. Lett.*65(1990), 690.
17) E. Chiavassa et al., Laboratoire National Saturne - C.E.N. Saclay, Proposal n. 237.

18) E. Chiavassa, G.Dellacasa, N.DeMarco, F.Ferrero, M.Gallio, A.Musso, A.Piccotti and E.Vercellin, F.Brochard, J.M.Durand, R.Bertini *Proc. of 5th International Conference on Nuclear Reaction Mechanisms, Varenna, June 1988, ed. by E. Gadioli, Ricerca Scientifica ed Educazione Permanente (Università degli Studi di Milano)*, Suppl. n.66 (1988) 633.

19) E. Vercellin et al., submitted for publication.

Common Problems and Trends of Modern Physics, pp. 269–275
ed. T. Bressani, S. Marcello and A. Zenoni
©1992 World Scientific Publishing Company

ON THE INTERACTION CROSS SECTIONS OF LIGHT NUCLEI AT LOW ENERGIES

A.BERTIN, N.SEMPRINI-CESARI, M.PICCININI, and A.VITALE
Dipartimento di Fisica dell'Università di Bologna, and
Istituto Nazionale di Fisica Nucleare, Sezione di Bologna

1. Introduction

The experimental information on the interactions of light nuclei at low energy is rather incomplete. In the range of thermal energies for instance, only neutron scattering lengths (n-n, n-d, n-He, n-Li) were measured in the different spin states. The corresponding data for the proton are obtained by assuming charge symmetry of strong interactions, while for more complex nuclei the interaction cross sections (like the d-d one, for instance) are determined by extrapolating data obtained at higher energies.

On the one hand, a comparison of the binding energies of mirror nuclei clearly indicates that the validity of charge symmetry of strong interactions should be rediscussed; on the other, as it will be seen in the following, recent experimental results[1] set some doubts on the reliability of extrapolating cross sections from higher to lower energies. In the present report, we will discuss how this lack of experimental knowledge demands to reexamine several important problems, like for instance (besides charge symmetry in strong interactions) the relative abundance of light nuclei within stellar matter and the problem of solar neutrinos. We will also show that experiments on muonic molecules formed with light nuclei may provide quite significant information also in this direction.

2. Hints from Cluster-Impact Fusion

In the experiment by Beuhler et al.[1] molecular clusters of $D_2 0$, accelerated in the energy range 200-325 keV, were impinging on a deuterated titanium (TiD) target. The protons produced, which were attributed to the fusion reaction

$$d+d \to p+t+4.03\,MeV \tag{1}$$

were measured as a function of the energy of the incoming molecules. The fusion rate R ($cm^{-3}\, s^{-1}$), in these conditions, would be given by the relation

$$R=\eta^2\bar{\sigma}\bar{v} \tag{2}$$

were η is the deuteron number density, $\bar{\sigma}$ is the cross section, $\bar{v}$ is the deuteron velocity averaged over the velocity distribution in the collision spike.

The cross section σ, instead, is expressed as

$$\sigma(E)=(S(E)/E)\,e^{-31.28/E^{1/2}} \tag{3}$$

where E is given in keV, and S(E) is derived from data on higher-energy D-D reactions.[2]

From the experimental data obtained by Beuhler et al.[1] one gets a value of σ more than 10 orders of magnitude larger than that computed (for 300 eV D) using the value of S(E) = 55 x 10^{-24} cm^2 keV. This fact suggests that the d-d cross section at low energy may have a resonant character, which may be a significant feature for several problems, besides the ones quoted at the end of Section 1: among the others, for instance, nuclear cold fusion.

3. The Problems of Light Elements in Stellar Matter and of Solar Neutrinos

The synthesis of light elements in stellar matter occurs mainly through the proton-proton cycle:

$$p+p \rightarrow d+e^{+}+\nu \tag{4}$$

$$p+e^{-}+p \rightarrow d+\nu \tag{5}$$

$$d+p \rightarrow {}^{3}He+\gamma \tag{6}$$

$${}^{3}He+{}^{3}He \rightarrow {}^{4}He+2p \tag{7}$$

$${}^{4}He+{}^{3}He \rightarrow {}^{7}Be+\gamma \tag{8}$$

$${}^{7}Be+e^{-} \rightarrow {}^{7}Li+\nu \tag{9}$$

$${}^{7}Li+p \rightarrow 2\,{}^{4}He \tag{10}$$

$${}^{7}Be+p \rightarrow {}^{8}B+\gamma \tag{11}$$

$${}^{8}B \rightarrow {}^{8}Be^{*}+e^{+}+\nu;\ {}^{8}Be^{*} \rightarrow 2\,{}^{4}He\ . \tag{12}$$

The relative abundances of light nuclei and the neutrino fluxes generated from this cycle depend on the cross sections of the various processes. Since directly measured data are missing, in all models of the stellar nuclei extrapolated cross section values are used, like the ones provided by the relation

$$\sigma(E)=(S(E)/E)\exp-(2\pi e^{2}Z_{1}Z_{2}/\hbar v) \tag{13}$$

where E is the c.m. energy of the interacting nuclei, v is their relative velocity and S(E) takes into account the energy behaviour of the cross sections at low energy. For the cases of reactions (7) and (8), at E = 0, we have S(0) = $(4.7^{+0.8}_{-0.7})$ MeV b and S(0) = (0.52 ± 0.15) keV b. If one follows this method, however, one obtains systematically underestimated predictions for the relative abundances of light nuclei, as well as overestimated predictions for the fluxes of produced neutrinos. In the past, these discrepancies were attributed to inadequacy of the models, so that a number of alternative models were developed and are available in the literature.

However, the possibility that such discrepancy may not be due to the chosen model, but to the behaviour of the cross sections for the interactions of light nuclei at low energies has seriously to be taken into account. It was shown by our group, in fact[3] that an increase in the cross section for the weak process (4), which might be due to an increase in the effective value of the axial-vector weak interaction coupling constant g_A, can reduce or ultimately eliminate the solar neutrino problem. In a similar way, a resonant behaviour of the interaction cross sections for light nuclei at low energies might explain as well the relative abundances of light nuclei in stellar matter.

4. Charge Symmetry in Strong Interactions

A significant experimental test on the validity of charge symmetry in strong interactions is missing, since the relevant information on the scattering cross sections of protons on light nuclei (p-p, p-d, p-^{3}He) is incomplete. Breaking of charge symmetry in strong interactions would imply e.g. interaction potentials between nucleons different from the ones which are currently considered and changements in the mass difference between light quarks. We shall see in the following that the present experimental picture is in favour of electromagnetic mixing mechanisms between mesons (like η^0 and π^0) which might be responsible of charge asymmetric potentials.

An important consequence of this mixing mechanism was pointed out by our group[4] and it concerns small violations of G-parity conservation in the $\bar{n}-p$ annihilation.

The validity of charge symmetry in strong interactions is generally tested by two independent methods.

One method consists in comparing the n-n and p-p scattering lengths,[5]

$$a_{nn} = -(18.8 \pm 0.3)\ F,\ a_{pp} = -(17.3 \pm 0.4)\ F \tag{14}$$

the difference may be explained with mechanisms of mixing of mesons.

The second method consists in comparing the binding energies of mirror nuclei (i.e. having an equal number of nucleons, but different proportions of neutrons and protons). If charge symmetry holds, the difference in binding energies must have electromagnetic origin, and can be calculated. In the case of the mirror nuclei ^{3}H and ^{3}He, the experimental value for the difference in the binding energies is of about 0.76 MeV, whereas the Coulomb energy of the two protons in ^{3}He turns out to be in the range between 0.62 and 0.66 MeV, according to the chosen model. There are then about 100 keV in the quoted difference of binding energies which cannot be attributed to the different Coulomb interaction and might point at a charge dependence of the interaction between nucleons.

Several attempts were proposed[5,6] to explain these facts. In the first place, one tried to attribute the 100 keV discrepancy to the mass difference between the proton and the neutron; in this way, however, only about 30 keV would be justified.

Another possible explanation considered possible radiative corrections (see Fig.1) which would be present only in the p-p interaction (therefore being absent in the n-n one). The relevant correction, however, would contribute for about 8 keV instead of the quoted 100.

A third model takes into account a modification in the charge distribution of the proton due to the strong interactions. Under this hypothesis, the charge distribution of the proton would not have a spherical symmetry, and this would affect also the Coulomb potential energy within the ^{3}He nucleus. In order to construct in this way a significant effect, however, it becomes necessary to assume the a_{nn} and a_{pp} scattering lenths out of the experimental errors.

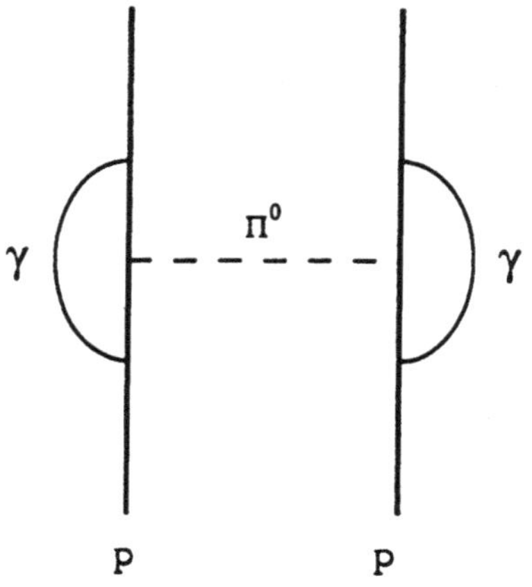

Fig. 1 - Radiative correction to p-p scattering.

The only mechanism which might explain in a convincing way the experimental facts is an isospin mixing induced by the electromagnetic interactions (see Fig.2). One supposes in this case that π^0 and η^0 (isospin eigenstates), are mixed between them, in this case the physical mesons $\bar{\pi}^0$ and $\bar{\eta}^0$ are expressed as:

$$\bar{\pi}^0 = \pi^0 + R_\eta \eta^0 \qquad (15)$$

$$\bar{\eta}^0 = \eta^0 - R_\pi \pi^0 \qquad (16)$$

were R_η and R_π are the mixing ratios.

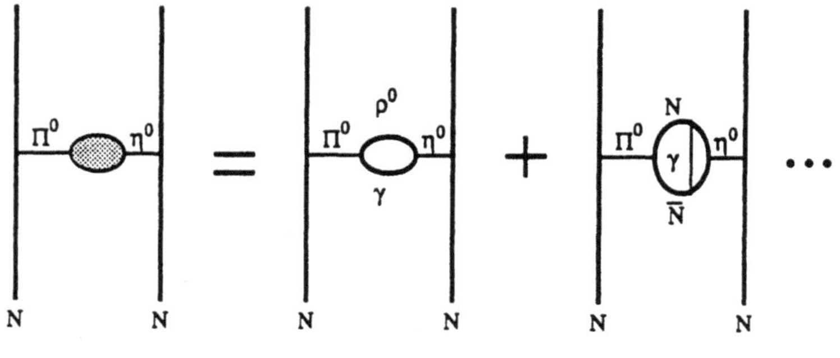

Fig. 2 - Diagram for π-η mixing in the nucleon-nucleon scattering.

The potential due the exchange of such $\bar{\pi}$ and $\bar{\eta}$ is the following:

$$U_{CA}(\pi\eta) = -(\tau_1^z+\tau_2^z)\,\frac{g_\pi g_\eta}{4\pi}\,[R_\eta U(m_\pi) - R_\pi U(m_\eta)] \tag{17}$$

where g_π and g_η are the coupling constants of π and η with the nucleons, and

$$U(m) = \left(\frac{m}{2M}\right)^2 \frac{e^{-\mu r}}{r} \qquad \left(\mu = \frac{mc}{\hbar}\right)$$

where m and M are meson and nucleon masses.

By assuming $R_\pi = R_\eta = 0.01$, one succeeds in this way in explaining a 0.14 MeV gap in the binding energies of the quoted mirror nuclei. Furthermore, one foresees in this way an alteration of the scattering lengths which is in the experimental errors. In such a way, the mixing assesses the whole experimental frame connected to the problem of charge symmetry.

5. Muonic Molecules: an Indirect Insight

As was already mentioned, technical difficulties make practically impossible direct determinations of the interaction cross sections for light nuclei in the energy range between eV and keV. An alternative, indirect approach to solve this problem is provided however by muonic molecules.

It is worth while recalling here that, when a negative muon is slowed down (e.g., in a hydrogen target) to the energy of a few eV he forms a μp muonic atom through the Auger process

$$\mu p + H \rightarrow (\mu p) + e^- . \tag{18}$$

The μp muonic atom is electrically neutral and has unusually small atomic dimensions, since its Bohr radius is

$$a_\mu = a_0\, m_e/m_\mu = 250\ F \tag{19}$$

where a_0 is the electronic Bohr radius of the hydrogen atom, m_e and m_μ the electron and muon mass, respectively.

The μp muonic atom is therefore rather similar to a slow neutrons, and can esaily diffuse throughout the surrounding matter.

One of the most important reaction which hydrogen (or deuterium or tritium) muonic atoms may undergo, if the medium is contaminated by an even small admixture of an inpurity element Y is the transfer process of the muon from the lighter muonic atom to the Y nucleus, following the irreversible reactions

$$\mu p + Y \rightarrow \mu Y + p \tag{20}$$

$$\mu d + Y \rightarrow \mu Y + d \tag{21}$$

$$\mu t + Y \rightarrow \mu Y + t \tag{22}$$

Processes (18) to (20) represent an alternative procedure to produce the μY muonic atoms with respect to the current method of stopping negative muons directly within a target made of the element Y. In particular, once the μY atoms are formed in this way, it can subsequently get bound to one more proton, deuteron or triton forming a $p\mu Y$, $d\mu Y$, or $t\mu Y$ muonic molecule, by a process which can be sketched like

$$\mu Y + p \rightarrow p\mu Y \tag{23}$$

$$\mu Y + d \rightarrow d\mu Y \tag{24}$$

$$\mu Y + t \rightarrow t\mu Y \tag{25}$$

The most known muonic molecules are those which are formed from μp, μd or μt atoms with hydrogen isotope nuclei, following a processes like for instance

$$\mu p + p \rightarrow p\mu p \quad or \quad d\mu + p \rightarrow p\mu d \tag{26}$$

These molecules can be treated in analogy to the H_2^+ molecular ion, getting for the internuclear equilibrium distance R_0 and for the relevant vibrational energy E_{vib} the values

$$R_0 = 500\ F \tag{27}$$

$$E_{vib} = 250\ eV \tag{28}$$

The effective value of density corresponding to this physical situation can be evaluated by the relation

$$\begin{aligned} d_u &= d_{liq.hydrogen}\ (R_0(liq.hydrogen)/R_0(muonic\,hydrogen)) = \\ &= d_{liq.hydrogen}\ (m_\mu/m_e)^3 = 10^7 d_{liq.hydrogen} \end{aligned} \tag{29}$$

The corresponding temperature, in its turn, can be evaluated through the relation

$$T = E_{vib}/K = 3\ 10^6\ K \tag{30}$$

which corresponds to conditions similar to the ones of stellar nuclei.

In a plasma, the rate of fusion reactions per unit volume is given by

$$N = n_1 n_2 \int_0^\infty \sigma(E) v P(E) dE \tag{31}$$

where n_1 and n_2 are the concentrations of different nuclei, v the relative velocity, and P(E) the probability distribution of energies.

In a muonic molecule (where the fusion reaction is catalyzed by the muonic bond, which keeps the nuclei at exceptionally small distances with respect to the ordinary electronic molecules) one has $n_1 = n_2 = 1$, so that the fusion rate for such a systems is given by

$$N = \int_0^\infty \sigma(E) v P(E) dE\ . \tag{32}$$

This is a parameter which can be measured. If for instance one observes an anomalous rate in the fusion of a given molecule, this might be attributed to a resonant behaviour of the interaction cross section of the involved nuclei in the energy range which is typical for muonic molecules, namely of the order of a few hundreds of eV (see Equation (26)). By observing then the fusion rates of muonic molecules like the pμ Li, dμ Li, and tμ Li ones, one would for instance obtain information on the cross section for reaction (10), which was seen to play a significant role in the proton-proton cycle for the synthesis of light nuclei in stellar matter.

6. Concluding Remarks

As a conclusion to this report, we should just like to summarize that three experimental lines are relevant to the subjects dealt in the previous Sections:

1) a new measurement of the neutron-neutron scattering length in process where there is no interaction in the final state, such as e.g. muon nuclear capture by deuterons

$$\mu + d \rightarrow n + n + \nu_\mu \tag{33}$$

2) a study of G-parity conservation in the annihilation reaction

$$\bar{n} + p \rightarrow \pi^+ \pi^0 \tag{34}$$

which in case of isospin symmetry violation could give information on the quark mass difference $m_u - m_d$.

3) The study of catalyzed fusion of muonic molecules formed with nuclei different from hydrogen isotopes which could give information on characteristics of fusion reaction of light nuclei in the region of energy of interest for the discussed problems in astrophysics.

REFERENCES

1. R.J.Beuhler et al., Phys. Rev. Lett., **63**, 12 (1989) 1292; R.J.Beuhler et al., Phys. Rev. Lett., **68**, 13 (1992) 2108.
2. D.D.Clayton, *Principles of Stellar Evolution* (Mc Graw-Hill, New York, 1968), p. 294.
3. A.Bertin et al., Phys. Lett. **B 199**, 1 (1987) 108.
4. A.Bertin et al., Phys. Lett. **B 244**, 3, 4 (1990) 519.
5. B.M.K.Nefkens, Comm. Nucl. Part. Phys. **20** (1992) 221.
6. K.Okamoto and C.Pask, Annals of Physics **68** (1971) 18.

Common Problems and Trends of Modern Physics, pp. 277–295
ed. T. Bressani, S. Marcello and A. Zenoni
©1992 World Scientific Publishing Company

New Developments of the Superradiant Nuclear Model

P.G. Ratcliffe

INFN, Sezione di Torino, via P. Giuria 1, 10125 Torino, Italy

ABSTRACT

In the framework of a novel approach to nuclear dynamics due to Preparata, based on a superradiant pion-nucleon-delta interaction, we discuss various phenomena of nuclear physics which have so far eluded convincing explanations. The central idea that the dynamics of the nucleus is dominated by a coherent π-N-Δ interaction leads to a lowering of the effective nucleon mass by about 60 MeV and the presence of a rather peculiar pion condensate coupled to a large isobar content of the nucleus. A theory of the EMC effect is then presented derived from this mass-shift in a convolution model. It is further shown that the shadowing effect of the small-x_B region can be explained via a single-diffractive mechanism interfered upon by Pauli exclusion. The reduced nucleon content of this nuclear ground state leads to the experimentally observed suppression of the so-called Coulomb sum rule in quasi-elastic scattering. And the completely determined pion condensate permits the calculation of the hypernuclear potential, predicting the depths and decay widths of lambda and sigma hypernuclei compatible with those measured experimentally.

1. Introduction

In this lecture I discuss various experimental aspects of nuclear dynamics and their explanation in terms of a novel approach to nuclear dynamics developed within the framework of the theory known as Superradiance (SR). The structure of the lecture is as follows: section 2 provides a brief outline of the ideas behind the SR solutions to many-body problems and their application to the case of nuclear structure. The resulting explanation of the EMC effect and related nuclear DIS phenomena is presented in Section 3. A somewhat similar phenomena, the experimental quenching of the Coulomb sum rule, in quasi-elastic scattering is discussed in section 4. In section 5 the interesting, if somewhat controversial, problem of the existence and properties of hypernuclei is considered and finally I close in Section 6 with some comments and conclusions.

2. Coherent Dynamics of the Nucleus and Superradiance

Let us begin by outlining the dynamical model of the nucleus with which the above-mentioned effects may be explained. The mechanism of SR has been applied for some time now with success to various aspects of condensed matter physics [1]. The philosophy underlying the development of this approach is that short-range forces cannot solely be responsible for large-scale structures and that one must seek long-range forces able to maintain coherence over large regions. In other words the picture of Fig. 1 (a), in which the matter fields behave incoherently and only interact when in direct collision, is to be replaced by that of Fig. 1 (b), in which communication via a background (radiation) field locks the phases of the matter fields and renders the dynamics coherent. The basic point is that (generic) dipole oscillations generate a radiation field which in turn influences the same dipoles, the non-linearity of the equations gives rise to stable non-zero configurations, *i.e.*, non-trivial vacuum structure.

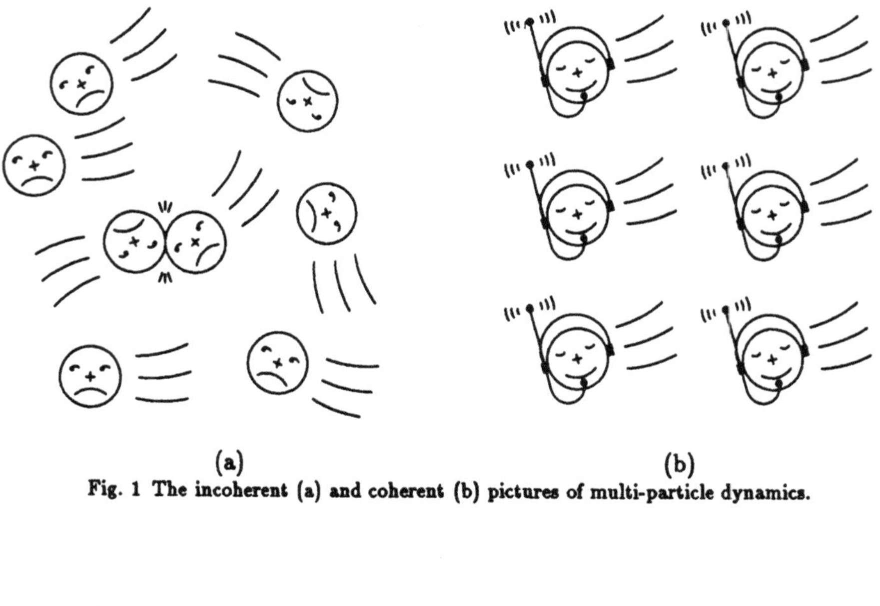

(a) (b)

Fig. 1 The incoherent (a) and coherent (b) pictures of multi-particle dynamics.

In condensed-matter physics the obvious candidate as a background field is the electromagnetic field with its potentially infinite range. While one would normally expect screening to kill any long-range effect it is possible that under conditions of sufficient density even the relatively weak QED can develop classical solutions with phase coherence over large regions or coherence domains (CD) in which the matter fields oscillate in phase with the background radiation field. The operation of the SR mechanism thus requires the presence of a two-level system whose transitions are determined by interaction with a potentially long-range radiation field. The Nucleon-Delta-Pion system is just such a case; the N-Δ two levels have a suitably small energy (mass) difference of ~ 292 MeV and the π (radiation) field, on the scale of nuclear structure, has an effectively long range. For detailed derivation of the SR solutions see refs. [2,3].

From refs. [2,3] the nucleon is described in terms of

(i) the nucleon and delta wave fields:

$$\Psi_N(\mathbf{x},t) = N_{st}\chi_s\eta_t \quad \text{and} \quad \Psi_\Delta(\mathbf{x},t) = \Delta_{ST}\chi_S\eta_T, \qquad [2.1]$$

where $\chi_{s,S}$ and $\eta_{t,T}$ are Pauli spinors for spin and isospin degrees-of-freedom, $s,t = \pm\frac{1}{2}$ and $S,T = \pm\frac{1}{2}, \pm\frac{3}{2}$;

(ii) the pion radiation field:

$$\phi_b(\mathbf{x},t) = \frac{1}{\sqrt{V}}\sum_{\mathbf{q}}\left[a_{b\mathbf{q}}e^{-i(\omega_q t-\mathbf{q}.\mathbf{x})} + a^\dagger_{b\mathbf{q}}e^{i(\omega_q t-\mathbf{q}.\mathbf{x})}\right], \qquad [2.2]$$

where $b = 0, \pm 1$, $\omega_q = \sqrt{(\mathbf{q}^2 + m_\pi^2)}$ and the operators $a_{b\mathbf{q}}$ and $a^\dagger_{b\mathbf{q}}$ obey the usual equal-time commutation relations. The SR state (the true nuclear ground state in this theory) fixes the dominant mode to be $\omega_q = m_\Delta - m_N = 292\,\text{MeV}$ and thus $q = 256\,\text{MeV}$. The interaction Hamiltonian is then given by

$$H_I = \frac{g}{\sqrt{4m_N m_\Delta}}\int d^3x\, \Delta^\dagger_{ST}N_{st}C^i_{Ss}C^b_{Tt}\nabla_i\phi_b \quad + \quad \text{h.c.}, \qquad [2.3]$$

where the C^i_{Ss} (C^b_{Tt}) are the spin (isospin) Clebsch-Gordan coefficients for coupling spin-$\frac{3}{2}$ to spin-$\frac{1}{2}$ plus spin-1 and the coupling g is entirely determined by the $\Delta \to N\pi$ decay width. The kinetic Hamiltonians are of the canonical form and we work in the non-relativistic limit.

The salient features of the solutions are that:

(i) the SR nuclear ground state shifts the nucleon mass from its free value by $\sim 60\,\text{MeV}$, the depth of the *self-consistent potential*,

(ii) the effective number of pions is $n_\pi \simeq 0.63$ per nucleon;

(iii) the "matter" content of the nucleus is $\sim 80\%$ nucleons and $\sim 20\%$deltas.

3. The Application of Superradiance to Nuclear DIS and the EMC Effect

The long-established vision of hadronic scattering at high energies embodied in the Quark-Parton Model [4] (QPM) requires that quarks, though confined inside a nucleon, behave at short distances (*i.e.*, when probed at high energies) as if free. Such a picture is supported by the coupling constant (α_S) behaviour of PQCD, known to decrease as the energy scale (for which I adopt the generic notation Q^2) increases and asymptotically to vanish. Thus the early belief was that since the quarks are effectively free inside the nucleon they should be unaware of the environment of the nucleon itself, *i.e.*, whether it is in turn bound inside a nucleus. One must factor out purely kinematical effects such as Fermi motion, however, the important point is though that no dynamical difference was expected.

Thus, one of the most striking results of the research programme carried out by the European Muon Collaboration (EMC) [5] at CERN was the discovery that DIS of leptons off nuclei does not behave as the incoherent sum of individual nucleon scattering cross-sections [6,7]. This deviation from the expectations has been fully confirmed in various other experiments [8,9].

In the QPM, lepton-nucleon DIS is described as the convolution of a point-like hard scattering of a photon off a free, on-shell quark (including PQCD corrections to the desired accuracy) and a long-distance, incalculable, nucleon quark distribution, see Fig. 2. The hadron-photon piece of such an interaction is described in terms of structure functions (SF), F_1 and F_2 (neglecting parity violation and polarisation effects for which further structure functions are necessary):

$$\begin{aligned} W^{\mu\nu} &= \frac{1}{\pi}\,\mathrm{Im}\int d^4z\, e^{iq.z}\,\langle p|\,[J^\mu(z),J^\nu(0)]\,|p\rangle \\ &= \left[\frac{q^\mu q^\nu}{q^2}-g^{\mu\nu}\right]F_1(x_B,Q^2)+\left[p^\mu-\frac{p.q}{q^2}q^\mu\right]\left[p^\nu-\frac{p.q}{q^2}q^\nu\right]\frac{1}{p.q}F_2(x_B,Q^2), \end{aligned} \qquad [3.1]$$

where p^μ, q^μ are the nucleon, virtual-photon four-momentum respectively for which $q^2 = -Q^2 < 0$ and defining ν, the energy transfer of the photon in the nucleon rest frame, with m the nucleon mass then $p.q = m\nu$, J^μ is the electromagnetic current and the ellipsis represents the other Lorentz covariant structures, governed by F_3, G_1, G_2 etc. At high energies the SF's depend only on the Bjorken scaling variable $x_B = Q^2/2p.q = Q^2/2m\nu$, and logarithmically (according to PQCD) on Q^2. The main purposes then of DIS experiments is to measure these distributions.

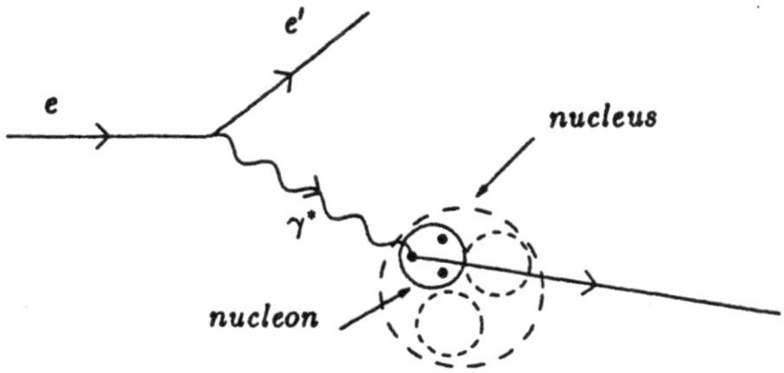

Fig. 2 The QPM picture of lepton-nucleon DIS.

In the *naïve* QPM F_2, for example, can be expressed in terms of quark distributions, denoted $q_f(x_B)$, which are interpreted as the probability of finding a quark (flavour f) carrying a fraction x_B of the parent momentum:

$$F_2^p(x) = x\left[\tfrac{4}{9}u(x) + \tfrac{1}{9}d(x) + \tfrac{1}{9}s(x)\right], \qquad [3.2]$$

where u, d and s are the *up*, *down* and *strange* quark distributions inside the proton and the factors $\frac{4}{9}$ *etc.*, are their charges squared in proton-charge units. From the kinematics x_B lies in the interval [0,1]. In Fig. 3 an example is displayed of the valence distribution relevant to an isoscalar target such as deuterium.

Since the x_B-dependence of the distributions is incalculable, one actually studies, in the case of nuclear DIS, the ratio of F_2 for a nucleus, of atomic number A, to that for deuterium $R(x) = F_2^A(x)/F_2^D(x)$. Considering F_2 as normalised to A, $R(x_B)$ is just the effective fraction of nucleons as a function of x_B. *Naïvely* then this ratio is expected to be unity. The behaviour of this quantity, displayed in Figs. 4,5, exhibits the following prominent features:

(i) in the region of x_B near to one a rise above unity in $R(x)$, due to the bound nucleon Fermi motion;

(ii) a pronounced dip near $x_B \sim 0.6$, where $R(x)$ shows a large defect of $\sim 20\%$;

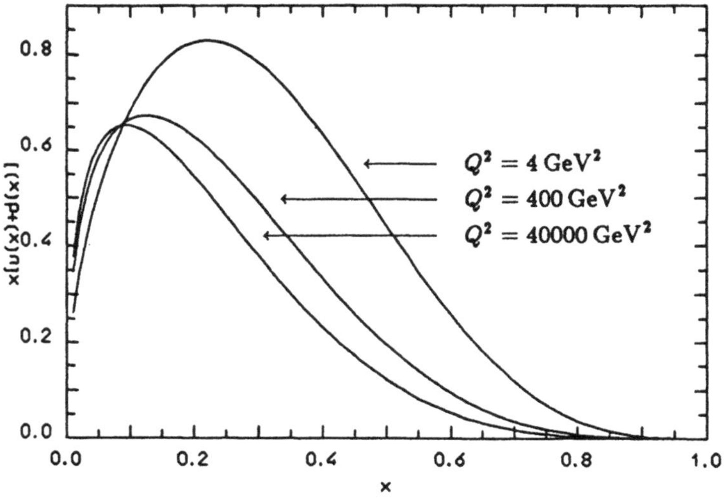

Fig. 3 **The valence-quark distributions relevant to an isoscalar target as a function of x_B and for various values of Q^2.**

(iii) a rise above unity for $x_B \sim 0.2$, more pronounced at higher beam energies, an effect termed *anti-shadowing*;

(iv) the so-called *shadowing* effect at very small x_B, where $R(x)$ again falls below unity by up to 30%.

There are several surprising aspects to this phenomenon. The scaling nature of the phenomenon (apart from the anti-shadowing effect with its strong Q^2 dependence), realised very precociously (*i.e.*, for very small values of Q^2), is all the more puzzling if interpreted as demonstrating that quarks scatter the highly virtual photon incoherently whereas nucleons do not. Furthermore, owing to the above-mentioned precocity, what is controlling the physics of the small-x_B region is not true shadowing as such a process is expected to have strong Q^2 dependence [10] (known as higher-twist). Although there exist in the literature descriptions of the small-x_B behaviour exhibiting slow Q^2 dependence [11], there is no generally accepted explanation and some parametrisation and phenomenological fitting to the basic effect itself as input is necessary. Many explanations of the phenomena outlined above have been proposed and there is no space here to discuss them; the

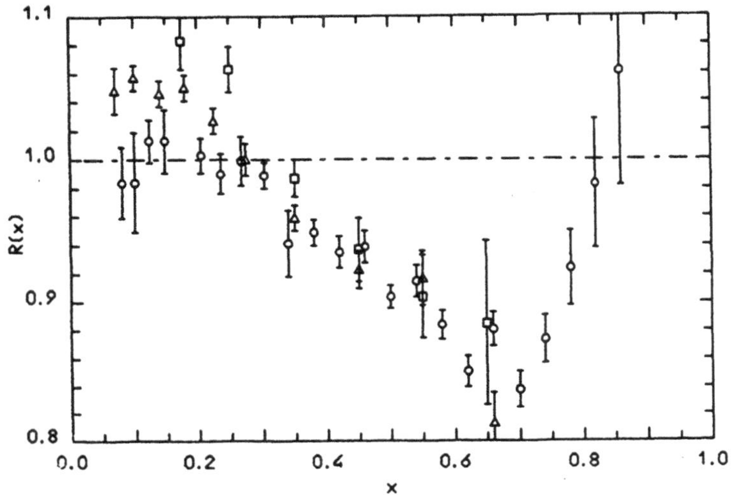

Fig. 4 $R(x_B)$ versus x_B as compared with the iron data from SLAC [8] (○), BCDMS [9] (△) and EMC [6,7] (□).

interested reader is referred to ref. 12.

Based on the model presented $R(x)$ can now be computed by describing it in terms of additive DIS of virtual photons off A nucleons of reduced effective mass $\widetilde{m} \simeq m - 60\,\text{MeV}$, populating the Fermi sphere of the nucleus and, in addition, off n_π pions. In the small-x_B shadowing region, our theory is a generalisation to massive, space-like photons of single-diffractive (SD) scattering, which provides an excellent description of hadron-nucleus scattering [13].

Leaving aside the small-x_B region for a moment, the nucleon contribution $F_2^{A,N}$ to F_2^A is given by the following convolution integral of the deuteron SF, F_2^D:

$$F_2^{A,N}(x_B) = \frac{3}{p_F^3} \int_0^{p_F} p^2\, dp \int_{-1}^{1} \frac{dz}{2} \int_0^1 d\xi\, F_2^D(\xi)\, \delta(\xi - x_B^N), \qquad [3.3]$$

where the integration variable $p^\mu = (E, p\sin\theta, 0, p\cos\theta)$, the struck nucleon momentum, ranges over the Fermi sphere: $|\vec{p}| \leq p_F$ and $z = \cos\theta$. The re-scaled, effective Bjorken scaling variable x_B^N for the bound nucleon is given by

$$x_B^N \simeq x_B \frac{m}{\widetilde{m}} \left[1 + \frac{p^2}{2\widetilde{m}^2} + \frac{zp}{\widetilde{m}}\right]^{-1}. \qquad [3.4]$$

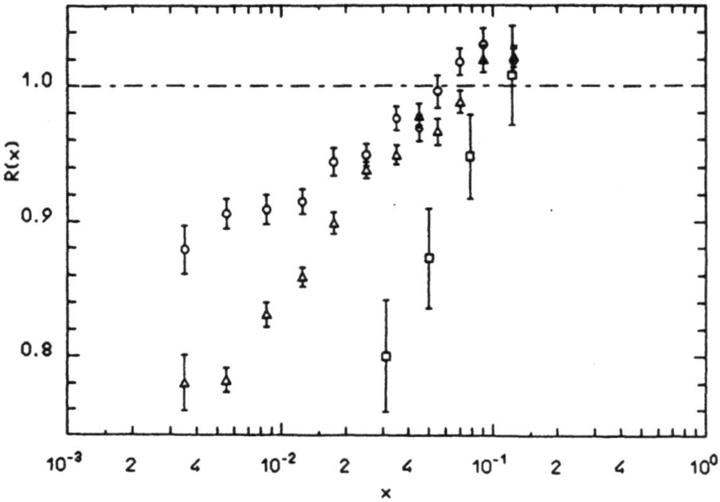

Fig. 5 $R(x_B)$, the F_2 ratio with respect to deuterium, in the small-x_B shadowing region as compared with the NMC data for carbon, $Q^2 = 0.7 - 11.0\,\mathrm{GeV}^2$ (○); calcium, $Q^2 = 0.6 - 10.0\,\mathrm{GeV}^2$ (△); and tin, $Q^2 = 4 - 14\,\mathrm{GeV}^2$ (□).

For the deuteron SF's the parametrisation of Duke and Owens [14] was used. One sees from eq. [3.4] that the result is to shift the effective x_B to a larger value and examining Fig. 3 one realises that for $0.2 < x_B$ such a shift is to a lower cross-section, hence the suppression.

As for the pion contribution to F_2^A, one writes analogously:

$$F_2^{A,\pi}(x_B) = \frac{n_\pi}{A} \int_{-1}^{1} \frac{dz}{2} \int_0^1 d\xi\, F_2^\pi(\xi)\, \delta(\xi - x_B^\pi). \qquad [3.5]$$

The same arguments as above lead to a similar re-scaling of x_B to

$$x_B^\pi = x_B \frac{m}{(\omega + kz)}, \qquad [3.6]$$

where the reduced pion frequency $\omega = 0.28\omega_0$, the momentum of the dominant pion mode $k = \sqrt{\omega_0^2 - m_\pi^2}$ and ω_0, the resonating π-mode frequency, is just the proton-delta mass difference. The pion SF was taken from the Owens parametrisation, extracted from J/Ψ and muon-pair production [15].

It is clear from Eqs. [3.5,3.6] that the pion contribution is important only for small x_B. Moreover, the experimentally relevant lower bound on the produced

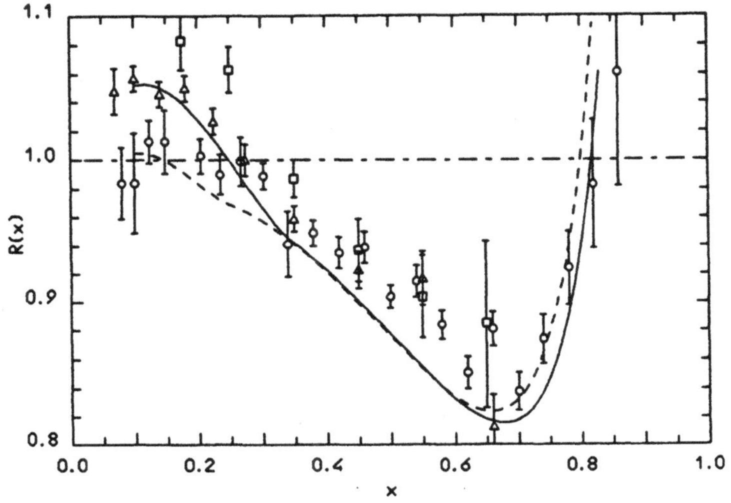

Fig. 6 The data are as in Fig. 4, the solid and dashed curves correspond to our calculation for $Q^2 \simeq 14-200$ and $2-10\,\text{GeV}^2$ respectively, where Q^2 varies with x_B according to the relevant experimental setup.

mass W^2 of $\sim 2\,\text{GeV}^2$ limits its contribution to high values of the incident lepton momentum, *i.e.*, rather above the original SLAC energy region. In Fig. 6 the result of the calculation is reported which, it should be noted, from the foregoing has absolutely no free parameters. For moderate values of x_B (where F_2 is rapidly decreasing) one sees the expected suppression due to the shift in the effective x_B to larger values. The suppression is eventually more than compensated for by the Fermi-motion effect. For $x_B \sim 0.1$ the extra contribution coming from the pion content leads to an enhancement, provided Q^2 is sufficiently high.

The Regge picture of high-energy behaviour implies that one diagram for DIS in the small-x_B region is that representing the SD process displayed in Fig. 7. Let us suppose this diagram to be indeed dominant here, then its diffractive nature implies the outgoing nucleon to be a proton of the same spin and approximately the same momentum as the initial proton. This kinematical contiguity of the nucleon is of no consequence in DIS off a free nucleon, but when the scattering occurs off a nucleon bound inside a heavy nucleus a large volume of final-state phase-space is unavailable, due to the Pauli exclusion principle. One is thus lead to a

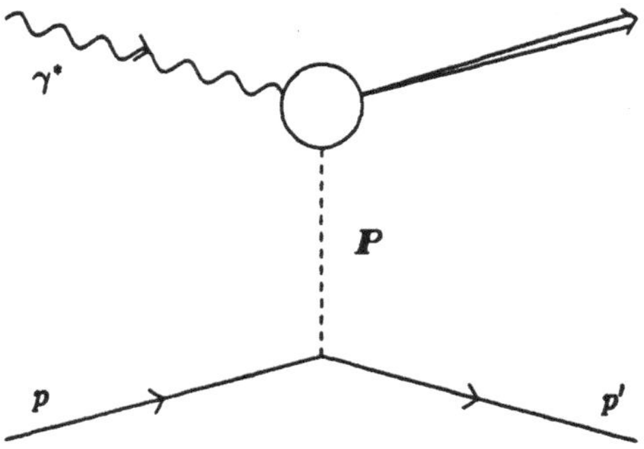

Fig. 7 The single-diffractive mechanism for the small-x_B shadowing region, the Regge exchange is dominated by the Pomeron, P. Definitions and relationships of the different variables are given in the text.

simple *Pauli-blocking* mechanism suppressing the cross-section as compared with the spectator-model predictions, which is as observed experimentally. The size and A-dependence of such a suppression are strongly dependent on the dominance of this diagram and the quality of its diffractive nature. A rather accurate analysis carried out in ref. 13 does in fact show that the fundamental SD mechanism by far dominates high-energy hadron-hadron scattering in the relevant kinematical region. Using the parameters obtained from phenomenological fits [13] one estimates that the SD mechanism should account for $\geq 75\%$ of the cross-section.

In order to calculate the diagram of Fig. 7 it is necessary to extend the triple-Regge formula to the case of a virtual photon of space-like momentum. Generalising the standard Regge SD cross-section formula [16] to a virtual photon of space-like momentum, $q^2 < 0$, for the process shown in Fig. 7 one derives

$$F_2(x_B) = \int \frac{d^3\vec{p}'}{2E'}\, g(t)\, (\cos\theta_t)^{2\alpha_P(t)}\, (P^2)^{\alpha_P(0)}, \qquad [3.7]$$

where $g(t)$, the Regge residue function, is parametrised as $g(t) = g_0\, e^{bt}$ (see ref. 13, in which $b = 5.81\,\mathrm{GeV}^{-2}$ was obtained) and here g_0 is fixed from the single-nucleon scattering cross-section; the primes refer to the outgoing proton and $P = p + q - p'$ is the momentum of the remaining outgoing system, $\alpha_P(t)$ is the Pomeron trajectory. To a good approximation one has

$$\cos\theta_t \simeq \frac{1}{2}\frac{[1 + x_P(1 - x_B)]}{[1 - x_P(1 - x_B)]} \qquad \text{and} \qquad P^2 \simeq Q^2\, \frac{(1 - x_B)(1 - x_P)}{x_B}, \qquad [3.8]$$

where x_F is the usual Feynman momentum fraction for the outgoing proton.

This reveals two important features: for very small x_B the outgoing nucleon has (i) x_F close to one and (ii) its transverse momentum kept small by the sharply falling t-dependence of $g(t)$. Thus one expects substantial suppression owing to the Pauli exclusion of the portion of outgoing momentum states lying below the Fermi surface, *i.e.*, the region of integration in Eq. [3.7] is restricted to $|\vec{p}'| \geq p_F$.

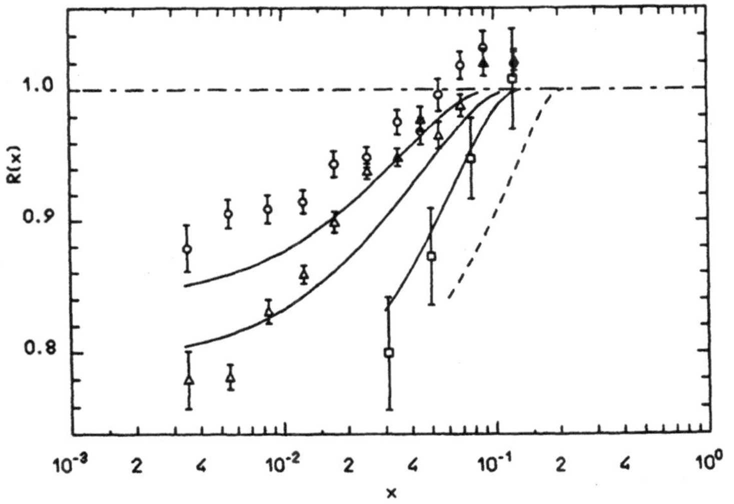

Fig. 8 The data are as in Fig. 5, the upper, middle and lower solid curves correspond to p_F = 198, 225 and 240 MeV respectively. The dashed curve gives the limiting case of infinite volume and ignoring the nucleon density reduction referred to in the text, for which p_F takes its *naïve* value of 264 MeV.

Owing to finite-size effects and to the reduced nucleon density coming from the model [2,3] there is a reduction in the Fermi momentum, more pronounced for lighter nuclei, from the infinite nuclear matter value of $p_F^{(0)} = 264$ MeV. According to the model the density of nucleons is reduced by a factor of 82%, see above, which in turn reduces p_F by $0.82^{1/3}$, this reduction is discussed in detail in ref. 12. The calculation yields, for the region $x_B \leq 0.1$, the results reported in Fig. 8, where they are compared with a collection of existing data. The solid curves in Fig. 8 refer to $p_F = 198$, 225 and 240 MeV; values appropriate for carbon, calcium and tin respectively, the dashed curve corresponds to the infinite nuclear matter value of

264 MeV. Thus shadowing is more conspicuous and persists to higher x_B in heavier nuclei, as indeed observed experimentally. Note the results are independent of Q^2 in this kinematical range, in accord with the recent, precise NMC data [17].

4. Quasi-Elastic Scattering and the Coulomb Sum Rule

It is commonly held that an effective way to count the protons in a nucleus is to consider the longitudinal response function (R_L) in the quasi-elastic scattering (QES) of electrons off the nucleus. In a range of nuclear models one derives a sum rule (analogous to the Coulomb sum rule):

$$C'(q) = \int_{\omega_{el}^+}^{\omega_{max}} d\omega \, R_L(q, \omega), \qquad [4.1]$$

where ω_{el}^+ signifies integrating from ω just above the elastic peak and ω_{max} is the maximum value for which R_L is non-zero.

The sum rule is expected to measure an effective nucleon charge form factor; in contrast the experimental data [18–20], summarized in fig. 9, show a clear discrepancy between theory and experiment.

From the discussion in section 2 on the nuclear model one may write, for the baryon wave-function in the nucleus:

$$|B\rangle_{\text{nucleus}} = \cos\theta \, |N\rangle + \sin\theta \, |\Delta\pi\rangle. \qquad [4.2]$$

The peculiar wave-function (3) derived in our model shows that the number of protons counted in the Coulomb sum rule will be reduced from Z by precisely a factor $\cos^2\theta \simeq 0.82$ [2,3]. One might worry that the contribution of the remaining Δ-π content of the nuclear wave-function should also be considered. However, both the pion and Delta fields of the coherent ground-state are much further off the mass-shell (by about 200–300 MeV) and thus their ejection is kinematically forbidden in the experimental range of ω. Moreover, since the Δ-N transition is known to be purely magnetic, one expects this contribution to the Coulomb sum-rule to be suppressed. Including the correction factor of 0.82 and taking into account the kinematical effects of the mass shift our prediction for the uncorrected Coulomb sum rule, displayed in fig. 9, is seen to describe the data well.

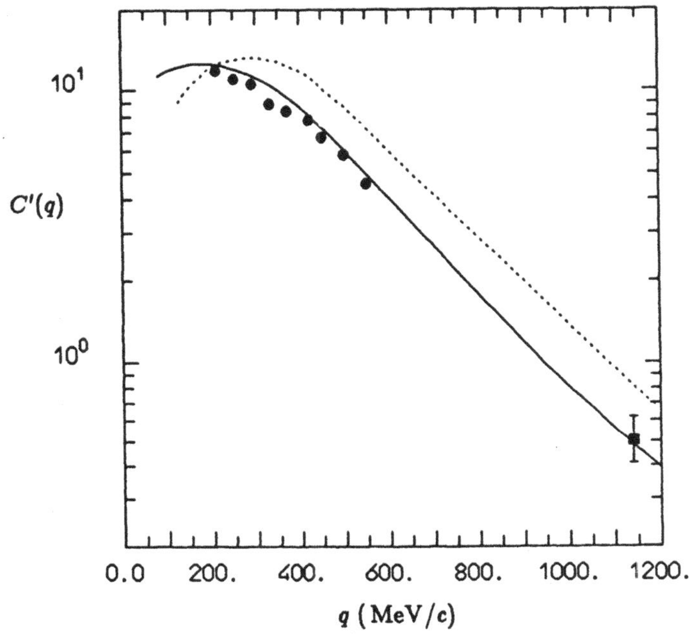

Fig. 9 $C'(q)$ for ^{56}Fe. The circle is the NE9 data [18], the squares are Saclay [19] and Bates [20] data. The solid curve is our result and the dashed curve is a Fermi-gas calculation.

The Coulomb sum rule is defined by dividing out the nucleon charge form factor with a relativistic correction [21]:

$$C(q) = \int_{\omega_{el}^{+}}^{\omega_{max}} d\omega \, \frac{R_L(q,\omega)}{Z\left[\overline{G}_E(Q^2)\right]^2}, \tag{4.3}$$

the effective nucleon charge form factor has been taken to be

$$\left[\overline{G}_E(Q^2)\right]^2 = \left\{\left[G_E^p(Q^2)\right]^2 + (N/Z)\left[G_E(Q^2)\right]^2\right\} \times \left[\frac{1+Q^2/4M_N^2}{1+Q^2/2M_N^2}\right], \tag{4.4}$$

where M_N is the nucleon mass, Z and N are the numbers of protons and neutrons, respectively. Taking into account smearing (due to Fermi motion) and the potential well of our coherent nuclear model we obtain for R_L:

$$R_L = \frac{3}{p_F^3}\int_0^{p_F} p^2 dp \int_{-1}^{1} \frac{d\cos\theta}{2}\, 2p_0'\, \delta(p'^2 - m^2), \tag{4.5}$$

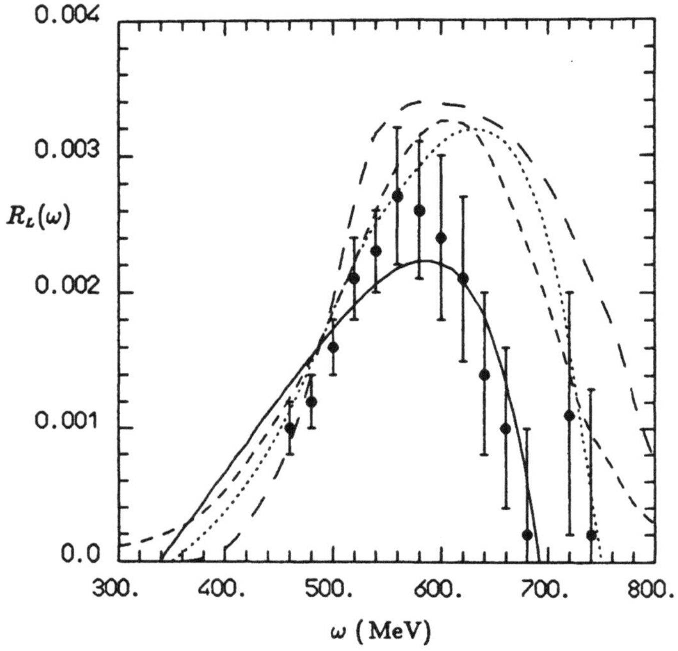

Fig. 10 The longitudinal response function for ^{56}Fe for an effective three-momentum transfer of $q = 1.14\,\mathrm{GeV}/c$. The data are from the NE9 collaboration [18], the curves are: our calculation (solid), a Fermi-gas model [22] (dashed), a random-phase approximation [23] (chain-dashed) and a plane-wave Born approximation [24] (dotted).

where p and p' are the intial and final nucleon momenta respectively and in p_0 we include the energy shift due to the coherent ground-state potential. For the proton form factor, G_E^p, a dipole form is used while for the neutron, one has $G_E^n = .0$. The use of other parametrisations has little effect on the Coulomb sum rule. We compare the experimental sum rule with our calculation, the results of a Fermi-gas model [22] and a random-phase approximation [23]:

$$\begin{aligned} C(q) &= 0.76 \pm 0.23 \quad &&(\text{expt. q=1.14 GeV/c}) \\ &= 0.82 &&(\text{our calculation}) \\ &= 1.10 &&(\text{Fermi gas}) \\ &= 1.16 &&(\text{random phase}). \end{aligned} \qquad [4.6]$$

Using the nucleon mass shift and the Fermi momentum of our model [25,26], in the case of ^{56}Fe $p_F = 230\,\mathrm{MeV}$, we obtain $R_L(\omega)$ plotted in fig. 10 together with data and previous calculations. One sees that the data are well represented by the results of our calculation, which we note has a $\chi^2 = 11.5$ for 14 data points.

5. Hypernuclei and the Hypernuclear Potential

One of the most interesting results of kaon-nucleus physics has certainly been the discovery of hypernuclei [27,28]. The extant phenomenology of the spectrum and properties of hypernuclei shows that lambda and sigma hyperons within a nucleus behave rather like the nucleons themselves, being arranged in typical shell-model levels associated with a potential of approximately half the depth of the standard nuclear potential and which displays no significant spin-orbit splitting [29]. Note that whereas there exists a quark model explanation of the smallness of the spin-orbit interaction in lambda hypernuclei the picture fails in the case of the sigma [30]. Furthermore, the very existence of sigma hypernuclei, with widths of the order of a few MeV, is surprising since the commonly accepted model would predict fast conversion of $\Sigma \to \Lambda$ through interaction with the nuclear environment and hence widths of a few tens of MeV. These then seem to be the challenges posed by established phenomenology to nuclear model building. Thus in this the final part of this lecture I present a *parameter-free* calculation of the hypernuclear potential that successfully meets these challenges.

According to our view it is a highly correlated physical system that a lambda or sigma hyperon encounters once produced in a kaon-nucleus interaction; the question then is: how do they interact with it?

Applying second-order perturbation theory, fully justified *a posteriori* from the point of view of the results, the effective hyperon-nucleus interaction is

$$V_{\rm eff} = \frac{-i}{4m_Y} \int_{-\infty}^{\infty} dt \, \langle Y|\, T(H_I(t/2)H_I(-t/2))\, |Y\rangle_{CD}, \qquad [5.1]$$

where Y is either lambda or sigma and the interaction Hamiltonian, H_I, will be given explicitly later and the expectation value is evaluated over a coherence domain.

We insert into eq. [5.1] a complete sum over intermediate states:

$$V_{\rm eff} = \frac{-i}{2m_Y} \sum_{Y^*} \int_0^{\infty} dt \, \langle Y|\, H_I(\mathbf{x}, t/2)\, |Y^*\rangle \frac{1}{2m_{Y^*}} \langle Y^*|\, H_I(\mathbf{x}', -t/2)\, |Y\rangle, \qquad [5.2]$$

where the intermediate states, Y^* of relevance to our calculation are the Σ^*, the Σ and Λ themselves; the $\Lambda(1405)$ and higher mass states give negligible contributions for reasons which will become clear later. A pictorial representation of eq. [5.2] is given in fig. 11.

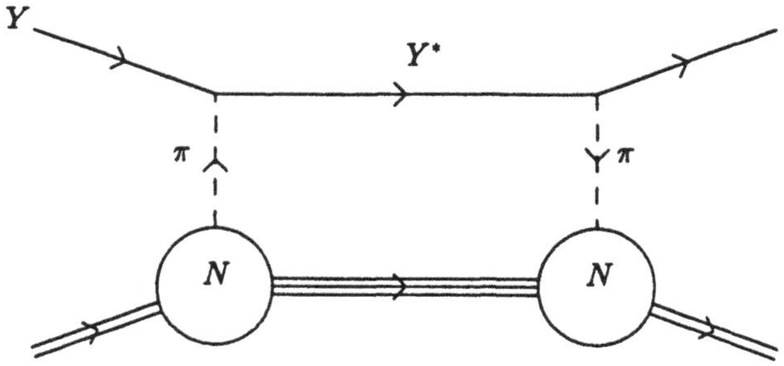

Fig. 11 The hyperon-nucleus interaction via the SR pion condensate, to this should be added the contribution with the pion lines crossed.

The energy denominators coming from the t integral provide an effective cut-off on the intermediate state momentum of the order of several hundred MeV, much larger than the impulse corresponding to the size of a coherence domain (~44 MeV), thus the space integrals in the expectation value of eq. [5.2] are restricted to the region $\mathbf{x} = \mathbf{x}'$. The resulting expression for the effective potential for a hyperon inside the nuclear medium is

$$V_{\text{eff}} = \frac{4}{3} G^2_{\pi YY^*} \omega_q^2 \left\langle |\alpha_\pi(\mathbf{x})|^2 \left[\frac{1}{m_{Y^*} - m_Y - \omega(\mathbf{x})} + \frac{1}{m_{Y^*} - m_Y + \omega(\mathbf{x})} \right] \right\rangle_{CD}, \quad [5.3]$$

where α_π is the pion field amplitude, $\omega(\mathbf{x}) \sim$ 90MeV is the renormalised pion frequency inside the SR nuclear medium and m_{Y^*} implicitly includes a possible imaginary part coming from the finite width in the case of the Y^*. The two different energy denominators correspond respectively to the diagram of fig. 11 and that with the pion lines crossed. The dimensionless coupling constant $G^2_{\pi YY^*}$ is of order unity and is defined by analogy with refs. 2,3 as

$$G^2_{\pi YY^*} = \frac{8\pi^2}{9} \frac{\tilde{g}^2 q^2}{4 m_Y m_{Y^*}} \frac{\rho_{\text{nucl}}}{\omega_q^3}, \quad [5.4]$$

with q and ω_q as in eq. [2.2], ρ_{nucl} is the density of normal nuclear matter and $\tilde{g}$ is now fixed (analogously to g above) by the $\Sigma^* \to \Lambda(\Sigma)\pi$ decay width.

The reasons why the higher mass strange baryonic resonances (including the $\Lambda(1405)$) give negligible contributions are twofold: i) suppression due to the energy denominator of eq. [5.3] and ii) the couplings (as deduced from the widths) are very small. Since the energy denominators for the case $Y^* = \Sigma$ have opposite signs, the resulting cancellation also renders this contribution negligible. Thus by far the dominant coupling is precisely $\Lambda\Sigma^*\pi$ for the lambda potential and $\Sigma\Sigma^*\pi$ for that of the sigma.

In the previous problems we have systematically neglected any spatial dependence inside a CD, however, we have recently made progress in this direction [31] and have now obtained the radial dependence of α_π and ω displayed in fig. 12. The relationship between $\dot{\phi}$ the SR pion-field phase velocity and ω is $\omega = (1 - \dot{\phi})\omega_q$ [2,3].

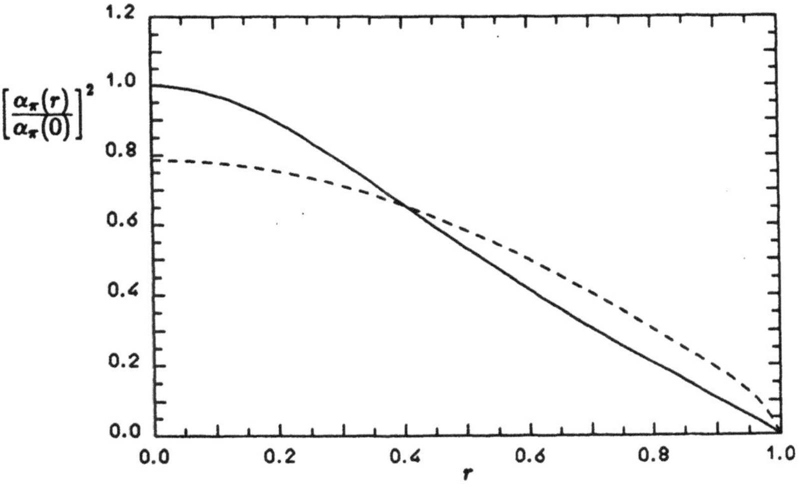

Fig. 12 The radial dependence of the SR pion field, the solid curve represents the pion-field density, $[\alpha_\pi(r)/\alpha_\pi(0)]^2$, and the dashed curve is $\dot{\phi}(r)$; r is in units of $R_{CD} = 4.5\,\text{fm}$.

With this information (and $G^2_{\pi YY^*} \sim 1.2$, $\alpha^2_\pi(0) \sim 0.1$ and $\omega(0) \sim 90\,\text{MeV}$) we obtain the following results:

(i) the hypernuclear potential has the same shape as the nuclear potential with a depth of $\sim 30\,\mathrm{MeV}$ for both the Λ and Σ (slightly shallower in the case of the Σ);

(ii) the width of the states (the imaginary part coming from the Σ^* width) calculated from eq. [5.3] is $\sim 8\,\mathrm{MeV}$ for Σ hypernuclei, while for Λ the system is below the threshold for Σ^* decay and thus has a weak-interaction lifetime.

(iii) due the size of the CD we find that the spin-orbit coupling is suppressed by a factor $\sim (R_{Y^*}/R_{CD})^2$ (R_{Y^*} is the resonance wavelength).

6. Conclusions and Comments

We regard our explanation of the unexpected behaviour of the EMC effect and the longitudinal response function in quasi-elastic scattering as significant confirmation of the correctness and potentialities of the picture of nuclear structure afforded by the coherent π-N-Δ interaction introduced in refs. 2,3.

The undeniable success obtained in this simple, preliminary, calculation of the hypernuclear interaction shows that the universality of shell-model levels for baryons inside a nucleus is profoundly linked to a collective dynamics of nucleons interacting with a coherent pion field inside the nuclear medium. Indeed it is the universality of the structure of such a coherent pion field that is able to guarantee the universality of the shell-model states. The large spin-orbit coupling our model predicts for nucleons in the nucleus [2,3], being an important feature of the *collective nucleon* $\pi N\Delta$ interaction, is not present in the hypernuclear potential for in this case there do not exist such collective effects at the hyperon level and our calculations bear this out.

Looking toward the future then, our approach to the collective dynamics of the nucleus clearly opens the way to simple and straightforward descriptions of many diverse nuclear phenomenon. The developments we foresee in the near future regard the spatial dependence of the SR solutions, which could in particular lead to an understanding of the instability and decay of large nuclei and, on the other hand, to a treatment (not yet possible) of smaller nuclei where finite size effects are likely to become important. It goes without saying that application of the model to all forms of nuclear scattering is expected to lead to a substantially clearer picture of such processes.

REFERENCES

1. G. Preparata, in *Problems of Fundamental Modern Physics*, eds. R. Cherubini, P. Dalpiaz and B. Minetti (World Scientific, 1990), p. 303.
2. G. Preparata, *Nuovo Cim.* **A103** (1990) 1213.
3. G. Preparata, in *Common Problems and Ideas of Modern Physics*, eds. T. Bressani, B. Minetti and A. Zenoni (World Sci., 1992) p. 3.
4. F.E. Close, *An Introduction to Quarks and Partons* (Academic Press, 1979).
5. T. Sloan, G. Smadja and R. Voss, *Phys. Rep.* **162** (1988) 45.
6. EMC collab., J.J. Aubert *et al.*, *Phys. Lett.* **B123** (1983) 275.
7. EMC collab., M. Arneodo *et al.*, *Nucl. Phys.* **B333** (1990) 1.
8. SLAC E-139, R.G. Arnold *et al.*, *Phys. Rev. Lett.* **52** (1984) 727.
9. BCDMS collab., A.C. Benvenuti *et al.*, *Phys. Lett.* **B189** (1987) 483.
10. S.J. Brodsky, F.E. Close and J.F. Gunion, *Phys. Rev.* **D6** (1972) 177.
11. J. Qiu, *Nucl. Phys.* **B291** (1987) 746.
12. P.G. Ratcliffe, in *Common Problems and Ideas of Modern Physics*, eds. T. Bressani, B. Minetti and A. Zenoni (World Sci., 1992) p. 71.
13. L. Angelini, L. Nitti, M. Pellicoro and G. Preparata, *Phys. Rev.* **D41** (1990) 2081.
14. D.W. Duke and J.F. Owens, *Phys. Rev.* **D30** (1984) 49.
15. J.F. Owens, *Phys. Rev.* **D30** (1984) 943.
16. P.D.B. Collins and A.D. Martin, *Hadron Interactions* (Adam Hilger, 1984), p. 81.
17. NMC collab., P. Amaudruz *et al.*, CERN preprint PPE/91-52 (1991).
18. J.P. Chen *et al.*, *Phys. Rev. Lett.* **66** (1991) 1283.
19. Z.E. Meziani *et al.*, *Phys. Rev. Lett.* **52** (1984) 2130; *Phys. Rev. Lett.* **54** (1985) 1233.
20. R.M. Altemus *et al.*, *Phys. Rev. Lett.* **54** (1980) 965.
21. T. de Forest, Jr., *Nucl. Phys.* **A414** (1984) 347.
22. J.W. Van Orden, Ph.D. thesis, Stanford University, 1978.
23. X. Ji, *Phys. Lett.* **B219** (1989) 143.
24. D.B. Day, in *Proceedings of the Workshop on Momentum Distributions*, ed. R.N. Silver and P.E. Sokol (Plenum, New York, 1989).
25. G. Preparata and P.G. Ratcliffe, *A Novel Approach to Nuclear Shadowing*, Milano preprint MITH 91/13.
26. G. Preparata and P.G. Ratcliffe, *Phys. Lett.* **B276** (1992) 219.
27. R. Bertini, *Europhys. News* **22** (1991) 115.
28. T. Bressani, in *Common Problems and Ideas of Modern Physics*, eds. T. Bressani, B. Minetti and A. Zenoni (World Sci., 1992) p. 211.
29. W. Brückner *et al.*, *Phys. Lett.* **B79** (1978) 157.
30. H.J. Pirner, *Phys. Lett.* **B85** (1979) 190.
31. G. Preparata and P.G. Ratcliffe, work in progress.

Common Problems and Trends of Modern Physics, pp. 297–306
ed. T. Bressani, S. Marcello and A. Zenoni
©1992 World Scientific Publishing Company

Σ-Hypernuclei: an overview.

Simonetta Marcello
Istituto Nazionale di Fisica Nucleare, Via P. Giuria 1, 10125 Torino, Italy

ABSTRACT

Hyperons embedded in the nucleus can be very useful probes to reveal quark degrees of freedom in nuclei. Furthermore they are distinguishable from other nucleons, because the hyperon shell-model orbits are not Pauli blocked, allowing to observe all the levels down to the ground state. The current puzzle of Σ-hypernuclei is described and a new experiment with stopped K^- is suggested for the facility DAΦNE, the new Frascati φ factory.

1. Introduction

Since 1953 when the hypernuclei were discovered by M.Danysz and J.Pniewski in photografic emulsions exposed to the cosmic rays, the experimental efforts were devoted to study the Λ-hypernuclei using K^- beams and a lot of data have been produced[1].

On the other hand, few data with poor statistics exists on Σ-hypernuclei and not all of them confirm the existence of such a nuclear system.

The strength of the Λ-N interaction is of the same order of magnitude than that of the N-N interaction, then one or more nucleons in the nucleus can be replaced by one or more Λ particles forming a nuclear system known as hypernucleus. In principle, we can expect the same process occuring for other hyperons (Σ, Ξ, etc.). Such a nuclear system offers the possibility to investigate the effect of quark degrees of freedom in nuclei, in particular the hyperon-nucleon spin-orbit interaction can reveal such effects, due to its short range nature. The Λ spin-orbit interaction has been found weak ($V^{ls} \approx 0$), as expected by the additive quark model [2], on the contrary the Σ spin-orbit splitting predictions are strikingly different.

Moreover the hyperons in a nucleus are distinguishable from other nucleons and the shell model orbits are not Pauli blocked allowing to observe all the levels down to the

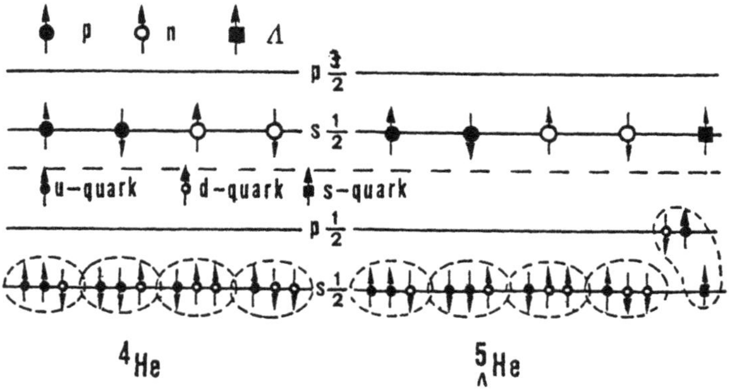

Fig.1. Baryon (upper row) and quark (lower row) configuration for the ^{4}He and $^5_\Lambda$He systems.

ground state; at the moment the observation of the lowest-lying shell-model states is possible only in light nuclei.

On the other hand Pauli blocking at the quark level could occur in a hypernucleus, due to the overlap hyperon-nucleons, and we should be able to reveal a shift in the single particle energies in heavy hypernuclei.

In fig. 1 two possible configurations for ^{4}He nucleus and $^5_\Lambda$He hypernucleus are shown: in the upper one the quark components of the baryons are fully confined; in the lower one the quarks are deconfined. For the ^{4}He nucleus the differences between the two descriptions are not substantial; on the contrary for the $^5_\Lambda$He all the baryons can occupy the lowest energy state ($s_{1/2}$) in the traditional description, while the u and d quarks need the second energy state ($p_{3/2}$) in the quark description.

This effect at quark level can be responsible for the anomalous small Λ binding energy in ^{5}He.

2. Σ-hypernuclei puzzle

The mean life of Λ-hyperon is ~ 10^{-10}s, greater than the typical nuclear interaction time (~ 10^{-22}s), then the Λ-hypernuclear system is stable enough to study the quantum-mechanical states of the Λ into the nucleus.

Even if the mean life of Σ particle is the same, it was a surprise to discover signatures of Σ-hypernuclei in 1979 at CERN[3] in the (K^-,π^-) reaction on ^{9}Be at $p_K = 720$ MeV/c and $\vartheta = 0°$.

Fig. 2 shows the spectrum plotted as a function of the transformation energy $M_{Hy} - M_A = (m_{hy} - m_N) - (B_{hy} - B_N)$, where M_{Hy} is the mass of the hypernuclear system and M_A the target mass, m_{hy} and m_n are respectively the hyperon mass and the nucleon mass, B_{hy} and B_N the binding energy. The Λ binding energy scale is given at the bottom. The two higher peaks stem from the recoilless Λ production and the two smaller ones, shifted by roughly the Λ-Σ mass difference (~ 80 MeV), were ascribed to Σ-hypernuclear states. Their width is about 8 MeV.

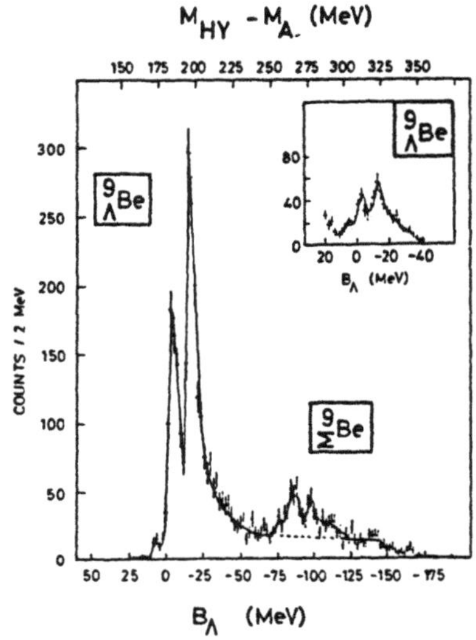

Fig. 2. Spectrum of the (K^-,π^-) reaction on ^{9}Be at a k momentum of 720 MeV/c (Ref. 3)

The existence of such narrow Σ-hypernuclear states was unexpected because, unlike the Λ-hypernuclear states, the conversion reaction $\Sigma + N \rightarrow \Lambda + N$, by strong interaction, is allowed in the nucleus. Then the life time of a Σ in the nuclear matter was thought to be of the order of 10^{-22}s. An energy width of about 30 MeV was expected, due to the Heisemberg principle, for a Σ-hypernuclear bound state.

Since this first evidence of Σ-hypernuclear states, several experiments tried to confirm again their existence. Two main methods were used: with K^- in flight and with K^- at rest.

2.1 Experiments with K^- in flight

We can obtain Σ-hypernuclei by using K^- beams with the interactions:

$$K^- + {}^AZ \rightarrow {}^A_{\Sigma^+}(Z-1) + \pi^- \qquad (1)$$

$$K^- + {}^AZ \rightarrow {}^A_{\Sigma^0}Z + \pi^- \qquad (2)$$

$$K^- + {}^AZ \rightarrow {}^A_\Sigma(Z\text{-}1) + \pi^+ \qquad (3);$$

corresponding respectively to the elementary processes:

$$K^- + p \rightarrow \Sigma^+ + \pi^- \qquad (1bis)$$
$$K^- + n \rightarrow \Sigma^0 + \pi^- \qquad (2bis)$$
$$K^- + p \rightarrow \Sigma^- + \pi^+ \qquad (3bis)$$

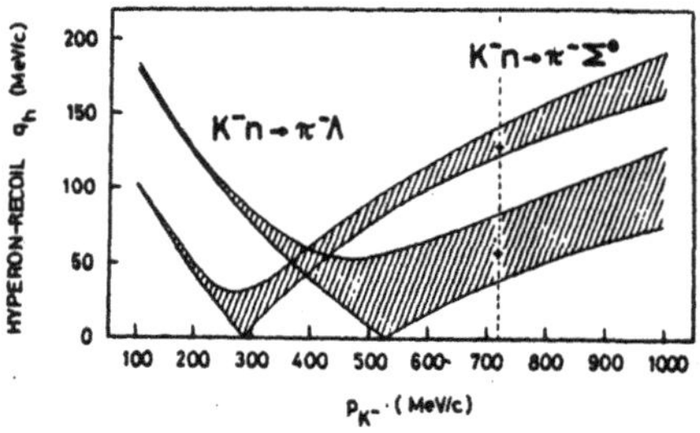

Fig. 3. Kinematics in the forward direction for the Λ and Σ^0 production.

Fig. 3 shows the recoil momentum for Λ and Σ as a function of the incoming K^- momentum. It's worth to notice the existence of a "magic" value of the K^- momentum (~ 280 MeV/c), for which the Σ is produced at rest and the π^- carries a momentum equal to the incident one.

At this "magic" momentum the hypernucleus is produced in a substitutional excited state, that means the nucleon of the target nucleus is replaced by a Σ, without any change in the spatial and spin wave function.

In order to reduce the background, due to quasifree Σ-production, the CERN experiment used a K^- beam at lower momentum (400 - 450 MeV/c) [4,5], close to the "magic" momentum on C, O and Li as targets.

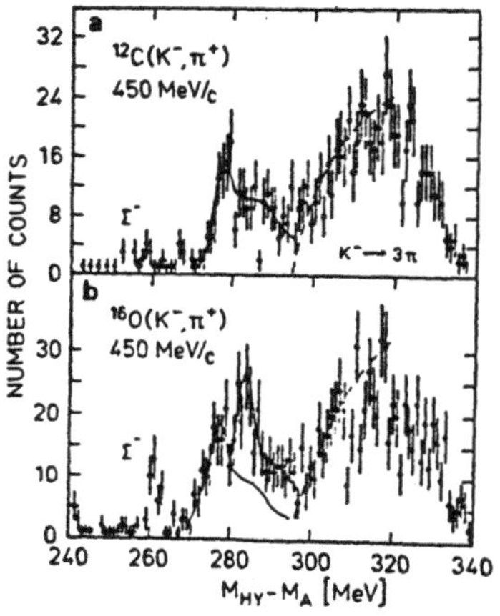

Fig.4. Σ hypernuclear spectra of the (K^-,π^+) reaction from ^{12}C (a) and ^{16}O (b) at 450 MeV/c (Ref. 5).

Some results concerning C and O are shown in Fig.4. The oxygen spectrum shows two peaks corresponding to the Σ production on nucleons of the $p_{1/2}$ and $p_{3/2}$ shells.

In order to determine isospin dependence of the Σ-nucleon interaction, the measurements have been performed at different incident K^--momenta. As it's shown in

Fig. 5 the free K⁻ - N cross sections are strikingly different for the Σ^0 and Σ^+ production at different momenta.

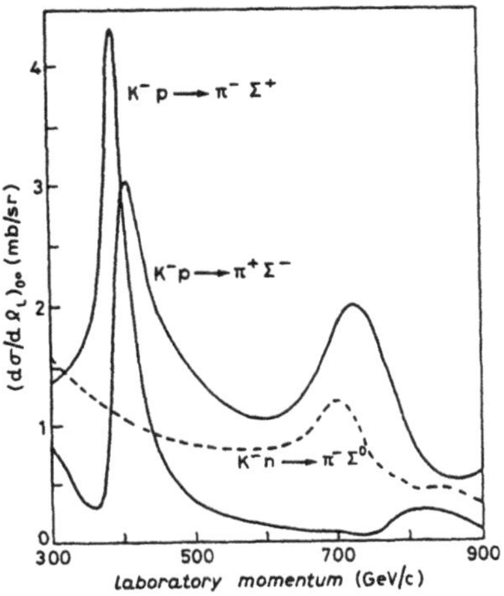

Fig. 5. The free $(K^-, \pi^\pm)$ cross-section as a function of the K^- momentum.

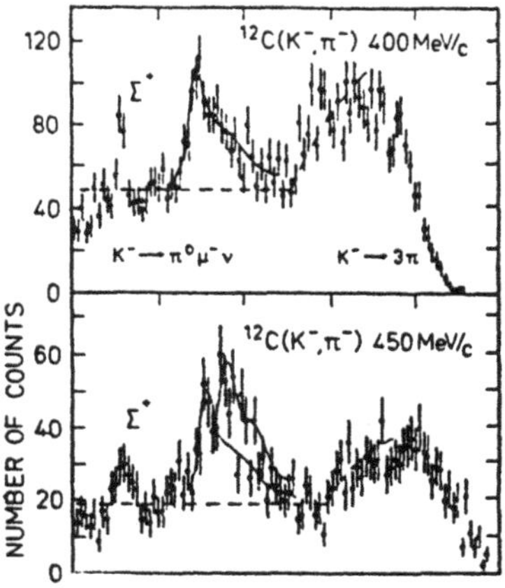

Fig. 6. M_{hy} - M_A for the (K^-,π^-) reaction on ^{12}C at 400 (top) and 450 (bottom) MeV/c.

At 400 MeV/c the production of Σ^+ should dominate, at 450 MeV/c the two cross sections for Σ^+ and Σ^0 production are almost the same. The measured missing mass spectra for the (K^-, π^-) reaction on ^{12}C are shown in Fig. 6. At 400 MeV/c the (K^-, π^-) reaction on ^{12}C populates one state at M_{Hy} - M_A = 270 MeV, whereas at 450 MeV/c two states are produced one at 270 MeV and the second at 275 MeV. The difference of the two spectra was interpreted as the production of Σ^+ alone and Σ^0 and Σ^+, respectively.

Another experiment [6,7], looking for Σ bound states, has been performed at BNL, using K^- in flight at p_K = 713 MeV/c and measuring the (K^-, π^+) reaction on several targets: ^{16}O, ^{12}C, ^6Li and ^7Li.

For the ^6Li target two states similar to the ones of the Λ were measured, but they were explained as a reflection of the Λ-hypernuclear states, $(1p^{-1}, 1p)_{p\Lambda}$ and $(1s^{-1}, 1s)_{n\Lambda}$.

No enhancement has been found for the O and C targets probably due to the overwhelming background, because of the quasi free Σ-production. Only a special tagging method allowed to obtain a small signal in the same region of the peak measured at CERN.

2.2 Experiments with stopped K⁻

In 1985 another experimental method, using K^- at rest, was performed by a group working at KEK[8]. This method was already used for the study of Λ hypernuclei in 1973.

There are two main advantages on using stopped K^-. First there is a high production rate of hyperons, that means it's possible to achieve high statistics in a few time. Second the large momentum transfer in the reaction allows the population of all the states and the detection of the ground state of Σ hypernuclei is more efficient. Then the Σ binding energy can be readly obtained.

New techniques of tagging were developed in order to overcome the large background due to the quasi-free Σ production. The basic idea was to identify the Λs, coming from the conversion reaction $\Sigma + N \rightarrow \Lambda + N$ of the Σ trapped in the nucleus, by detecting the decay products $p\pi^-$ or $n\pi^0$ in the counters around the target and to identify the interaction point in an active target of $(CH)_n$.

The (K^-, π^+) reaction on ^{12}C was studied and after the analysis three peaks stood out above the suppressed quasi-free continuum in the π^0-tagged spectrum, as shown in Fig. 7. Two of them at $M_{Hy} - M_A = 277.2 \pm 0.3$ MeV and $M_{Hy} - M_A = 281.8 \pm 0.5$ MeV were interpreted like $\Sigma^-(p_{1/2})$ and $\Sigma^-(p_{3/2})$ levels and the spin-orbit splitting of Σ^- was deduced to be 5 MeV. The interpretation of the third peak was uncertain and it was left unexplained.

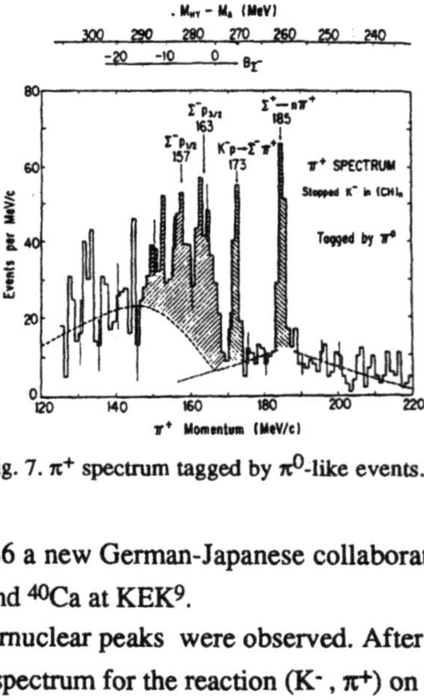

Fig. 7. π^+ spectrum tagged by π^0-like events.

The results were so surprising that in 1986 a new German-Japanese collaboration started data taking stopping K^- in ^{12}C, ^{16}O, ^{6}Li and ^{40}Ca at KEK[9].

In the inclusive spectra no narrow Σ-hypernuclear peaks were observed. After the tagging analysis a peak was found (Fig.8) in the spectrum for the reaction (K^-, π^+) on ^{12}C in the same position of the hypernuclear state observed (Fig.4a) with 450 MeV/c K^- beam at CERN[4,5], but the statistical significance was low.

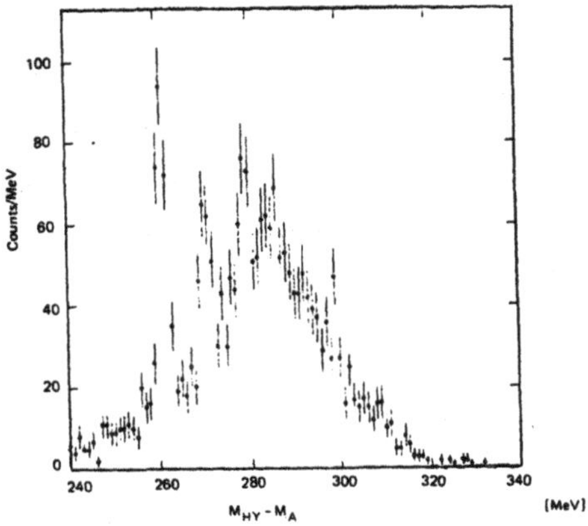

Fig. 8. Neutral tagged spectrum from Ref. 9.

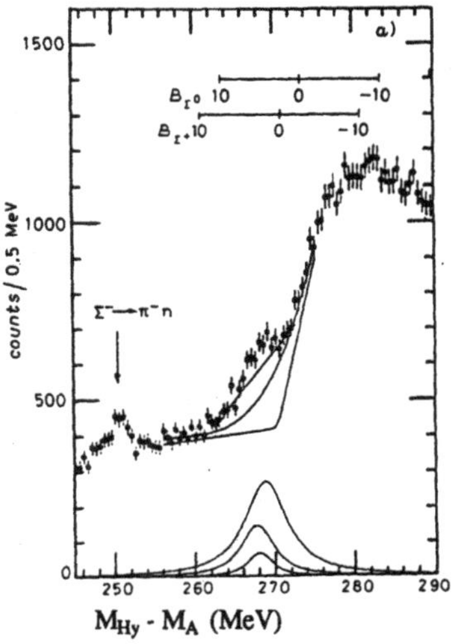

Fig. 9. π^- spectrum, three different background assumptions and corresponding peak shapes are shown (Ref. 11).

A theoretical paper[10], predicting the existence of only one bound state of I = 1/2 and S = 0 with a narrow width in ${}^4_\Sigma$He, led the Japanese group to look for Σ bound states with kaons at rest on a liquid helium target[11].

In the (K^-, π^-) spectrum a shoulder-like structure around ΔM ≈ 268 MeV in the Σ^+ bound region was observed. On the other hand the (K^-, π^+) spectrum didn't have any enhancement in the Σ^- bound region.

Assuming three different behaviours for the background, they found different position, width and height of the fitted peak, as shown in Fig. 9, and in all cases the statistical significance was good.

The peak was assigned to the formation of the ground state of ${}^4_\Sigma$He with a position of 267.7 ± 0.3 MeV and a width of 4.6 ± 0.5 MeV.

The difference in the π^- and π^+ spectra suggested that there is an isospin dependence in the Σ-nucleus interaction. The (K^-, π^+) reaction can produce I = 3/2 states, while the (K^-, π^-) reaction can produce both I = 1/2 and I = 3/2 states. Considering the two spectra, the absence of peak in the (K^-, π^+) spectrum led to assign an isospin 1/2 to the bound state in the (K^-, π^-) spectrum as foreseen by Akaishi et al.[10].

3. FiNuDA Experiment

This short review shows that the experimental situation until now is controversial. The existent data have been used to extract the Σ-nucleus potential depth, and in particular the spin-orbit potential, but no unique information is available and the question is still open.

The theoretical interpretations are based essentially on the assumption that the two-body Σ-N potential is strongly spin-isospin dependent[12], then strong energy shifts are foreseen, due to the residual Σ particle-nucleon hole interaction, and the Σ-hypernuclear states can be produced.

In order to clarify the Σ puzzle, new experiments with more statistics are required.

The problem of low statistics for the experiment with K^- in flight will be probably solved in the next future by using the low momentum high intensity K^- beam at the new KAON facility at TRIUMF[13].

New possibilities for better understanding of the Σ spectroscopy will arise from the new Φ-factory in Frascati. The Φ meson will be produced in an e^+e^- collider with a c.m. energy of 1.02 GeV at a luminosity of 10^{33} $cm^{-2}s^{-1}$. The $\Phi \rightarrow K^+K^-$ decay provides an intense source (I = 2 kHz) of monoenergetic (E_{kin} = 16 MeV) K^-, which can be stopped in a thin target producing ~ 7000 hypernuclei per hour.

The new experiment FiNuDA[14] has been planed to study the hypernuclear spectroscopy with stopped K^-.

In spite of the large overwhelming background due to the quasi-free continuum using K^- at rest, the Σ-nucleus interaction could be studied because of the particular tagging capabilities of the FiNuDA detector.

For instance, let us examine how the Σ^+-bound state, coming from the interaction (1), could be discriminated from other processes in the π^- spectrum[15] shown in Fig. 10.

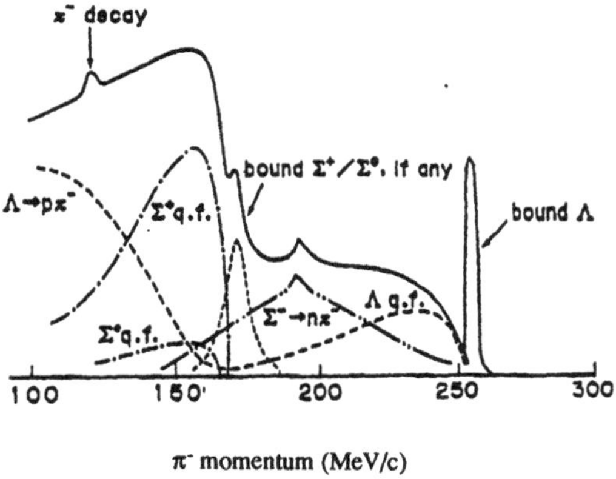

Fig. 10. Expected inclusive π^- momentum spectrum from nuclear absorption of K^- at rest (Ref. 15).

The detection at the trigger level[16] of μ^+, coming from $K^+ \to \mu^+\nu_\mu$, identifies unambigously the presence of K^- (Fig. 11).

The Σ^+ bound undergoes $\Sigma^+ + n \to \Lambda + p$, by requiring off line the Λ and the p on the reaction point, the background from $\Sigma^- \to n\,\pi^-$ can be supressed, the Σ^+ quasi-free part is not completely removed, because the Σ^+ can be absorbed by other nuclei, emitting Λ particles.

Probably in the future both methods, K^- in flight[13] and K^- at rest[14], will be able to throw light on the Σ puzzle.

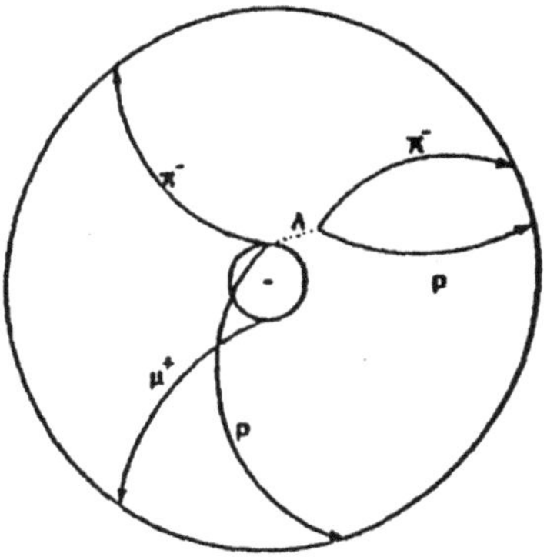

Fig. 11. The (K^-,π^-) reaction for Σ^+-bound production at p_π 170 MeV/c (Ref. 15).

1. References

1. T. Bressani, in *Proc. of "Hadronic Physics at Intermediate Energy"*, Folgaria, February 17-22, 1986, Eds. T. Bressani and R. A. Ricci, p. 259.
2. H. J. Pirner et al., *Phys. Lett.* **B85** (1979) 190.
3. R. Bertini et al., *Phys. Lett.* **B90** (1980) 375.
4. R. Bertini et al., *Phys. Lett.* **B136** (1984) 29.
5. R. Bertini et al., *Phys. Lett.* **B158** (1985) 19.
6. H. Piekarz et al., *Phys. Lett.* **B110** (1982) 428.
7. L. G. Tang et al., *Phys. Rev.* **C38** (1988) 846.
8. T. Yamazaki et al., *Phys. Lett.* **B54** (1985) 102.
9. S. Paul et al., *Nucl. Phys.* **A479** (1988) 137c.
10. Y. Akaishi et al., *Few-Body Systems.*, Suppl., **No. 1** (1986) 120 (*Proc. of the European Conference on Few-Body Systems*).
11. R. S. Hayano et al., *Phys. Lett..* **B231** (1989) 355.
12. C. B. Dover et al., *Phys. Rep.* **184, No.1** (1989) 1.
13. S. Paul et al., *Zeit. Phys..* **C46**, (1990) 51.

14. T. Bressani, in *Proc. of Whorkshop on Physics and Detectors for DAΦNE*, Frascati, April 9-12, 1991, Eds. G. Pancheri, p. 475.
15. T. Yamazaki, in *Proc. of Whorkshop on Physics and Detectors for DAΦNE*, Frascati, April 9-12, 1991, Eds. G. Pancheri, p. 315.
16. P. Gianotti and E. Rossetto, this volume.

Common Problems and Trends of Modern Physics, pp. 307–313
ed. T. Bressani, S. Marcello and A. Zenoni
©1992 World Scientific Publishing Company

Low-momentum K⁺-nucleus interaction

V. Lucherini
I.N.F.N. Laboratori Nazionali di Frascati - 00044 Frascati - Italy

ABSTRACT

The properties of low-momentum K^+-nucleus scattering to investigate the behavior of nucleons inside nuclei are reviewed. The existing experimental results below 800 MeV/c of K^+ momenta for the ratio R between the total K^+ cross section on Carbon and Deuteron are compared with a conventional calculation using optical potential and starting from the elementary K^+-nucleon amplitude. The deviations between such calculation and experimental results are discussed on the light of possible modifications of nucleon properties inside nuclei or direct contributions of exchanged pions. The possibility to pursue an experiment of K^+-nucleus scattering at the very low K^+-momentum of 110 MeV/c with DAΦNE at Frascati are briefly examined.

1. Introduction

Low-momentum K^+-nucleus scattering holds a very special position in investigating the dynamical behavior of nucleons embedded inside nuclei. K^+ of momenta less than 500 MeV/c have only a very "weak" interaction with nucleons, they have mean free path in nuclei larger than 6 fm (Fig.1) thus probing the entire nuclear volume, including neutron distributions.

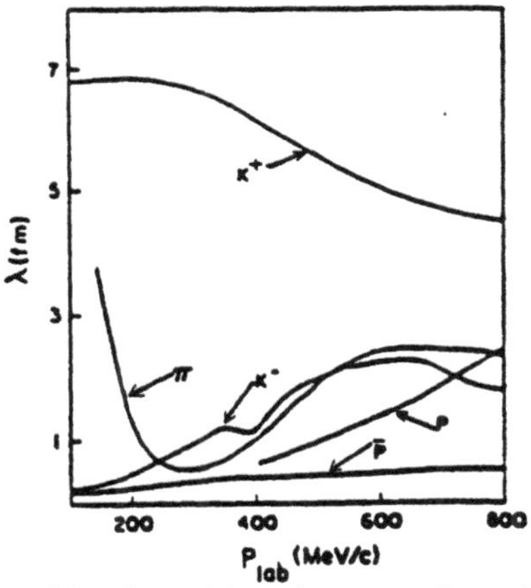

Fig.1 Mean free path in nuclear matter of strongly interacting particles.

Moreover, since single- and double-scattering dominate K^+-nucleus interactions, reliable calculations are possible for the K^+ interaction with nuclei, based on the elementary K^+-nucleon amplitude, and deviations of calculations from experimental results can be interpreted in a more straightforward way. These properties make the low-momentum K^+ the ideal hadronic probe of possible nuclear-medium effects on nucleons, and a useful low-momentum complement to high-momenta leptonic and electro-magnetic probes.

Given these properties of K^+ as a *clean hadronic probe*, it was surprising to find experimentally that the ratio R between the total K^+ cross section on Carbon and Deuteron (each one normalized to the respective number of nucleons, as defined by the following relation):

$$R = \frac{2\sigma_T(K^{+\,12}C)}{12\sigma_T(K^+d)}$$

was R≠1 (>1) for 450÷900 MeV/c K^+-lab momenta ("EMC-like effect"), as it can be seen in Fig.2.

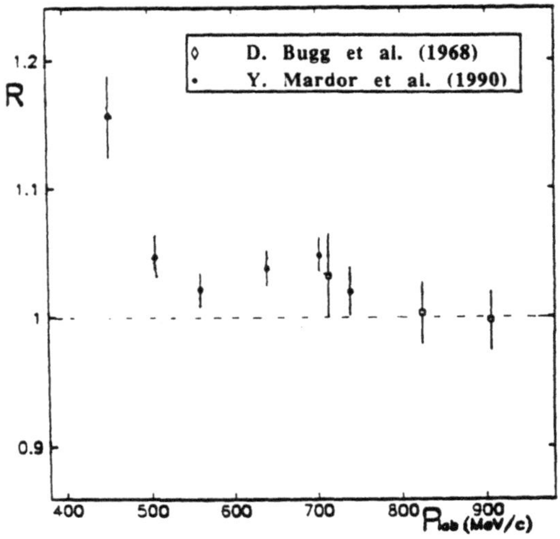

Fig.2 Ratio R between the total K^+ cross section on Carbon and Deuteron (each one normalized to the respective number of nucleons) measured in the transmission experiments of D. Bugg et al.[1] (open points) and Y. Mardor et al.[2] (closed points).

Accurate theoretical calculations based on conventional reaction mechanisms (optical model starting from the elementary K+-nucleon amplitude) suggest instead a moderate shadowing, at least at Lab. momenta > 600 MeV/c. For these reasons, possible non-conventional explanations of the experimental data have been explored, such as the "swelling" of nucleons inside nuclei, in analogy with some interpretations of the EMC effect. Also possible direct contribution of the exchanged pions inside nuclei has been considered. The construction at Frascati of the ϕ-Factory DAΦNE offers the possibility to extend the measurement of K^+-nucleon at the very low momentum of 110 MeV/c, at which no data exists and will be briefly examined in the last paragraph.

2. K^+-nucleon scattering and K^+-nucleus optical potential

In a series a papers P. Siegel, W.B. Kaufmann and W.R. Gibbs[3,4] address the questions:
a) How serious is the deviation of expectations from experiment? **b**) Can nuclear effects account for the discrepancy? In the conventional view of nuclear physics nucleons properties are unchanged within the nucleus. On a deeper level, however, the nucleon is a dynamic quantity which should respond to its surroundings: but before to assess any such conclusion the point **b**) must be fully exploited.

The K^+-nuclear scattering has been calculated with microscopic optical potential, incorporating: nucleon binding, recoil and Pauli blocking, with a three-body model of meson, struck nucleon and core. To "detect" exotic contributions to the total K^+-nucleus cross section several "traditional" effects have been carefully examined: i) change in the total cross section due to the oblateness of ^{12}C; ii) effects of second order scattering due to nucleon-nucleon correlation; iii) double spin flip scattering; iv) relativistic off-shell corrections; v) nuclear density effects; vi) sequential double charge exchange: $K^+n_1 \rightarrow K^o p_1$; $K^o p_2 \rightarrow K^+ n_2$. The final result of these calculations for the ratio R is showed in Fig.3 as a band representing the possible uncertainties in the conventional theoretical predictions. As seen, it is not possible to explain the obtained experimental data.

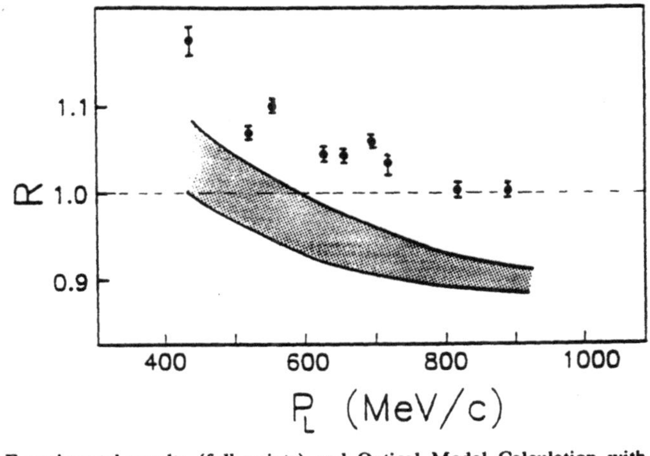

Fig.3 Experimental results (full points) and Optical Model Calculation with its theoretical uncertainties (gray band) for the normalized ratio R of the total K^+ cross section on Carbon and Deuteron

Following W.Weise[5], a simplified eikonal approximation helps to understand the Optical Model results. The forward scattering of a K^+ with impact parameter b from a N=Z nucleus is described by the eikonal phase $\chi(b)$, in terms of the elementary amplitudes f on proton and neutron:

$$\chi(b) = \frac{\bar{\sigma}}{4}(i+\beta)\int_{-\infty}^{\infty} dz\, \rho(z)$$

where:

$$\beta = \frac{\mathrm{Re}\, \bar{f}(\theta=0)}{\mathrm{Im}\, \bar{f}(\theta=0)}$$

$$\bar{f}(K^+\mathcal{N}) = \frac{1}{2}[f(K^+p)+f(K^+n)]$$

$$\bar{\sigma}(K^+\mathcal{N}) = \frac{1}{2}[\sigma(K^+p)+\sigma(K^+n)]$$

If the nucleus has a Gaussian density distribution $\rho(r)$ with r.m.s. radius:

$$\langle r^2 \rangle^{\frac{1}{2}} = r_0 A^{\frac{1}{3}}$$

the eikonal phase has the closed expression:

$$\chi(b) = \frac{\bar{\sigma}(\beta+i)}{4}\exp\left[-\frac{b^2}{2r_0^2 A^{\frac{2}{3}}}\right]$$

In term of the eikonal phase $\chi(b)$ the total K^+-nucleus cross section is:

$$\sigma(K^+A) = 4\pi\chi(b)\int_0^{\infty} bdb\{1-e^{-2\mathrm{Im}\chi(b)}\cos(2\mathrm{Re}\chi(b))\}$$

Expanding the "exact" eikonal K^+-nucleus cross section up to second order in $\chi(b)$ i.e. including only single and double scattering (respectively, 90% and 8% of σ_{tot}), we obtain:

$$\sigma(K^+A) \cong A\bar{\sigma}(K^+\mathcal{N})\left[1-\frac{3}{8}\frac{\bar{\sigma}(K^+\mathcal{N})}{\pi r_0^2}A^{\frac{1}{3}}(1-\beta^2)+\ldots\right]$$

This expression qualitatively explains the theoretical behavior shown in Fig.3 for the ratio R. The relevant observation is that: β decreases in magnitude from low momenta to $|\beta| \approx 1$ at $p_L(K^+)=(500 \div 600)$ MeV/c. Hence the ratio β determines the degree of shadowing (or antishadowing) in K^+-nucleus cross section.

3. K^+-nucleus scattering and non-conventional effects

Since , within the classical picture of unchanged nucleons inside nuclei, the reconstructed (with great reliability) K^+-nucleus interaction is unable to reproduce the experimental data, various attempts have been made to reconcile experimental data with theoretical (not traditional) calculations. A list of the followed approaches includes: i) there is nucleon swelling;[4] ii) density dependence of vectors mesons exchanged.[6] The eventual presence of NN bubbles in nuclear ground state has also been considered. In a more traditional approach, the explicit contribution of exchanged pions has been taken into account:[7] this last point is interesting in view of the predictions of pions in nuclei of the Superradiance theory.[8]

The results of these calculations show a better agreement with data, as can be seen in Fig. 4a and 4b.

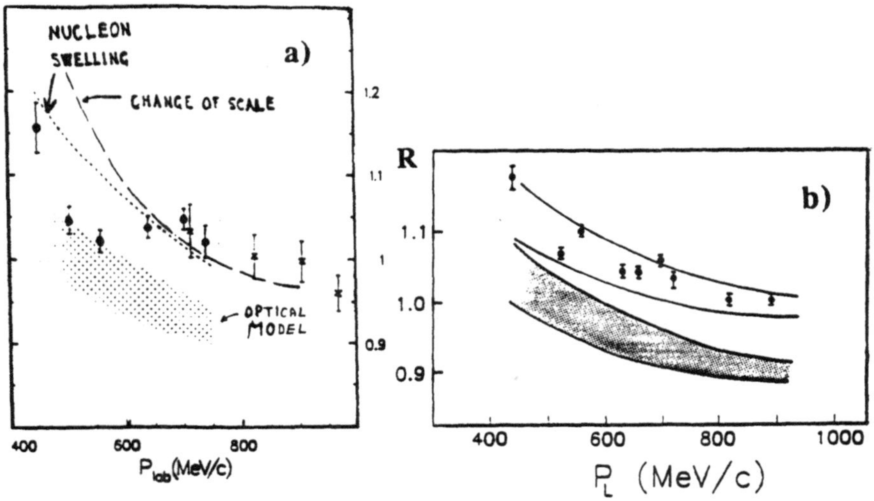

Fig.4 Experimental results (full points) and calculation for the normalized ratio R of the total K^+ cross section on Carbon and Deuteron: **a)** dashed curve indicates the effects of "nucleon swelling",[4] consistent with the parameters chosen by Close et al.[9] to explain the EMC effect; the dot-dashed curve is the result of calculation of ref.6; **b)** the two full curves show the predictions of direct pion contribution.[7] The shaded area indicates the prediction of the optical model.

Clearly, before any definite conclusion could be taken, it is very important to confirm the (rather scarce) set of experimental data, both studying other nuclei and extending at lower energies such measurements.

4. Experimental aspects of K^+-nucleus scattering

Experimentally, the two experiments up to now done have been typical transmission experiments, in which the total K^+-nucleus cross section is obtained in a two step procedure (target In and Out) and using an extrapolation at zero degrees of quantities measured at larger angles.[1,2]

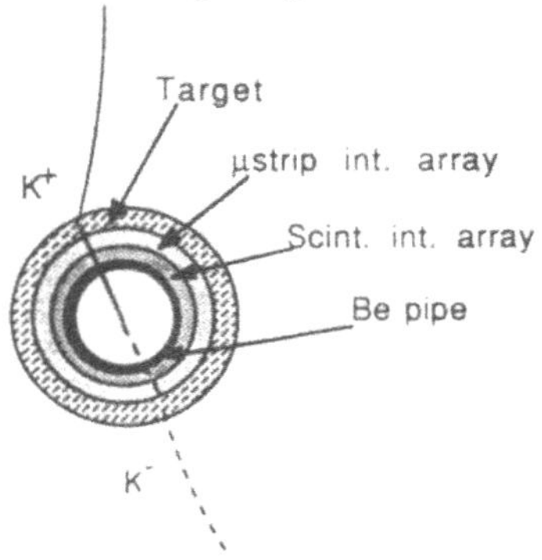

Fig.5 A sketch of the interacting region with the target and the inner detectors.

The planned construction at Frascati of the Φ-factory DAΦNE,[10] would allow a direct K^+-nucleus total cross section measurement at the very low K^+-momentum of 110 MeV/c, using the K^+ from Φ decay. The monochromatic (127 MeV/c) K^+s from Φ decay, after traversing the Be pipe, and a thin internal scintillator and µstrip array (Fig.5), scatter (at 110 MeV/c) with the nuclei of the selected target, cylindrically shaped, around the pipe.

The incoming K^+ direction can be determined using the internal µstrip information of the impact points of both K^+ and K^- (tagging), while the fast scintillator signals may be used as trigger and for dE/dx.

The direction of the scattered K^+ can be accurately determined with the trajectory reconstruction given by the same set of drift chambers in magnetic field to be installed for the study of hypernuclei formation and decay (see P. Gianotti, contribution to this Course). From an experimental point of view, the use of such low energy K^+ beam imposes severe constraints on the choice of the target thickness, and, in general of all interposed thicknesses, to minimize energy losses and multiple scattering.

The following Table 1 compares a typical transmission-like experiment as that of ref.2 with a possible experiment with DAΦNE. The predominant contribution at this low momenta of Rutherford scattering also at angles of several degrees may, however, be the most severe limitation to the effective useful solid angle, and hence, to the counting rate.

In conclusion, the study of K^+-nucleus scattering at low momenta allows to test several interesting and fundamental questions about the behavior of nucleons inside nuclei, and the use of DAΦNE in this respect (at the planned maximum luminosity) seems feasible.

Table 1. Characteristics of a typical low-momentum K^+ scattering experiment with an extracted K^+ beam and a possible experiment at DAΦNE .

Characteristic	Extracted beam	DAΦNE (at $L=10^{33}$ cm^{-2}s^{-1})
K^+ lab. mom. (MeV/c)	450	110 (after 200μm Be, 1mm scint. and 200 μm Si
Background	$\pi^+/K^+ = 10$	Free (tagging!)
Intrinsic beam angular resolution (mrad)	35	5
Extrapolation of exp. data	Yes	No
Target thickness (C,g/cm^2)	3.6	0.025 (to have similar target effects)
Counting rate (s^{-1})	4.4	0.15 (assumed overall efficiency: 50%)

References

1) D. Bugg *et al.*, Phys. Rev. C **25**, 2619 (1982).
2) Y. Mardor *et al.*, Phys. Rev. Lett. **65**, 2110 (1990).
3) P.B. Siegel *et al.*, Phys. Rev. C **30**, 1256 (1984).
4) P.B Siegel *et al.*, Phys. Rev. C **31**, 2184 (1985).
5) W. Weise, Il Nuovo Cimento **120A**, 265 (1989).
6) G.E. Brown *et al.*, Phys. Rev. Lett. **60**, 2723 (1988).
7) S.V. Akulinichev, Phys. Rev. Lett. **68**, 290 (1992).
8) G. Preparata, Il Nuovo Cimento **103A**, 1213 (1990).
9) F.E. Close *et al.*, Phys. Rev. D **31**, 1004 (1985).
10) Proposal for a Φ-Factory, Frascati Internal Report **LNF-90/031** (1990).

Common Problems and Trends of Modern Physics, pp. 315–330
ed. T. Bressani, S. Marcello and A. Zenoni
©1992 World Scientific Publishing Company

SEMICLASSICAL DESCRIPTION OF HEAVY ION COLLISIONS

ALDO BONASERA

INFN, Sezione di Catania, c.so Italia 57
95129 Catania, Italy

and

FRANCESCA GULMINELLI

Dipartimento di Fisica and INFN, Sez. Milano, v.Celoria 16
20133 Milano, Italy

ABSTRACT

Semiclassical transport equations are presented as a general tool to analyze nuclear dynamics in different energy regimes, from low energy to relativistic heavy ion collisions. In particular the relevance of high order correlations in heavy ion collisions is discussed. An extension of the Boltzmann Nordheim Vlasov (BNV) equation to include three body collisions is proposed and compared to Enskog's theory. Then a general method to evaluate fluctuations of the semiclassical one body Wigner function is presented both in the equilibrium and non equilibrium framework. The model is valid far away from critical points, where BNV represents a good approximation of the average Wigner function. Different applications (collective flow, particle production, relaxation times of collective modes) are studied and compared to experimental data.

1. Introduction

The aim of nuclear physics is to understand the complex behaviour of atomic nuclei. Heavy ion physics at intermediate energy uses collisions of nuclei from 50 to 1000 MeV/nucleon to probe the properties of nuclear matter and contribute to this understanding. It is hoped that through a detailed explication of the experimental data and the theoretical models, one can accurately describe the ephemeral nucleus-nucleus interaction.

As a matter of fact, it is not possible to extract the nuclear equation of state (EOS) directly from experimental data, since the EOS is a property of equilibrated nuclear matter, and in the highly non equilibrium situation of a heavy ion collision, portions of equilibrated nuclear matter, if any, survive only for a very short time compared to the reaction time. It is only with microscopic phase space simulations that it is possible to disentangle the non equilibrium from the equilibrium component, that is to keep track of the nucleon-nucleon collisional dynamics which determines the degree of thermalization. Thus a comparison of the kinetic simulations with experimental data can provide unique informations on the high density, high temperature zone of the nuclear equation of state, and on the strong interaction vertices at high density.

Moreover, an interesting point in a kinetic description of intermediate energy heavy ion collisions is the quantum many body problem itself. At low energies, the non equilibrium finite size fermion gas is in first approximation fully degenerate, and a fairly accurate description is possible in terms of TDHF theory. At energies comparable to the Fermi energy and higher, this is not true any more, and higher order correlations have to be taken into account. So intermediate energy heavy ion physics is a suitable field to test many body theories and experimentally explore the boundary between an explicit dynamical treatment and statistical theories.

The most reliable transport equation that applies in the nuclear physics domain is the Boltzmann Nordheim Vlasov equation (BNV)[1–4], also known in the literature as Boltzmann-Uehling-Uhlenbeck (BUU)[2,4], Vlasov-Uheling-Uhlenbeck (VUU)[1] or Landau-Vlasov (LV) [3].

This equation has originally been proposed by Nordheim [5], and later on by Uehling and Uhlenbeck [6] for weakly interacting fermion systems as an extension of the Boltzmann equation. While its derivation has commonly been based on perturbative arguments [7] or simply by physical intuition [2], Botermans and Malfliet showed [8] that it may consistently be derived also in a nonperturbative framework. In particular, the time dependent mean field and the effective nucleon-nucleon cross section, that determine the nuclear dynamics, are in principle uniquely extracted from the bare interaction. In section 2 we shall introduce the BBGKY hierarchy and show the truncation scheme that leads to the BNV equation. In section 3 we shall show some extensions of the BNV formalism to include triple collisions and fluctuations. Conclusions are finally given in section 4.

2.The Boltzmann Nordheim Vlasov Equation

Let's consider [9] an isolated system with hamiltonian H of n non relativistic particles interacting via a two body force; the dynamical description of the nuclear many body problem is based on the Liouville - Von Neumann equation for the

n-body density matrix ρ_n, i.e.

$$i\hbar\partial_t\rho_n(1,\ldots,n) = \mathcal{L}_n(1,\ldots,n)\rho_n(1,\ldots,n) \tag{2.1}$$

where $\mathcal{L}_n$ is the Liouvillian of the system, that can be separated in a free kinetic part and in an interaction part

$$\mathcal{L}_n = [H,\ ] = \mathcal{L}_n^0 + \sum_{i<j}^{n} \Theta_{ij} \tag{2.2}$$

In terms of the density matrix, the expectation value of any observable A reads

$$< A >= Tr(1,\ldots,n)A\rho_n \tag{2.3}$$

Since most of the physical observables depend only on a restrict number of degrees of freedom, it is useful to define reduced density matrixes of rank k ($k < n$) as

$$\rho_k(1,\ldots,k) = \frac{n!}{(n-k)!}Tr(k+1,\ldots,n)\rho_n \tag{2.4}$$

In terms of the reduced density matrices the Liouville - Von Neumann equation can be recast in a hierarchy of equations (BBGKY)

$$i\hbar\partial_t\rho_k = \mathcal{L}_k^0\rho_k + \sum_{j=1}^{k}\sum_{i<j}\Theta_{ij}\rho_k + \sum_{i=1}^{k}Tr(k+1)\Theta_{i,k+1}\rho_{k+1} \tag{2.5}$$

where ρ_k is coupled to the matrix of higher rank ρ_{k+1}. In particular, the first two equations of the hierarchy are

$$i\hbar\partial_t\rho_1 = \mathcal{L}_1^0\rho_1 + Tr(2)\Theta_{12}\rho_2 \tag{2.6a}$$
$$i\hbar\partial_t\rho_2 = \mathcal{L}_2^0\rho_2 + \Theta_{12}\rho_2 + Tr(3)(\Theta_{13}+\Theta_{23})\rho_3 \tag{2.6b}$$

All the kinetic equations that are used in the nuclear transport theory correspond to different truncation schemes of the BBGKY hierarchy, i.e. different approximations to the higher order density matrices [4].

The simplest possible approximation corresponds to retaining the first equation of the hierarchy only, that is truncating at the two body level. If one assumes that the only correlations are represented by the antisymmetrization of the density matrix (that is the wave function is a Slater determinant)

$$\rho_2(1,2) = \mathcal{A}_{12}\rho_1(1)\rho_1(2) \tag{2.7}$$

the first equation 2.6a of the hierarchy reduces to the Hartree Fock approximation

$$i\hbar\partial_t\rho_1 = \mathcal{L}_1^0\rho_1 + U_{HF}\rho_1 = \left[H_{HF}, \rho_1\right] \tag{2.8}$$

where $\mathcal{A}$ is the antisymmetrization operator and

$$U_{HF}(1) = Tr(2)\Theta_{12}\mathcal{A}_{12}\rho_1(1)\rho_1(2) \tag{2.9}$$

is the Hartree Fock mean field.

TDHF can be written in the phase space representation by means of a Wigner transform, defined for a generic one body operator $\mathcal{A}$ as

$$A_W(\mathbf{r},\mathbf{p}) = (2\pi\hbar)^{-3/2}\int ds \quad exp(\frac{-i\mathbf{p}\mathbf{s}}{\hbar}) < \mathbf{r}+\mathbf{s}/2|\mathcal{A}|\mathbf{r}-\mathbf{s}/2 > \tag{2.10}$$

The Wigner transform of the one body density matrix ρ_1 is the quantum phase space distribution function $f(\mathbf{r},\mathbf{p},t)$. The utility of f lies in its similarity to a classical distribution function, as far as the computation of expectation values is concerned. The Wigner transform of eq.2.8 results

$$\partial_t f + \frac{2}{\hbar}H_W sin\left[\frac{\hbar}{2}\Lambda\right]f = 0 \tag{2.11}$$

where $\Lambda = \nabla_r\nabla_p - \nabla_p\nabla_r$ and H_W is the Wigner transform of the Hartree Fock Hamiltonian

$$H_W = \mathbf{p}^2/2m + U(\mathbf{r},\mathbf{p},t) \tag{2.12}$$

The semiclassical approximation corresponds to keeping only the lowest order in $\hbar$ and gives

$$\partial_t f + \frac{\mathbf{p}}{m}\nabla_r f + \nabla_p U\nabla_r f - \nabla_r U\nabla_p f = 0 \tag{2.13}$$

which is known in the literature as the Vlasov equation.

The Vlasov equation is a useful tool at low energy, where most of the nucleon-nucleon collisions are Pauli blocked. At intermediate energies however it exhibits a transparency behaviour similar to TDHF, which can be removed by consistently taking into account two particle scatterings, i.e. truncating the hierarchy at the successive order. The truncation scheme is the following

$$\rho_2(1,2) = \mathcal{A}_{12}\rho_1(1)\rho_1(2) + c_2(1,2) \tag{2.14a}$$

$$\rho_3(1,2,3) = \sum_{i<j<k} \mathcal{A}_{ij}\rho_1(i)\rho_1(j)\rho_1(k) + c_2(i,j)\rho_1(k) \tag{2.14b}$$

The correlated part of the two body density matrix $c_2(1,2)$ accounts for the residual interaction. Eqs.2.14 have to be substituted in the second equation of the hierarchy eq.2.6b, which in turn will give a kinetic equation for the one body density matrix.

With the extra assumptions of a local approximation for the density matrix (quasi-homogeneity of the f) and a weak time dependence (Markov approximation) it turns out [4] that the semiclassical limit of the kinetic equation has a Boltzmann structure

$$\partial_t f + \frac{\mathbf{p}}{m} \cdot \nabla_r f - \nabla_r U \cdot \nabla_p f + \nabla_p U \cdot \nabla_r f = I_2[f] \tag{2.15}$$

$$\begin{aligned} I_2[f] = \int & d\mathbf{p}_2 d\mathbf{p}'_1 \sigma |v_1 - v_2| \cdot \\ & \Big[f(1')f(2')(1-f)(1-f(2)) - ff(2)(1-f(1'))(1-f(2')) \Big] \cdot \\ & \delta(\mathbf{p}_1 + \mathbf{p}_2 - \mathbf{p}'_1 - \mathbf{p}'_2) \end{aligned} \tag{2.16}$$

where the self consistent nuclear mean field $U(\mathbf{r},\mathbf{p},t)$ and the nucleon- nucleon scattering cross section $\sigma(E,\mathbf{q})$ are in principle univocally given in terms of the G matrix

$$U(\mathbf{r},\mathbf{p},t) \propto \int d\mathbf{r}' d\mathbf{p}' Re[G(E,\mathbf{r}-\mathbf{r}',\mathbf{p}-\mathbf{p}')] f(\mathbf{r}',\mathbf{p}',t) \tag{2.17a}$$

$$\frac{d\sigma}{d\Omega}(E,\mathbf{q}) \propto G(E,\mathbf{q}) G\dagger(E,\mathbf{q}) \tag{2.17b}$$

where $\mathbf{q}$ is the momentum transfer in a nucleon nucleon collision.

In eq. 2.16 only binary scatterings are considered, i.e. the mean collision time was supposed much shorter than the average time between collisions (dilute gas approximation). The application of the BNV equation in the intermediate energy regime, and its extension in the relativistic regime, has greatly clearified the HI fenomenology. We have understood very well the dynamical transition between mean field dynamics and two body collisions, and most one body observables have been qualitatively, and in many cases quantitatively, reproduced [1–4]. With the availability of new HI experimental facilities at intermediate and relativistic energy, it is now possible to build up portions of nuclear matter far away from equilibrium, at a density which is easily twice the saturation density and an excitation energy that can easily reach the binding energy per nucleon. In such a situation the incoherent binary scattering approximation given by the Boltzmann collision integral can be fairly inadequate and multiple scatterings have to be taken into account. Moreover in this dynamical regime critical phenomena are expected, and it is probable that high order correlations and the dynamical propagation of fluctuations in the nuclear medium can lead to critical behaviours, like nuclear multifragmentation and phase transitions.

In the following we would like to briefly discuss both the inclusion of triple scatterings into the collision integral, and the evaluation of the fluctuations of the distribution function with respect to the average behaviour given by the BNV equation.

3. Higher Order Correlations in BNV

3.1 Triple Collisions

As we have seen, one of the assumptions to be done in order to derive the BNV equation is that the gas is very rarified, so that all correlations except local incoherent two body scatterings can be neglected. This can be physically expressed by the condition

$$d/r_0 \ll 1 \tag{3.1}$$

As a rough estimation, $r_0 \approx (3/4\pi\rho)^{1/3}$ and $d \approx \sqrt{\sigma/\pi} \approx \sqrt{\sigma_f \cdot < (1-f_1)(1-f_2) > /\pi}$ where σ_f is the free nucleon-nucleon cross section $\sigma_{free} \approx 40$ mb and $< (1-f_1)(1-f_2) >$ is an average Pauli blocking factor. At beam energies of about 200 MeV/u Pauli blocking is almost negligeable and so $d/r_0 \approx 1$ already at $\rho = \rho_0$, and becomes greater with increasing density. So we expect three body collisions to be very important in this dynamical regime.

A more quantitative estimation of the relative contribution of two body and three body scatterings can be obtained in the simplified situation of a semiclassical quasi-homogeneous gas. In this case, an explicit form for the three body collision integral can be worked out. The truncation scheme will correspond to an explicit inclusion of three particle correlations

$$\rho_3(1,2,3) = \sum_{i<j<k} \mathcal{A}_{ij}\rho_1(i)\rho_1(j)\rho_1(k) + c_2(i,j)\rho_1(k) + c_3(1,2,3) \tag{3.2}$$

and the dilute gas approximation has to be relaxed. Then the three body collision integral reads [10,11]

$$I_3 = \int d(2)d(3)\Theta_{12}\mathcal{R}(123)f(1)f(2)f(3) \tag{3.3}$$

where 1,2,3 are abbreviations for the momentum and configuration $\vec{p}_1\vec{x}_1; \vec{p}_2\vec{x}_2; \vec{p}_3\vec{x}_3$ of particles 1,2,3; $\mathcal{R}(123) = \mathcal{S}(123) - \mathcal{S}(12)\mathcal{S}(23) - \mathcal{S}(12)\mathcal{S}(13) + \mathcal{S}(12)$ and $\mathcal{S}(1,\ldots,n)$ is the substitution operator defined by

$$\mathcal{S}(1,\ldots,n) = \lim_{t\to\infty} exp(-t\mathcal{L}_n)exp(t\mathcal{L}_n^0) \tag{3.4}$$

The effect of the substitution operator $\mathcal{S}(1,\ldots,n)$ on the n-body distribution function f^n is

$$\mathcal{S}(1,\ldots,n)f^n(1,\ldots,n,t) = f^n(1',\ldots,n',t) \tag{3.5}$$

where $1'$ is the value that the variables of particle 1 must have at a time before the collision in order to be at 1 at time t, according to the n-body kinematics of the collision. Note that $\mathcal{R}(123)$ is zero for a sequence of two body collisions, i.e. it isolates the contribution of actual triple collisions.

Eq.3.3 is still very involved; however, it can be further simplified in the case of a hard sphere gas, i.e. by assuming that the successive collisions do not overlap in time.

Then the time integral can be immediately worked out giving [11–13]

$$I_3 = \frac{4}{3\sqrt{\pi}} \int d\vec{p}_2 d\vec{p}_3 |v_{12}| \sigma^{5/2} \Big[\mathcal{U}(123) - \sum_{\alpha>\beta} \mathcal{U}(\alpha)\mathcal{U}(\beta)\Big] f_1(1) f_1(2) f_1(3) \tag{3.6}$$

where the Ursell operators $\mathcal{U}$ are defined by

$$\mathcal{U}(ij) = \mathcal{S}(ij) - 1 \quad ; \quad \mathcal{U}(ijk) = \mathcal{S}(ijk) - \mathcal{S}(ij) - \mathcal{S}(ik) - \mathcal{S}(jk) + 2 \tag{3.7}$$

The effect of the Ursell operators is to transform the distribution function in the distribution function itself prior the collision, minus the distribution function after the collision. In this way all the dependence on the detailed dynamics during the collisions is washed out, and the kinetic equation displays a Markovian character.

In an abbreviated form we can write

$$I_3 = \frac{4}{3\sqrt{\pi}} \int d\vec{p}_2 d\vec{p}_3 \sigma^{5/2} |v_1 - v_2| \cdot$$
$$\sum_k \Big[f_1^k f_2^k f_3^k (1-f_1)(1-f_2)(1-f_3) - f_1 f_2 f_3 (1-f_1^k)(1-f_2^k)(1-f_3^k)\Big] \tag{3.8}$$

where the variables with a k refer to momenta evaluated after the collision and the sum is formally taken over all the possible kinematical diagrams included in $\mathcal{R}(123)$.

It is important to stress that in the derivation we have neglected all contributions arising from a genuine three-body force, that is in our approach a three body scattering is any succession of three 2-body encounters between particles laying inside the same interaction sphere, as it is implied by the form of the $\mathcal{R}(123)$ operator [10]. The quantum nature of the system is revealed by the occupation factors on final states, but all higher order correlations arising from off-shell propagation and Pauli blocking on intermediate states are neglected.

The numerical tecnique that we use to simulate eq.3.8 is explained in detail in ref.[14], where a comparison to an analytical result is also presented. The algorithm is based on a generalization to three body correlations of the concept of mean free path. In the spirit of Enskog's virial theory [15] the effect of many body correlations on mean free path can be expressed by an expansion in powers of density, which at first order reads

$$\lambda = \frac{c_1}{\rho\sigma + c_2\rho^2\sigma^{5/2}} \tag{3.9}$$

with c_1, c_2 constants.

Furthermore, since λ is inversely proportional to the collision probability Π, and the probabilities Π_2, Π_3 of two and three body collisions are additive, we get

$$\begin{aligned} \Pi = \Pi_2 + \Pi_3 &\propto \lambda^{-1} \\ &\propto \rho\sigma + c_2\rho^2\sigma^{5/2} \\ &\propto \lambda_2^{-1} + \lambda_3^{-1} \end{aligned} \tag{3.10}$$

In this way the mean free path can be divided in a two body (λ_2) and a three body (λ_3) component, that are used to decide the collision probabilities in the numerical simulation of the two and three body collision integrals [14]. The close analogy of our formalism to Enskog's theory is more clearly seen by computing the number of collisions occurring in a classical gas at temperature T. It is well known that in an Enskog's gas at a constant density ρ and cross section σ the number N_3 of three body collisions is simply proportional to the number N_2 of the two body ones [15]

$$N_3 = c_2\rho\sigma^{3/2}N_2 \tag{3.11}$$

a feature which is correctly reproduced in our calculations.[14] The topic is discussed in more detail in ref.13,14.

In order to get a feeling of how three nucleon interactions influence the dynamics, we have calculated the transverse flow $< p_x >$ versus rapidity y for the Ca(400

MeV/nucleon)+Ca reaction at b=3 fm. In Figure 1 the result of this calculation (full line) is compared to a calculation with only two body correlations (dashed line) and with a simulation of Enskog's kinetic equation[15] (dashed dotted line). In all cases a soft (K=200 MeV) equation of state with no momentum dependence has been used. It is seen that our model is in good agreement with Enskog's theory and that the inclusion of three body collisions increases the transverse momentum transfer of about 20%. This result is also in reasonable agreement with the data [16] but, of course, the problem to be studied is which equation of state and which in medium nucleon-nucleon cross section should be used.

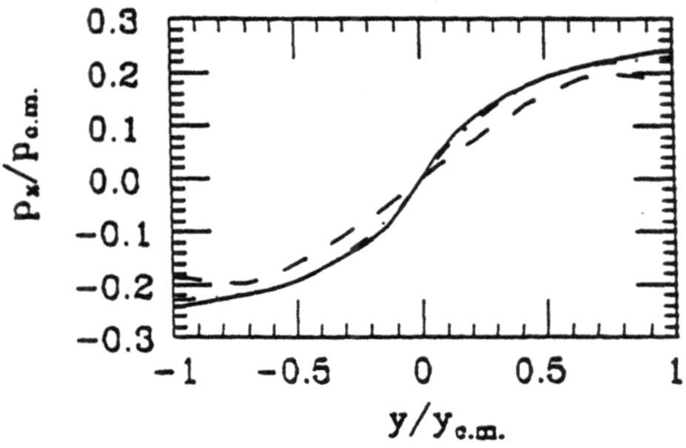

Fig.1 Transverse flow versus rapidity for the Ca+Ca system with a soft equation of state. Lines:see text.

Apart from the influence on the average dynamics of the nucleus-nucleus collisions, three body correlations can be very important for subthreshold production of energetic particles like pions or hard photons as pointed out in ref.[17–18] In fact, incoherent nucleon-nucleon collisions are responsable for some of the high energy particle production. But, when the total energy of the produced particle is very close to the kinematical threshold, the incoherent scatterings are not sufficient to explain the data and other mechanisms have to be invoked. One such mechanism could be the cooperation of three nucleons which carry a lot of energy and momentum available for the production of a new particle. To elucidate these ideas, we have plotted in Figure 2 the proton production differential cross section for a O+Al collision at 94 MeV/nucleon in comparison with experimental data [19]. The dashed line gives the ordinary BNV calculation while the full line includes the contribution of three body collisions. The low energy part of the distribution is about the same, but a high energy tail is seen in the three body case, which agrees with the data.

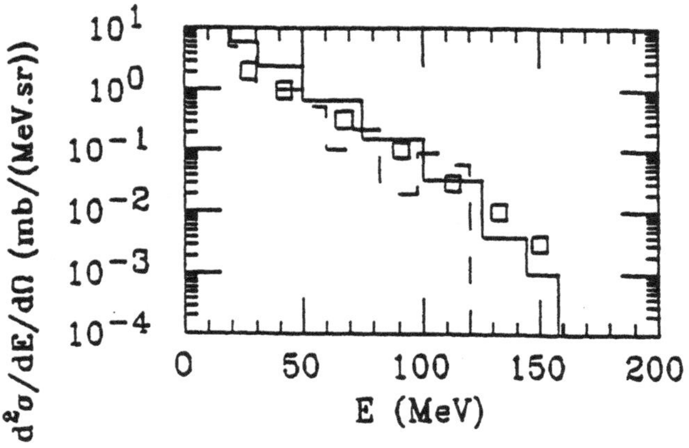

Fig.2 Nucleon production differential cross section versus nucleon energy in the c.m. system. Lines: see text

3.2 Fluctuations of the One Body Distribution Function

The recent experimental findings about fragmentation and multifragmentation have raised a great interest about the problem of fluctuations in nuclear physics [20–28]. In principle the kinetic equations represent an ideal tool to analyze microscopically and self-consistently the dynamical evolution of a heavy ion collision, and an appealing possibility is given by the BNV equation eq.2.16. However, as we have already stressed, the BNV equation presents serious drawbacks for an application in the critical regime. As we have seen, in order to derive the BNV equation from the BBGKY hierarchy, the dynamical correlations caused by collisions are supposed to be very well localized both in space and in time. Then the two body distribution function is completely uncorrelated except at the very occurrence of a collision. Such position implies that the one body distribution function gives only the average behaviour at the one body level. As a consequence, the fluctuations induced by the scattering processes are averaged out before they can be propagated by the mean field, and the BNV equation cannot describe the approach to critical conditions.

Even if in an actual HI collision no critical phenomenon is happening, it would be very useful to know the fluctuations of the exact microscopic distribution function, with respect to the Boltzmann approximation. As a matter of fact, in the computation of one body observables, the average distribution function $\overline{f}$ allows us to reproduce only mean values, while no informations on the variances of the observables is achieved. We start [10]by defining the correlation function of the fluctuations as

$$\sigma(\vec{r}_1,\vec{r}_2,\vec{p}_1,\vec{p}_2,t_1,t_2) = \langle \delta f(\vec{r}_1,\vec{p}_1,t_1)\delta f(\vec{r}_2,\vec{p}_2,t_2)\rangle \tag{3.12}$$

with $\delta f(\vec{r},\vec{p},t) = f(\vec{r},\vec{p},t) - \overline{f}(\vec{r},\vec{p},t)$.

Here, f is the exact microscopic distribution function which in principle should be obtained by solving the complete Liouville equation, and $\overline{f}$ is the average distribution function, solution of the Boltzmann equation eq.2.16. The average $\langle\rangle$ has to be interpreted as a time average, and is equivalent to an ensamble average in the ergodic limit.

If the fluctuation δf is small and the system is in a non equilibrium state represented by the small addition δf to the average distribution distribution $\overline{f}$, then we can use the theory of quasi-stationary fluctuations [10,25]. The process of relaxation of the system to equilibrium is described by a linearized Boltzmann equation

$$\frac{\partial}{\partial t}\delta f(\vec{r},\vec{p},t) + \frac{\vec{p}}{m}\cdot\nabla_r\delta f(\vec{r},\vec{p},t) - \nabla_r U\cdot\nabla_p\delta f(\vec{r},\vec{p},t) = I_l\cdot\delta f(\vec{r},\vec{p},t) \tag{3.13}$$

with I_l linearized collision integral. Since δf is a small quantity and $\langle\delta f\rangle = 0$, the same equation is solved by the correlation function eq.3.12. This means that the correlation function is easily obtained once the initial condition, i.e. the equal time correlation function, is known.

If we are dealing with an homogeneous gas in equilibrium, the correlations between simultaneous fluctuations are propagated at distances of the order of the nucleon-nucleon interaction, which is regarded as zero in BNV theory (recall that the collision integral in eqs.2.16, 3.8 is a local quantity). As a consequence, the equal time correlation function is a δ-function whose coefficient is the mean square fluctuation at the given phase space point. As it is well known [25], in an ideal gas this is equal to the mean value of the distribution itself. The initial condition for a Maxwell gas reads

$$\sigma(\vec{r},\vec{p}_1,\vec{p}_2,0) = f_0(\vec{r},\vec{p}_1)\delta(\vec{r})\delta(\vec{p}_1-\vec{p}_2) \tag{3.14a}$$

In the Fermi Dirac case non zero range fluctuations appear as a consequence of antisymmetrization, that creates correlations among particles. However, the expression for the equal time correlation function at $\vec{r}=0$ is similar [25] to eq.(3.14a)

$$\sigma(0,\vec{p_1},\vec{p_2},0) = f_0(0,\vec{p_1})\big(1-f_0(0,\vec{p_1})\big)\delta(\vec{p_1}-\vec{p_2}) \tag{3.14b}$$

The simultaneous solution of eqs.3.13, 3.14 solves completely the problem of the determination of the correlation function. In particular, the behaviour of σ versus time is expected [24] to be exponentially decreasing

$$\sigma(t) = \langle \delta f^2 \rangle exp(-t/\tau) \tag{3.15}$$

Here τ represents the relaxation time and for simplicity we have integrated over space and momentum coordinates.

The numerical procedure is straightforward: we start with the initial condition given by an equilibrated homogeneous gas. The distribution function is initialized with the test particle method [2] and the Boltzmann equation eq.2.16 is solved in the standard way [1-4,14]. The knowledge of $\overline{f}$ allows us to compute the fluctuation δf from eq.3.13, and this in turn gives at each time step the correlation function via eq.3.12.

In ref.[27] we have shown that both eqs. 3.14 and 3.15 are correctly reproduced by our method, and that for model cases in momentum space the relaxation times are in very good agreement with some analytical result from Fermi liquid theory [26].

The same reasoning applies to the non equilibrium case, but in this case the initial condition is given by [10]

$$\sigma_{12}(0) = \langle \overline{f}_{12} \rangle - \langle \overline{f}_1 \overline{f}_2 \rangle + \langle \overline{f}_1 \rangle \delta(\vec{r_1}-\vec{r_2})\delta(\vec{p_1}-\vec{p_2}) \tag{3.16}$$

where we have used the abbreviated notation $\overline{f}_i(t) = \overline{f}(\vec{r_i},\vec{p_i},t_i)$ and $\sigma_{ij}(t) = \langle \delta f_i(t) \delta f_j(0) \rangle$. Eq.3.15 explicitely contains the average two body distribution function $\overline{f}_{12}$, thus a solution of the second equation of the BBGKY hierarchy is required. It is possible to show [10] that the correlated part of the two body distribution function $c_{12} = \overline{f}_{12} - \overline{f}_1 \overline{f}_2$ satisfies a Boltzmann- like equation

$$\frac{D}{Dt} c_{12} = I_l(1)c_{12} + I_l(2)c_{12} + I_{col}(\overline{f})\delta(\vec{r_1}-\vec{r_2}) \tag{3.17}$$

In eq.3.17 D/Dt is the total time derivative, $I_l(i)$ are linearized collision operators, and I_{col} is the usual Boltzmann collision integral from eq.2.16. The first two terms on the r.h.s. of eq.3.17 describe the collisions of particles in the phase space cells $d\vec{r}_i, d\vec{k}_i (i = 1,2)$ with some other particles in different cells. The third term takes into account the possibility that the two particles 1 and 2 occupy the same cell in $\vec{r}$ space: then they can collide with each other. This third term is the source of the simultaneous correlations which are absent in the equilibrium case. Since in the general case the solution of eq.3.17 is not feasible, it is necessary to find some physically reasonable prescription for the initial condition eq.3.16. One very promising possibility [21–23] is to make use of the fluctuation-dissipation theorem and to treat the time evolution of the fluctuations as a Poisson process, under the action of a random force as a proper generalization of eq.3.13.

Here we would like to examine, as a zero order approximation, the minimum fluctuations contained in the equation, without adding any extra physical ingredient to the simple Boltzmann evolution for the one body distribution function. This means we put $\langle \overline{f}_{12} \rangle = \langle \overline{f}_1 \overline{f}_2 \rangle$ neglecting all the simultaneous correlations between different phase space points. According to the above discussion, this corresponds to assume small deviations of the real distribution function with respect to the Boltzmann solution. This assumption is not at all valid near to a critical point, when long range correlations play an important role and even the Boltzmann average distribution function $\overline{f}$ is incorrect. However, our method could give an indication if such critical points are reached in a HI collision.

In ref.27 we have shown that this is a good approximation to study small oscillations around equilibrium.

The method is suitable to describe both mean field fluctuations (from the Vlasov equation eq.2.13) and fluctuations coming from the collision integral (from the homogeneous Boltzmann equation, eq.2.16 where $U = 0$). An example is given in Fig. 3 for a quadrupole deformation of a Ca nucleus. The full line represents the correlation function of fluctuations integrated over momentum space in the homogeneous case, i.e. neglecting the mean field. The amplitude of the deformation is extracted from GQR phenomenology[29] and the details of the calculation are chosen such to compare to the analytical calculation ref.26. The slope of the correlation function, $\tau = 130$ fm/c, gives the collisional relaxation time of the the deformation of the Fermi sphere, while the analytical estimate [26] gives $\tau = 142$ fm/c.

On the other hand, the dashed line in Fig.3 refers to a Vlasov calculation (mean field only) for the same initial condition of the previous one. The correlation function of fluctuations has been integrated both over $\vec{r}$ and $\vec{p}$ space. In this case the slope results $\tau = 375$ fm/c and is connected to one body dissipation mechanisms, the so called Landau damping.

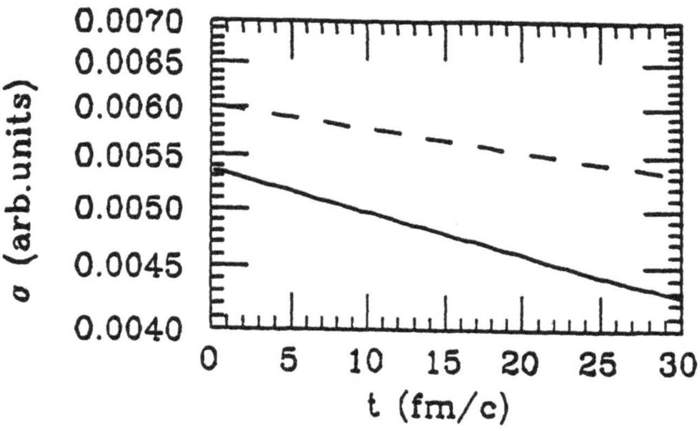

Fig.3 Correlation function versus time for a GQR excitation of a Ca nucleus. Dashed line: mean field only. Full line: collision integral only.

Many different applications are possible along this line. In ref.27 we have studied in some detail the Landau Damping and its connection with the escape width of collective modes. In ref.28 the interplay between mean field and collision integral is discussed, together with the temperature effects on collective modes.

4. Conclusions

In this contribution we have recalled some aspects of the kinetic theory with the purpose of application in the high density, high temperature regime of HI collisions, where the BNV equation treatment does not represent an adequate approximation of the dynamics any more. First of all we have presented an extension of the BNV equation including three-body scatterings. The equation is numerically solved with an algorithm based on the concept of the mean free path. We have shown that the three body term has already an important effect on observables at energies as low as 94 MeV/nucleon. This is due to the high densities reached during the heavy ion collision and to the decreased importance of Pauli blocking. The results presented here indicate that multiple scatterings must be included in future transport equation calculations at high and intermediate incident energies.

We have also presented a new method to evaluate fluctuations in the time evolution of the average one body distribution function, both in the classical and in the quantum case. Only two full ensamble BNV runs are needed to compute the correlation function of fluctuations. We have applied our model in the equilibrium and near to equilibrium situation, where it shows a good agreement with the analytical results [10,25]from fluctuation theory.

Fluctuations given by the collision integral and mean field fluctuations are treated unambigously on the same ground. The model is particularly suitable to describe relaxation times of collective modes, while it breaks down if the fluctuations get too large, i.e. approaching a critical region. A comparison with experimental widths of the distributions of one body observables will be very interesting, since it could give informations about the approach to critical phenomena.

References

1. H.Stocker,W.Greiner,*Phys.Rep.* **137** (1986) 277.
2. G.F.Bertsch and S.Das Gupta,*Phys.Rep.* **160** (1988) 190.
3. P.Schuck, R.W. Hasse, J.Jaenicke, C. Gregoire, B. Remaud, F. Sebille, and E. Suraud,*Progr. in Part. and Nucl. Phys.* **22** (1989) 181.
4. W.Cassing and U.Mosel, *Progr. in Part. and Nucl. Phys.* **25** (1990) 235.
5. L.Nordheim, *Proc.Roy.Soc.Lond.* **119** (1928) 689.
6. E.A.Uehling, G.E.Uhlenbeck, *Phys.Rev.* **43** (1933)552.
7. J.H.Irving, R.W.Zwanzig, *J.Chem.Phys.* **19** (1951) 1173.
8. W.Botermans, R.Malfliet, *Phys.Lett.* **B171** (1986) 22.
9. see f.i. P.Ring, P.Schuck, *The Nuclear Many Body Problem* (Springer Verlag, 1980).
10. E.M. Lifshits and L.P. Pitaevskii,*Physical Kinetics* (Pergamon press, New York, 1981).
11. M.S. Green,*Phys. Rev.* **A136** (1964) 905.
12. J.V.Sengers, in *Lectures in Theoretical Physics, Kinetic Theory, vol.IX C*, ed. W. E. Brittin (Gordon and Breach Publishers, NY 1967) p.335.
13. A.Bonasera, F.Gulminelli, in preparation
14. A.Bonasera, F.Gulminelli, *Phys.Lett.* **B259** (1991) 399.
    A.Bonasera, F.Gulminelli, *Phys.Lett.* **B275** (1992) 24.
15. S.Chapman and T.G.Cowling, *The Mathematical Theory of Non-Uniform Gases*, (Cambridge University Press, London 1970), chapter 16 and references therein.
16. H.A. Gustaffson, H.H. Gutbrod, B. Kolb, H. Loehner, B. Ludewigt,

A.M. Poskanzer, T. Renner, H. Riedesel, H.G. Ritter, A. Warwick, F. Weik, and H. Wieman, *Phys. Lett.* **B142** (1984) 141.
17. R.Shyam and J.Knoll, *Nucl. Phys.* **A426** (1984) 606.
18. A.Adorno, A.Bonasera, F.Gulminelli, in *Proc. of the VIII Winter Workshop on Nuclear Dynamics, Jackson Hole (WI), 1992*, ed. W.Bauer.
19. D.Durand et al., *Nucl.Phys.* **A511** (1990) 442.
20. W.Bauer, G.F.Bertsch, S.Das Gupta, *Phys.Rev.Lett.* **58** (1987) 863.
21. S.Ayik, C.Gregoire, *Nucl.Phys.* **A513** (1990) 187.
22. J.Randrup, B.Remaud, *Nucl.Phys.* **A514** (1990) 339.
23. P.Chomaz, G.F.Burgio, J. Randrup, *Nucl.Phys.* **A529** (1991) 157.
24. E.Suraud, S.Ayik, J.Stryjewski, M.Belkacem, *Nucl.Phys.* **A519** (1990) 171c.
25. L.D.Landau and E.M.Lifshits, *Statistical Physics*(Pergamon Press, New York, 1981) chap.12.
26. G.F.Bertsch, *Z.Phys.* **A289** (1978) 103.
27. A.Bonasera, F.Gulminelli, P.Schuck, *Phys.Rev.* C, in press.
28. A.Bonasera, F.Gulminelli, P.Schuck, in *Proc.of the Int. Workshop on Dynamical Fluctuations, Aussois, 1992*, to be published in *Nucl.Phys.* **A.**
A.Bonasera, F.Gulminelli, P.Schuck, to be published.
29. A.Bonasera, M.Di Toro, F.Gulminelli, *Phys.Rev.* **C42** (1990) 966.

Common Problems and Trends of Modern Physics, pp. 331–340
ed. T. Bressani, S. Marcello and A. Zenoni
©1992 World Scientific Publishing Company

RESULTS OF COLD FUSION EXPERIMENTS ON Ti/D_2 AND Pd/D_2 SYSTEMS WITH GAS LOADING

ELENA BOTTA and DANIELA CALVO

Istituto Nazionale di Fisica Nucleare, sezione di Torino, via P. Giuria n. 1
10125 Torino Italy

ABSTRACT

Measurements of neutron emission from gas (H_2, D_2) loaded metals (Ti, Pd) have been carried out in the TOFUS experiment. A result of two years ago was a 2.5 σ signal for the emission of 2.5 MeV neutrons from a Ti/D system, while a result with a better statistical significance (~ 5 σ) has been obtained recently. The result on the Pd/D system does not show a statistically significant signal (less ~ 2 σ).

1. Introduction

The D_2 gas loading technique in Ti and Pd metals had its birth and growth in parallel with the electrochemical loading technique in Cold Fusion experiments. The methodology developed for this technique is the answer to the request for fast metal loading in order to achieve the maximum atomic ratio D/Pd (Ti).

Furthermore, it gives the opportunity of investigating the nuclear origin of the Cold Fusion phenomena by the detection of 2.5 MeV neutrons, as an unambiguous signature of fusion events. In fact, using this method it is possible to evaluate with high accuracy the energy of the emitted particle, while this result is practically impossible with electrolytical cells, where the water thermostatic bath acts as a neutron moderator.

Other significant features of the gas loading technique are:

1) good cleaning of the sample obtained by degassing;
2) the sample is not in contact with the atmosphere since the circuit is completely sealed;
3) good monitoring of the thermodynamic variables, like pressure and temperature.

This latter information allows evaluation of the time evolution of the loading (defined as the ratio between the number of monoatomic gas moles in the metal and the metal moles) and to follow the time evolution of the Ti(Pd)/D system in the phase diagram

In general, the method of pressure and temperature variations allows one to investigate the phase transition regions and, in particular, the α <-> β transition for the

Pd/D and β <-> δ transition for the Ti/D system. In fact, during these phase transitions, depletion regions are produced, and just at their interfaces, between two regions, the first with a higher D concentration and the second with a lower one, large strains induce dislocations and cracks which can increase the D-D tunneling probability (coherence effects [1]).

2. The TOFUS Experimental Set-up

In the TOFUS experiment, carried out at INFN Laboratorio Tecnologico in Torino, the gas loading technique has been applied in a Cold Fusion cell located in front of a neutron detector. This detector has been described previously [2,3,4,5,6]. Its main performances are: neutron energy reconstruction based on the measurement of the neutron time of flight between two blocks of NE110 plastic scintillators (double scattering technique) and of the impact position on the scintillator slabs. The resolution of the reconstructed neutron energy is ~ 1 MeV FWHM for neutrons of 2.5 MeV, as estimated by means of a Monte Carlo simulation, while the overall efficiency, measured by using an Am-Be neutron source, is $2.5\ 10^{-4}$ for the geometry used; the maximum event acquisition rate is 250 hz, but the maximum neutron counting rate (without reconstruction) is of the order of 5 Mhz, while the background is 200 trigger/hour (±8%) reduced to ~ 68 events/hour (±10%) after kinematical cuts. Now the neutron detector can reconstruct the energy spectrum for measurement of Cold Fusion neutron emission from D/Ti (Pd) system, but it is also necessary to measure the background that has to be subtracted from the first.

In particular it was chosen to operate with thermal cyclical variations on the metal/gas system in order to investigate the dependence of the neutron emission, if any, upon the thermodynamic conditions. Now, to take data in the same macroscopical conditions (mainly the temperature) as in the thermal cycles on Ti (Pd)/D system, several cycles were performed by filling the metal with H_2 instead of D_2: in this way the influence on the detector of every macroscopical effect was exactly the same for background and Cold Fusion measurements.

Two sets of measurements were performed in two years with different Cold Fusion devices: the first in 1990 with Ti and the second in 1991 with Ti and then Pd. The first Cold Fusion device consisted of a stainless steel cell of cylindrical shape (3 mm thickness, 20 cm^3 total volume), containing 3 gr of metallic Ti shavings. The pressure in the cell could be controlled between 1.3 10^{-8} bar and 2 bar. The temperature was controlled from 25 °C to 540 °C by means of a small heater put under the cell, in contact with the base, and was measured by means of a thermocouple inserted into a copper ring surrounding the basis of the cell: the temperature of the Ti shavings in the cell was estimated to be ~ 40 °C lower (500 °C instead of 540 °C).

From this experiment it was evident that the gas loading technique shows a high complexity due to several features as the temperature and pressure measurements, the loading, the treatment used to clean the Ti (Pd) sample surfaces, the reproducibility, etc....

To overcome such problems we have designed, realized and utilized in the 1991 experiments [7] the gas loading system shown in Fig.1.

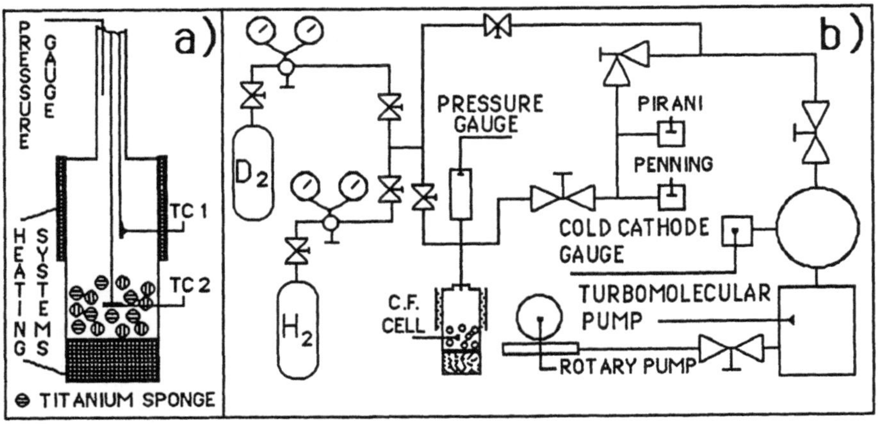

Fig.1 Cold Fusion cell (a) and gas loading system (b) scheme.

First of all the system allows clean experimental conditions during the tests, second the temperature and pressure monitoring for the metal lattice and the gas is satisfactory and allows evaluation, with filling of a known volume, of the amount of gas introduced and thus the average value of the atomic ratio in the sample.

In particular, a cylindrical cell of 44 mm diameter and 102 mm height contains the metal while a heating system is positioned in contact with its lower basis.

Two K-type thermocouples, the first one embebbed in the metal and the second one lying in the upper internal part of the cell, allow to monitor simultaneously the temperature of the metal and of the surrounding gas, while the pressure of the gas is monitored by means of a piezoresistive pressure gauge located in the upper part of the cell.

3. Description of the Thermal Cycles

Initially, the sample needs surface cleaning performed by means of a careful degassing procedure at high temperature: after this treatment the surface is very active with complete gas absorptive capacity. Then a known amount of D_2 was dosed at room temperature.

In the 1990 tests, the cell was degassed in a vacuum for 24 hours at 540 °C [8], then a series of measurements with the cell heated and cooled alternately were performed. During the heating phase the heater was kept on for a period of ~ 2 hours (~ 1 hour in a temperature ramp up from 25 °C up to 540 °C and ~ 1 hour at steady temperature); during the cooling phase the heater was kept off for a period of ~ 22 hours (~ 4 hours in a temperature ramp down from ~ 540 °C down to ~ 25 °C, ~ 18 hours at steady room temperature)

From the volume and pressure measurements the average atomic ratio D/Ti was estimated ~ 0.32; however, this value is totally inconsistent with the Ti/D phase diagram. In fact, in the range of the temperature and pressure scanned the average atomic ratio expected should be ~ 1.8: a possible explanation is that the surface condition of the Ti shavings was such as to avoid a complete filling of Ti with D near the value expected from the phase diagram and that only a reduced portion of the sample could reach the equilibrium conditions expected. However, no macroscopic changes were visible in the Ti shavings when the experiment was stopped.

The 1991 tests were performed first with 20 gr of Ti sponge and then with 54 gr of metallic Pd: the degassing was at 600 °C and the final vacuum inside the cell was $\sim 10^{-10}$ bar. During the immission step, the pressure of the gas inside the cell rose suddenly and then decreased slowly due to Ti deuteride formation. Since the absorption is an exothermic reaction, this leads to an increase of the metal temperature: in Fig.2 the gas pressure and the metal temperature are reported as a function of time during a two step immission experiment.

After D_2 immission, the system was submitted to a number of thermal cycles

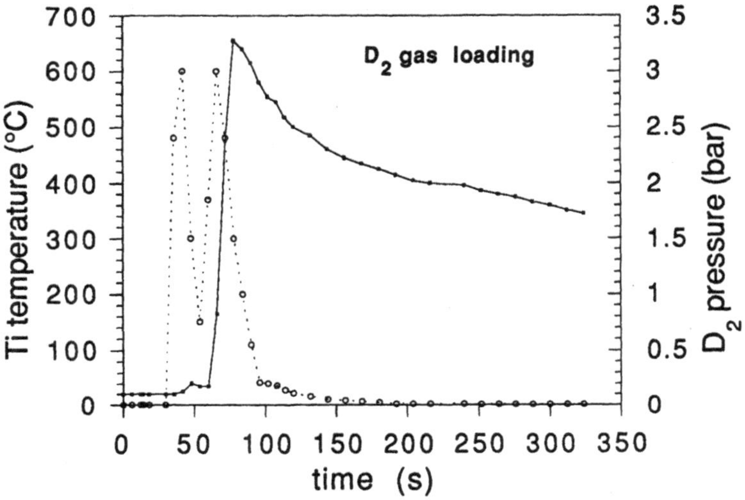

Fig.2 Ti temperature (continuous line) and D_2 pressure (dotted line) as a function of time during a gas loading phase.

consisting of a heating step from room temperature (~ 25 °C) up to 540°C at least, for Ti and 350°C for Pd followed by a cooling step down to the room temperature. The thermal cycles for the Ti were performed at two average atomic ratios, 0.7 and 1.8. For the lower loading a β<->δ phase transition occurs around 270°C: in Fig.3 (curve a) the presence of the phase transition is evident from the approximate constancy of the temperature; in the higher loading (curve b) the cycles in the Ti deuteride phase diagram were such that only the δ phase was concerned and therefore no phase transition was observed.

Moreover in the case of Pd the system crosses a phase transition: in particular from β to α during the heating and from α to β during the cooling (both at ~ 120°C, Fig.4); the average atomic ratio at room temperature was ~ 0.7.

However in both cases the morphology of the sample changes gradually during the repeated cycles: Ti from sponge to powder and the Pd from small cylinders of 1 mm diameter and 2 mm length to small spheres. This is due to the large strains associated with the hydride formation and phase transformations, in particular to the rearrangement of atomic positions and to the volume variation of the elementary cell caused by gas flowing in the sample, which produce internal cracks and dislocations with consequent macroscopic changes.

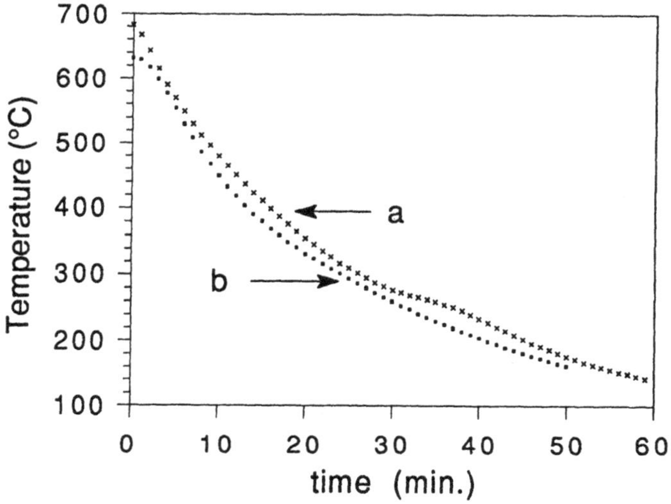

Fig.3 Lattice temperature as a function of the time during cooling steps of the Ti/D system (the average atomic ratio is 0.7 for the curve a and 1.8 for the curve b).

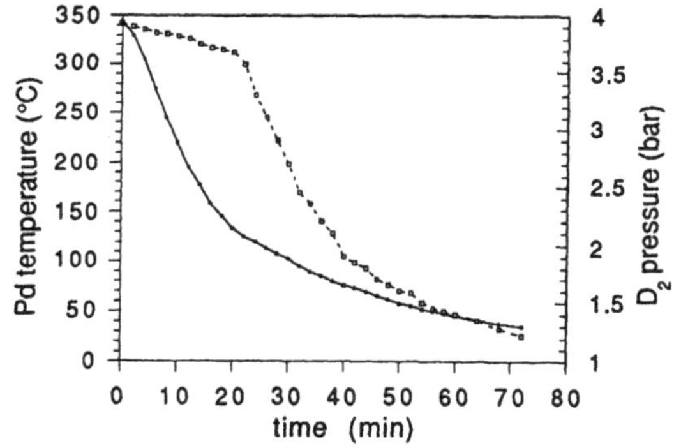

Fig.4 Lattice temperature (continuous line) and D_2 pressure (dotted line) of the Pd/D system as a function of time (initial average atomic ratio ≈ 0.7)

4. Neutron Emission Measurements

The result of the neutron measurements [9] in the 1990 is shown in Fig.5: the spectrum, obtained by subtracting the sum of runs in the heating step normalized in time

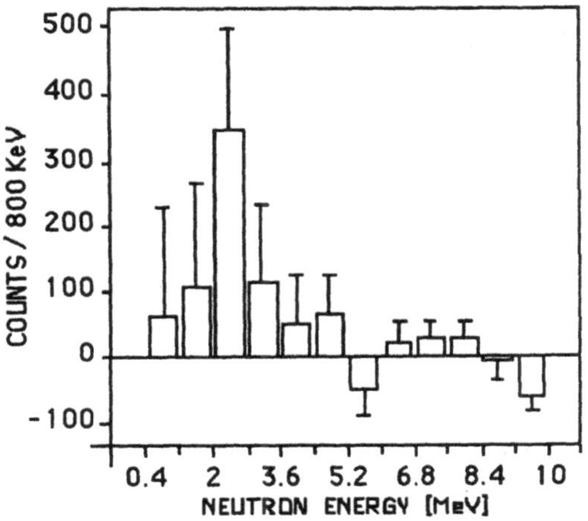

Fig.5 Neutron emission spectrum from the Ti/D system. The error bars indicate the statistical error and are to be understood as plus and minus.

to the sum of the runs in the cooling step, presents in the range 2-2.8 Mev a peak with a statistical significance of 2.5 standard deviations. This particular subtraction was necessary since the spectra corresponding to the D_2 and H_2 filling were taken with about a 3 week delay: since the average room temperature was different then the photomultipliers noise was different, slightly larger in the H_2 filling measurements.

The neutron emission measured in this first experiment was ~ 1.3 ± 0.5 n s^{-1} g^{-1}.

It should be noticed that this value has to be considered as a lower limit for these measurements, just due to the subtraction technique.

The final result of the 1991 measurements on the Ti/D system is shown in Fig.6, and is obtained from a subtraction of a background (H_2), normalized in time, from the Cold Fusion measurements.

In this spectrum one can see that the energy range between 2 and 3 MeV contains ~ 377 counts with a significance of ~ 3.9 standard deviations, while a subtraction of two sample of background measurements, chosen at random, gives a result, Fig.7, with a

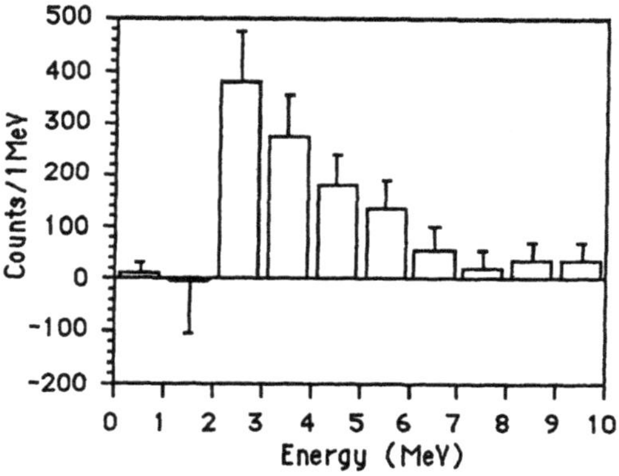

Fig.6 Neutron emission spectrum from the Ti/D system after the background subtraction. The error bars indicate the statistical error and are to be understood as plus and minus.

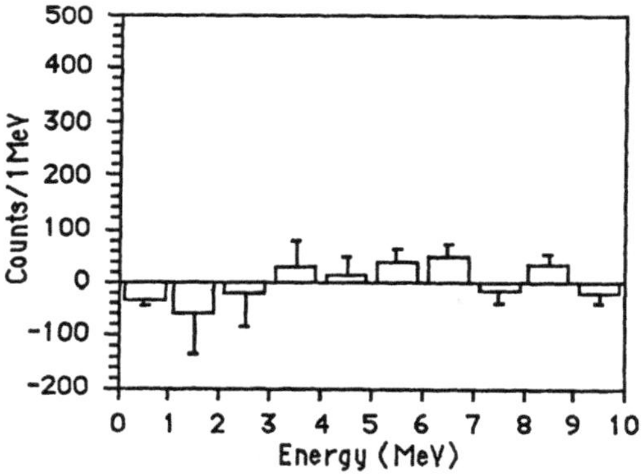

Fig.7 The result of the background subtraction procedure applied to two different sets of background (Ti/H system) measurements. The error bars indicate the statistical error and are to be understood as plus and minus.

statistical fluctuation consistent with zero.

However, another way of searching for neutron excesses is that of subtracting from each run with D_2 filling the total spectrum obtained with H_2 filling, properly normalized in time: the counts in each channel were obtained as the weigthed mean of the values obtained for each D_2 run, while the error was calculated as the standard deviation. This latter result is shown by Fig.8 and no substantial difference is apparent between the two methods apart the reduction of the errors: the channel between 2 MeV and 3 MeV is again the most populated, at a 5.4 σ level.

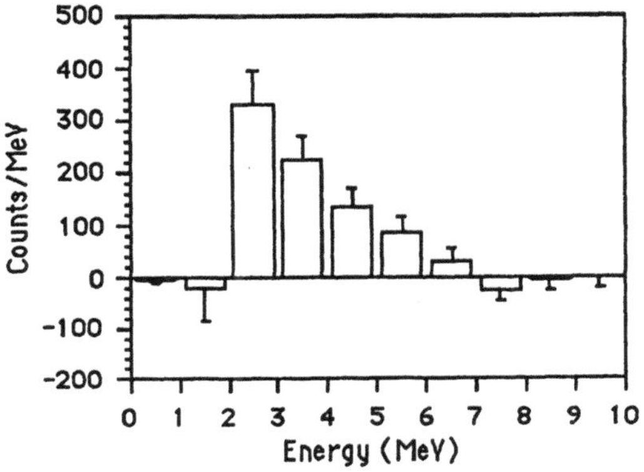

Fig.8 Spectrum of neutrons emitted from the Ti/D system calculated by the analysis of the single runs as described in the text.

An estimate of the neutron emission for unit mass and time, assuming that the neutron production rate was independent of time, was ~ 0.11 ± 0.03 neutrons $g^{-1}s^{-1}$.

For the Pd/D experiment the final spectrum obtained is shown in Fig.9: in the 2-3 MeV channel the statistical significance is small, less than 2 standard deviations.

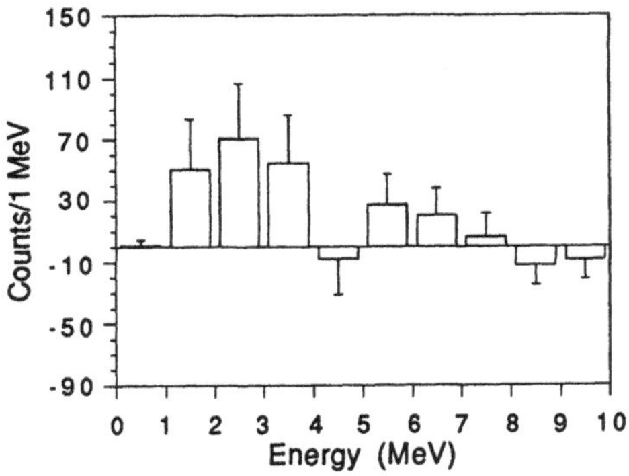

Fig.9 Neutron emission spectrum from the Pd/D system after the background subtraction. The error bars indicate the statistical error and are to be understood as plus and minus.

5. Conclusions

Both the measurements of 1990 and 1991 on the Ti/D system give a positive result for 2.5 MeV neutron emission. The neutron emission rate difference may be due either to the different forms of the Ti used (metal Ti shavings in the 1990 experiment, Ti sponge in the second one) or to the different integration time or to both effects. No significant neutron emission was observed for the Pd/D system, again submitted to the thermodynamic cycles.

Nevertheless, due to reduced statistics, it is not possible to attempt a statistically significant correlation between the neutron emission and the thermodynamic conditions.

Experimentally it seems reasonable that the neutron emission appears to be statistically distributed along the full duration of the runs, or to occur in many smaller bursts, again distributed along all the runs, but not concentrated in a few bursts.

6. References

1. T. Bressani et al., in *"Understanding Cold Fusion Phenomena"*, Conf. Proc. **Vol. 24**, SIF, Bologna (1989) p. 213.
2. G. C. Bonazzola et al., in *"Understanding Cold Fusion Phenomena"*, Conf. Proc. **Vol. 24,** SIF, Bologna (1989) p. 313.
3. G. C. Bonazzola et al., *NIM* **A299** (1990), 25.
4. M. Agnello et al., " The Trigger of the TOFUS Detector " accepted for publication on *IEEE Trans. Nucl. Sci.* **Vol. 39** (1992).
5. M. Agnello et al., " Performances of a Method..." presented at the 1991 Nuclear Science Symposium and submitted to IEEE Trans N. S. for publication.
6. M. Agnello et al., in *"The Science of Cold Fusion"*, Proc. of the II Annual Conference on Cold Fusion, Como, 29 June-4 July 1991, Eds. T. Bressani, E. Del Giudice and G. Preparata, Conf. Proc. **Vol. 33,** SIF, Bologna, 1991, cited in the following as Como '91, p. 249.
7. E. Botta et al., accepted by *Il Nuovo Cimento* for publication.
8. T. Bressani et al., *Il Nuovo Cimento* **Vol. 104 A** N. 9 (1991), p 1413.
9. T. Bressani et al., in Como '91, p. 105.

Common Problems and Trends of Modern Physics, pp. 341–347
ed. T. Bressani, S. Marcello and A. Zenoni
©1992 World Scientific Publishing Company

STATUS OF THE BOLOGNA EXPERIMENT ON NUCLEAR COLD FUSION AT THE GRAN SASSO LABORATORY

A.BERTIN, U.MARCONI, M.PICCININI, A.VITALE and G.ZAVATTINI
Dipartimento di Fisica dell' Università di Bologna, and
Istituto Nazionale di Fisica Nucleare, Sezione di Bologna

1. Introduction

The absorption of hydrogen, deuterium or tritium in palladium (Pd) or titanium (Ti) samples takes place via a process of interstitial solid solution: the nuclei of each hydrogen isotope, in practice, can settle in the interstices between the atoms of the host material, forming with them a palladium or titanium hydride.

Within the metal lattices loaded with hydrogen isotopes, the generation of suitable conditions to the fusion of two of the guests nuclei is currently expected on the basis of two main possibilities:

i) the creation (via different techniques) of *non-equilibrium conditions* during - or successively to - the loading procedure.

ii) the action of the physical frame constituted by the metal lattice and by microscopic and collective interactions, where the lattice itself plays a fundamental role, which in their turn can explain some features of the problem which are unexpected on the basis of conventional nuclear physics.[1]

In the present report, we shall discuss some methodological points concerning experimental research on nuclear cold fusion in general (Section 2) and the Bologna experiment at the Gran Sasso Laboratory (Section 3). Unusually but intentionally, no conclusions will be drawn in the present paper.

2. General Remarks on the Present Status of Experimental Research on Nuclear Cold Fusion

In the present Section, we shall formulate some synthetic considerations on the recent evolution of experimental research on nuclear cold fusion (NCF). For less recent results, survey reports presented in the last two years at the Folgaria Winter School of Hadronic Physics are quoted.[2,3]

As is well known, the nuclear synthesis reaction on which the main experimental interest has been focussed in the field of NCF so far is the one which involves two deuterium nuclei, which occurs through the channels

$$d + d \rightarrow {}^3He + n + 3.27\ MeV \quad (1)$$

$$\rightarrow t + p + 4.03\ MeV \quad (2)$$

$$\rightarrow {}^4He + \gamma + 24\ MeV \quad (3)$$

Channels (1) and (2) "in vacuo" occur with similar probabilities (about 50%), whereas channel (3) takes place which a branching ratio very much reduced in size.

It is known as well that experimental work on the production of these processes following the immission of deuterons within Pd or Ti samples has been developing along two fairly independent and complementary directions, namely:

a) by detecting one or more reaction products;

b) by looking at the emission of excess heat (with respect to the energy expenditure for the deuterium-loading operation).

Deuterium loading of the crystal lattice is currently performed by electrolysis or pressure injection, while non-equilibrium conditions are obtained by several techniques, among which submitting the loaded metal sample to thermal cycles, using pulsed electrolysis or making low-energy molecular clusters to impinge on deuterided metals.

Experimental work begun on these lines in 1989, yielding fairly controversial and hardly reproducible results, some features of which can be summarized as follows:

i) *Excess heat* production by several orders of magnitude larger than the energy associated to the observed neutron emissions.

ii) *Neutron emission (Reaction (1))*: a significant fraction of the positive results seem to agree on very low neutron emission levels. Neutron emission shows up with different features (i.e. occuring in bursts or in uncorrelated way; different time connections are also experienced with respect to the electrochemical and thermodynamical operations performed on the physical systems under observation).

iii) *Tritium production (Reaction (2))*: The observation of tritium production within electrolytic cells with Pd or Ti cathodes was by far the most significant experimental result on NCF obtained during the second year of work. Similar importance had the observation, recorded by different groups, of the smallness of the ratio between the recorded quantities of neutrons and of tritons (which would mean that process (1) occurs within the metal lattices with a probability by several orders of magnitude smaller than process (2)).

iv) *Observation of ^4He.* Significant quantities of ^{4}He nuclei within the gas produced by D_2O electrolysis with palladium cathodes, correlated with excess heat quantities, (although not with the emission of gamma radiation, see process (3)) were observed[4] recently. This result was undoubtedly the most significant experimental novelty of the third year of research on NCF.

Experimental work on NCF is then continuing, yielding interesting results in a progressively widening horizon. The scientific work should be acknowledged of theorists[1,5] who supported by their research and their interest a difficult field, cold-shouldered by many scientific establishments, and opening beyond the conventional programs of financing Agencies. Under this respect, an analogous acknowledgment should be attributed to the III INFN Scientific Committee, who kept an open-minded attitude with repect to supporting experimental research in this field in Italy.

3. The Bologna Experiment at the Gran Sasso Laboratory: Criteria and Evolution

The experiment is mainly aiming at the detection of neutrons released following the immission of deuterons into Ti and Pd electrodes.

Working at the underground Gran Sasso Laboratory is a choice which is founded on the fact that the neutron background within the tunnel is extremely reduced, and the expected neutron emission due to reaction (1) is a low-level one.

Aiming at the detection of these neutrons has the advantage that they are the only uncharged massive products of reactions (1) - (3), so that they may easily leave the material in which they have been produced, without loosing a significant fraction of their initial energy. The experiment has so far gone through the following two steps:

1) *Preliminary measurements (1989).* These ones were performed in collaboration with S.E.Jones and his group of the Brigham Young University, using small cathodes of *fused* Ti, and the same electrolyte of different salts dissolved in D_2O which was used by the Jones group in their pioneer work.[6]

In this attempt already, the neutron detectors used allowed to discriminate neutrons from the local radioactivity gamma-rays by means of pulse-shape discrimination (PSD). These detectors provided for monochromatic neutrons a continuous spectrum, which had to be interpreted by Monte Carlo calculation to extract out the energy information.[7] The residual gamma-background, however,[8] was significant (about a factor of 5 larger than the signal). It was also clearly preferable to work with a neutron-detecting apparatus which directly should provide a spectrometric energy information (amplitude peak for monoenergetic neutrons), and which should prove to be more effective in the rejection of gamma-rays.

2) The coincidence neutron spectrometer, background and stability measurements (1990/91). The group has therefore developed a dedicated apparatus under a low-radioactivity lead shield, which decreased the local gamma-radioactivity within the Gran Sasso tunnel by a factor 7-8. The apparatus has first been focussed on the construction of a novel coincidence neutron spectrometer, the performances of which have been described elsewhere, with regard to its gamma-ray rejection, spectrometric operation, and background measurements in the Gran Sasso tunnel.[9] It is worth while mentioning here that the gamma-ray background is thereby reduced by a factor of 1.5×10^5, while the detected neutron background is of a very low level (some 10^{-6} neutrons s^{-1}). The stability of the apparatus proved to be excellent over periods of measurements of the order of monthes.

4. Present Status of the Experiment

While the Bologna group was preparing the measurements with the renewed set-up, experimental research in the field of NCF prosecuted on a world scale, with alternant successes and failures, and opening to new methods of investigation. The situation which was summarized by us one year ago by laying side by side the data "as received" from different groups on the neutron emissions observed following electrolytic loading is recalled in Fig.1.[4] This Figure was presented by the authors with one *caveat* on the reliability of such a straightforward comparison: these data were in fact referring to electrodes differing for structure, purity and treatment of the electrodes, and to measurements little homogeneous as far as duration, experimental conditions and reproducibility were concerned. A naive attempt which was performed[10] to normalize these data to the unit volume and to the unit surface of the electrodes yielded frustrating results: the wide spreading of the (even positive) experimental results was practically unaffected by this ingenuous treatment.

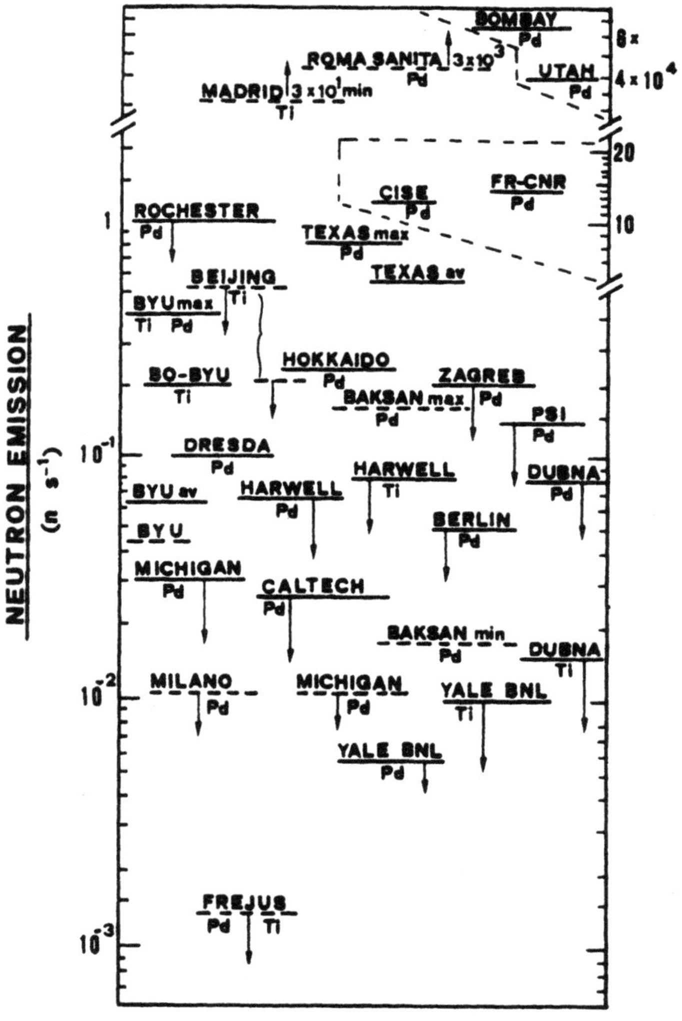

Fig. 1 - Neutron emission observed in different experiments using palladium (Pd) and Titanium (Ti) cathodes in electrolytic D_2O cells. The results underlined with full lines were reported by us in Reference 7. The values underlined with dashed lines are those discussed in the present work. The arrows pointing downwards indicate upper limits. (From Reference 4)

A little more reassuring picture came out from the comparison of the results "as received" on the neutron emission observed following pressure loading of Ti or Pd samples, (see Fig.2).[4]

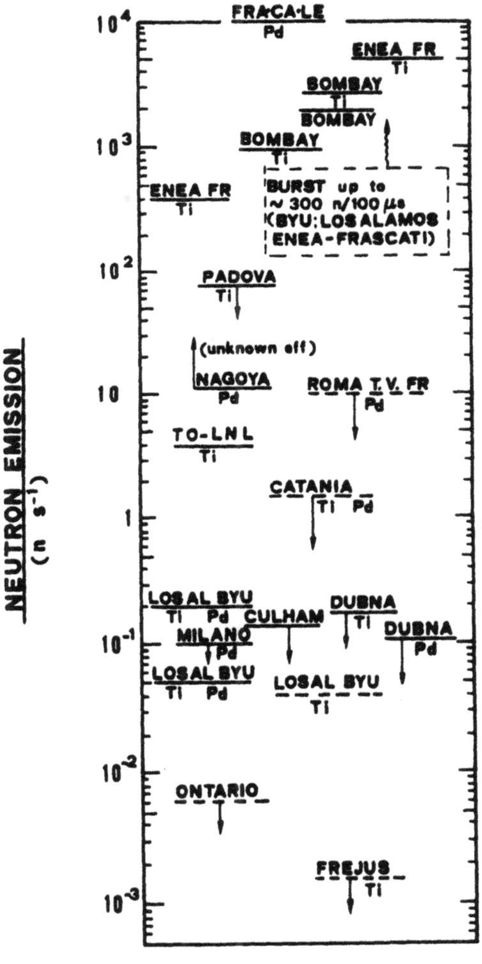

Fig. 2 - Neutron emissions observed in different experiments following deuterium pressure-loading of Pd and Ti samples. Arrows, dashed and continuous underlying are set as for Fig.1. (From Reference 4; TO-LNL from Reference 11)

These difficulties confirmed to us that one essential figure to be controlled - which might well be the only significant normalization parameter for non-homogenous measurements - is the *loading fraction* (i.e. the stoichiometric ratio (*deuterium atoms*)/(*Pd or Ti atoms*)) obtained in the chosen experimental conditions: a parameter which is not easily determined, although experienced groups try to be more and more precise on this point.[11]

Despite these difficulties, we decided to prosecute the search for neutrons emitted from electrolitically-loaded Ti or Pd samples. Openings to the pressure-loading method are also foreseen. The updating aims of this work are the following:

i) to carry out systematically accurate measurements in order to establish the low-neutronic (or a-neutronic) character of the fusion reaction of deuterium nuclei occurring within a metal lattice.

ii) priority will be given to the measurements with the electrolytic loading, in order to complete the experimental program started in 1989 [6], following the methodology which first was used[12], to explore the possibilities of NCF.

To fulfill the purpose of point i) implies nowadays, given the above-mentioned difficulties, continuous corrections to the measurement perspective, which makes pale the ingenuous original expectation of producing NCF via *bread and butter* experiments. In our case, aside of the improvements in the detecting apparatus and of the background, stability and blank measurements already mentioned, the following improvements have also been introduced:

1) In parallel to the coincidence neutron spectrometer mentioned in Sect.2, a system of four conventional NE 213 neutron counters, with neutron-gamma PSD, has been set up in order to detect the possible emission of *neutron bursts*. While the spectrometer is constitutionally dedicated to the detection of uncorrelated, or *continuous*, neutron emission, the 4 neutron detectors are aiming to detect the emission of neutrons in time coincidence or correlated in time within a short time interval (of the order of 1 ms, with a minimum delay of few tenths of ns between one neutron and the subsequent one).

The method, which in practice is based on the detection of the distribution of the time intervals occurring between two subsequent neutron signals, has the advantage of defining the emission intensity corresponding to one neutron burst in an analogous way to the one by which the decay rate of a radioactive source is given. The background measurements in the Gran Sasso tunnel yield a rate of two-neutron events of the order of a few units per more than 100 hours, with a flat time distribution, which represents the sensitivity limit for this type of measurements.

2) A significant effort has been devoted to develop electrolytic cells in which electrodes and electrolytes have more controlled characteristics than the ones used in the past. Aside of the original Jones-like salt mixture, we are presently working with a 0.1M LiOD solution in D_20, with one thermostatic cell, the temperature and current of which are under on-line control. We are also working in order to control the actual loading fraction achieved in each experiment, which can be determined both by degassing (once the measurements are concluded), and by destructive analysis of the electrode. One more technique which is also under development by our group is the measurement of the gas volumes developed during the electrolysis. The methods are giving some consistent results, so that we now have rasonable hopes of having achieved a substantial step to keep under control also this impor-

tant parameter.

This effort is considered as one contribution to the objective of getting truly quantitative and reproducible, data in the field of NCF.

REFERENCES

1. T.Bressani, E.Del Giudice, and G.Preparata, *Il Nuovo Cimento* **A 101** (1989) 845.
2. A.Bertin and A.Vitale, in *Problems of Fundamental Modern Physics II*, R.Cherubini, P.Dalpiaz and B.Minetti Editors, World Scientific (1991) p. 257.
3. A.Bertin and A.Vitale, in *Common Problems and Ideas of Modern Physics*, World Scientific (1992), T.Bressani, B.Minetti and A.Zenoni Editors, p. 257.
4. B.F.Bush, J.J.Lagowski, M.H.Miles and G.S.Ostrom, J.Electroanal. Chem. **304** (1991) 271.
5. J.Schwinger, in *Proc. of the First Annual Conference on Cold Fusion*, National Cold Fusion Istitute, Salt Lake City, Utah (1989) 130.
6. S.E.Jones, E.P.Palmer, J.B.Czirr, D.L.Decker, G.L.Jensen, J.M.Thorne, S.F.Taylor, and J.Rafelski, Nature **338** (1989) 737.
7. A.Bertin, M.Bruschi, M.Capponi, S.De Castro, U.Marconi, C.Moroni, M.Piccinini, N.Semprini-Cesari, A.Trombini, A.Vitale, A.Zoccoli, S.E.Jones, J.B.Czirr, G.L.Jensen and E.P.Palmer, Il Nuovo Cimento **101 A** (1989) 997.
8. A.Bertin, M.Bruschi, M.Capponi, S.De Castro, U.Marconi, C.Moroni, M.Piccinini, N.Semprini-Cesari, A.Trombini, A.Vitale, A.Zoccoli, S.E.Jones, J.B.Czirr, G.L.Jensen and E.P.Palmer, J.Fusion Energy **9** (1990) 209.
9. A.Bertin, M.Bruschi, D.Bulgarelli, M.Capponi, I.D'Antone, S.De Castro, D.Galli, U.Marconi, M.Morganti, C.Moroni, M.Piccinini, M.Poli, N.Semprini-Cesari, M.Villa, A.Vitale, G.Zavattini, and A.Zoccoli, *Muon Catalyzed Fusion* **5/6** (1990/91) 396; S.Affatato, A.Bertin, M.Bruschi, D.Bulgarelli, V.M.Bystritsky, M.Capponi, I.D'Antone, S.De Castro, D.Galli, U.Marconi, I.Massa, M.Morganti, C.Moroni, M.Piccinini, M.Poli, N.Semprini-Cesari, M.Villa, A.Vitale, G.Zavattini, and A.Zoccoli, Il Nuovo Cimento **104 A** (1991) 437; A.Bertin, M.Bruschi, D.Bulgarelli, B:M.Bystritsky, M.Capponi, I.D'Antone, S.De Castro, D.Galli, I.Massa, C.Moroni, M.Piccinini, M.Poli, N.Semprini-Cesari, R.Spighi, V.A.Stolupin, S.Vecchi, M.Villa, A.Vitale, J.Wozniak, G.Zavattini and A.Zoccoli, *Neutron Spectrometry by Means of a Two-Stages Neutron-Gamma Pulse-Shape Discriminating Apparatus*, Il Nuovo Cimento **105 A** (1992) to be published.
10. A.Bertin e A.Vitale, in *Energia*, **n. 1** (1990) p. 2.
11. T.Bressani, D.Calvo, A.Feliciello, C.Lamberti, F.Iazzi, B.Minetti, R.Cherubini, A.M.I.Haque and R.A.Ricci, Il Nuovo Cimento **104 A** (1991) 1413.
12. M.Fleischmann and S.Pons, J.Electroanal. Chem. **261 N** (1989) 301.

IV. PHYSICS AT DAΦNE

Common Problems and Trends of Modern Physics, pp. 351–367
ed. T. Bressani, S. Marcello and A. Zenoni
©1992 World Scientific Publishing Company

PHYSICS WITH DAΦNE

LUCIANO MAIANI
Dipartimento di Fisica, Università di Roma "La Sapienza"
and
Istituto Nazionale di Fisica Nucleare, Sezione di Roma

Introduction

The DAΦNE accelerator complex has been approved by the INFN board in June 1990. It consists of a double ring e^+e^- collider with two intersecting regions, to be installed in the INFN Laboratories of Frascati, inside the ADONE Hall. The luminosity is optimized at the ϕ peak (E=1019 MeV), project luminosity is of $10^{33} cm^{-2}s^{-1}$. Commissioning is foreseen for mid '95, with initial luminosity in excess of 2.5 $10^{32} cm^{-2}s^{-1}$.

DAΦNE will be a very powerful ϕ-factory. In corrispondence to the decays (branching ratios in parenthesis):

$\phi \to K^+K^-$ (0.495)
$\phi \to K_L K_S$ (0.344)
$\phi \to \gamma\eta$ (1.28 10^{-2})
$\phi \to \gamma\eta'$ (< 4.1 10^{-4})

DAΦNE will be also a remarkably good K-factory, a good η-factory and, hopefully, an interesting source of η' and gluons. Neutral Kaons are produced (irrespectively of C or CP conservation) in a pure quantum state, consisting of an antisymmetric K_LK_S pair.

From the outset, the physics program of DAΦNE has been focussed on the observation of direct CP-violation in K_L decays, i.e. the measuremente of ε'/ε, with accuracy in the 10^{-4} range. After more detailed studies, it has become increasingly evident that there are, in addition, several interesting issues in low energy hadron dynamics, gamma-gamma physics, low energy meson spectroscopy and hypernuclear physics that can be studied thoroughly with this machine.

In this Seminar, I will illustrate a number of selected topics of the physics at DAΦNE. More details and information can be found in several studies on ϕ-factories done recently in Italy[1,2,3] and elsewhere[4,5]. A description of the machine is given by G.

Vignola in ref.[3]. The results of a theoretical Workshop on DAΦNE physics are going to appear soon as a special Report, ref.[6]

In the following, event rates will correspond to a luminosity of 5 10^{32} cm^2s^{-1} integrated over an effective year of 10^7 seconds. The ϕ peak cross section being:

$$\sigma_{peak} = \frac{12\pi}{M^2} \frac{\Gamma_e}{\Gamma} = 4.4\ 10^{-30}\ cm^2$$

the above integrated luminosity corresponds to the yearly production rates:

2.2 10^{10}	ϕ
1.1 10^{10}	K^+K^- pairs
7.6 10^9	K_LK_S pairs
2.8 10^8	η

The possibility of tagged decays is very important in many instances. Given the average decay lenghts:

$(\gamma\beta\tau)_{K_S} = 0.59$ cm;

$(\gamma\beta\tau)_{K_L} = 343$ cm;

$(\gamma\beta\tau)_{K^\pm} = 95$ cm

the tagging efficiency of K_S and $K^\pm$ depends directly from the detector size. Assuming a detector of a similar size ($\approx$150cm) and efficiency as the proposed KLOE detector, the annual rate of tagged Kaons can be given as:

1.7 10^9 tagged K_S
1.1 10^9 tagged K_S
9.0 10^9 tagged $K^\pm$ (using all $K^\pm$ decays as tag)
1.9 10^9 tagged $K^\pm$ (using $K^\pm \to \pi^\pm\pi^0$ decays, for a precision tag)

The possibility of an intense beam of tagged K_S is a most typical and interesting property of the ϕ-factory.

The summary of the Seminar is as follows:

1. CP violation
2. Chiral dynamics
3. Low-energy γ-γ pion production
4. Hadronic contributions to the muon g-2
5. Spectroscopy below the ϕ
6. Dalitz pair decays of vector mesons
7. Conclusions

1. CP violation

No doubt, the most important task of DAFNE is the measurement of ε'/ε, from the comparison of different CP-violating decay modes of the K_L-K_S pair produced in φ-decay.

The real part of ε'/ε can be extracted from the distance distribution of the correlated decays:

$$\phi \rightarrow K_L K_S \rightarrow \pi^+ \pi^- (d_1) + \pi^0 \pi^0 (d_2) \quad (1.1)$$

where $d_{1,2}$ are the distances of the decay point from the interaction region[7,5,8].

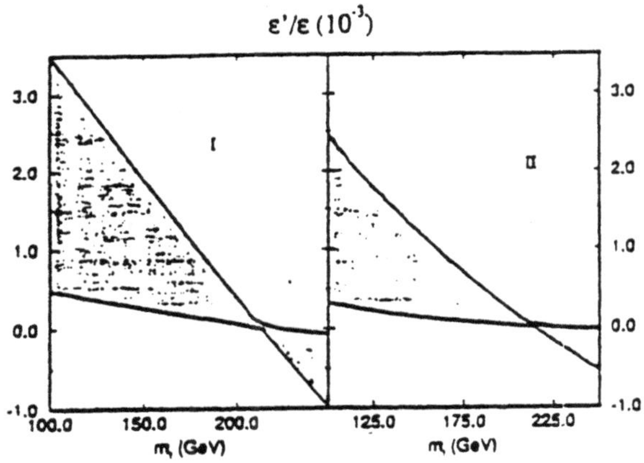

Fig.1. Upper and lower limits of e'/e vs. m_t, in the first and second quadrant of the CP-violating phase, d, ref.[9]. Parameter ranges used: $0.09 \leq |V_{ub}/V_{cb}| \leq 0.17$, $0.036 \leq |V_{cb}| \leq 0.046$, $0.1\text{GeV} \leq L_{QCD} \leq 0.3\text{GeV}$, $125 \leq m_s \leq 200\text{MeV}$. Other quantities are taken in the leading 1/N expansion.

Figs. 1 and 2 show the theoretical prediction of ε'/ε, as function of the t-quark mass, from two different, recent, papers. The analysis of ref.[9] is based on "continuum physics" estimates of the various weak matrix elements involved (the so-called B-factors). The authors of ref.[10] use, instead, lattice QCD determinations of the B-factors.

Considering that the two calculations are affected by very different systematic errors they are in quite good agreement with each other. Both estimates show a decrease of ε'/ε at the increase of the t-quark mass. The range of m_t presently preferred by the electroweak data indicates a rather small value of ε'/ε.

The figures make it clear that theory is not particularly good at predicting ε'/ε. We need this quantity because we want to know if CP-violation is indeed inside the weak interaction, more than to use it as a test of the Standard Theory, or of any other model.

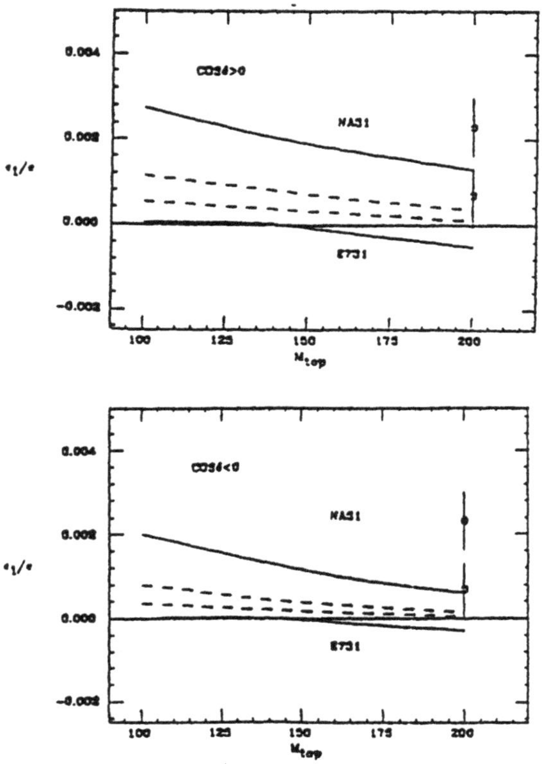

Fig.2. Upper and lower limits of e'/e vs. m_t, in the first and second quadrant of the CP-violating phase, d, ref.[10]. Parameter ranges used: $0.08 \leq |V_{ub}/V_{cb}| \leq 0.14$, $0.042 \leq |Vcb| \leq 0.051$, $0.1\text{GeV} \leq L_{QCD} \leq 0.3\text{GeV}$, $140 \leq m_s \leq 200\text{MeV}$. Other quantities are taken from lattice QCD calculations.

Should ε'/ε be too small to be detected at DAFNE, or with the experiments now in preparation at CERN and FNAL, B-physics would be the last resort to distinguish a superweak origin of CP-violation from the milliweak, Kobayashi-Maskawa, alternative.

The search of CP-violating asymmetries in charged Kaon decays can also be pursued with DAFNE. There are indications[11,12] that the optimistic estimate of the asymmetries given in ref.[13] is, indeed, quite optimistic. A considerable improvement of the present limits can be achieved, at any rate, and it would represent a significant progress.

The tagged K_S beam allows the leptonic asymmetry of K_S to be measured, which will provide an interesting test of the CPT symmetry[5,14]. Finally, the ϕ-factory should permit the first observation of order 10 events of the purely CP-violating decay:

$$K_S \rightarrow \pi^0 \pi^0 \pi^0 \qquad (1.2)$$

2. Chiral dynamics

As Weisskopf said once, physicists belong to two categories, those who study the radial and those who study the angular part of the wave-function. In field theory there are two different kinds of "angular wave-functions", corresponding to the two ways a continuous symmetry can be realized: in the Wigner or in the Goldstone mode.

Assume a symmetric lagrangian plus a small, symmetry breaking perturbation:

$$L = L_0 + L_{SB}$$

In the Wigner mode, the vacuum (ground) state is approximately symmetric, exactly so in the limit $L_{SB} \to 0$. The implications of the approximate symmetry are given by (correspondingly approximate) relations between masses, amplitudes, etc., which follow from the Wigner-Eckart theorem. Typical cases are isospin relations of the kind:

$$A(\Delta^+ \to P\,\pi^0) = \sqrt{\frac{2}{3}}\,A_0 \tag{2.1}$$

$$A(\Delta^+ \to N\,\pi^+) = \sqrt{\frac{1}{3}}\,A_0, \text{ etc.} \tag{2.2}$$

A_0, the "reduced matrix element" depends from the dynamics while the factors in front, the Clebsch-Gordan coefficients, are purely group-theoretical quantites.

In the Goldstone, spontaneously broken, mode, the stable vacuum is unsymmetric, even for $L_{SB} \to 0$. In this limit, one massless spinless particle, a Goldstone boson, is present for each broken generator. The Wigner-Eckart theorem is replaced by the representation of low-energy amplitudes given by the effective lagrangians[15].

In QCD, the chiral $SU(3)_L \times SU(3)_R$ symmetry is spontaneously broken down to approximate $SU(3)_{flavour}$, L_{SB} is the light-quark mass matrix, pions and Kaons play the role of (approximate) Goldstone bosons.

The low-energy self-interactions of pions and Kaons are described by the effective chiral lagrangian:

$$L_{eff} = \frac{f_\pi^2}{8}\,Tr(\partial_\mu U \partial^\mu U^\dagger) + \frac{v}{8}\,Tr(UM + MU^\dagger) \tag{2.3}$$

where:

$$U = \exp(i\frac{2\pi}{f_\pi}) \tag{2.4}$$

$$\pi = \Sigma_i\, \pi_i \frac{\lambda^i}{\sqrt{2}} \tag{2.5}$$

with π_i the pseudoscalar octet fields, λ^i the Gell-Mann matrices and M the quark mass matrix appearing in L_{SB}. The pion decay constant, $f_\pi \approx 132$ MeV, plays the role of a "reduced matrix element", i.e. it is determined by the underlying dynamics. All the rest, apart from small perturbations due to L_{SB}, is determined by the symmetry, like the Clebsch-Gordan coefficients in (2.1-2).

Goldstone bosons decouple at zero energy. Therefore, it is possible to perform a systematic expansion around this point, the Goldstone boson momenta acting as small, perturbative, parameters. Similarly, the effect of L_{SB} can also be power-expanded, at least for quark masses which are small with respect to the "typical QCD mass scales", e.g. the proton mass. The resulting scheme is often called Chiral Perturbation Theory (χPT, for short).

Chiral dynamics was much studied in the sixties[16], as a tool to determine the symmetries of hadron dynamics. It has been reconsidered more recently, after the work of Gasser and Leutwyler[17] and many others, who have carried out the expansion in external momenta to the next order, in many interesting instances.

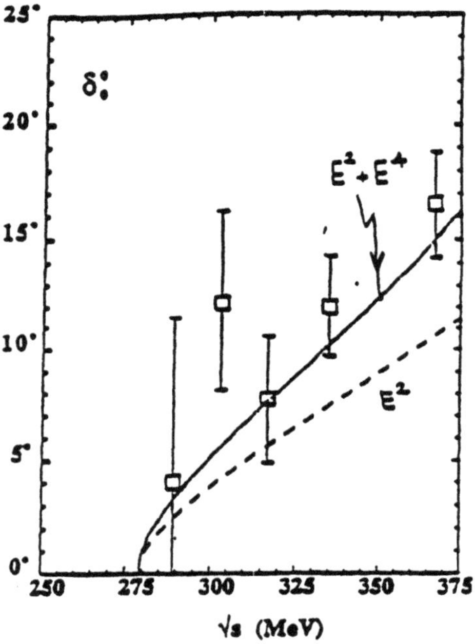

Fig.3 π-π I=J=o phase shifts vs energy. Dashed curve: lowest order prediction; solid line: one loop-corrected chiral prediction[17] (see the talk of J. Gasser in ref.[3] for more details and references).

Of course, to the next level of approximations, new "reduced matrix elements" appear, to be determined from the experiment. However, there remain enough testable predictions to obtain a meaningful test of the chiral dynamics from a systematic comparison with data. In addition, the coefficients of the effective lagrangian can be computed theoretically in some cases, e.g. with numerical methods in lattice QCD. The comparison with the experimentally determined coefficients provides a rigorous test the fundamental theory of strong interaction, in a fully non-perturbative regime.

Fig.3 gives a nice illustration of how all this is supposed to work. The I=J=0 π-π experimental phase-shift[17] is compared to the lowest order prediction, obtained by Weinberg and others from the effective lagrangian (2.3), dashed line, and to the one-loop corrected prediction, solid line. Another example is given[18] by the $\eta \to \pi^+\pi^-\gamma$ decay rate:

$$\Gamma(\eta \to \pi^+\pi^-\gamma) = 35 \text{ eV (lowest order)}$$
$$= 47 \text{ eV (including one-loop)}$$
$$= (53 \pm 10) \text{ eV (Particle Data Book}^{19})$$

Not everything works so nicely, however, as exemplified by the $\eta \to 3\pi$ decay rate:

$$\Gamma(\eta \to \pi^+\pi^-\pi^0) = 66 \text{ eV (lowest order)}$$
$$= (160 \pm 50) \text{ eV (including one-loop}^{17})$$
$$= (280 \pm 14) \text{ eV (Particle Data Book}^{19})$$

To obtain a better understanding of this theoretical method, one of the very few available to describe a field theory in the strong-interaction regime, it would be important to map systematically the regions of agreement and of discrepancy. DAFNE can contribute significantly to this program, with a systematic exploration of processes determined by the chiral dynamics, particularly in K-decays, but also in other areas (γ-γ physics is discussed in the next Section). Tab.1, taken from the second paper of ref. [17] shows DAFNE remarkable potential to improve on our present knowledge of semileptonic K-decays. Other interesting cases[6] include:

- radiative leptonic decays (e.g. $K \to \mu \nu \gamma$)
- radiative non-leptonic decays (e.g. $K_S \to \nu \gamma$)
- non- leptonic decays (e.g. $K \to 3\pi$).

Low energy 3π , or $\pi \gamma$ production by a virtual photon is also an interesting process, from the point of view of chiral dynamic. Detailed theoretical predictions, including one-loop and vector meson exchange corrections, are availables, as discussed in ref.[18].

Tab.1

Process	# events Part.DataBook	DAFNE(1year)	improvement
$K^0_{\mu 3}$	$4\ 10^6$	$3\ 10^8$	70
$K^+_{\mu 3}$	10^5	$3\ 10^8$	$3\ 10^3$
$K^+ \rightarrow \pi^0\pi^0 e^+\nu_e$	<50	$2\ 10^5$	$> 4\ 10^3$
$K^+ \rightarrow \pi^+\pi^- e^+\nu_e$	$3\ 10^4$	$3\ 10^5$	10
$K^0_L \rightarrow \pi^0\pi^\pm e^\mp\nu_e$	16	$7\ 10^4$	$4\ 10^3$

3. Low-energy γ-γ pion production

Fig. 4 shows the cross-section of the process:

$$\gamma\gamma \rightarrow \pi^0\pi^0 \tag{3.1}$$

around threshold, as determined by the Crystal Ball Collaboration (see ref.[20] for more details). χPT predicts this process to arise at the one-loop level. The loop amplitude, which arises from charged-pion exchange, is finite and calculable, simply because there is no allowed counterterm that could absorb an eventual divergence. The chiral prediction[21] is shown in Fig.4 as a dashed curve. The addition of vector meson exchange corrections has also been considered[20], and it does not modify appreciably the prediction, except in the high energy side. The disagreement of the data with the chiral prediction, if it is indeed real, is quite surprising.

The band in Fig.4 the most recent calculation by Pennington[22] based on dipersion relations, elastic unitarity and the experimental π-π phase shifts, (previous analysis by Morgan and Pennington is reported in ref.[23]). It agrees much better with the present data, particularly at low energy. How can this be, since, as we have just seen, chiral symmetry is quite proficient in predicting π-π phase-shifts at threshold?

I think the problem deserves further study, but I can suggest a possible way out.

Consider, for the sake of the argument, the amplitude of the crossed, Compton scattering, process:

$$\gamma\,\pi^0 \rightarrow \gamma\,\pi^0 \tag{3.2}$$

rather than that of (3.1). In a development of the amplitude in powers of the photon momentum, the constant and linear terms are easily shown to vanish, because of gauge invariance. By the Low theorem, these terms are proportional to the target electric charge and magnetic moment, which both vanish for π^0, so that the aplitude starts at least like E^2. This constraint is correctly taken into account in ref.[22].

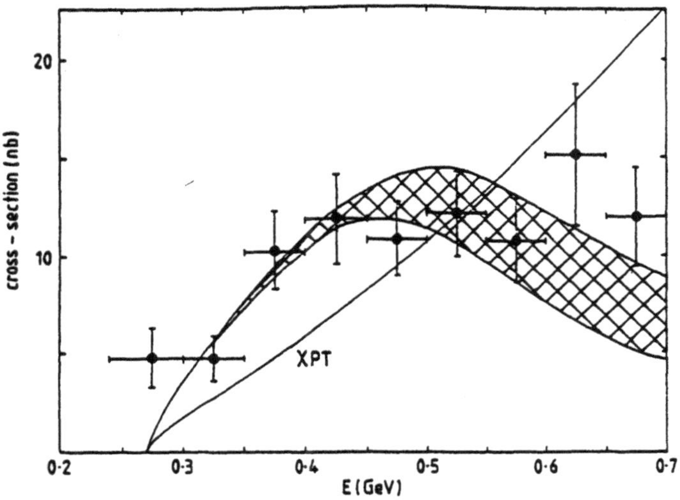

Fig.4 The integrated cross-section for $\gamma\gamma\to\pi^0\pi^0$ vs c.o.m. energy, around threshold, Crystal Ball data. Dashed curve: chiral dynamics prediction, ref.[21]. The band gives the (twice-subtracted) dispersive prediction of Pennington et al., ref.[22].

A second order contribution in the external momenta would arise from the gauge invariant effective coupling:

$$L_{int}=g\ (\pi^0)^2F_{\mu\nu}F^{\mu\nu} \tag{3.3}$$

However, chiral symmetry tells us that π^0 is a Goldstone boson and it must appear only with derivative couplings. So, the amplitude of (3.2) must vanish also to order E^2, and start with terms of the order of E^4, such as those generated by the effective interaction:

$$L_{int}=g_1\ (\partial^\mu\pi^0)(\partial_\lambda\pi^0)F_{\mu\rho}F^{\rho\lambda}+g_2\ (\partial^\mu\pi^0)(\partial_\mu\pi^0)F_{\alpha\rho}F^{\rho\alpha} \tag{3.4}$$

which respects all constraints. There is an exception to this rule, namely the amplitude due to the charged pion loop, for which the infrared singularity (in the $m_\pi\to0$ limit) may compensate one power of the energy in the numerator, giving an order E^2 amplitude. This is, of course, the reason why the charged pion loop dominates the amplitude in lowest order in E, i.e. it determines the slope of the χPT curve in Fig.4.

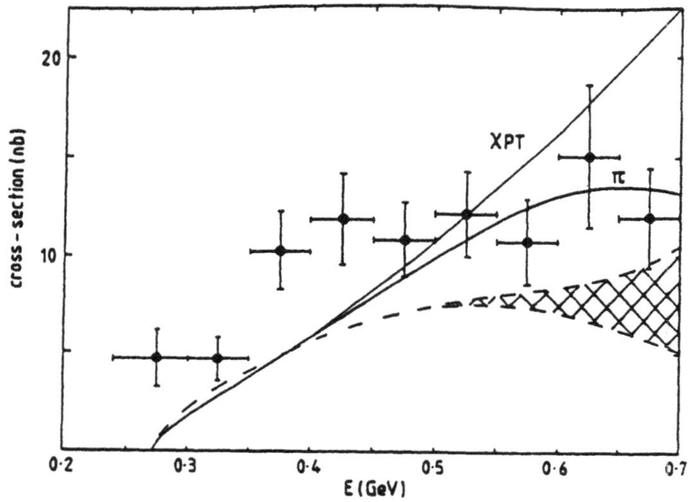

Fig.5 Integrated cross-section for $\gamma\gamma \to \pi^0\pi^0$ vs c.o.m. energy. The band gives the dispersive prediction of Pennington et al., ref.[22], where the lowest-order (Weinberg) $\pi\pi$ scattering phases are used as input, in the low-energy region. The dispersive prediction agrees now with chiral perturbation theory, around threshold.

A first consequence of these considerations is that a dispersive calculation should use a twice-subtracted form, to enforce correctly the Goldstone property of the pion. Truly enough, if we would treat the intermediate states *exactly*, we would automatically recover the behaviour implied by (3.4). However, any truncation of the intermediate states is likely to spoil the delicate cancellations implied by current algebra and give rise to an unreliable result. This second subtraction has also been performed in the calculation of ref.[22].

However, the dispersive calculation uses the experimentally determined $\pi\pi$ phase shifts. Fig.5 shows what one obtains using the lowest order, Weinberg, phase-shifts. The new result tends nicely to the χPT curve as we go toward threshold. This behaviour strongly suggests that failure of χPT to reproduce the data is mainly due to the departure of the lowest order $\pi\pi$ phase-shifts from the experimental ones. A much more precise prediction should obtain at the next order of approximation, in line with what happens to the $\pi\pi$ scattering phases.

A two-loop calculation of the amplitude (3.1) is strongly called for. In the meanwhile, DAΦNE can improve quite considerably the $\gamma\gamma$ cross-sections at low energy, for both neutral and charged pions, see Tab.2. The possibility of detecting both final electrons, so as to measure the azimuthal dependence of the $\gamma\gamma$ cross-section, has been stressed by Courau[24] as an important advantage of the lower energy of DAFNE.

Tab.2

$e^+e^- \to e^+e^- X$	# events/year			
	$M_{\pi\pi}< 0.45$ GeV		$M_{\pi\pi}< 1.0$ GeV	
	(total yield)	$(8.5<\theta_{LAB}<171.5)$	(total yield)	$(8.5<\theta_{LAB}<171.5)$
$X=\pi^0\pi^0$	$1.7\ 10^4$	$1.5\ 10^4$	$3.8\ 10^4$	$3.5\ 10^4$
$X=\pi^+\pi^-$	$2.2\ 10^6$	$1.8\ 10^6$	$2.8\ 10^6$	$2.3\ 10^6$

4. Hadronic contributions to the muon g-2

Consideration of hadron production at low energy brings me to the subject of the corrections to the muon g-2. The obvious target, in this domain, is the detection of the electroweak corrections due to vector and Higgs boson exchange. The present experimental value of the muon anomaly is[25]:

$$a_\mu(\text{expt}) = \frac{\alpha}{2\pi} + \ldots =$$
$$= 116593700 \pm 1200 \times 10^{-11} \ (\mu^-)$$
$$= 116591100 \pm 1100 \times 10^{-11} \ (\mu^+) \qquad (4.1)$$

with the electroweak contribution predicted to be:

$$a_\mu(\text{EW}) = (195 \pm 1)\ 10^{-11} \qquad (4.2)$$

To identify the effect (4.2), we need to reduce the experimental error in (4.1) by more than one order of magnitude, which does not seem impossible, but also reduce the error in the theoretical prediction of the other contributions to a_μ.

The main source of the latter error is the hadronic vacuum polarization correction, Fig.6(a), in turn given by an appropriate integral over the cross-section of $e^+e^- \to$hadrons. With the present data, one has[26,27]:

$$a_\mu(\text{had, vac-pol}) = (7030 \pm 59 \pm 164)\ 10^{-11} \qquad (4.3)$$

(the first is the statistical, the second the systematic error).

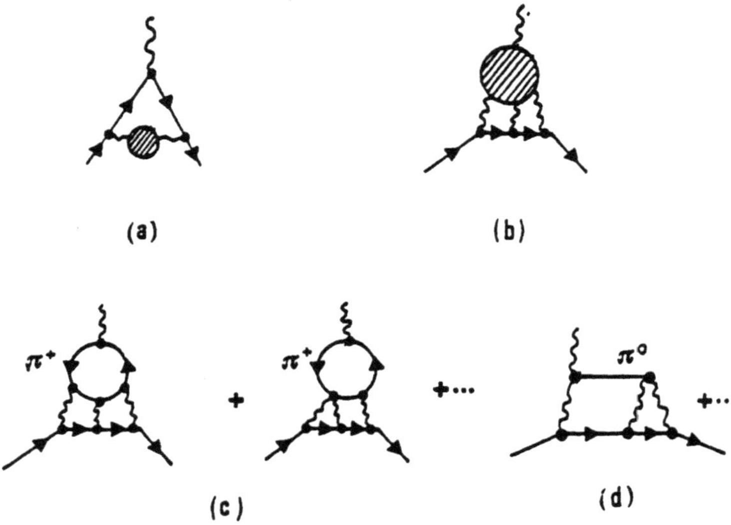

Fig.6 Feynman diagrams representing hadronic corrections to the muon anomaly. (a) lowest order contribution due to vacuum polarization; (b) next order contribution from light-light scattering; (c) charged and (d) neutral pion contributions to the corrections 5 (b).

Tab.3

Analysis[26] of the present statistical and systematic error of the hadronic contribution to the muon anomaly, in units of 10^{-11}. Totals are obtained in quadrature.

energy region:	ρ	ω	φ	high en.	tot
stat. error	21	48	18	21	59
syst. error	150	15	13	63	164
total				67	174

The error in (4.3) is mainly due to the imperfect knowledge of $\sigma(e^+e^- \to \text{hadrons})$ at low energy, as shown in Tab.3 [26].

DAFNE can improve considerably on the statistical and systematic errors in the ρ, ω and φ regions, reducing the overall error to something like 70 10^{-11}, sufficient to identify the bulk of the electroweak correction.

This is not the full story, however. Hadronic corrections, to the next order, enter via the light-light scattering diagram, Fig.6(b), and we have to make sure that the error on this contribution is also sufficiently small. This question was considered in ref.[26], where various estimates are given, see Tab.4, replacing the hadron bubble in Fig.6(b) with a poinlike quark loop, or with a charged pion loop and adding the contribution arising from the $\pi^0 \to \gamma\gamma$ coupling, see Fig.6(c). For the charged pion loop, the pointlike case has been considered, as well as the case where ρ-dominated form factors are associated with the γ-$\pi\pi$ vertices (form factor softening is always necessary for the neutral pion contribution, which would be UV divergent otherwise).

The quark loop result, with the masses indicated in the Table, agrees remarkably well with the sum of the contributions of the neutral and charged pions (with form factors)leading the authors to quote, as best value:

$$a_\mu(\text{had, light-light}) = 49 \pm 5 \times 10^{-11} \qquad (4.4)$$

The error is their estimate of the model dependence of the result.

Following ref.[2,6], I think that the error may have been quite underestimated. For one, the pion result arises from the cancellation of different (gauge invariant) contributions, each of which is larger than the electroweak correction itself, see Tab.4 and ref.[26], so that the uncertainty of the result is unlikely to be as small as quoted in (4.4). Moreover, the quark model result is very sensitive to the values of the assumed light quark masses (it goes approximatly like m^{-2}) and the agreement of the quark and meson results looks suspiciously close to being accidental. A quark loop estimate of a_μ(had, vac-pol) , with the quark masses given in Tab.4 yields:

$$a_\mu(\text{had, vac-pol})_{\text{quark loop}} \approx 3000\ 10^{-11}$$

To obtain the experimental result (4.3), one would rather need a light quark mass ≈ 0.19 GeV. If we take the latter mass for $m_{u,d}$, we would get: $a_\mu(\text{had, light-light})_{\text{quark loop}} \approx 150\ 10^{-11}$, a much less favorable result.

I think it is fair to say that the problem is still open. More thinking is needed to estimate the size of γ-γ scattering. In the meanwhile, a good measurement of the hadron cross-section at low energy, with DAFNE, cannot be but welcome.

Tab.4

Contributions to the muon anomaly, in units of 10^{-11}, from differents estimates of ligh-light scattering[26], see Fig.6. Numbers in parenthesis are the contributions from two different, gauge invariant set of diagrams, see ref.[26].

quark loop (m_u=m_d=0.3, m_s = 0.5, m_c=1.5 GeV)	charged pion loop (pointlike)	charged pion loop (ρ form-factor)	neutral pion (ρ form factor)
60	-48 (=-521+473)	-16(=-339+323)	+66

5. Spectroscopy

Hadron spectroscopy today means, by and large, trying to go beyond the standard quark model, with $q\bar{q}$ mesons and qqq baryons. Candidates for exotic states are glueballs, in the first place, and also four body, $q\bar{q}q\bar{q}$, states.

There are a few but significant issues in this area spectroscopy about which DAFNE may give a contribute significantly (see also ref.[28,29]).

i. Glue fraction of η'. Glueball states are definitly predicted to exist, in QCD, above 1 GeV. A light 0^- state could mix significantly with the isoscalar pseudoscalar mesons, η and η'. Writing the η or η' wave-function according to:

$$|P^0\rangle = N\,|(q\bar{q})_{I=0}\rangle + S\,|s\bar{s}\rangle + G\,|glue\rangle \tag{5.1}$$

$$|(q\bar{q})_{I=0}\rangle = \frac{1}{\sqrt{2}}(|u\bar{u}\rangle + |d\bar{d}\rangle) \tag{5.2}$$

we can determine the coefficients N and S from the transitions: $P^0 \to \gamma V$ (or $V \to P^0\gamma$) and $P^0 \to \gamma\gamma$ and check if N^2+S^2 differs from unity, thereby providing limits on (or evidence for) the fraction of glue, G in (5.1).

Present data on $\phi \to \eta\gamma$, $J/\psi \to \eta\gamma$, $\eta \to \rho\gamma$ and $\eta \to \gamma\gamma$ are compatible[28], within 20% errors, with $N^2+S^2=1$.

For η', the rates of $\eta' \to \rho\gamma$ and $\eta' \to \gamma\gamma$ are known, but $\phi \to \eta'\gamma$ is lacking. Observation of this decay would provide the missing information to check if η' has a significant glue component. Data on $\psi \to \eta'\gamma$ and $\eta' \to \rho\gamma$ may provide room for suspicion[28].

The branching ratio for $\phi \to \eta'\gamma$ is expected at the 10^{-4}-10^{-5} level, corresponding to 10^6-10^5 η' /year, and should be seen without question.

ii. The nature of the I=0 and I=1 scalar mesons f_0 (once, S) and a_0 (once, δ) has been debated since a long time.

The small coupling of f_0 to pions (it decays comparably into the $K\bar{K}$ channel which is almost completely below threshold) would suggest a strong $s\bar{s}$ component. On the other hand, f_0 and a_0 are almost degenerate, more appropriate to a ρ-ω like situation.

In addition, in the standard $q\bar{q}$ model, 0^{++} scalar mesons belong to the L=1 supermultiplet (L=orbital angular momentum) together with 1^{++} and 2^{++} mesons. While mass differences of 1^{++} and 2^{++} meson are well described by the spin-orbit coupling, the latter would predict 0^{++} mesons to lie definitely above 1 GeV (by 100-200 MeV), again suggesting that the observed a_0 and f_0 are not $q\bar{q}$ states.

The $q\bar{q}q\bar{q}$ assignement:

$$a_0 = \frac{1}{\sqrt{2}} (s\bar{s}) (u\bar{u} - d\bar{d})$$
$$f_0 = \frac{1}{\sqrt{2}} (s\bar{s}) (u\bar{u} + d\bar{d})$$

or a description in terms of $K\bar{K}$ molecule, have been invoked to solve the puzzle. The observation of the radiative decays:

$$\phi \rightarrow f_0/a_0\, \gamma \qquad (4.1)$$

will certainly help to clarify the situation[28].

6. Dalitz pair decays of vector mesons

The vector meson decays:

$$\phi \text{ (or } \rho, \text{ or } \omega) \quad \rightarrow \eta\, e^+e^-, \pi^0\, e^+e^- \qquad (6.1)$$

$$\phi \text{ (or } \rho, \text{ or } \omega) \quad \rightarrow \eta\, \mu^+\mu^-, \pi^0\, \mu^+\mu^- \qquad (6.2)$$

as well as the continuum reactions:

$$e^+e^- \rightarrow \gamma^* \rightarrow V\, \eta, V\, \pi^0 \qquad (6.3)$$

($V=\rho$, ω, ϕ), give direct information on the vector form factor of the transition: Vector→Pseudoscalar, in the time-like region. As estimated by Eidelmann[30], at DAFNE one could get as much as $5\ 10^3$ and $6\ 10^2$ events/year for $\eta\, e^+e^-$ and $\pi^0\, e^+e^-$, respectively, enough for reliable branching ratio and form factor measurements.

Results could be compared to predictions from Vector Meson Dominance, or from lattice QCD calculations[31], which are quite within present possibilities.

7. Conclusion

The observation of direct CP-violation is still the DAΦNE "plat de resistence". New, interesting physics issues have emerged, however, in the studies of the last years, as I hope to have shown in this Seminar. I think more arguments will come.

The construction of the machine and of adequate detectors is a considerable challenge to the national community and, in particular, to the Frascati Laboratory. I am convinced the enterprise is worth the effort and that its success will be of great importance for the future of Particle Physics in our country.

8. References

1. *Physics Program at a Φ - Factory*, ed. by M. Giorgi, M. Greco, M. Piccolo, LNF-90/031(R), 1990.
2. R. Barbieri et al., *Report of the Φ - Factory Working group* , LNF-90/041(R), May 29, 1990; LNF-91/020(R), May 3, 1991.
3. Proc. of the *Workshop on Physics and Detectors for DAΦNE*, Frascati, April 9-12, 1991, ed. by G.Pancheri, 1991.
4. *Physics at a Φ - Factory*, ed. by M. Fukawa et al., KEK Report, Tsukuba, 1990.
5. C.Buchanan, R.Cousins, C.Dib, R.Peccei, J.Quackenbush, UCLA/91/TEP/44.
6. *The DAΦNE Physics Handbook*, ed. by L.Maiani, G.Pancheri and N.Paver, in preparation.
7. P.Franzini, Proc. of the DAΦNE Workshop, ref[3].
8. V.Patera, A.Pugliese in ref.[6].
9. G.Buchalla, A.J.Buras and M.K.Harlander, *Nucl.Phys.* **B337** (1990) 313; A.J.Buras, M.K.Harlander, MPI-PAE/PTh 1/92, to appear in the Review Volume on Heavy Flavours, ed. by A.J.Buras and M.Lindner, World Scientific Publishing Co., Singapore.
10. M. Lusignoli, L. Maiani, G. Martinelli, L. Reina, *Nucl.Phys.* **B369** (1992) 139.
11. J. Donoghue, in ref.[3].
12. G.Isidori, L.Maiani, A.Pugliese, Univ. of Roma"La Sapienza", preprint n.848, 1991, to appear on Nucl. Phys.B.
13. A.Belkov, G.Bohm, D.Ebert and A.Lanjov, *Phys. Lett.* **232B** (1989) 118.
14. L.Maiani, in ref.[6].
15. S. Coleman, J. Wess, B. Zumino, *Phys. Rev.* **177** (1969) 2239; C.G. Callan, S. Coleman, J. Wess, B. Zumino, *Phys. Rev.* **177** (1969) 2247.
16. see, e.g., S.Weinberg, *Physica* **96A**, 327 (1979); S.Coleman in *Aspects of Symmetry*, Cambridge University Press, 1985.
17. J.Gasser, H.Leutwyler, *Ann.Phys.* (NY) **158** (2984) 142; *Nucl.Phys.* **B250** (1985) 465. For a recent discussion and summary of the relevant literature, see J.Bjinens, G.Ecker, J.Gasser, in ref.[6].
18. A.Bramon, in ref.[3]; A.Bramon, A.Grau, E.Pallante, G.Pancheri, R.Petronzio, in ref.[6].
19. Particle Data Book, *Phys. Lett.* **B239** (1990) 1.
20. S. Bellucci, in reff.[3, 6].
21. J.F.Donoghue, B.R.Holstein and Y.C.Lin, *Phys. Rev.* **D37** (1988) 2423.
22. M.R.Pennington, in ref.[6].
23. D.Morgan, M.R.Pennington, *Phys. Lett.* **192B** (1987) 207; *Z.Phys.* **C37** (1988) 431; *Z.Phys.* **C39** (1988) 590.
24. A.Courau, in ref.[3].
25. Bailey et al., *Nucl. Phys.* **B150** (1979) 1.
26. T.Kinoshita, B.Nizic and Y.Okamoto, *Phys.Rev.* **D31** (1985) 2108.

27. L.Martinovic and S.Dubnicka, preprint Dubna E2-89-144, 1989.
28. F.Close, in ref.[3]; F.Close, N.Brown, in ref.[6]..
29. N.N.Achasov, A.A.Ivanchenko, *Nucl. Phys.* **B315** (1989) 465; see also he contributions of N.N.Achasov and M.R.Pennington to the Proceedings in ref.[3].
30. S.Eidelman,in ref.[3].
31. M.Crisafulli, V.Lubicz, in ref.[6].

Common Problems and Trends of Modern Physics, pp. 369–385
ed. T. Bressani, S. Marcello and A. Zenoni

©1992 World Scientific Publishing Company

VECTOR MESON RECURRENCIES BELOW 2 GeV

R. BALDINI
Laboratori Nazionali dell'INFN, CP 13, I-00044 Frascati, Italy

and

D. BISELLO
Dipartimento di Fisica e Sezione INFN, via Marzolo 8, I-35131, Padova, Italy

Abstract

The production of vector meson recurrencies with masses up to 2 *GeV* is shortly reviewed. The importance of a DAΦNE contribution in this field is enphasized.

1. Introduction.

The important multihadron production above the ϕ, first observed at ADONE[1] twenty years ago, is usually interpreted as the contribution of the recurrencies of the ρ, ω, ϕ mesons. Nevertheless the observation of such recurrencies is still controversial.

Although the direct observation as a resonant production in final states of e^+e^- annihilation remains the best signature for a new vector meson, data are also available from diffractive photoproduction, from phase shift analysis of many pions production in high statistics hadronic interactions, and from τ decays into an even number of pions according to the CVC hypothesis. However photoproduction results also depend on large inelastic contributions which may account for the observed discrepancies with corresponding e^+e^- data. On the other hand, at present, e^+e^- data, when they do not suffer of low statistical significance, are hard to interpret because of large interference effects.

2. Isovector Sector.

The production of isovector mesons represents the largest contribution to the total hadronic e^+e^- cross section.

Ten years ago, before the DM2 experiment at the DCI, the salient experimental features were the presence of a very large and smooth bump[2] in the $e^+e^- \rightarrow \pi^+\pi^-\pi^+\pi^-$ channel, peaked at 1.5 GeV with a width of the order of 0.4 GeV (Fig. 1a). The same bump was observed with a smaller width in $\pi^+\pi^-\pi^+\pi^-$ and $\pi^+\pi^-$ photoproduction[3] (Fig. 2a). The other relevant process, $e^+e^- \rightarrow \pi^+\pi^-\pi^0\pi^0$ (including $\omega\pi^0$) showed a sharp rise in the cross section at 1.1 GeV followed by a slow fluctuating decrease[4] (Fig. 1b). This result was roughly consistent with $\pi^+\pi^-\pi^0\pi^0$ diffractive photoproduction[5] taking in mind the large background coming from the photoproduction of the axial meson B(1.25).

The $4\pi^\pm$ data were interpreted[6] as evidence of a ρ recurrency, the $\rho'(1.6)$. On the other hand, to explain the $\pi^+\pi^-\pi^0\pi^0$ behaviour, the ρ tail contribution was added[7] to

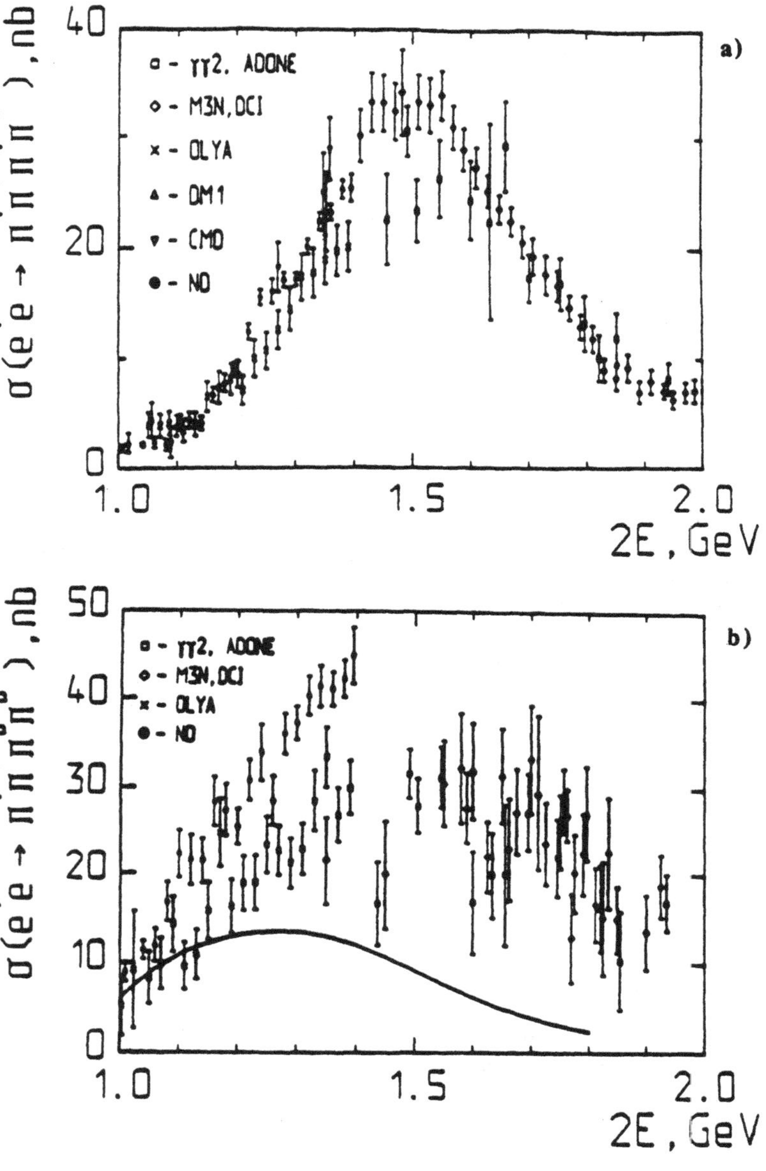

Fig. 1 $e^+e^- \to \pi^+\pi^-\pi^+\pi^-$ (a) and $e^+e^- \to \pi^+\pi^-\pi^0\pi^0$ (b) cross sections before DM2.

the $\rho'(1.6)$ via the process $\rho \to \omega\pi^0$ with coupling constants coming from $e^+e^- \to \rho$ and $\omega \to \rho\pi$.

The total hadronic cross section was assumed to be mainly due to the contributions of the even ρ recurrencies[8], according to the Veneziano formula. This model gave an asymptotic value of $\sigma_{tot}/\sigma_{\mu\mu}$ consistent with the quark model prediction[9] by assuming $\Gamma_{ee}(\rho_n) \propto 1/M(\rho_n)$ and $\Gamma_{tot}(\rho_n) \propto M(\rho_n)$, in agreement with the $\rho'(1.6)$ parameters. In this scenario a $\rho'(1.3)$ and orbital ρ excitations were disregarded.

The naive expectation of a single $\rho'(1.6)$ has been ruled out by the DM2[10] measurement of the pion form factor from $e^+e^- \to \pi^+\pi^-$ (Fig. 2b). The interference pattern requires at least two resonances, a $\rho'(1.45)$ and a $\rho'(1.7)$[11]. On the other hand, in $\pi^+\pi^-$ diffractive photoproduction a bump is observed at 1.6 GeV in the $\pi\pi$ invariant mass, corresponding to a deep in the pion form factor. To preserve the vector meson dominance model of diffractive photoproduction very large inelastic contributions must be taken into account[12], spoiling in part the strength of this model. Furthermore data on $e^+e^- \to \rho\eta$[13], as well as data on phase shifts analysis[14] on $\pi^-p \to \rho\eta n$, are in favour of the two resonances hypothesis, but the statistical accuracies are not very compelling (Fig. 4a,b).

On the other hand the smooth bump in $e^+e^- \to \pi^+\pi^-\pi^+\pi^-$, peaked at $m = 1.5\ GeV$, has been confirmed by DM2[15] (Fig. 3a). It has been shown[11] that this pattern is not in disagreement with the above-mentioned two superimposed resonances.

DM2 data[15] on $e^+e^- \to \pi^+\pi^-\pi^0\pi^0$ are also consistent with the two ρ' hypothesis. Actually the sharp drop of the $e^+e^- \to \omega\pi^0$ cross section at 1.5 GeV (Fig. 5), in good agreement with data on $\tau \to \omega\pi^0$ [16], is evidence for an interference between a ρ tail and a $\rho'(1.4)$. On the other hand a different bump in $e^+e^- \to \pi^+\pi^-\pi^0\pi^0$, peaked at $m = 1.7\ GeV$, has been put in evidence, once $e^+e^- \to \omega\pi^0$ has been removed (Fig. 3b). Unfortunately high statistics Novosibirsk[17] data at lower energies (Fig. 1b), already inconsistent with data at the same energies extracted from $\tau \to \pi^-\pi^+\pi^-\pi^0$ [18], do not match with those of DM2.

In conclusion the DM2 data show evidence for two isovector resonances at $\simeq 1.4$ and $\simeq 1.7\ GeV$. Nevertheless such results cannot be taken as conclusive.

Actually a renewed evidence for the old $\rho'(1.3)$, with mass $m = 1.30 \pm 0.03\ GeV$ and width $\Gamma = 0.14 \pm 0.05\ GeV$, has emerged in the $\pi^+\pi^-$ phase shifts analysis in $K^-p \to \pi^+\pi^-\Lambda$ from LASS[19] (Fig. 6a,b). The statistical evidence is not overwhelming, however a check of this experiment is that a clear $\rho - \omega$ interference effect is observed (Fig. 6c), with branching ratio and phase in agreement with storage ring measurements.

MARK III[20] measuring the pion form factor from $J/\psi \to \pi^+\pi^-\pi^0$ claims to be consistent with LASS (Fig. 2c). However no real evidence is found for a $\rho'(1.3)$ in the pion form factor or in the 4π channels in the DM2 and photoproduction data [21] . Relevant channels in which this resonance should be seen are $K\bar{K}$ and $\omega\pi^0$, but for the latter one should get rid of the not resonant background coming from the ρ tail.

Several years ago an experiment on e^+e^- pairs photoproduction on proton[22] carried out at DESY gave a clear indication of a ρ', with mass and width consistent with the LASS result (Fig. 7a). In such a reaction a Bethe-Heitler C-even, real amplitude and a Compton C-odd amplitude, real if resonant, contribute. The events were weighted in a suitable way to extract the interference between these two amplitudes. In fact their interference is not

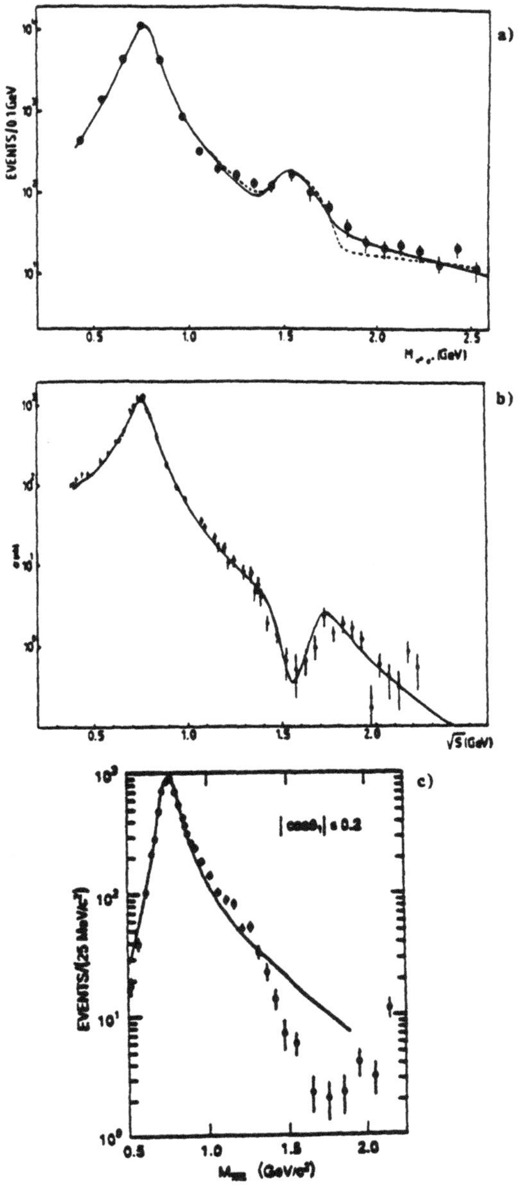

Fig. 2 $\pi^+\pi^-$ diffractive photoproduction cross section (a); $e^+e^- \to \pi^+\pi^-$ cross section from DM2 (b) and $|F_\pi|^2$ from MARK III (c).

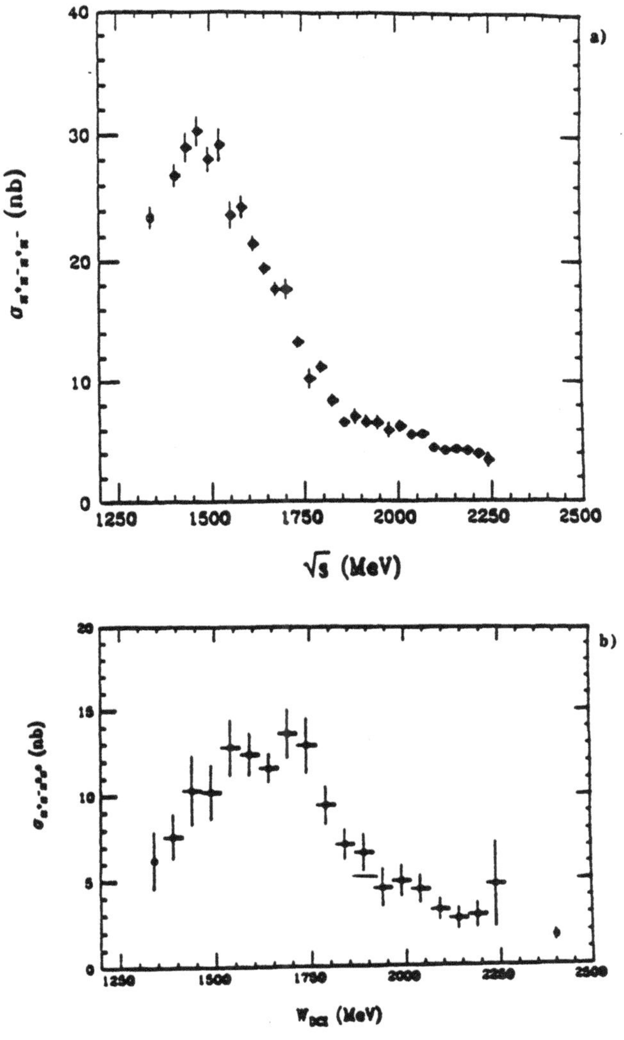

Fig. 3 DM2 results on $e^+e^- \to \pi^+\pi^-\pi^+\pi^-$ (a) and on $e^+e^- \to \pi^+\pi^-\pi^0\pi^0$ (b).

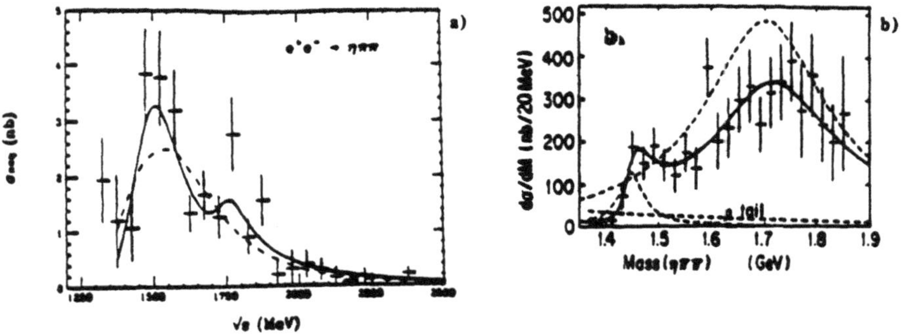

Fig. 4 DM2 results on $e^+e^- \to \rho\eta^{(11)}$ (a) and 1^{--} contribution from $\rho\eta$ phase shift analysis of $\pi^- p \to \rho\eta n$ (b).

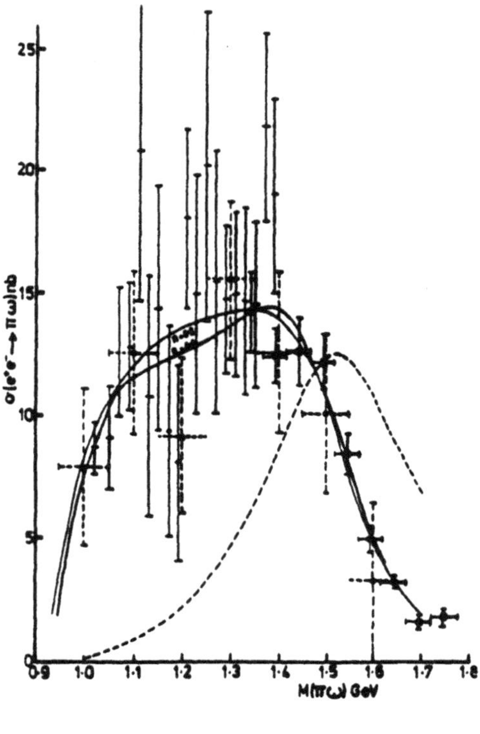

Fig. 5 DM2 reesults on $e^+e^- \to \omega\pi^0$ cross section.

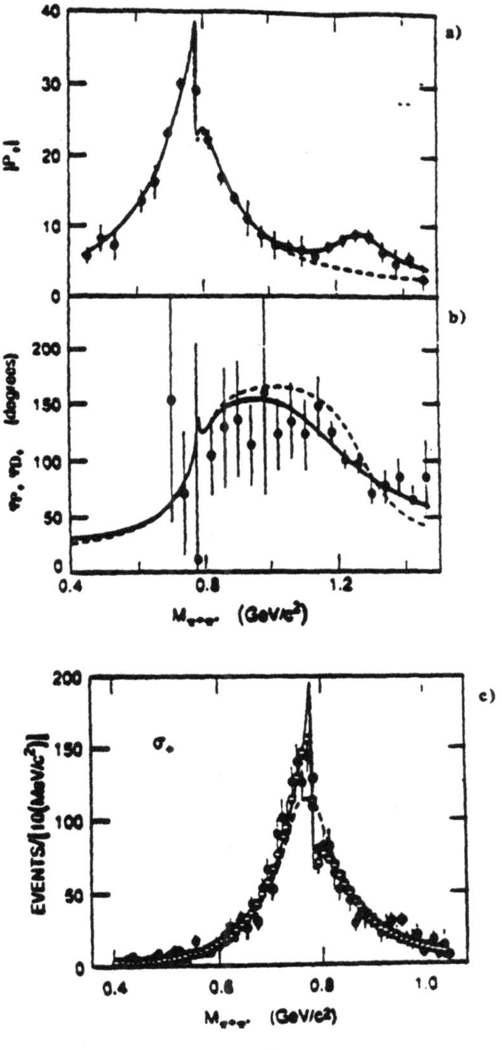

Fig. 6 1^{--} contribution from $\pi^+\pi^-$ phase shift analysis of $K^-p \to \pi^+\pi^-\Lambda$ (a), phase difference between 1^{--} and 2^{++} (b) and evidence for the $\rho - \omega$ interference (c).

vanishing only if there is a resonant amplitude, which in turn is amplified .

It has to be noted that this experiment presented also ρ and ϕ cross sections in good agreement with the expectations according to the vector dominance model of diffractive photoproduction on proton (Fig. 7b). Nevertheless at that time this experiment was disregarded because the ρ' photoproduction cross section on proton would have implied a too high e^+e^- cross section, still according to the vector dominance model.

This result could be revived if there is an anomalous $\rho'(1.3)$ coupling to a baryon, like for instance a I=1 $q^2\bar{q}^2$ interpretation of the $\rho'(1.3)$.

Currently a serious candidate for an isoscalar $q^2\bar{q}^2$ or baryonium state has been observed by various high statistics experiments[23], with a mass $m = 1.52$ GeV and $J^{PC} = 2^{++}$. In the near future these $q^2\bar{q}^2$ or baryonium mesons should be identified in the $p\bar{p}$ annihilation at threshold at LEAR. It has been pointed out[24] that an isoscalar $J^{PC} = 1^{--}$ is expected exactly at $m = 1.1$ GeV if these states lie on a Regge trajectory with the universal slope 0.9 GeV^{-2}. Actually the DESY experiment showed also another unexpected narrow resonance at $m = 1.10$ GeV, which could be just this I=0 partner of the $\rho'(1.3)$.

3. Isoscalar Sector.

The situation is still more confused in the isoscalar sector because of the smaller statistics obtained in e^+e^- experiments, roughly 1,500 events in front to $\simeq 40,000$ with positive G-parity.

In the past, the DM1 experiment showed evidence of a state around 1650 GeV/c^2 in the $e^+e^- \to K\bar{K}$, $K^0_S K^\pm \pi^\mp$ and $\omega\pi^+\pi^-$ channels. The lack of signal in the $\rho\pi$ reaction strongly supported the interpretation[25] of this state as a ϕ recurrency, a $\phi'(1.68)$. Moreover the existence of an ω' degenerate in mass and width with the $\rho'(1.6)$ had been supposed.

However a peak in $\pi^+\pi^-\pi^0$ was subsequently observed in photoproduction[26] at the same mass, mostly suggesting an ω' assignment for the observed state. The same experiment claimed for a ϕ' state observed in its K^+K^- decay at 1.76 GeV[27].

The kaon form factor has been extensively studied at the ϕ mass in the charged and neutral modes. Above this energy data from Novosibirsk[28] and Frascati[29] suggested that $|F_{K^\pm}|$ could be larger than the contribution of ρ, ω and ϕ tails. The DM1 experiment[30] concluded that the experimental behaviour of $|F_{K^\pm}|^2$ was not compatible (Fig. 8b) with the VDM hypothesis because of an enhancement in the 1.35 $\div$ 1.70 GeV energy range which has been related to the contribution of higher vector mesons like the $\rho'(1.6)$. On the contrary the $e^+e^- \to K^0_S K^0_L$ cross section[31] was generally depressed but showed a resonant behaviour at $\simeq 1.65$ GeV/c^2 (Fig. 8c). This effect and the fast drop of both form factors near 1.7 GeV cannot be explained by the $\rho'(1.6)$ alone and an interference with an isoscalar vector meson is needed, identified[25] later as the $\phi(1.68)$.

DM2[32] has confirmed that ρ, ω, ϕ and $\rho'(1.6)$ are not sufficient in order to explain the $|F_{K^\pm}|$ behaviour and the contribution of a further amplitude, compatible with the $\phi(1.68)$, is the most consistent with the data. In particular its branching ratio into $K\bar{K}$ width seems too large for an ω' assignment. On the other hand the contribution of a narrow ϕ' state at 1.76 GeV[27], alone or associated to an $\omega'(1.6)$, is rejected by DM2.

A resonant signal at $\simeq 1.67$ GeV was also first observed by DM1 (Fig. 8d) in the

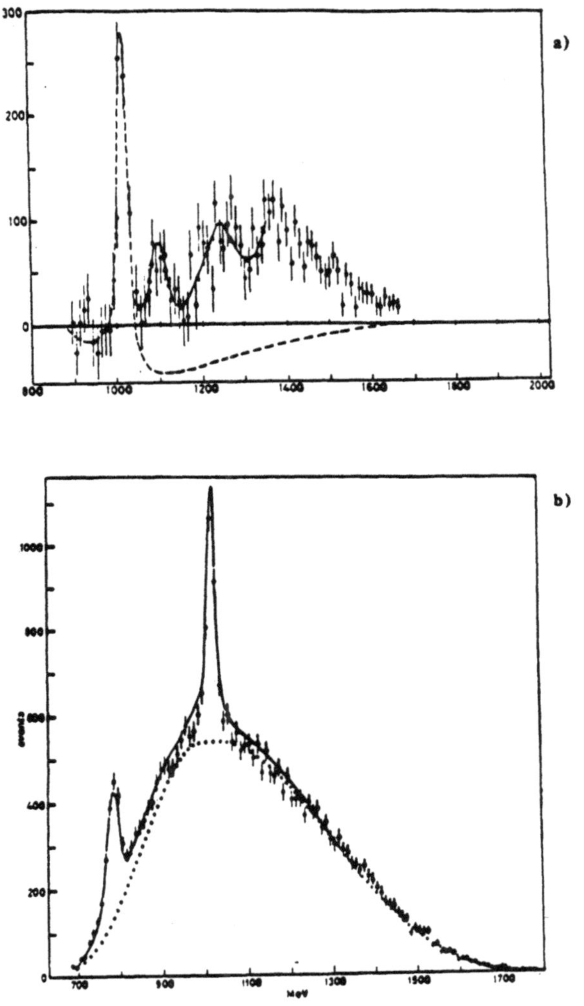

Fig. 7 Interference between C-even and C-odd amplitudes in e^+e^- photoproduction (a) and e^+e^- photoproduction cross section on protons (b).

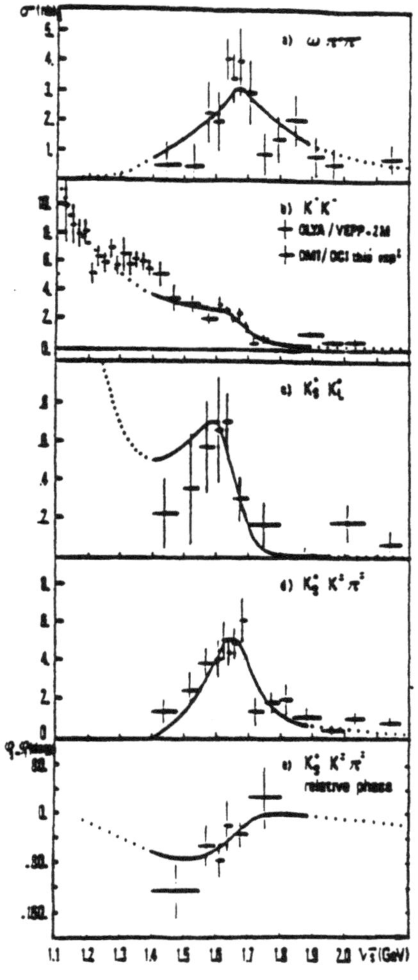

Fig. 8 Summary of DM1 results: cross section of $\omega\pi^+\pi^-$ (a), K^+K^- (b), $K^0_S K^0_L$ (c), $K^0_S K^\pm\pi^\pm$ (d) and relative phase between isoscalar and isovector production in $K^0_S K^\pm\pi^\mp$. The curves refer to the global fit described in Ref.[25].

$K_s^0 K^\pm \pi^\mp$ channel[33]. The related Dalitz plot shows a clear K^*K dynamics with K^{*0} dominance. This effect indicates an interference between isovector ($\rho - like$) and isoscalar ($\phi - like$) amplitudes. A Dalitz plot analysis shows a resonant behaviour for the isoscalar amplitude whose phase crosses 90^0 on the resonance peak (Fig. 8e) whereas the isovector cross section is largely suppressed.

DM2[34] has confirmed these results with a larger statistical significance (Figs. 9,10). Furthermore a measurement of the $e^+e^- \to K^+K^-\pi^0$ cross section is well compatible with the one expected (Fig. 11) by SU(3) relations for the coupling constants which imply a destructive interference between ρ-like and ϕ-like resonances for this channel.

Concerning the final states with an odd number of pions, in the past results were available from DM1[35] at Orsay and ND[17] at Novosibirsk. The former experiment measured the $e^+e^- \to \omega\pi^+\pi^-$ channel giving evidence of a large bump centered at $\simeq 1.68\ GeV/c^2$, $\simeq .23\ GeV/c^2$ wide (Fig. 8a). The second showed that the $e^+e^- \to \pi^+\pi^-\pi^0$ cross section is larger than a VDM extrapolation from the $\omega - \phi$ peaks, but does not show structures.

Subsequent DM2 results[36] show that the cross section for $e^+e^- \to \pi^+\pi^-\pi^0$, which is dominated by the $\rho\pi$ dynamics, remains constant up to $\simeq 1.62\ GeV$ where it goes suddendly to zero (Fig. 12). On the other hand the cross section for $e^+e^- \to \omega\pi^+\pi^-$, is dominated by the same wide bump reported by DM1. An analysis[37] of the available $\pi^+\pi^-\pi^0$ and $\omega\pi^+\pi^-$ data, including preliminary DM2 results, has proposed the existence of two ω' states with $m_{\omega'_1} \simeq 1390\ GeV/c^2$ and $m_{\omega'_2}$ ranging in the $1.6 - 1.7\ GeV/c^2$ interval depending on the parametrization model. The existence of a further ω' state at higher mass has been also suggested[38].

Final DM2 results[36] (Fig. 13) are well compatible with two contributing states, $\omega'(1.42)$ and $\omega'(1.66)$. The parameters of the $\omega'(1.42)$ which is not coupled to the $\omega\pi\pi$ final state can undergo significant changes if the ND data[17] are taken into account. On the other hand the parameters of the $\omega'(1.63)$ are largely compatible with those of the isoscalar resonance observed in the $K\bar{K}\pi$[33,34] reaction. However the identification of these two signals with a single state is not at all automatic.

The large $\omega\pi\pi$ production would suggest an ω' nature for this state, but then SU(3) invariance would imply a $\rho\pi$ cross section even higher than the K^*K one. On the other hand strict application of the OZI rule for a pure $s\bar{s}$ state would suppress both $\omega\pi\pi$ and $\rho\pi$ decays relatively to K^*K. Therefore the mixing between ω' and ϕ' could be different from the ideal value.

Anyway the single $\omega'(1.63)$ production cannot reproduce all the resonant pattern we observe in this mass interval. The large K^*K production can be explained only if the contribution of the $\phi'(1.68)$ is taken into account. The existence of both such states has been recently suggested by an analysis of the DM2 data[39] and also proposed[40] on an independent basis.

The experimental situation, including the isovector data, is summarized in Tab. 1. The two ρ' are in accord with quark model predictions for a 2 S and a 1 D excitations of the ρ[41]. On the other hand, although two ω' and one ϕ' states seems to be raisonably observed, it is not clear yet at which recurrencies they correspond and thus if space is available for exotic states which can be foreseen.

Concerning exotic states, the existence of a vector meson decaying into $\phi\pi$ has been

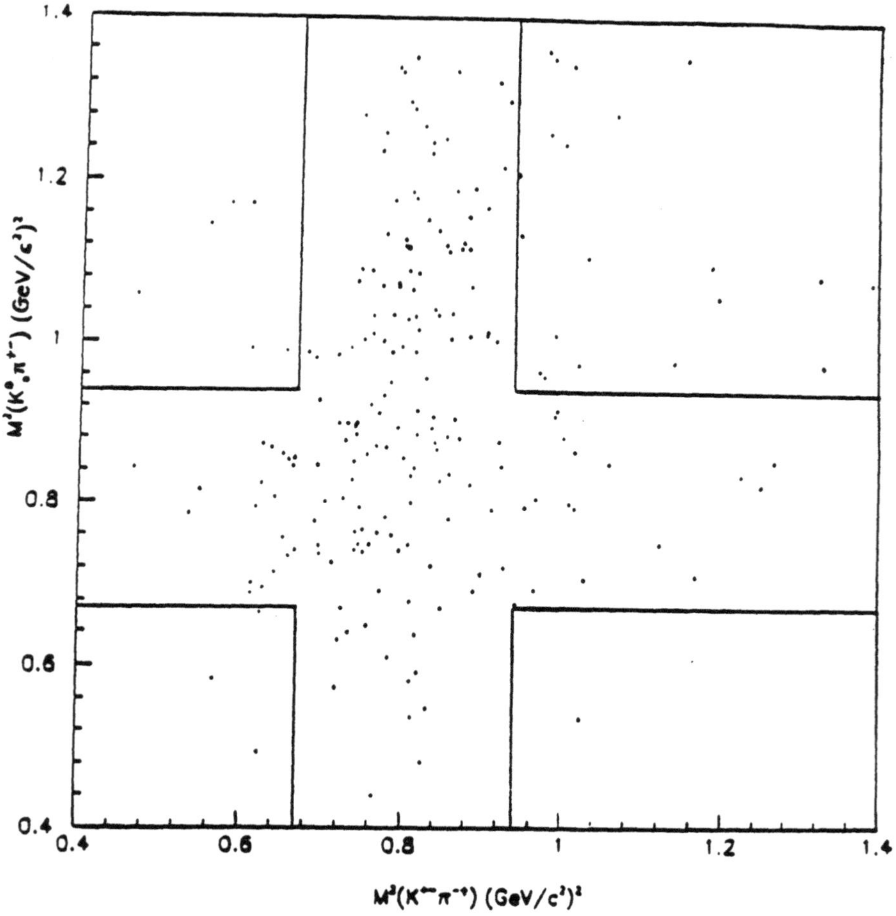

Fig. 9 Dalitz plot of $K_S^0 K^{\pm}\pi^{\mp}$ events (DM2).

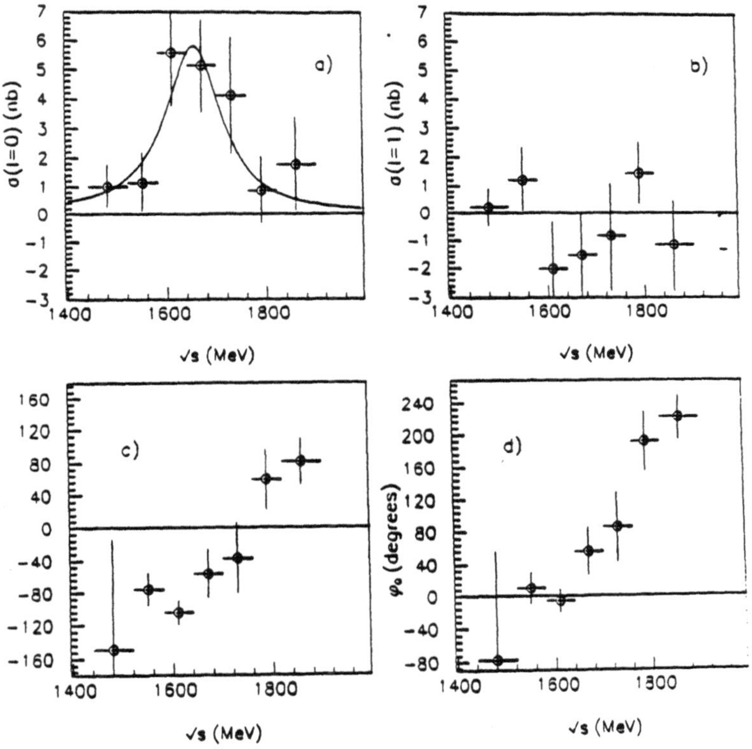

Fig. 10 Isoscalar (a) and isovector (b) cross sections, relative (c) and absolute (d) isoscalar phase in $K_S^0 K^{\pm}\pi^{\mp}$ (DM2).

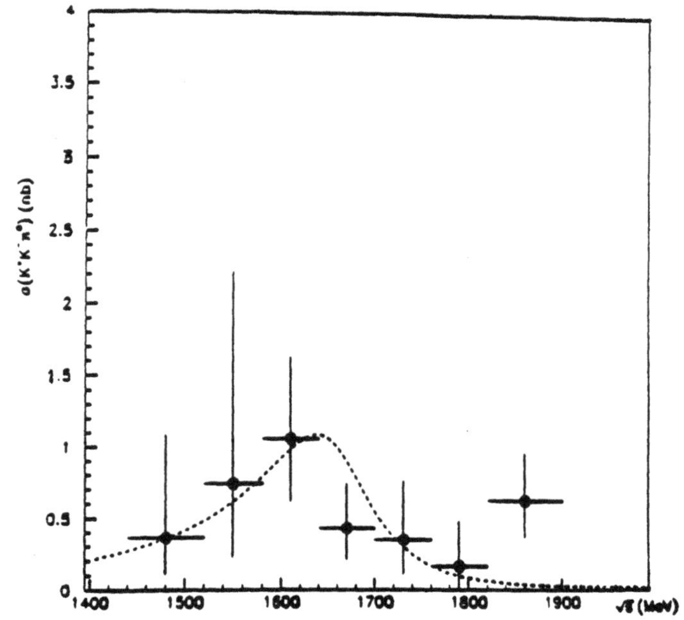

Fig. 11 $K^+K^-\pi^0$ cross section compared with the expectation from the $K_S^0K^\pm\pi^\mp$ analysis (DM2).

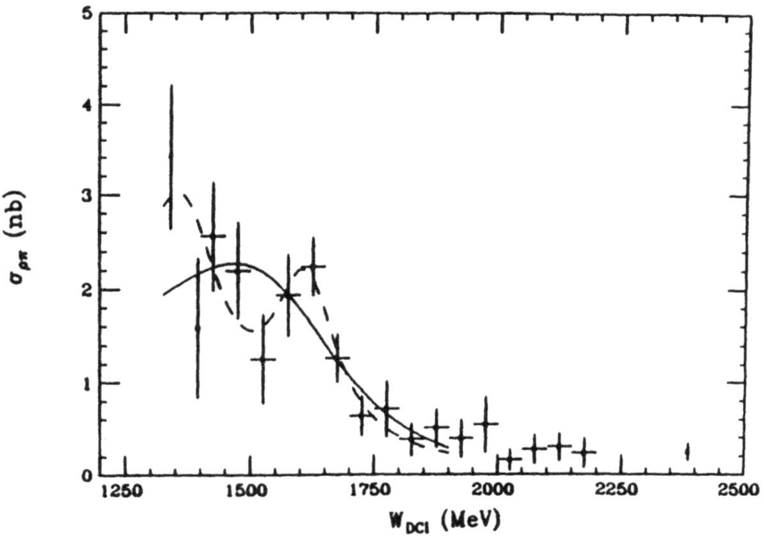

Fig. 12 DM2 results on $e^+e^- \to \rho\pi$. Lines correspond to the fits with a single or two Breit Wigner functions.

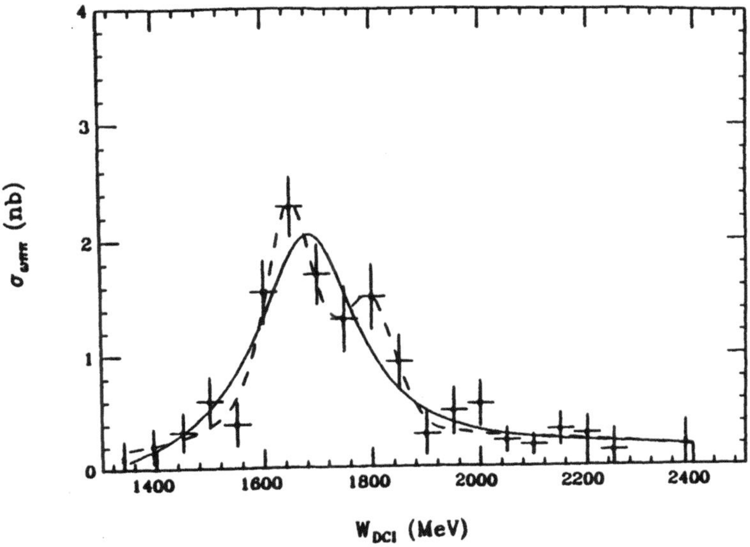

Fig. 13 DM2 results on $e^+e^- \to \omega\pi\pi$. Lines correspond to the fits with a single or two Breit Wigner functions.

Tab. 1 DM1 and DM2 results on isoscalar resonant productions.

channel [*mass/width* (MeV/c^2)]	*DM1*	*DM2*
$K\bar{K}$	~ 1670 ~ 100	1655 ± 17 207 ± 45
$K\bar{K}\pi$	1653 ± 18 113 ± 55	1657 ± 27 146 ± 55
$\omega\pi\pi$	1657 ± 13 136 ± 46	(1*BW*) 1682 ± 12 226 ± 44
$\rho\pi$		(1*BW*) 1540 ± 35 505 ± 26

proposed by a hadroproduction experiment[42] at $\simeq$ 1.48 GeV. No evidence for such a state has been still observed in e^+e^- experiments[34] but the evaluable limit is not stringent because of the low integrated luminosity.

3. Conclusion.

In the study of hadron spectroscopy DAΦNE can make definitive contributions thanks to its high luminosity. Moreover it is the first machine which should be able to cover the whole energy interval of interest if its maximum energy would be raised to at least 1.9 GeV. In this case the contributions of both radial and orbital recurrencies could be precisly measured, including mixing angle values in the isovector sector.

However a maximum energy in the center of mass of 1.5 GeV, as scheduled for the moment, should be very useful to settle at least the $\rho'(1.3)$ and $\rho'(1.1)$ puzzles. It must to be stressed that any present interpretation strongly suffers of large systematic errors because no experiment has completely covered this energy interval. A careful study of the hadron form factors and of the presently flat $e^+e^- \to \rho\pi$ cross section could reveal contributions from unseen narrow resonances. The actual production of the exotic state around 1.48 GeV should be also evidenced.

4. References

1. G. Barbiellini et al. and B. Bartoli et al, *XVth Int. Conference on High Energy Physics*, Kiev (1970).
2. A. Cordier et al., *Phys. Lett.* **B109** (1982) 129.
   L.M. Kurdadze et al., *JETP Lett.* **47** (1988) 512.
3. D. Aston et al., *Nucl. Phys.* **B189** (1981) 15.
   D. Aston et al., *Phys. Lett.* **B92** (1980) 215.
4. C. Bacci et al., *Nucl. Phys.* **B184** (1981) 31.
5. M. Atkinson et al., *Nucl. Phys.* **B243** (9184) 1.
   M. Atkinson et al., *Z. Phys.* **C26** (1985) 499.
6. G. Barberino et al., *Lett. N. Cim.* **3** (1972) 689.
7. F.M. Renard, *N. Cim.* **A64** (1969) 979.
8. M. Greco, *Phys. Lett.* **B70** (1977) 441.
9. A. Bramon, E. Etim and M. Greco, *Lett. N. Cim.* **12** (1975) 91.
10. D. Bisello et al., *Phys. Lett.* **B220** (1989) 321.
11. A. Donnachie and H. Mirzaie, *Z. Phys.* **C33** (1987) 407.
12. A. Donnachie and A.B. Clegg, *Z. Phys.* **C42** (1989) 663.
13. A. Antonelli et al., *Phys. Lett.* **B212** (1988) 133.
14. S. Fukui et al., *Phys. Lett.* **B202** (1988) 441.
15. L. Stanco et al., *HADRON 91*, Maryland Univ., Aug. 1991 and LAL 91-64 .
16. H. Albrecht et al., *Phys. Lett.* **B185** (1987) 223.
17. S.I. Dolinsky et al., *Phys. Rep.* **202** (1991) 99.
18. C. Kiesling, DESY 91-067 .
19. D. Aston et al., *Nucl. Phys.* **B21** (1991) 105.
20. L.P. Chen et al., SLAC-PUB-5674 (1991).

21. A. Donnachie, *Phys. Lett.* **B269** (1991) 450.
22. S. Bartalucci et al., *N. Cim.* **A49** (1979) 207.
23. B. May et al., *Z. Phys.* **C46** (1990) 191.
E. Aker et al., *Phys. Lett.* **B260** (1991) 249.
E-760 Collab., *HADRON 91*, Maryland Univ., Aug. 1991.
24. L.A. Kondratyuk et al., private communication
25. J. Buon et al., *Phys. Lett.* **B118** (1982) 221.
26. M. Atkinson et al., *Phys. Lett.* **B127** (1983) 132.
27. M. Atkinson et al., *Z. Phys.* **C27** (1985) 233.
28. P.M. Ivanov et al., *Phys. Lett.* **B107** (1981) 297.
29. B. Esposito et al., *Phys. Lett.* **B67** (1977) 279.
B. Esposito et al., *Lett. N. Cim.* **28** (1980) 337.
30. B. Delcourt et al., *Phys. Lett.* **B99** (1981) 257.
31. F. Mane et al., *Phys. Lett.* **B99** (1981) 261.
32. D. Bisello et al., *Z. Phys.* **C39** (1988) 13.
33. F. Mane et al., *Phys. Lett.* **B112** (1982) 178.
34. D. Bisello et al., *Z. Phys.* **C52** (1991) 227.
35. A. Cordier et al., *Phys. Lett.* **B109** (1982) 129.
36. A. Antonelli et al., *Z. Phys.* **C**, in press, and LAL 92-08.
37. A. Donnachie and A.B. Clegg, *Z. Phys.* **C42** (1989) 663.
38. A.B. Govorkov, *Sov. J. Nucl. Phys.* **48** (1988) 237.
39. A. Donnachie, *HADRON 91*, Maryland Univ., Aug. 1991.
40. D.C. Peaslee, Maryland Univ. Pubbl. 92-109.
41. S. Godfrey and N. Isgur, *Phys. Rev.* **D32** (1985) 189.
42. S.I. Bityukov et al., *Phys. Lett.* **B188** (1987) 383.

Common Problems and Trends of Modern Physics, pp. 387–407
ed. T. Bressani, S. Marcello and A. Zenoni
©1992 World Scientific Publishing Company

Low–Energy KN Interactions at a ϕ–Factory

Paolo M. Gensini
Dip. di Fisica dell' Università di Perugia, Perugia, Italy, and
I.N.F.N., Sezione di Perugia, Italy.

1. Introduction.

We intend to illustrate in this lecture the possibilities opening up at machines planned for the nineties for low–energy kaon–nucleon interactions, keeping the focus on the theoretical problems they should help to solve.

We shall deal here only with the ϕ–factory DAΦNE[1] and consider, for brevity, interactions with light, gaseous targets, using gaseous H_2 as a benchmark for which to estimate the rates to be expected in a typical apparatus.

The interest in this field is of a systematic rather than exploratory nature: information on low–energy kaon–nucleon interactions is scarce and of a poor statistical quality, when compared to the corresponding pion–nucleon one. As an example, just take a look at the two pages dedicated by the PDG booklet[2] to $K^{\pm}p$ and $K^{\pm}d$ total and elastic cross sections: other data do not present a rosier perspective[3].

The low quality of low–momentum elastic and inelastic scattering data reflects in turn on our knowledge of the "elementary" parameters of the KN interaction, remarkably poorer than in the $SU(3)_f$–related πN case[3]. On top of this sorry situation, one must add the still unsolved mystery of kaonic–hydrogen level–shifts and widths, whose experimental determinations are in total disagreement with theoretical expectations[4].

Data at very low momenta and at rest are essential to clarify many of the above–mentioned problems[3]; however, experiments of this kind pose formidable problems at conventional fixed–targed machines, some of which can be circumvented at a ϕ–factory. For instance, at the KAON factory planned for TRIUMF[5], beams in the lowest momentum range (from 400 to 800 MeV/c) have intensities of

10^6 - 10^8 K^-s^{-1}, with K^+ beams about twice more intense. Already the purity of these beams is limited by $K^\pm$ decays in flight: to experiment at momenta below 400 MeV/c one has to use moderators, which at the same time decrease the kaon intensity, degrade the beam resolution, and increase enormously the beam contamination at the final target. All these effects make the experiments much more complex, overturning all the advantages offered by the higher initial beam intensities.

2. Capabilities of a KN–scattering experiment at DAΦNE.

DAΦNE is the ϕ–factory (the acronym stands for "Double Annular Φ–factory for Nice Experiments"), due to replace the Adone colliding–beam machine in the same experimental hall of the I.N.F.N. National Laboratories in Frascati. From its expected commissioning luminosity[6] of 5×10^{32} $cm^{-2}s^{-1}$, and an annihilation cross section of about 5 μb at the ϕ–resonance peak, one can see that its *two* interaction regions will be the sources of $\simeq 1.2 \times 10^3$ $K^\pm s^{-1}$, at a central momentum of 126.9 MeV/c, with the momentum resolution of $\simeq 1.1 \times 10^{-2}$ due to the very small energy spreads in the beams, as well as of $\simeq 850$ $K_L s^{-1}$, at a central momentum of 110.1 MeV/c, with the slightly poorer resolution of $\simeq 1.5 \times 10^{-2}$.

Both $\pi^\pm$'s and leptons coming out of the two sources are easy–to–control backgrounds: the first because the $\pi^\pm$'s, though produced at a rate of about 380 $\pi^\pm s^{-1}$ (not counting those from K_S decays), come *almost all* from events with three or more final particles and can thus be suppressed by momentum and collinearity cuts; the second, as well as collinear pions from $e^+e^- \to \pi^+\pi^-$, produced at much lower rates of order 0.75 s^{-1} (the leptons) or 0.25 s^{-1} (the pions), are eliminated by a momentum cut, having momenta about four times those of the $K^\pm$'s.

The *two* interaction regions are therefore *small–sized* sources of *low–momentum, tagged* $K^\pm$'s and K_L's, with *negligible contaminations* (after suitable cuts on angles and momenta of the particles are applied *event by event*), in an environment of very low background radioactivity: this situation is simply unattainable with *conventional* technologies at fixed–target machines[7], where the impossibility of placing experiments too close to the production target limits from below the charged–kaon momenta, and kaon decays in flight always contaminate the beams:

low–momentum experiments are thus possible only with the use of moderators, with a huge beam contamination at the target, as well as a large final–momentum spread due to straggling phenomena.

It is therefore of interest to consider the feasibility of low–energy, $K^{\pm}N$ and K_LN experiments at DAΦNE, with respect to equivalent projects at machines such as, *e.g.*, KAON at TRIUMF[8] (or to ideas advanced for the sadly aborted EHF project[9]).

We shall, in this part, try and give an evaluation of rates to be expected in a simple apparatus at DAΦNE. We shall assume cylindrical symmetry, with a toroidal target fiducial volume, limited by radii a and $a + d$ and of length L (inside and outside of which one can imagine a tracking system, surrounded on the outside by a photon detecting system – *e.g* lead–Sci–Fi sandwiches – and a superconducting, solenoidal coil to provide the moderate magnetic field **B** needed for momentum measurements), filled with a gas at moderate pressure.

One must convert the usual, fixed–target expression for reaction rates to a spherical geometry, and also include kaon decays in flight, getting (for $\mathbf{B} = 0$ or K_L^0's: the general case can be easily treated with slight modifications)

$$dN_r = [\frac{1}{\rho^2}(\frac{3}{8\pi})\ (L\sigma_\phi B_\phi)\sin^2\theta e^{-\rho/\lambda}]\sigma_r\rho_t(\rho^2 d\rho\sin\theta d\theta d\phi)\ , \tag{1}$$

with ρ, θ and ϕ spherical coordinates (with the z–axis oriented along the beam direction), L the machine luminosity, σ_ϕ the annihilation cross section at the ϕ–resonance peak, B_ϕ the ϕ branching ratio into the desired mode (either K^+K^- or K_LK_S), σ_r the reaction cross section for the process considered, ρ_t the target nuclear density, and $\lambda = p_K\tau_K/m_K$ the decay length (respectively of 0.954 m for $K^{\pm}$'s and of 3.429 m for K_L's) at the ϕ–resonance momenta.

Integrating over the fiducial volume, the reaction rate can be cast into the simple formula

$$N_r = \frac{3\pi}{4}rd(L\sigma_\phi B_\phi)\rho_t\sigma_r\ , \tag{2}$$

with both geometrical acceptance and kaon decay in flight thrown into the reduction factor r, which we have estimated to take the values 0.50 for $K^{\pm}$'s and 0.72 for K_L's for a fiducial volume defined by $a = 10\ cm$, $d = 50\ cm$ and $L = 1\ m$, to represent a "person–sized" detector, fitting in DAΦNE's interaction region. A longer detector or a larger outer radius would not give any substantial improvement in the rates, due to the angular distribution of the produced kaons and to the value of the decay length λ for these low–momentum K's; besides, for $K^{\pm}$ r increases almost linearly but slowly with increasing field **B**, due to the interplay of the increased path length inside the fiducial volume on one side, and of the particle decays on the other.

This gives, for a target volume filled by a diatomic, ideal gas at room temperature, the rates for $K^{\pm}$–initiated processes

$$N_r = p(\text{atm}) \times \sigma_r(\text{mb}) \times (4.0 \times 10^4 \text{ events/y}) \ , \tag{3}$$

for a "Snowmass year" of 10^7 s (for K_L's the figure in eq. (3) is about the same, because of an approximate compensation between the variations in r and B_ϕ), or, with rough estimates of the partial K^-p cross sections at the ϕ–decay momenta, to about 10^7 two–body events per year in H_2 gas at atmospheric pressure, of which about 3.6×10^6 elastic scattering events, 2.4×10^6 $\pi^+\Sigma^-$ and about 10^6 for each of the remaining four two–body channels $\pi^0\Sigma^0$, $\pi^0\Lambda$, $\bar{K}^0 n$, and $\pi^-\Sigma^+$. The above rates are enough to measure angular distributions in all channels, and also the polarizations for the self–analyzing final–hyperon states, particularly for the decays $\Lambda \to \pi^- p$, $\pi^0 n$ (asymmetry $\alpha \simeq 0.64$) and $\Sigma^+ \to \pi^0 p$ ($\alpha \simeq -0.98$). One could also expect a total of about 10^4 radiative–capture events, which should allow a good measurement on the *absolute* rates for these processes as well.

Such an apparatus will need: tracking for incoming and outgoing charged particles, time–of–flight measurements (for charged–particle identification), a moderate magnetic field (due to the low momenta involved) for momentum measurements, and a system of converters plus scintillators for photon detection and subsequent geometrical reconstruction of π^0 and Σ^0 decays, amounting thus to a rather simple (on today's particle–physics scale), not too costly apparatus. Mentioning costs, we wish to point out that DAΦNE, though giving the experimenters a very small momentum range, saves them the cost of the *separate* tagging system needed to reject contaminations in a *conventional* low–energy, fixed–target experiment[7].

The above formulæ for $K^{\pm}$ rates do not include particle losses in the beam–pipe wall and in the internal tracking system, which were assumed sufficiently thin

(e. g. of a few hundred μm of low-Z material, such as carbon fibers or Mylar). We have indeed checked that, due to the shape of the angular distribution of the kaons, particle losses are contained (mostly at small angles, where K-production is negligible, and events would anyhow be hard to be fully reconstructed), and momentum losses flat around $\theta = \pi/2$ (where most of the $K^{\pm}$'s are produced): even for a total thickness of the above-mentioned materials of 1 mm, kaon momenta do not decrease below 100 MeV/c and losses do not grow beyond a few percents. Rather, one could exploit such a thickness as a low-momentum, thin moderator, to span the interesting region just above the charge-exchange threshold at $p_L(K^-) \simeq$ 90 MeV/c, measurements which would add precious, additional constraints on low-energy amplitude analyses[10].

We have presented the above simplified estimates to show that acceptable rates can be achieved, orders of magnitude above those of existing data at about the same momentum, *i.e.* to the lowest-energy points of the British-Polish Track-Sensitive Target (TST) Collaboration, taken in the mid and late seventies at the (R.I.P.) NIMROD accelerator[11].

Since losses do not affect K_L's, a detector of the kind sketched above, similar in geometry to the one proposed by T. Bressani[7] to do K^+-nucleus scattering and hypernuclear experiments, could be used without any problem to study low-energy $K_L \rightarrow K_S$ regeneration and charge-exchange in gaseous targets, providing essential information for this kind of phenomena.

We wish to add that a DAΦNE detector dedicated to kaon experiments on gaseous H_2 and D_2 can continue its active life, without substantial changes, to measure K^+-, K^--, and K_L^0-interactions on heavier gases as well (He, N_2, O_2, Ne, Ar, Kr, Xe), exploring not only the properly nuclear aspects of these interactions, such as nucleon swelling in nuclei[12], but also producing $\pi\Sigma$, $\pi\Lambda$ and $\pi\pi\Lambda$ systems at invariant masses below the elastic $\bar{K}N$ threshold in the so-called unphysical region, with statistics substantially higher than those now available[13], due to the $\simeq 4\pi$ geometry allowed by a colliding-beam-machine detector.

3. KN and $\bar{K}N$ amplitudes at low momenta: formalism and some problems.

Any $a_1(0^-, q) + B_1(\frac{1}{2}^+, p) \to a_2(0^-, q') + B_2(\frac{1}{2}^+, p')$ process is described in the c.m. frame by two amplitudes $G(w, \theta)$ and $H(w, \theta)$, when the T–matrix element $T_{\alpha\beta}$ is expressed in terms of two–component Pauli spinors χ_α and χ_β for the final and initial $\frac{1}{2}^+$ baryon as $T_{\alpha\beta} = \chi_\alpha^\dagger \mathbf{T} \chi_\beta$, where

$$\mathbf{T} = G(w, \theta) \times \mathbf{I} + iH(w, \theta) \times (\vec{\sigma} \cdot \hat{n}) \tag{4}$$

and $\hat{n}$ defines the normal to the scattering plane[14].

These amplitudes have a simple partial–wave expansion, given by

$$G_N(w, \theta) = \sum_{\ell=0}^{\infty} [(\ell + 1) T_{\ell+}(w) + \ell\, T_{\ell-}(w)] P_\ell(\cos\theta) \tag{5}$$

and

$$H_N(w, \theta) = \sum_{\ell=1}^{\infty} [T_{\ell+}(w) - T_{\ell-}(w)] P'_\ell(\cos\theta) \ , \tag{6}$$

where the subscript N indicates that only the nuclear interaction has been considered. To describe the data, they must include electromagnetism and be rewritten as

$$G(w, \theta) = \tilde{G}_N(w, \theta) + G_C(w, \theta) \tag{7}$$

and

$$H(w, \theta) = \tilde{H}_N(w, \theta) + H_C(w, \theta) \ , \tag{8}$$

where the tilded amplitudes differ from the untilded ones in the Coulomb shifts $\sigma_{\ell\pm}^{\mathrm{in(fin)}}$, non–zero only for *both* initial (final) charged particles, having been applied to each partial wave:

$$T_{\ell\pm} \ \to \ \tilde{T}_{\ell\pm} = e^{i\sigma_{\ell\pm}^{\mathrm{in}}} T_{\ell\pm}(w) e^{i\sigma_{\ell\pm}^{\mathrm{fin}}} \ . \tag{9}$$

The one–photon–exchange amplitudes G_C and H_C (absent for charge– or strangeness–exchange processes, but not for K_S regeneration, which at $t \neq 0$ goes also via one–photon exchange) are expressed in terms of the nucleon Dirac form factors as[14,15] (μ and m indicate respectively the meson and baryon masses)

$$G_C(w,\theta) = \pm e^{\pm i\phi_C} \cdot \{(\frac{2q\gamma}{t} + \frac{\alpha}{2w}\frac{w+m}{E+m}) \cdot F_1(t) +$$

$$+[w - m + \frac{t}{4(E+m)}] \cdot \frac{\alpha F_2(t)}{2wm}\} \cdot F_K(t) \tag{10}$$

and

$$H_C(w,\theta) = \pm \frac{\alpha F_K(t)}{2w \tan\frac{1}{2}\theta} \cdot \{\frac{w+m}{E+m} \cdot F_1(t) + [w + \frac{t}{4(E+m)}] \cdot \frac{F_2(t)}{m}\} \tag{11}$$

for $K^\pm$ interactions with nucleons, while for K_S regeneration one has to change the sign of the isovector part of the kaon form factor $F_K(t)$.

Here $\gamma = \alpha \cdot (w^2 - m^2 - \mu^2)/2qw$ and the Coulomb phase ϕ_C is defined as

$$\phi_C = -\gamma \log(\sin^2 \frac{1}{2}\theta) + \gamma \cdot \int_{-4q^2}^{0} \frac{dt}{t} \cdot [1 - F_K(t)F_1(t)] \tag{12}$$

for charged kaons scattering on protons, while it reduces to

$$\phi_C = -\gamma \int_{-4q^2}^{0} \frac{dt}{t} F_K(t)F_1(t) \tag{12'}$$

for processes involving K^0's and/or neutrons.

w and θ are the c.m. total energy and scattering angle, $q = [\frac{1}{4}w^2 - \frac{1}{2}(m^2+\mu^2) + (m^2-\mu^2)^2/4w^2]^{1/2}$ the c.m. momentum (in the initial state: for inelastic processes, including charge exchange, we indicate final–state quantities with primes), E the c.m. (initial) baryon energy $E = (w^2+m^2-\mu^2)/2w$, and t the squared momentum–transfer, $t = m^2 + m'^2 - 2EE' + 2qq'\cos\theta$. We shall also use the laboratory–frame, initial–meson momentum $k = \frac{1}{2}(\omega^2 - \mu^2)^{1/2}$ and energy ω, related to the c.m. total energy via $\omega = (w^2 - m^2 - \mu^2)/2m$, and, besides t, the two other Mandelstam variables $s = w^2$ and u, the square of the c.m. total energy for the crossed channel $\bar{a}_2(0^-) + B_1(\frac{1}{2}^+) \to \bar{a}_1(0^-) + B_2(\frac{1}{2}^+)$, obeying on *the* mass *shell* the indentity $s + t + u = m^2 + m'^2 + \mu^2 + \mu'^2$.

In terms of the amplitudes G and H the c.m. differential cross section for an unpolarized target takes the simple form

$$\frac{d\sigma}{d\Omega} = \frac{1}{2}\sum_{\alpha,\beta} |T_{\alpha\beta}|^2 = |G|^2 + |H|^2 \ . \tag{13}$$

The other observable accessible at DAΦNE, in the strangeness–exchange processes $\bar{K}N \to \pi\Lambda$, $\pi\Sigma$, is the polarization P_Y ($Y = \Lambda$, Σ) of the final hyperon, measurable through the asymmetry α of their weak, nonleptonic decays $\Lambda \to \pi^- p$, $\pi^0 n$ (both have an asymmetry $\alpha \simeq 0.64$), and $\Sigma^+ \to \pi^0 p$ (which has $\alpha \simeq -0.98$), while there is little chance to use the neutron decay modes $\Sigma^\pm \to \pi^\pm n$, which have $\alpha \simeq \pm 0.068$; we have for these quantities

$$P_Y \cdot (\frac{d\sigma}{d\Omega}) = 2\ \mathrm{Im}\ (G \cdot H^*)\ . \tag{14}$$

For an $(S + P)$–wave parametrization (adequate at such low momenta), while the integrated cross sections depend only quadratically on the P–waves, both the first Legendre coefficients of the differential cross sections

$$L_1 = \frac{1}{2}\int_{-1}^{+1} \cos\theta\ (\frac{d\sigma}{d\Omega})\ d\cos\theta = \frac{2}{3}\ \mathrm{Re}\ [T_{0+} \cdot (2T_{1+} + T_{1-})^* + \ldots] \tag{15}$$

and the polarizations

$$P_Y \cdot (\frac{d\sigma}{d\Omega}) = 2\ \mathrm{Im}\ [T_{0+} \cdot (T_{1+} - T_{1-})^* + 3T_{1-} \cdot T_{1+}^* \cos\theta + \ldots] \sin\theta \tag{16}$$

are essentially linear in the P–wave contributions, and give complementary information on these latter. It is perhaps not useless to remind that the low–statistics experiments performed only up to the late seventies have been able to put only rather generous upper bounds[11] on these parameters for the hyperon production channels.

We shall devote the last part of this section to show why this absence of *direct* information on the low–energy P–waves has been a serious shortcoming for $\bar{K}N$ amplitude analyses. Remember that from *production* experiments we know that the $I = 1$, $S = -1$ T_{1+} partial wave resonates *below* threshold at a c.m. energy around $w = 1385\ MeV$, the mass of the isovector member of the $J^P = \frac{3}{2}^+$ decuplet.

For any analytical extrapolation purpose, one has to turn from the Pauli amplitudes G and H to the invariant amplitudes $A(s,t)$ and $B(s,t)$, defined in term of four–component Dirac spinors as

$$2\pi w\ T_{\alpha\beta} = \bar{u}_\alpha(p')[A(s,t) + B(s,t) \cdot \gamma^\mu Q_\mu] u_\beta(p)\ , \tag{17}$$

where $Q = \frac{1}{2}(q + q')$, the average between incoming– and outgoing–meson c.m. four–momenta: these amplitudes obey simple crossing relations and are free of kinematical singularities, so that they are the ones to be used, rather than G and H. It is also customary to use the combination $D(\nu, t) = A(\nu, t) + \nu \cdot B(\nu, t)$, where $\nu = (s - u)/2(mm')^{1/2}$, which has the same properties as $A(\nu, t)$ under crossing and for elastic scattering obeys the optical theorem in the simple form

$$\text{Im } D(\nu, t = 0) = k \cdot \sigma_{tot} \ , \tag{18}$$

where of course all electromagnetic effects must be subtracted on both sides.

A and B can be rewritten in terms of G and H and reexpressed through the partial waves $T_{\ell\pm}$ by projecting eq. (17) on the different spin states: for elastic scattering one gets

$$A(\nu, t) = \frac{4\pi}{E + m}\{(w + m)G(w, \theta) +$$

$$+[(E + m)^2(w - m) + (\frac{1}{2}t + E^2 - m^2)(w + m)] \cdot \frac{H(w, \theta)}{q^2 \sin\theta}\} \ , \tag{19}$$

and

$$B(\nu, t) = \frac{4\pi}{E + m}\{G(w, \theta) - [(E + m)^2 - \frac{1}{2}t - E^2 + m^2]\frac{H(w, \theta)}{q^2 \sin\theta}\} \ . \tag{20}$$

The amplitudes become, leaving out D– and higher waves,

$$D(\nu, 0) = \frac{4\pi w}{m}[T_{0+} + 2T_{1+} + T_{1-} + \ldots] \tag{21}$$

and

$$B(\nu, 0) = \frac{4\pi w}{mq^2}[(E - m)T_{0+} - 2(2m - E)T_{1+} + (E + m)T_{1-} + \ldots] \ . \tag{22}$$

Introducing the (complex) scattering lengths $a_{\ell\pm}$ and (complex) effective ranges $r_{\ell\pm}$ one can expand up to $O(q^2)$, and obtain for the forward D amplitudes close to threshold,

$$D(q, 0) = 4\pi(1 + \frac{\mu}{m})\{a_{0+} + ia_{0+}^2 q + [2a_{1+} + a_{1-} - (a_{0+} + \frac{1}{2}r_{0+})a_{0+}^2 - \frac{a_{0+}}{2m\mu}]q^2 +$$

$$+ \ldots\} \ , \tag{23}$$

dominated by the S–waves, while for the B amplitudes the same approximations give

$$B(q,0) = \frac{2\pi}{m}(1+\frac{\mu}{m})[a_{0+} - 4m^2(a_{1+} - a_{1-}) + ia_{0+}^2 q + \ldots] , \qquad (24)$$

where the factor $4m^2 \simeq 90\ fm^{-2}$ enhances the low–energy P–waves (virtually unkown), rendering practically useless the unsubtracted dispersion relation for the better converging B amplitudes, so important for the πN case in fixing accurately the values of the coupling constant f^2 and of the S–wave scattering lengths[14].

4. Impact of DAΦNE on baryon spectroscopy: the states Λ(1405) and Σ(1385).

At low momenta, comparable to those of the kaons from DAΦNE, we have data from low–statistics experiments, mostly hydrogen bubble–chamber ones on K^-p (and K^-d) interactions[11,16] (dating from the early sixties trough the late seventies), plus scant data from K_L interactions and K_S regeneration on hydrogen[17].

The inelastic channels, open at a laboratory energy $\omega = \frac{1}{2}M_\phi$ (for $K^\pm$'s the value of ω at the interaction point has to include ionization energy losses as well), are the two–body ones $\pi\Lambda$ and $\pi\Sigma$ (in all possible charge states), plus the three–body one $\pi\pi\Lambda$ for K^- or K_L interacting with nucleons: K^+–initiated processes are (apart from charge exchange) purely elastic in this energy region.

For interactions in hydrogen, the c.m. energy is limited by momentum conservation to the initial one, equal (neglecting energy losses) to $w = (m_p^2 + \mu_K^2 + m_p M_\phi)^{1/2}$, or 1442.4 MeV for incident $K^\pm$'s and 1443.8 MeV for incident K_L's. Energy losses for charged kaons can be exploited (using the inner parts of the detector as a moderator) to explore K^-p interactions in a *limited* momentum range, down to the charge–exchange threshold at w = 1437.2 MeV, corresponding to a K^- laboratory momentum of about 90 MeV/c.

For interactions in nuclei, momentum can be carried away by spectator nucleons, and the inelastic channels can be explored down to threshold. The possibility of reaching energies below the $\bar{K}N$ threshold allows exploration of the unphysical region, containing two resonances, the $I = 0$, S–wave Λ(1405) and the $I = 1$, $J^P = \frac{3}{2}^+$ P–wave Σ(1385), observed mostly in production experiments (and, in the first case, in very limited statistics ones[13]): the information on their couplings to the $\bar{K}N$ channel relies entirely on extrapolations of the low-energy $\bar{K}N$ data. The coupling of the $\Sigma(1385)$ to the $\bar{K}N$ channel, for instance, can be determined via forward dispersion relations involving the total sum of data collected at $t \simeq 0$, but still with uncertainties which are, *at their best*, still of the order of 50% of the flavour-$SU(3)$ symmetry prediction[18]; as for the $\Lambda(1405)$, even its spectroscopic classification is an open problem, *vis-à-vis* the paucity and (lack of) quality of the best available data[19].

A formation experiment on *bound* nucleons, in an (almost) 4π apparatus with good efficiency and resolution for low-momentum γ's (such as KLOE[20]), can measure a channel such as $K^-p \to \pi^0\Sigma^0$ (above threshold), or $K^-d \to \pi^0\Sigma^0 n_s$ (both above and below threshold), which is pure $I = 0$: up to now all analyses on the $\Lambda(1405)$ have been limited to charged channels[13], and assumed the $I = 1$ contamination to be either negligible or smooth and non-interfering with the resonance signal. Since the models proposed for the $\Lambda(1405)$ differ mostly in the details of the resonance shape, rather than in its couplings, and it is precisely the shape which could be changed even by a moderate interference with an $I = 1$ background, such measurements would be decisive. Having in the same apparatus and at almost the same energy *tagged* K^- and K_L produced at the same point, one can further separate $I = 0$ and $I = 1$ channels with a minimum of systematic uncertainties, by measuring all channels $K_Lp \to \pi^0\Sigma^+$, $\pi^+\Sigma^0$ and $K^-p \to \pi^-\Sigma^+$, $\pi^+\Sigma^-$, besides, of course, the above-mentioned, pure $I = 0$, $K^-p \to \pi^0\Sigma^0$ one.

Another class of inelastic processes which are expected to be produced, at a much smaller rate, by DAΦNE's kaons are the radiative capture processes $K^-p \to \gamma\Lambda$, $\gamma\Sigma^0$ and $K_Lp \to \gamma\Sigma^+$ (both in hydrogen and deuterium), and $K^-n \to \gamma\Sigma^-$ and $K_Ln \to \gamma\Lambda$, $\gamma\Sigma^0$ (only in deuterium). Up to now only searches for photons emitted after stops of K^-'s in liquid hydrogen and deuterium have been performed with some success: the spectra are dominated by photons from unreconstructed π^0 and Σ^0 decays[21], and separating the signals from this background poses serious difficulties, since only the photon line from the $\gamma\Lambda$ final state falls just above the endpoint of the photons from π^0 decays in the $\pi^0\Lambda$ final state, while that from $\gamma\Sigma^0$ falls right on top of the latter. Indeed these experiments were able to produce only an estimate of the respective branching ratios[21].

The 4π geometry possible at DAΦNE, combined with the "transparency" of a KLOE–like apparatus[20], its high efficiency for photon detection and its good resolution for spatial reconstruction of the events, should make possible the full identification of the final states and therefore the measurement of the absolute cross sections for these processes, although in flight and not at rest.

Data[21] are presently indicating branching ratios around 0.9×10^{-3} for $K^- p \to \gamma\Lambda$ and 1.4×10^{-3} for $K^- p \to \gamma\Sigma^0$, with errors of the order of 15% on both: most models[23] give the first rate larger than the second, with both values consistently higher than the observed ones. Only a cloudy–bag–model[24] exhibits the trend appearing (although only at a 2σ–level, and therefore waiting for confirmation by better data) from the first experimental determinations, but this is the only respect in which it agrees with the data, still giving branching ratios larger than observations by a factor two.

Data are also interpretable in terms of $\Lambda(1405)$ electromagnetic transition moments[22]: this interpretation is clearly sensitive to the interference between the decay of this state and all other contributions. An extraction of the $\Lambda(1405)$ moments freer of these uncertainties would require measurements of $\gamma\Lambda$ and $\gamma\Sigma$ (if possible, in different charge states) over the unphysical region, using (gaseous) deuterium or helium as a target. Rates are expected to be of the order of 10^4 *events/y* only, but such a low rate would correspond to better statistics than those of the best experiment performed on the $\Lambda(1405) \to \pi\Sigma$ decay spectrum[13].

5. Description of coupled $\bar{K}N$, πY channels: the K–matrix.

A description of the low–energy $\bar{K}N$ partial waves must couple the two–body inelastic channels to each other and to the elastic one: the three–body channel $\pi\pi\Lambda$ is expected to be suppressed, for $J^P = \frac{1}{2}^-$, by the angular momentum barrier, but it could contribute appreciably to the $I = 0$, $J^P = \frac{1}{2}^+$ P–wave, due to the strong final–state interaction of two pions in an $I = 0$ S–wave. Most bubble–chamber experiments were unable to fully reconstruct events at the lowest momenta, and therefore assumed all *directly* produced Λ's to come from the $\pi\Lambda$ channel alone, neglecting the small $\pi\pi\Lambda$ contribution altogether.

The appropriate formalism is to introduce a K–matrix description (sometimes it is convenient to use, instead of the K–matrix, its inverse, *a.k.a.* the M–matrix), defined in the isospin eigenchannel notation as

$$\mathbf{K}_{\ell\pm}^{-1} = \mathbf{M}_{\ell\pm} = \mathbf{T}_{\ell\pm}^{-1} + i\,\mathbf{Q}^{2\ell+1}\ , \tag{25}$$

for both $I = 0, 1$ S–waves (and perhaps also for the four P–waves as well). The K–matrices, assuming $SU(2)$ symmetry, describe the S–wave data at a given energy in terms of *nine* real parameters (six for $I = 1$ and three for $I = 0$), while the experimentally accessible processes are described, with pure S–waves and in the same symmetry limit, by only *six independent* parameters, for which one can choose the two (complex) amplitudes $A_{0,1}$ for the elastic channel, the phase difference ϕ between the $I = 0$ and $I = 1$ $\pi\Sigma$ production amplitudes, and the ratio ϵ between the $\pi\Lambda$ production cross section and that for total hyperon production in an $I = 1$ state[25].

Thus a single–energy measurement does not allow a complete determination of the K–matrix elements. For a precise determination of the S–waves one should also subtract out the P–wave contributions to the integrated cross sections

$$\sigma = 4\pi L_0 = 2\pi \int_{-1}^{+1} \left(\frac{d\sigma}{d\Omega}\right) d\cos\theta = 4\pi[|T_{0+}|^2 + 2|T_{1+}|^2 + |T_{1-}|^2 + \ldots]\ , \tag{26}$$

which could be obtained either from L_1 alone for the elastic and charge–exchange channels, or from both L_1 and P_{Y} for the hyperon production channels. None of these quantities has been measured up to now: the TST Collaboration tried to extract L_1 from some of their low–statistics data, but found results consistent with zero within their obviously very large errors[11]. At the same level of accuracy, one should also be able to isolate out the $\pi\pi\Lambda$ channel as well. Note that an accurate analysis has also to include complete e.m. corrections[15,26]: up to now all $\bar{K}N$ analyses have relied on the old, approximate formulæ derived by Dalitz and Tuan for a pure S–wave scattering[27].

To fix the redundant K–matrix parameters different ways have been tried: some authors have used the data on the shape of the $\pi\Sigma$ spectrum from production experiments[28], others have constrained the amplitudes in the unphysical region by imposing consistency with forward dispersion relations for both $K^{\pm}p$ and $K^{\pm}n$

elastic–scattering D amplitudes[29], relying on the accurate total–cross–section data at higher energies. More recently, some attempts have been made to combine both constraints into a global analysis, but with no better results than each of them taken separately[30].

Unfortunately, neither of these methods has been very powerful because of the low statistics of the $\pi\Sigma$–production data on one side, and on the other because of the need to use for the dispersion relations the often inaccurate information (and particularly so for the $K^{\pm}n$ amplitudes) on the real–to–imaginary–part ratios.

We list below the constant K–matrices found by Chao *et al.* using the first method[28], which did not include the TST Collaboration data, and the more complex parametrization found by A.D. Martin using the second[29], and including the preliminary TST data. Note that to describe the data for $I = 0$ both above and below threshold A.D. Martin had to introduce a constant–effective–range M–matrix $\mathbf{M}^{(0)} = (\mathbf{K}^{(0)})^{-1} = \mathbf{A} + \mathbf{R}k^2$, so that to make the two analyses comparable we list separately his threshold K–matrix values.

Table I

Chao et al.	*A. D. Martin*		
$K^{(0)}_{NN} = -1.56 fm$	$A_{NN} = -0.07 fm^{-1}$	$R_{NN} = +0.18 fm$	$K^{(0)}_{NN}(0) = -1.65 fm$
$K^{(0)}_{N\Sigma} = -0.92 fm$	$A_{N\Sigma} = -1.02 fm^{-1}$	$R_{N\Sigma} = +0.19 fm$	$K^{(0)}_{N\Sigma}(0) = +0.16 fm$
$K^{(0)}_{\Sigma\Sigma} = +0.07 fm$	$A_{\Sigma\Sigma} = +1.94 fm^{-1}$	$R_{\Sigma\Sigma} = -1.09 fm$	$K^{(0)}_{\Sigma\Sigma}(0) = -0.15 fm$
$K^{(1)}_{NN} = +0.76 fm$			$K^{(1)}_{NN} = +1.07 fm$
$K^{(1)}_{N\Sigma} = -0.97 fm$			$K^{(1)}_{N\Sigma} = -1.32 fm$
$K^{(1)}_{N\Lambda} = -0.66 fm$			$K^{(1)}_{N\Lambda} = -0.30 fm$
$K^{(1)}_{\Sigma\Sigma} = +0.86 fm$			$K^{(1)}_{\Sigma\Sigma} = +0.27 fm$
$K^{(1)}_{\Sigma\Lambda} = +0.51 fm$			$K^{(1)}_{\Sigma\Lambda} = +1.54 fm$
$K^{(1)}_{\Lambda\Lambda} = +0.04 fm$			$K^{(1)}_{\Lambda\Lambda} = -1.02 fm$

The table shows that there is considerable uncertainty even on the $K^{(I)}_{NN}$ elements (the real parts of the corresponding scattering lengths): the data have been re–analyzed by Dalitz *et al.*[30], using both sets of constraints with different weigths and different parametrizations, and yielding a variety of fits, all of them of about the same quality of, but none of them improving very much over, the above ones.

To further highlight the difficulties met in fitting the data, we point out that A.D. Martin himself[29] found that including in his analysis a $\Sigma(1385)$ resonance with the width given by production experiments and the coupling to the $\bar{K}N$ channel dictated by flavour–$SU(3)$ symmetry, was worsening rather than improving the results obtained neglecting it altogether. He proposes therefore to consider the Σ Born–term contribution a superposition of the former and of that of the P–wave resonance: a rather unsavoury situation, considering the different J^P of the two states, which may raise questions about the applicability of his analysis away from $t \simeq 0$. An analogous superposition is considered in $K^{\pm}p$ dispersion relations, where one can not separate the Σ– from the Λ–pole, but here the two contribute to the same partial wave, and the Σ–pole can be extracted independently from data on $K^{\pm}n$ scattering and K_S regeration on protons[31,32].

In the analysis of the low–energy data collected in the past, a further difficulty comes from the large momentum spread of the low–energy kaon beams, for $K^{\pm}$'s because of the degrading in a moderator of the higher–energy beams needed to transport the kaons away from their production targets, for K_L's because of the large apertures needed to achieve satisfactory rates in the targets (typically bubble chambers): this made unrealistic the proposals (advanced from the early seventies) of better determining the low–energy K–matrices by studying the behaviour of the cross sections for K^-p–initiated processes at the $\bar{K}^0n$ charge–exchange threshold[10]. The high momentum resolution available at DAΦNE will instead make such a goal a realistically achievable one.

In this case $SU(2)$ can no longer be assumed to be a symmetry of the amplitudes: under the (reasonable) assumption that the forces are still $SU(2)$–symmetric, one can however retain the previous K–matrix formalism, but no longer decouple the different isospin eigenchannels[33]. Introducing the orthogonal matrix $\mathbf{R}$, which transforms the six isospin eigenchannels for $\bar{K}N$ $(I = 0, 1)$, $\pi\Lambda$ $(I = 1$ only) and $\pi\,\Sigma$ $(I = 0, 1, 2)$ into the six charge–states K^-p, $\bar{K}^0n$, $\pi^0\Lambda$, $\pi^-\Sigma^+$, $\pi^0\Sigma^0$ and $\pi^+\Sigma^-$, and the diagonal matrix $\mathbf{Q}_c$ of the c.m. momenta for these latter, one can rewrite the T–matrix in the isospin–eigenchannel space for the S–waves as

$$\mathbf{T}_I^{-1} = \mathbf{K}_I^{-1} - i\mathbf{R}^{-1}\mathbf{Q}_c\mathbf{R} \ , \tag{27}$$

where $\mathbf{K}_I$ is a box matrix with zero elements between channels of different isospin, and $\mathbf{R}^{-1}$ $\mathbf{Q}_c$ $\mathbf{R}$ is of course no longer diagonal.

Apparently this involves one more parameter, since it also contains the element $K^{(2)}_{\Sigma\Sigma}$: in practice, if one is interested in the cross sections only in the neighbourhood of the $\bar{K}N$ charge–exchange threshold, one can take the c.m. momenta in the three $\pi\Sigma$ channels as equal and decouple the $I = 2$, $\pi\Sigma$ channel from the $I = 0, 1$ ones, since the "rotated" matrix $\mathbf{R}^{-1}\mathbf{Q}_c\mathbf{R}$ has now only two non–zero, off-diagonal elements, equal to $\frac{1}{2}(q_0 - q_-)$ (the subscripts refer to the kaon charges), between the $I = 0$ and $I = 1$ $\bar{K}N$ channels, the diagonal ones being the same as in the $SU(2)$–symmetric case, if one substitutes for the $\bar{K}N$ channel momentum q the average over the two charge states $\frac{1}{2}(q_0 + q_-)$. Indeed $K^{(2)}_{\Sigma\Sigma}$ could only be important for an accurate description of the $\pi\Sigma$ and $\pi\Lambda$ mass spectra close to the $\pi\Sigma$ threshold.

6. Outline of a theoretical program for a modern KN amplitude analysis.

The measurements proposed for DAΦNE will provide data of the same statistical quality now available *only* for the πN system: theoretical tools for their analysis must thus be improved as well, to meet the standards required by this, long awaited for, "forward leap" in KN data. Since long, tools of just this level have been provided, for πN amplitude analysis, by the so–called "Karlsruhe–Helsinki collaboration" headed over the years by Prof. G. Höhler[14,34]: alas, their software can not be straightforwardly imported to do KN analyses, mainly because of the complicated analytic structure of the low-energy $\bar{K}N$ amplitudes.

Much for the same reason, the dispersive treatment of Coulomb corrections, developed at NORDITA[15,26] by Hamilton and collaborators, can not be immediately transferred to the strange sector. It has to be recalled that old data were always analyzed using, for these corrections, the approximate formula of Dalitz and Tuan[27], which considers only a pure–S–wave strong interaction, and furthermore might be inapplicable to a *strongly absorptive* interaction close to threshold[35].

Since the basic principles, on which both approaches are based, have to hold for the $\bar{K}N$ system as for the πN one, it remains only to work out the details of a partial–wave–analysis procedure, applicable to a system strongly absorptive at threshold, and possessing *ab–initio* the following requirements: i) consistency with both fixed–t and partial–wave dispersion relations; ii) crossing symmetry (and isotopic–spin symmetry as well, to describe *simultaneously* charge–exchange and regeneration data); iii) analyticity in t beyond the Lehman ellipses, with the *correct* low–mass, t–channel–cut discontinuities given by the $\pi\pi$ cut; iv) a *complete* treatment of radiative and Coulomb corrections.

The author and its collaborators (chiefly G. Violini of Università della Calabria, Cosenza, and C.I.F., Bogotá, and R.C. Barrett of the University of Surrey, Guildford) have in the past carried out parts of this program (as many others have done, more or less at the same epoch), but only for *limited purposes*, such as extrapolations either to the hyperon poles[31,36] or to the Cheng–Dashen point[37], or studies of Coulomb effects[35] and radiative capture at threshold[24]: what remains to be done is a merging together of all these techniques into a "global" analysis, on which work is presently under way.

In this perspective we have advanced the proposal to I.N.F.N. for a program of extensive collaboration, code–named KILN (for "Kaon Interactions at Low energies with Nucleons"), which has already received an initial, and thus limited, financial support. Participation in this collaboration is highly welcome, and we take this occasion for calling upon all theorists wich have been or wish to be active in this still open and very much alive (despite greatly exaggerated rumors on the contrary[38]) field of particle physics.

ACKNOWLEDGEMENTS

Exchanges of information with theorists and experimentalists involved in the planning of experiments at both machines, KAON and DAΦNE, is gratefully acknowledged by the author and his collaborators.

REFERENCES

1. See the Proceedings of the *"Workshop on Physics and Detectors for DAΦNE"*, ed. by G. Pancheri (I.N.F.N., Frascati 1991).
2. M. Aguilar–Benitez, *et al.* (Particle Data Group): *Phys. Lett.* **B 239** (1990) 1.
3. P.M. Gensini and G. Violini: *"Workshop on Science at the KAON Factory"*, ed. by D.R. Gill (TRIUMF, Vancouver 1991), Vol. II, p. 193; P.M. Gensini: *"Workshop on Physics and Detectors for DAΦNE"*, ed. by G. Pancheri (I.N.F.N., Frascati 1991), p. 453.
4. C.J. Batty and A. Gal: *Nuovo Cimento* **102 A** (1989) 255; C.J. Batty: *"First Workshop on Intense Hadron Facilities and Antiproton Physics"*, ed. by T. Bressani, F. Iazzi and G. Pauli (S.I.F., Bologna 1990), p. 117.
5. J. Beveridge: *"Workshop on Science at the KAON Factory"*, ed. by D.R. Gill (TRIUMF, Vancouver 1991), Vol. I, p. 19.
6. G. Vignola: *"Workshop on Physics and Detectors for DAΦNE"*, ed. by G. Pancheri (I.N.F.N., Frascati 1991), p. 11.
7. T. Bressani: *"Workshop on Physics and Detectors for DAΦNE"*, ed. by G. Pancheri (I.N.F.N., Frascati 1991), p. 475; *"Common Problems and Ideas of Modern Physics"*, ed. by T. Bressani, B. Minetti and A. Zenoni (World Scientific, Singapore 1992), p. 222, and the talk presented at this Folgaria Winter School.
8. See the Proceedings of the *"Workshop on Science at the KAON Factory"*, ed. by D.R. Gill (TRIUMF, Vancouver 1991).
9. *"Proposal for a European Hadron Facility"*, ed. by J.F. Crawford, report *EHF–87–18* (Trieste–Mainz, may 1987).
10. See the discussion on this point by D.J. Miller: *"Proc. of the Int. Conf. on Hypernuclear and Kaon Physics"*, ed. by B. Povh (M.P.I., Heidelberg 1982), p. 215.
11. TST Collaboration: R.J. Novak, *et al.*: *Nucl Phys.* **B 139** (1978) 61; N.H. Bedford, *et al.*: *Nukleonika* **25** (1980) 509; M. Goossens, G. Wilquet, J.L.

Armstrong and J.H. Bartley: *"Low and Intermediate Energy Kaon–Nucleon Physics"*, ed. by E. Ferrari and G. Violini (D. Reidel, Dordrecht 1981), p. 131; J. Ciborowski, *et al.*: *J. Phys.* **G 8** (1982) 13; D. Evans, *et al.*: *J. Phys.* **G 9** (1983) 885; J. Conboy, *et al.*: *J. Phys* **G 12** (1986) 1143. A good description of the experiment is in D.J. Miller, R.J. Novak and T. Tyminiecka: *"Low and Intermediate Energy Kaon-Nucleon Physics"*, ed by E. Ferrari and G. Violini (D. Reidel, Dordrecht 1981), p. 251.

12. E. Piasetzky: *Nuovo Cimento* **102 A** (1989) 281.
13. R.J. Hemingway: *Nucl. Phys.* **B 253** (1985) 742. Older data are even poorer in statistics: see ref. 27 for a comparison. See also, for formation on bound nucleons, B. Riley, I.T. Wang, J.G. Fetkovich and J.M. McKenzie: *Phys. Rev.* **D 11** (1975) 3065.
14. For conventions and kinematical notations we have adopted the same as: G. Höhler, F. Kaiser, R. Koch and E. Pietarinen: *"Handbook of Pion–Nucleon Scattering"* (Fachinformationszentrum, Karlsruhe 1979), and *"Landolt–Börnstein, New Series, Group I, Vol. 9b"*, ed. by H. Schopper (Springer–Verlag, Berlin 1983), which have become a "standard" for describing πN scattering.
15. B. Tromborg, S. Waldenström and I. Øverbø: *Ann. Phys. (N.Y.)* **100** (1976) 1; *Phys. Rev.* **D 15** (1977) 725; *Helv. Phys. Acta* **51** (1978) 584.
16. $K^{\pm}p$ data: W.E. Humphrey and R.R. Ross: *Phys. Rev.* **127** (1962) 1; G.S. Abrams and B. Sechi–Zorn: *Phys. Rev.* **139** (1965) B454; M. Sakitt, *et al.*: *Phys Rev* **139** (1965) B719; J.K. Kim: Columbia Univ. report *NEVIS–149* (New York 1966) and *Phys. Rev. Lett.* **14** (1970) 615; W. Kittel, G Otter and I. Waček: *Phys. Lett* **21** (1966) 349; D. Tovee, *et al.*: *Nucl. Phys.* **B 33** (1971) 493; T.S. Mast, *et al.*: *Phys. Rev.* **D 11** (1975) 3078 and **D 14** (1976) 13; R.O. Bargenter, *et al.*: *Phys. Rev.* **D 23** (1981) 1484. K^-d data: R. Armenteros, *et al.*: *Nucl. Phys.* **B 18** (1970) 425.
17. K_Lp data: J.A. Kadyk, *et al.*: *Phys. Lett.* **17** (1966) 599 and report *UCRL–18325* (Berkeley 1968); R.A. Donald, *et al.*: *Phys. Lett.* **22** (1966) 711; G.A. Sayer, *et al.*: *Phys. Rev.* **169** (1968) 1045.
18. G.C. Oades: *Nuovo Cimento* **102 A** (1989) 237.
19. A.D. Martin, B.R. Martin and G.G. Ross: *Phys. Lett.* **B 35** (1971) 62; P.N.

Dobson jr. and R. McElhaney: *Phys. Rev.* **D 6** (1972) 3256; G.C. Oades and G. Rasche: *Nuovo Cimento* **42 A** (1977) 462; R.H. Dalitz and J.G. McGinley: *"Low and Intermediate Energy Kaon–Nucleon Physics"*, ed. by E. Ferrari and G. Violini (D. Reidel, Dordrecht 1981), p. 381, and the Ph.D. thesis by McGinley (Oxford Univ. 1979); G.C. Oades and G. Rasche: *Phys. Scr.* **26** (1982) 15; J.P. Liu: *Z. Phys.* **C 22** (1984) 171; B.K. Jennings: *Phys. Lett.* **B 176** (1986) 229; P.B. Siegel and W. Weise: *Phys. Rev.* **C 38** (1988) 2221; R.H. Dalitz and A. Deloff: *J. Phys.* **G 17** (1991) 289.

20. A. Aloisio, *et al.* (The KLOE Collaboration): *"KLOE – A General Purpose Detector for DAΦNE"*, report *LNF–92/019 (IR)* (Frascati, April 1992).
21. B.L. Roberts: *Nucl Phys* **A 479** (1988) 75c; B.L. Roberts, *et al.*: *Nuovo Cimento* **102 A** (1989) 145; D.A. Whitehouse, *et al.*: *Phys. Rev. Lett.* **63** (1989) 1352.
22. See the review by J. Lowe: *Nuovo Cimento* **102 A** (1989) 167.
23. J.W. Darewich, R. Koniuk and N. Isgur: *Phys. Rev.* **D 32** (1985) 1765; H. Burkhardt, J. Lowe and A.S. Rosenthal: *Nucl Phys.* **A 440** (1985) 653; R.L. Workman and H.W. Fearing: *Phys. Rev.* **D 37** (1988) 3117; R.A. Williams, C.R. Ji and S. Cotanch: *Phys. Rev.* **D 41** (1990) 1449; *Phys. Rev.* **C 43** (1991) 452; H. Burkhardt and J. Lowe: *Phys. Rev.* **C 44** (1991) 607. For radiative capture on deuterons (and other light nuclei), see: R.L. Workman and H.W. Fearing: *Phys. Rev.* **C 41** (1990) 1688; C. Bennhold: *Phys. Rev.* **C 42** (1990) 775.
24. Y.S. Zhong, A.W. Thomas, B.K. Jennings and R.C. Barrett: *Phys. Lett.* **B 171** (1986) 471; *Phys. Rev.* **D 38** (1988) 837 (which corrects a numerical error contained in the previous paper).
25. See the review presented by B.R. Martin at the 1972 Baško Polje Int. School, published in: *"Textbook on Elementary Particle Physics. Vol. 5: Strong Interactions"*, ed by M. Nikolič (Gordon and Breach, Paris 1975).
26. J. Hamiltom, I. Øverbø and B. Tromborg: *Nucl. Phys.* **B 60** (1973) 443; B. Tromborg and J. Hamilton: *Nucl. Phys.* **B 76** (1974) 483; J. Hamilton: *Fortschr. Phys.* **23** (1975) 211.
27. R.H. Dalitz and S.F. Tuan: *Ann. Phys. (N.Y.)* **10** (1960) 307.

28. Y.A. Chao, R. Krämer, D.W. Thomas and B.R. Martin: *Nucl. Phys.* **B 56** (1973) 46.
29. A.D. Martin: *Phys. Lett.* **B 65** (1976) 346; *"Low and Intermediate Energy Kaon-Nucleon Physics"*, ed. by E. Ferrari and G. Violini (D. Reidel, Dordrecht 1981), p. 97; *Nucl. Phys.* **B 179** (1981) 33.
30. R.H. Dalitz, J. McGinley, C. Belyea and S. Anthony: *"Proc. of the Int. Conf. on Hypernuclear and Kaon Physics"*, ed. by B. Povh (M.P.I., Heidelberg 1982), p. 201.
31. G.K. Atkin, B. Di Claudio, G. Violini and N.M. Queen: *Phys. Lett.* **B 95** (1980) 447; *"Low and Intermediate Energy Kaon–Nucleon Physics"*, ed. by E. Ferrari and G. Violini (D. Reidel, Dordrecht 1981), p. 131.
32. J. Antolín: *Phys. Rev.* **D 43** (1991) 1532.
33. G.C. Oades and G. Rasche: *Phys. Rev.* **D 4** (1971) 2153; H. Zimmermann: *Helv. Phys. Acta* **45** (1973) 1117; G. Rasche and W.S. Woolcock: *Fortschr. Phys.* **25** (1977) 501.
34. R. Koch: *"Low and Intermediate Energy Kaon-Nucleon Physics"*, ed. by E. Ferrari and G. Violini (D. Reidel, Dordrecht 1981), p. 1; see also the update by G. Höhler: *πN Newsletter* **2** (1990) 1.
35. P.M. Gensini and G.R. Semeraro: *"Perspectives on Theoretical Nuclear Physics"*, ed. by L. Bracci, *et al.* (ETS Ed., Pisa 1986), p. 91.
36. B. Di Claudio, G. Violini and N.M. Queen: *Nucl. Phys.* **B 161** (1979) 238; G.K. Atkin, B. Di Claudio, G. Violini, J.E. Bowcock and N.M. Queen: *Z. Phys.* **C 7** (1981) 249.
37. P.M. Gensini: *J. Phys. G* **7** (1981) 1177; *Nuovo Cimento* **84 A** (1984) 203.
38. The same also happened to Mark Twain, from which we took the liberty of borrowing the pun.

Common Problems and Trends of Modern Physics, pp. 409–417
ed. T. Bressani, S. Marcello and A. Zenoni
©1992 World Scientific Publishing Company

FINUDA experiment at DAΦNE

Paola Gianotti
Istituto Nazionale di Fisica Nucleare, Via P. Giuria 1,
10125 Torino, Italy
and
Elena Rossetto
Università degli Studi di Torino, Via P. Giuria 1,
10125 Torino, Italy

ABSTRACT

DAΦNE, the Frascati ϕ factory, will be a source of low energy K^-. This will offer the opportunity of studying hypernuclear spectroscopy with K^- at rest. Here the project of a magnetic spectrometer (FINUDA), devoted to hypernuclear physics to be placed on one of the interaction regions of DAΦNE, is described . Also a general information about the machine is given.

1. Introduction

At the Frascati INFN laboratories, in 1993, will start the construction of DAΦNE (Double Anular ϕ-factory for Nice Experiments) machine, a e^+e^- high luminosity collider at center of mass energy of the ϕ (1020) resonance.

The ϕ resonance decays with a branching ratio of ~50% in K^+K^-, of ~35% in $K_L^0K_S^0$ and of ~15% in other channels (mostly 3π). Thus a ϕ-factory is a source of low energy (16.3 MeV), background free, $K\overline{K}$ pairs.

During 1991 DAΦNE workshop was pointed out[1,2] that K^- from ϕ decay can be stopped in a thin solid target to study hypernuclear spectroscopy. In fact K^- at rest can populate different hypernuclear states while, using K^- in flight around the "magic" moment, it is possible to obtain only those states for which the hyperon has the same spin and angular momentum as the struck nucleon.

Some brief remainds on hypernuclear physics will be given in Sec. 3, while in Sec. 2 the reader will find some details on DAΦNE machine.The description of an apparatus dedicated to hypernuclear physics is reported in Sec. 4.

2. The DAΦNE machine

DAΦNE will be the first of a new generation of high luminosity e^+e^- colliders and it will be located in the ADONE area of Frascati National Laboratory.

Its main characteristics are listed in Tab. I.

Tab. I DAΦNE machine parameters

Energy (MeV)	510
Circumference(m)	94.56
Luminosity($cm^{-2}s^{-1}$)	$10^{32} \div 10^{33}$
Number of bunches(max)	120
Number of particles per bunch(max)	$8.9\ 10^{10}$
Bunch dimensions: Δx(mm)	2.11
Δy(mm)	0.021
Δz(mm)	30
Bunch crossing(MHz)	70

The basic idea of building a ϕ-factory is to have an intense source of $K\overline{K}$ pairs from the decay of the ϕ resonance and, to achieve this goal, it has been designed a collider with two separated rings, one for e^+ and another for e^-. The solution of having two rings to increase machine luminosity has been chosen because in this way it is possible to put a very large number of particle bunches inside each ring (120). This is impossible in a machine in which the two beams circulate in the same ring because of beam-beam interaction. From what said before it is easy to see that the maximum number of particles in each ring will exceed 10^{13}; thus to maintain the injection time of the order of 10 minutes it will be necessary to have a small storage ring serving as an accumulator between the positron/electron LINAC and the main ring. Thus particles accelerated from the LINAC will be injected into the accumulator, then extracted and reinjected in the main ring filling one bucket at a time.

More details about DAΦNE can be found in Ref. 3.

3. Some basic concepts on hypernuclear physics

A hypernucleus is a many-body system made of conventional (nonstrange) nucleons with one or more hyperons[4, 5, 6, 7] (Λ, Σ etc.). The hyperon inside the nuclear matter is a probe of nuclear structure. Because the Λ has isospin 0 and the nucleons N isospin 1/2, the ΛN interaction can't arise from a π exchange (the isospin of the π is 1). Thus there isn't a dominance of OPE force as for NN interaction. This ensures that shorter range forces must be involved in ΛN interaction.

Another interesting feature of hypernuclear physics is the non-mesonic decay of Λ hyperon[3,4] when bound in nuclear matter. The reason of Λ non-mesonic decay, $\Lambda N \rightarrow NN$, is to be found in Pauli principle that reduces the phase-space for $\Lambda \rightarrow N\pi$ reaction, but the interest on it is above all due to the large momentum transfer in this process that can probe short distances associated with quark/gluon degrees of freedom.

The ways in which hypernuclei are produced are essentially two:

$$K^- + n \rightarrow \Lambda + \pi^- \qquad (1)$$

$$\pi^+ + n \rightarrow K^+ + \Lambda \qquad (2)$$

In reaction (1) the momentum transfer is low (0 if $P_K = 505$ MeV/c) so the Λ in the final state has the same spin and orbital wave function of the transformed neutron. In reaction (2) the momentum transfer to the Λ is high (> 250 MeV/c) so it is possible to excite high spin hypernuclear states. The cross-section of reaction (2) is two orders of magnitude lower than that of reaction (1), but this disadvantage is compensated by the higher intensity of π^+ beams.

A way of combining both advantages of reaction (1) and (2), high production rate and wide spectrum of hypernuclear final states, is that of working with K^- at rest. In this way the momentum transfer is high and the production rate of order of 10^{-3} per stopped K^-. This technique will be used at DAΦNE.

4. FINUDA detector

FINUDA (FIsica NUcleare a DAΦNE) will be the spectrometer designed to study hypernuclear spectroscopy at DAΦNE.

At DAΦNE kaons will be emitted over 4π sr with an angular distribution proportional to $\sin^2\theta$.

With a luminosity of the machine of 10^{32} cm^{-2}s^{-1} one can expect a rate of ϕ of the order of 480 s^{-1}. This will give a rate of K^- of about 240 s^{-1} for $8.5^o < \theta < 117.5^o$ that, with a rate of production of hypernuclei of $10^{-3}/K^-$ a solid angle for the detection of the π^- of about 0.6π sr and an efficiency of the whole detector of 50%, will give a rate of "good" events of 30 ev/h. This number will be multiplied by a factor 10 at the top machine luminosity of 10^{33} cm^{-2}s^{-1}. However, if one wants to tag the events using the K^+ emitted in the ϕ decay the rate of events will be reduced to one half.

Taking into account what has been said for the machine and the ϕ resonance, to optimize the events collection FINUDA detector will be a magnetic spectrometer of cylindrical simmetry with :

- wide angular acceptance (> 2π sr);
- good momentum resolution (~ 0.2% at 270 MeV/c) over a wide range of momentum values (100 ÷ 300 MeV/c);
- capability of detecting slow charge particles form decays (above all π and p arising from hypernucleus decays).

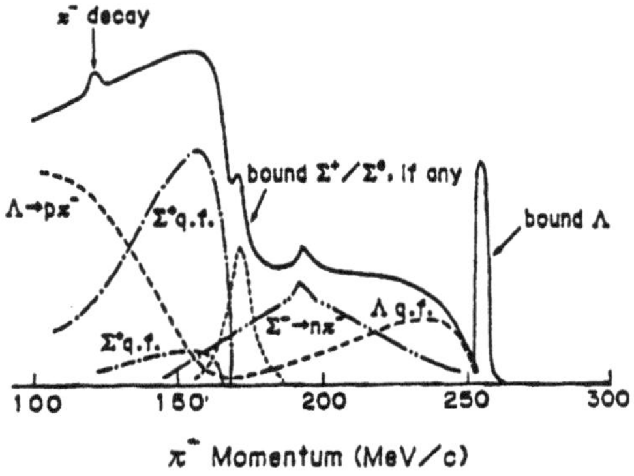

Fig. 1 Expected inclusive π^- spectrum, which is composed of various components as indicated [Ref. 2]

The need of a magnetic spectrometer with good performances is indispensable either to select hypernucleus events from other K^-N interaction channels or to identify hypernucleus state (see Fig. 1 and Fig. 2). In particular if one expects a separation of 1 MeV between the different energetic levels of an hypernucleus, it is necessary a $\Delta p/p$ of 0.4% FWHM.

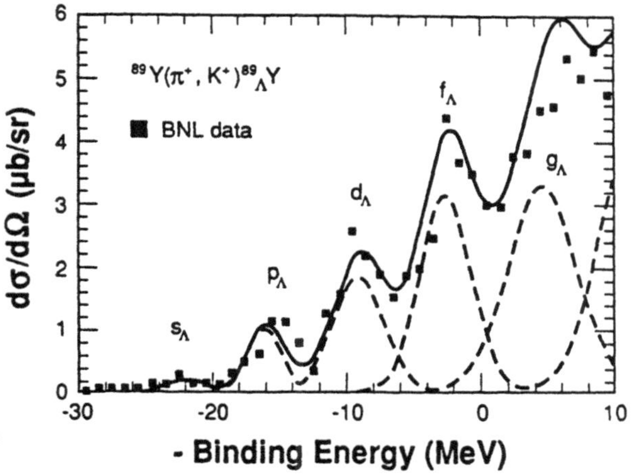

Fig. 2 Experimental $^{89}Y(\pi,K)$ spectrum showing the Λ shell structure of the residual hypernucleus [Ref. 8].

To satisfy all the requirements mentioned above we have designed the detector shown by Fig. 3.

Moving from the beam pipe one finds:

- an array of scintillators;
- a 0.5 g/cm^2 solid target;
- a microstrip array;
- a system of drift chambers;
- a second array of scintillators.

Scintillators, that are fast detectors, will be used above all at the trigger level and for Time Of Flight (TOF) purposes.

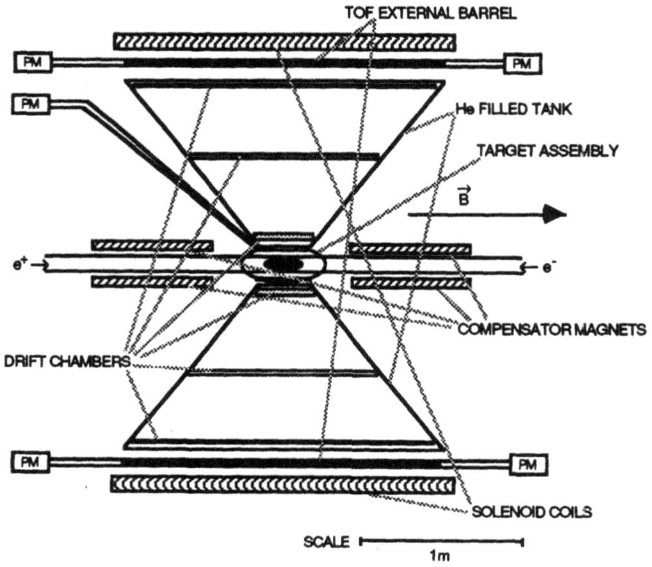

Fig. 3 Layout of FINUDA magnetic spectrometer for hypernucler spectroscopy.

A fast first level trigger, considering the typical hypernuclear event sketched in Fig. 4, could ask for multiplicity 2 in coincidence with a well defined $\Delta E/\Delta x$ in internal TOF, in order to eliminate Bhabha events, and multiplicity 3 or 4 in the external TOF. It is also possible to use the signal given by the μ^+, from the decay of the K^+, to make a delayed coincidence between the signal arising from a scintillator slab of internal TOF and another from the external ones. A scheme of a trigger like that is given in Fig. 5.

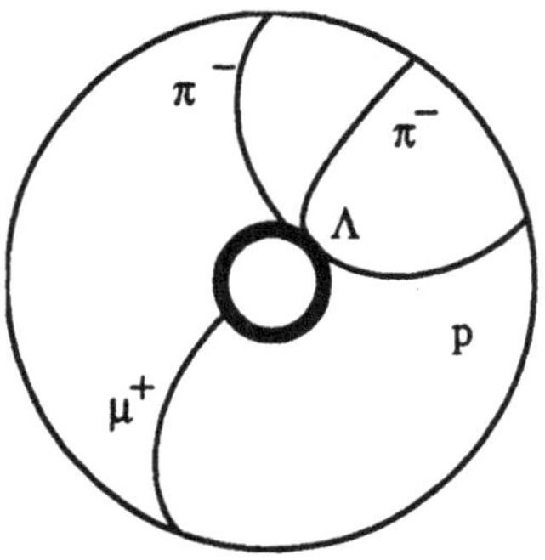

Fig. 4 Topology, of a typical hypernuclear event, inside detector FINUDA.

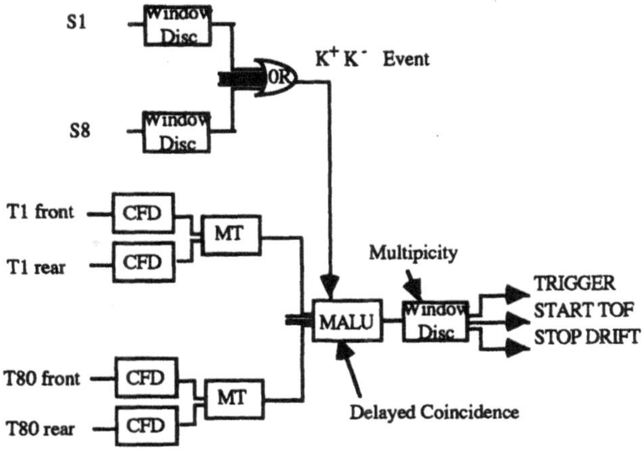

S1....S8 : Signals from inner scintillator barrel
T1....T80 : Signals from outer scintillator barrel
MALU : Majority Logic
DISC : Discriminator
MT : Mean Timer Circuit
CFD : Constant fraction

Fig. 5 Block scheme of FINUDA trigger.

The drift chambers, together with the microstrip array, will give the momentum of charge particles. In order to minimize the Coulomb multiple scattering, the total thickness of the material of the chambers must be kept as thin as possible. The chambers will be filled with the standard Ar + C_2H_6 mixture and will be put in a tank filled with Helium. The separation between the two mixtures will be done with thin mylar sheets.

In order to understand whether such a detector is feasible, we are performing Monte Carlo simulations based on GEANT3 package. As a first step we are above all interested in understanding which will be the best momentum resolution that may be achieved with an apparatus such that described. Tab. II reports the results we have obtained with different experimental set-ups.

Tab. II First results of FINUDA Monte Carlo simulation

magnetic field [T]	mylar window thickness [μm]	$\Delta x = \Delta y = 0$ mm $\Delta z = 0$ mm	$\Delta x = \Delta y = 0.12$ mm $\Delta z = 11.8$ mm	$\Delta x = \Delta y = 4.72$ mm $\Delta z = 47.2$ mm
1.0	10	2.03 ‰	2.55 ‰	10.10 ‰
1.0	20	---	2.97 ‰	---
1.0	30	---	3.33 ‰	---
1.5	10	1.35 ‰	1.58 ‰	7.46 ‰
1.5	20	---	1.82 ‰	---
1.5	30	---	2.02 ‰	---

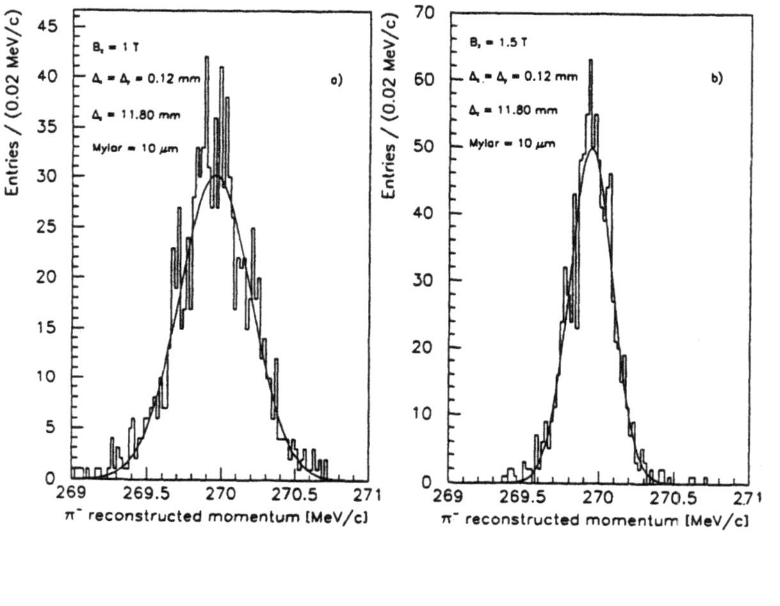

Fig. 6 Histograms of the reconstructed momentum distributions from Monte Carlo simulation. Monoenergetic π^- of 270 MeV/c, isotropically emitted from the center of the spectrometer were tracked in the apparatus by means of the GEANT3 package. The distributions have been obtained for the cases of a magnetic field intensity of, respectively, 1.0 T (a) and 1.5 T (b), a thickness of the mylar window of 10 μm and a nominal spatial resolution (FWHM) $\Delta x = \Delta y = 0.12$ mm, $\Delta z = 11.8$ mm for the drift chambers.

All the results are very good even if our simulation is not completely realistic: in fact the non uniformity of magnetic field have not been taken into account. Nevertheless we are optimistic and we think that an apparatus like that will be very suitable in studying hypernuclear spectroscopy. In Fig. 6 are reported two spectra of momentum resolution obtained from Monte Carlo simulation with different configurations of the detector.

1. References

1. T. Bressani, in *Proc. of Whorkshop on Physics and Detectors for DAΦNE*, Frascati, April 9-12, 1991, Eds. G. Pancheri, p. 475.
2. T. Yamazaki, in *Proc. of Whorkshop on Physics and Detectors for DAΦNE*, Frascati, April 9-12, 1991, Eds. G. Pancheri, p. 315.
3. G. Vignola, in *Proc. of Whorkshop on Physics and Detectors for DAΦNE*, Frascati, April 9-12, 1991, Eds. G. Pancheri, p. 11.
4. B. F. Gibson, in *Proc. of International Symposium on Hypernulear and Low-Energy Kaon Physics*, LNL Padova, 12-16 September, 1988, Eds. T. Bressani, F. Cannata, J. Lowe, R. A. Ricci, p.367.
5. J. Cohen, in *Progress in Particle and Nuclear Physics* **vol. 25**, (Pergamon 1990), p. 139.
6. T. Bressani, this volume.
7. S. Marcello, this volume.
8. B.F. Gibson, in *PILAC Users Group report on the Physics with PILAC* **LA-UR-92-150**, 15 october 1991

Common Problems and Trends of Modern Physics, pp. 419–431
ed. T. Bressani, S. Marcello and A. Zenoni
©1992 World Scientific Publishing Company

Phebo: a proposed detector for DaΦne.

G.Batignani, F.Forti, M.A.Giorgi, P.S.Marrocchesi, G.Rizzo and G.Triggiani.
INFN and University of Pisa

July 30, 1992

Abstract

A general purpose detector for the DaΦne Φ-factory is proposed. Emphasis is put on principles rather than on technical details. The Phebo magnetic detector is based on a silicon vertex detector, a gaseous TPC for tracking and a liquid argon tracking calorimeter, that together form a system capable of measuring the CP violating parameters in the $K\bar{K}$ system. We show that with such a system the systematics can be kept as low as the statistical error expected with $10^{40}\mathrm{cm}^{-2}$ integrated luminosity.

1 Introduction

The violation of CP invariance was first observed in 1964 in the two pions decay of the long lived neutral kaon and since then no other particle system has shown measurable effects of CP violation. The violation can be parametrised[1] with two complex quantities, ϵ and ϵ', which are connected with two different physical processes ($\Delta S = 1$ and $\Delta S = 2$), both foreseen by the standard model. The $\Delta S = 2$ violation is present in $K^0 \leftrightarrow \bar{K}^0$ transitions and translates to the fact that the mass eigenstates K_L,K_S are different from the CP eigenstates $K_S = K_1^{CP+} + \epsilon K_2^{CP-}$ and $K_L = K_2^{CP-} + \epsilon K_1^{CP+}$. The $\Delta S = 1$ CP violation corresponds to direct 2π decays of the $CP-$ state K_2 and its relative magnitude is given by the parameter ϵ' which is a function of the decay amplitudes $\langle 2\pi\!:\!I|T|K^0\rangle$ and $\langle 2\pi\!:\!I|T|\bar{K}^0\rangle$ where I is the isospin state of the two π's. The connection of these parameters to observable quantities comes from the relations:

$$\begin{aligned} \eta_{+-} &\equiv \frac{\langle \pi^+\pi^-|T|K_L\rangle}{\langle \pi^+\pi^-|T|K_S\rangle} &\approx \epsilon + \epsilon' \\ \eta_{00} &\equiv \frac{\langle \pi^0\pi^0|T|K_L\rangle}{\langle \pi^0\pi^0|T|K_S\rangle} &\approx \epsilon - 2\epsilon' \end{aligned} \tag{1}$$

The $\Delta S = 2$ violation is well measured, and the present world average of the magnitude and phase of ϵ are[2]:

$$|\epsilon| = (2.265 \pm 0.023) \times 10^{-3}, \phi_\epsilon = 45.9^\circ \pm 1.4^\circ. \tag{2}$$

On the other hand, there is no clear evidence of the $\Delta S = 1$ violation, the present measurements of $\Re(\epsilon'/\epsilon)$ being compatible with sero at the 3σ level: $((2.3 \pm 0.7) \times 10^{-3})$[3] and $((0.6 \pm 0.7) \times 10^{-3})$[4]. The standard model of electroweak interactions foresees a value for $\Re(\epsilon'/\epsilon)$ compatible with the present experimental results, its exact figure being dependent on the (unknown) top quark mass and QCD corrections. DaΦne, the Frascati Φ-factory, which produces Φ's at rest in the laboratory from e^+e^- annihilation, offers a unique laboratory to study CP-violation, providing us with plenty of coherently produced K_LK_S pairs. In Table 1 we summarize the most relevant parameters of the machine.

Circumference	94.56 m
Beam energy	510 MeV
Luminosity	$\approx 10^{33}$ cm^{-2}sec^{-1}
Interaction point spread	2 mm $\times$ 20 μm $\times$ 3 cm
Number of bunches	$1 \div 120$
Crossing half angle	10 mrad
$K_L K_S$ rate	5 KHz

Table 1: Principal parameter of DaΦne.

At DaΦne, with an optimistic, but not irrealistic, integrated luminosity of $\int \mathcal{L} = 10^{40}\text{cm}^{-2}$, the statistical error on $\Re(\epsilon'/\epsilon)$ could be as low as 0.16×10^{-3}. Besides, $\Im(\epsilon'/\epsilon)$ is poorly measured at present ($(-1.7 \pm 12) \times 10^{-3}$), and its error could be reduced by a factor 10 at DaΦne.

In this paper we propose a detector which in our opinion is able to perform these measurements keeping systematic errors as low as the previously quoted statistical one. In section 2 and 3 we describe various methods to measure ϵ'/ϵ and we analyze the sources of errors and the requirements on the experimental setup to keep them under control. In section 4 we describe a detector for DaΦne based on a silicon/gas tracking system and on a liquid Argon TPC. In section 5 we present the results of our simulation of the detector focused on the problem of $\pi\mu\nu$ background rejection.

2 Measurement of ϵ'/ϵ.

There are several ways to measure ϵ'/ϵ at DaΦne, some of which depend on the fact that the initial state is an antisymmetric superposition of K_L and K_S thanks to the C conservation in the strong decay $\phi \rightarrow K^0\bar{K}^0$. This initial state coherence causes quantum mechanical interference patterns in the decay products that can be used to extract CP (as well as CPT or T) violating parameters[1]. The K_L and K_S decays can be treated separately, as in the double ratio method, or can be correlated, requiring a specific final state (for instance $\pi^+\pi^-\pi^0\pi^0$) as in the time and final state asymmetry methods. We assume in all calculations an integrated luminosity of $\int \mathcal{L} = 10^{40}\text{cm}^{-2}$, which corresponds to one year (actually 10^7 s) running at full luminosity and we consider 0.1×10^{-3} as limit for systematic errors on ϵ'/ϵ.

2.1 *Double ratio*

The double ratio (R) is defined as:

$$R = \left|\frac{\eta_{00}}{\eta_{+-}}\right|^2 = \frac{BR(K_L \rightarrow \pi^0\pi^0)BR(K_S \rightarrow \pi^+\pi^-)}{BR(K_S \rightarrow \pi^0\pi^0)BR(K_L \rightarrow \pi^+\pi^-)} = 1 - 6\Re(\epsilon'/\epsilon) \tag{3}$$

Its sensitivity to $\Re(\epsilon'/\epsilon)$ has been exploited in fixed target experiments[6] where K_S and K_L were collected in an uncorrelated fashion at different times and/or different places. At DaΦne, this is the method that in principle could provide the best statistical error, because

- no correlation is required between the K_L and K_S decay modes and all events are acceptable;
- the fortunate factor 6 in the dependance of $\Re(\epsilon'/\epsilon)$ on R yields a free error reduction factor.

The expected statistical error as a function of fiducial radius is plotted in Fig. 1 , showing that with a fiducial radius of 60 cm a statistical error of 0.14×10^{-3} could be obtained.

Unfortunately at DaΦne this method has two important drawbacks. In the first place, as the K_L and K_S are produced and measured at the same time, two distinct fiducial zone must be defined

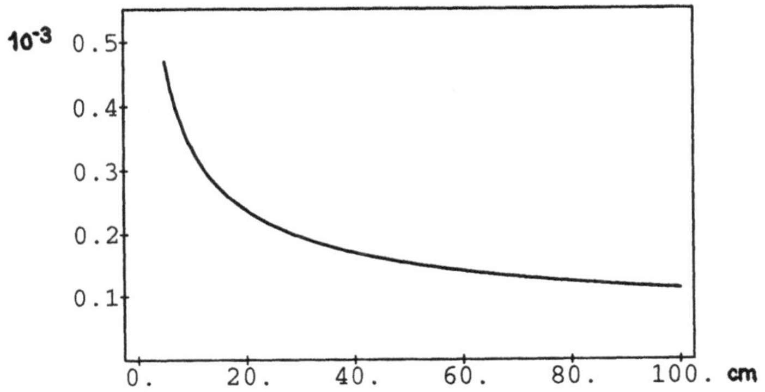

Figure 1: Statistical error on $\Re(\epsilon'/\epsilon)$ vs. fiducial radius for the double ratio measurement

to assign final 2π states to either K_L or K_S. This implies a 10% worsening of the statistical error (0.16×10^{-3} with a 10 cm inner fiducial zone) and of course some more difficulty in the systematic errors control. In the second place, R is sensitive to the trigger and reconstruction efficiency of totally neutral events (i.e. $K_L K_S \to n\pi^0$) versus events with at least on charged particle. Calling ϵ_N and ϵ_C these two efficiencies, the ratio between the measured R and the ideal one can be written as:

$$\frac{R_{meas}}{R_{id}} = \frac{B_S^N \epsilon_N + B_S^C \epsilon_C}{B_L^N \epsilon_N + B_L^C \epsilon_C} \tag{4}$$

where $B_L^{N,C}(B_S^{N,C})$ are the $K_L(K_S)$ branching fractions in neutral and charged final states. If these branching fractions are measured well enough in the same experiment, the requirement on $\Delta_{sys}[\epsilon'/\epsilon] < 0.1 \times 10^{-3}$ translates in a $\Delta[\epsilon_N/\epsilon_C] < 5 \times 10^{-3}$ which may not be easy to obtain. To circumvent this problem one can select events where at least one the kaons undergoes a charged decay, with an approximate 30% ($= BR(K_S \to \pi^0\pi^0)$) loss in statistics on the K_L CP-violating decays counting. In this case the the measured R compares to the ideal one as:

$$\frac{R_{meas}}{R_{id}} = \frac{B_S^C}{B_L^C} \tag{5}$$

requiring a control as accurate as 0.5×10^{-3} on the branching fractions ratio. The attainable statistical error in this hypothesis is $0.18 \div 0.2 \times 10^{-3}$.

2.2 *Final state asymmetry*

One way to overcome the two fiducial regions problem and to exploit the initial state coherence is to construct the quantity[7]:

$$\Omega(X_1, X_2) \equiv \frac{N(X_1, X_2)}{BR(K_S \to X_1)BR(K_S \to X_2)} \tag{6}$$

where X_1, X_2 are any final states and $N(X_1, X_2)$ is the number of events in which both X_1 and X_2 originated in a particular time interval. The time interval is not indicated in the formula because the measurable asymmetries:

$$\mathcal{A}_F^{(+-+-,0000)} \equiv \frac{\Omega(\pi^+\pi^-,\pi^+\pi^-)}{\Omega(\pi^0\pi^0,\pi^0\pi^0)} = 1 + 6\Re(\epsilon'/\epsilon) \tag{7}$$

$$\mathcal{A}_F^{(+-+-,+-00)} \equiv \frac{\Omega(\pi^+\pi^-,\pi^+\pi^-)}{\Omega(\pi^+\pi^-,\pi^0\pi^0)} = 1 + 3\Re(\epsilon'/\epsilon) \tag{8}$$

are independent of it. Here the requirement of a specific final state reduces the statistics, but only one fiducial sone is needed because the final state is considered globally and not divided in K_L and K_S decay products. For eq. (7) there applies the same argument on detection efficiency of totally neutral events, worsened by the fact that now these events constitute the main signal. The limit on $\Delta_{sys}[\epsilon'/\epsilon]$ requires in this case $\Delta[\epsilon_N/\epsilon_C] < 10^{-3}$. With the same 60 cm fiducial volume as before $\Delta[\Re(\epsilon'/\epsilon)] = 0.23 \times 10^{-3}$. For eq. (8) the neutral events can be disregarded altogether by evaluating the K_S branching fractions only on events with charged K_L decays. Of course the loss in statistics on K_S decays doesn't affect the error, which is mainly determined by K_L CP-violating decays counting. In this case the statistical error becomes 0.27×10^{-3}.

2.3 *Time asymmetry*

Here we can make full use of the coherence of the initial state analysing its time evolution. Let's consider the final state $\pi^+\pi^-\pi^0\pi^0$ and let t_{+-}, t_{00} be the charged and neutral decay time respectively. The probabibilty density of having such an event turns out to be the product of a function depending on the sum of the two times and a function depending only on their difference. It is proportional to[8]

$$P(t_{+-}, t_{00}) \propto e^{\frac{-\Gamma t}{2}} \left|3\epsilon'/\epsilon \cos(\Delta t \Delta\lambda) + i(2 - \epsilon'/\epsilon)\sin(\Delta t \Delta\lambda)\right|^2 \tag{9}$$

where $\Gamma = \Gamma_S + \Gamma_L$; $\Delta\lambda = M_S - M_L - i(\Gamma_S - \Gamma_L)$; $t = t_{+-} + t_{00}$ and $\Delta t = t_{+-} - t_{00}$. Integrating in t as far as the fiducial radius allows, we obtain $f(\Delta t) = \int_{|\Delta t|}^{\mathrm{Rfid}} P(t_{+-}, t_{00})dt$ which is plotted in Fig. 2 for two values of ϵ'/ϵ and has the characteristic dip at $\Delta t = 0$. There are two ways to extract ϵ'/ϵ from the measurement of Δt: either with the integral of $f(\Delta t)$, or fitting the curve shape. The latter method gives information both on the real part and on the imaginary part of ϵ'/ϵ.

Integrating over positive and negative Δt one can construct a time asymmetry which is sensitive to $\Re(\epsilon'/\epsilon)$:

$$\mathcal{A}_T \equiv \frac{N(\Delta t > 0) - N(\Delta t < 0)}{N(\Delta t > 0) + N(\Delta t < 0)} = 3\Re(\epsilon'/\epsilon) \tag{10}$$

The statistical error can be computed by considering that, except for the interference region near the origin which has a small weight in the integral, $N(\Delta t > 0) \approx N(K_L \to \pi^+\pi^-, K_S \to \pi^0\pi^0)$ and the analogue for $N(\Delta t < 0)$. With 60 cm fiducial volume $\Delta[\epsilon'/\epsilon] = 0.2 \times 10^{-3}$.

Fitting the curve provides another method of measurement which is sensitive to both $\Re(\epsilon'/\epsilon)$ and $\Im(\epsilon'/\epsilon)$. For the latter, only the region near the origin is relevant for the measurement, whose sensitivity is dominated by the resolution on the neutral vertex reconstruction which determines the precision of our knowledge of Δt. As can be seen from Fig. 3 , a 5 mm neutral vertex resolution is needed to obtain a sensitivity to the imaginary part (2×10^{-3}) close to the best possible value obtainable with this experiment. Besides the vertex resolution, the error on Δt is also affected by the beam spread that, in the plane of the machine, is about 2 mm in x and 3 cm in z, implying an angular cut around the horizontal plane. It is crucial, that the y beam spread is actually the design 20 μm value. For the real part, the vertex resolution is not relevant, and with a fiducial radius of 60 cm we obtain an error of 0.2×10^{-3} which is comparable with the one obtained with the other methods.

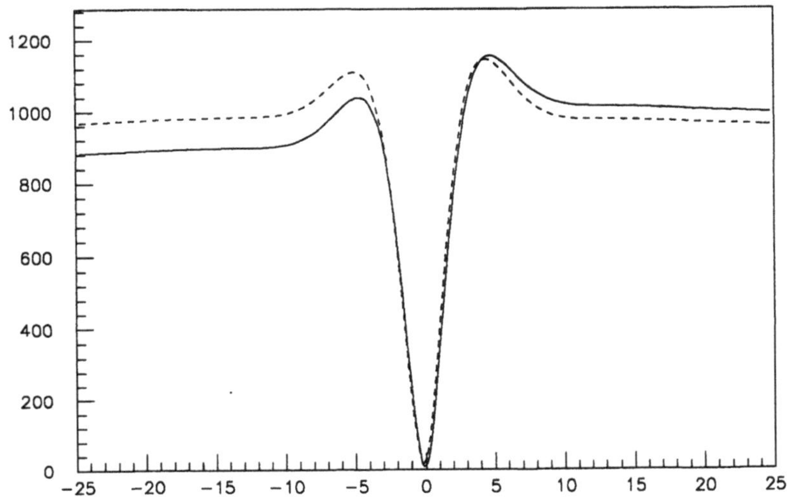

Figure 2: $f(\Delta t)$ with Δt in units of the K_S lifetime τ_S for two cases: $\epsilon'/\epsilon = 0.02$(continuous line) and $\epsilon'/\epsilon = 0.05i$ (dashed line).

3 Sources of systematic errors

It is extremely important that experiments at DaΦne keep systematic errors under control. Failure to do so may spoil completely the precision on the measurement of ϵ'/ϵ. In this section we will go through the sources of systematics errors we have spotted and discuss how they could be dealt with in the experiment. The detector uniformity and calibration play an obvious but never overestimated role. In the detector proposed in this paper we tried to use continuos media whenever possible, to avoid fragmentation and cracks.

3.1 Regeneration

Regeneration of K_L traversing part of the detector material is an important source of background for CP-violation measurement. The overall effect is a dilution of the signal that must be corrected for, thus requiring a good knowledge of the regeneration cross section in the various materials, which is *not* available at this energy and must be measured in the same apparatus. The size of the effect is determined by the ratio between the K_L decay length ($\lambda_L = 342$ cm) and the regeneration length $\lambda_R(x) = (\sigma n)^{-1}$ where σ is the regeneration cross section and n is the volume density of scattering centers. The position dependance of λ_R represents the distribution of material inside the experimental setup. Indicating with $N(x)$ the number of K_L surviving at a certain distance x from the origin, the number of final states of type α (i.e. $\pi^+\pi^-$ or $\pi^0\pi^0$) coming from K_L decaying or regenerating in the fiducial region (extending from the origin to R_F) can be written as:

$$N_L^\alpha = \int_0^{R_F} N(x) B_L^\alpha \frac{dx}{\lambda_L} \times \left[1 + \underbrace{\frac{\lambda_L}{\lambda_R}\frac{B_S^\alpha}{B_L^\alpha}}_{\kappa^\alpha}\right] \approx N^\alpha_{L(\text{vacuum})}(1+\kappa^\alpha) \qquad (11)$$

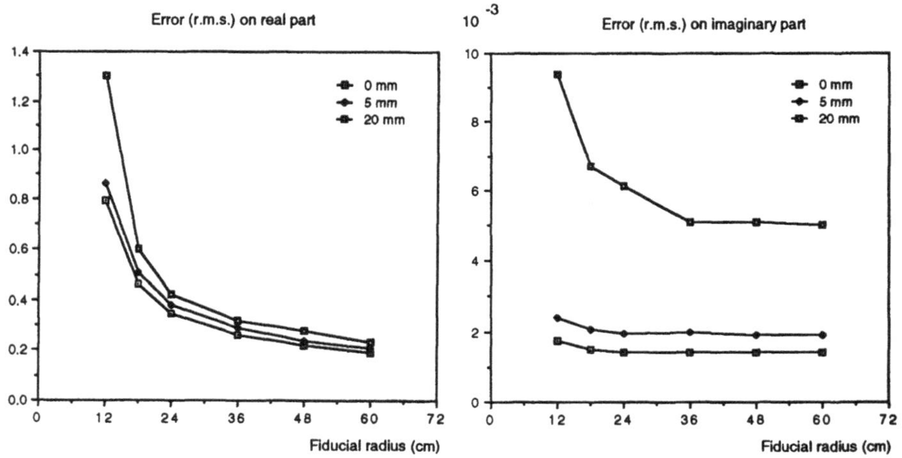

Figure 3: Statistical errors on $\Re(\epsilon'/\epsilon)$ (a) and $\Im(\epsilon'/\epsilon)$ (b) obtained fitting the asymmetry function as a function of the fiducial radius and the resolution on the neutral vertex reconstruction.

where $B^{\alpha}_{(L,S)}$ are the branching ratios of $K_{(L,S)}$ to a final state α and

$$\frac{1}{\overline{\overline{\lambda_R}}} = \int_0^{R_F} \frac{N(x)}{\lambda_R(x)} dx \Big/ \int_0^{R_F} N(x) dx \tag{12}$$

is the average inverse regeneration length weighted with the number of K_L actually crossing a given material. In first approximation and for small distances this is just the average inverse regeneration length. The approximation in eq. (11) amounts to neglecting in $N(x)$ the K_L's disapperaring because of regeneration respect to those decaying, which can be done as long as light and thin materials are used in the experiment. If we now insert eq. (11) into the double ratio expression and expand all terms up to first order in ϵ'/ϵ, we end up with the following relations, where a † indicates quantities measured with regeneration present, and a $\Delta[...]$ represents the statistical error on a quantity:

$$\Re(\epsilon'/\epsilon)^\dagger = (\kappa+1)(1-R^\dagger) \tag{13}$$
$$\Delta[R^\dagger] = \Delta[R]/\sqrt{\kappa+1} \tag{14}$$
$$\Delta[\Re(\epsilon'/\epsilon)^\dagger] = \sqrt{\kappa+1}\Delta[\Re(\epsilon'/\epsilon)] \tag{15}$$
$$\Delta_{sys}[\Re(\epsilon'/\epsilon)^\dagger] = \Re(\epsilon'/\epsilon)\Delta[\kappa]/(\kappa+1) \tag{16}$$

Here κ stands for κ^{+-}.

The apparent statistical improvement on $\Delta[R^\dagger]$ is due to the increased number of events recorded. Of course the error on ϵ'/ϵ, which takes into account the dilution factor is worse than for vacuum. These results imply that:

- regeneration must be kept small ($\kappa \leq 0.5$) to avoid blowing up the statistical error.
- the regeneration factor must be well known ($\Delta\kappa/(\kappa+1) \leq 0.1$) to avoid systematic effects.

Material	Berillium	Silicon	Neon(21°C)
λ_R(cm)	300	230	6.6×10^5

Table 2: Approximate values of K_L regeneration length for materials of interest.

An estimate of $\overline{\lambda_R}$ is not trivial because data at a kaon momentum of 110 MeV/c are not currently available. A very crude extrapolation of present data to the low energy limit ($\sigma(K_L\,p \rightarrow K_S\,p) \approx 3$ mb) and from proton to nuclei ($\sigma \propto A$) gives the results shown in Table 2 where a factor 2 uncertainty is more than likely. For the experimental setup proposed in this paper $\kappa \approx 0.1$, while the presence of localised layers of material (the silicon vertex and the TPC inner field cage) allows the measurement of the regeneration cross section by counting the excess of CP-violating decays coming from these layers.

3.2 Backgrounds

Three-body decays of the K_L where one of the final particles is undetected constitute the most serious background to the measurement[8]. Processes like $K_L \rightarrow 3\pi$ where one π^0 is missed because of detector acceptance and efficiency, or $K_L \rightarrow \pi l \nu$ where the lepton fakes a pion, are not CP-violating and their rate is much higher than the 2π decay rate, asking for a rejection factor that ranges from 10^{-4} to 10^{-5} to have a corresponding systematic error, due to the background subtraction, in the range $0.5 \div 1\times 10^{-4}$. This rejection factor is particularly hard to obtain for the $K_{\mu 3}$ decay because at these low energies the positive identification of μ's is not possible with ordinary absorption tecniques used in high energy experiments. Of course, the three-body decay has a fairly different kinematics with respect to the 2π decay, thus allowing the use of invariant mass and opening angle cuts to reject muons. Although these methods are quite effective, they pose severe requirements on the tracking system, and anyway don't seem to be able to do better than a 10^{-3} rejection factor (see section 5). As a matter of fact, we believe that a hardware identification of μ's is needed, thus our detector includes a liquid argon tracking calorimeter (see section 4.3) which is able to give a 10^{-2} additional rejection factor which is completely independent from the kinematical cuts.

3.3 Absolute scale

We have seen that the determination of the charged and neutral decay vertices of the kaons is crucial both to measure Δt and to ensure containment inside the fiducial volume. Because the two vertices are measured in utterly uncorrelated ways (one with the tracking devices and the other with the calorimeter), the problem arises of the systematic differences between the two measurements, which affect all methods of measurement we have examined so far. Let's consider the situation depicted in Fig. 4 in which the reconstructed position of the charged and neutral are pictorially shown systematically different from the real position of the vertex. Of course, in real life we do not expect an exactly linear relationship between the real and reconstructed vertices positions. For the double ratio method, the relevant quantities are the differences $\delta_{(1,2)}$ between the charged and neutral vertex reconstructed positions at the two fiducial radii R_1 and R_2 used to define the K_S and K_L zones, while for the time asymmetry method only δ_2 matters because only one fiducial zone is defined. For the double ratio we have to count the number of K_L that decay in 2π in the region $R_1 < r < R_2$, and this number is altered by the shift of the boundary induced by the uncertainty on the absolute scale. The double ration R is shifted by a quantity

$$\Delta[R] \approx \frac{\delta_2 - \delta_1(1+\frac{R_2-R_1}{\lambda_L})}{R_2 - R_1} \tag{17}$$

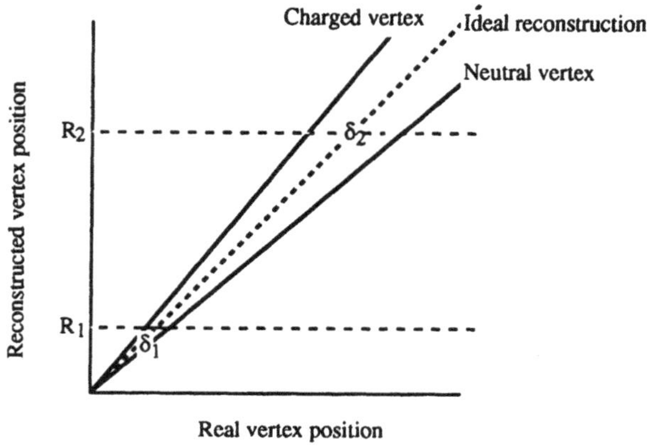

Figure 4: Pictorial representation of quantities used in the discussion of the absolute scale error.

while the time-asymmetry defined in section 2.3 is shifted by $\Delta[\mathcal{A}_T] \approx \frac{\delta_2}{R_2}$. For $R_1 = 10$ cm and $R_2 = 60$ cm and asking for a maximum 10^{-4} error on $\Re(\epsilon'/\epsilon)$ the requirements on $\delta_{(1,2)}$ become:

$$\delta_1 < 300\ \mu\text{m}$$
$$\delta_2 < \begin{cases} 350\ \mu\text{m} & \text{for double ratio} \\ 200\ \mu\text{m} & \text{for time asymmetry} \end{cases}$$

4 Proposed detector.

The requirements on the detector that emerge from the preceding discussion can be summarised as follows:

- The detector must have a homogeneous response and cover as much solid angle as possible to guarantee high acceptance and uniformity in charged and neutral measurements.
- Its fiducial region must be extremely transparent to avoid multiple scattering and keep regeneration under control.
- It must be able to reject backgrounds, in particular the $K_L \to \pi\mu\nu$ decay to a level of $10^{-4} \div 10^{-5}$.
- The absolute scale of the reconstructed vertices must be under control a level of about 200 μm

We propose a conceptual detector based on three main components: a silicon vertex detector, a gaseous TPC tracking system and a liquid Argon tracking calorimeter. In Fig. 5 a sketch of the detector used in the simulation is shown. Of course it is extremely schematic and does not contain engineering details.

In the rest of this section we will go through the three components of the detector illustrating their characteristics and functions. No attempt is made here to describe a trigger system, that could be based on ad hoc plastic scintillator counters placed outside the TPC and/or on the prompt scintilation light and electrical signal in the liquid argon TPC.

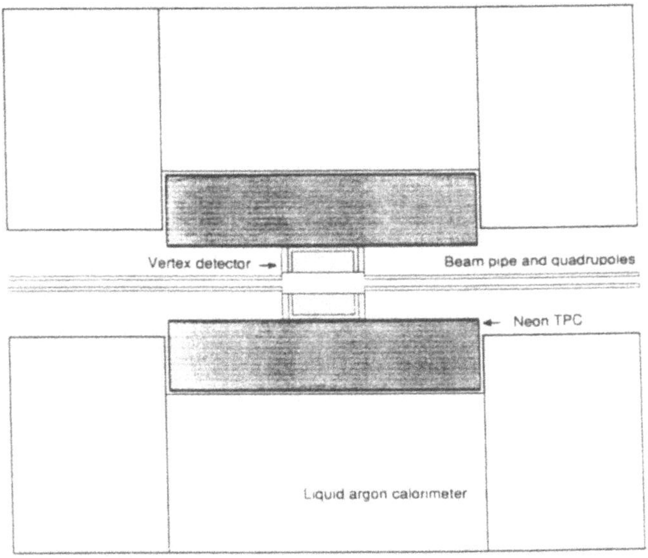

Figure 5: Proposed detector layout. The overall diameter is about 5 m not including the coil.

4.1 Interaction zone and silicon vertex

At the core of the detector, outside the beam pipe, we foresee a vacuum region of about 30 cm that acts as a superfiducial zone, meaning that the absence of interactions and regeneration allows a clean measurement of interference and rates. The silicon vertex, which has a $\approx$ 100% efficiency for charged particles detection, is inside this vacuum zone, determining very precisely whether a charged vertex is inside or outside the zone.

The silicon vertex is made of two layers of double side readout silicon strip detectors which allow for a measurement of two spatial points along the charged tracks' trajectories. The thickness of these layer is about 200 μm each and the single point resolution better than 50 μm. The additional multiple scattering introduced by this component is largely compensated by the improved track and vertex reconstruction.

4.2 The tracking system

Immediately outside the vertex detector, at a radius from 35 cm to about 1 m and for a length of about 3 m in z, it is placed the Neon Time Projection Chamber[9]. The main advantage of this tracking system is the absence of longitudinal wires, i.e. homogeneity and uniformity of response and thin endplates that need not stand the wire tension. Besides, a low-A material can be used, resulting in a good momentum resolution, which at these low energies is mainly limited by multiple scattering. Finally, the simultaneous measurement of the z coordinate is essential for the 3-d reconstruction of the charged vertex. On the cons side there is the intrinsic slowness of this kind of detector, due to the long drift time; this may create event pileup and poses problems with the gating needed to remove the space charge in the gas. Nevertheless, at a luminosity $\mathcal{L} = 10^{33}$ cm^{-2}sec^{-1} the expected event rate is of about 5 KHz that, with a 50 μs drift time, correspond to 25% duty cicle, which we think feasible[10].

4.3 The liquid Argon tracking calorimeter

The electromagnetic calorimeter plays a crucial part in an experiment for DaΦne. In the previous chapters we demonstrated that the statistical and systematic error limits on the ϵ'/ϵ measurement are dominated by the space resolution on the neutral decay vertex reconstruction. This translates into efficiency and resolution requirements on the calorimeter which are hardly met by any device commonly used in high energy physics experiments.

First of all, the choice of a continuously sensitive medium is mandatory, in order to have the best possible energy measurement. Fine grained readout is then necessary, allowing for a very good space resolution. We choose a solution based on a liquid noble gas TPC as developed (with argon) by the Icarus[11] collaboration at CERN. Such a detector acts as an electronic bubble chamber with astonishing resolution both in energy (4% at 1 MeV) and position ($600 \times 600 \times 60\ \mu m^3$). Another crucial feature of this device is the capability, using the range and the specific ionization for charged particles, to discriminate π from μ to a level better than 10^{-2} at the DaΦne energies. This feature allows one to push the overall $K_{\mu 3}$ event rejection (kinematical cuts included) to a suitable level for our experiment.

We found two items preventing us from developing immediately an actual design for this calorimeter, namely:

- with the smallest cell size we can imagine (15 cm. drift) the charge collection time for this device is $\approx 300\mu s$ in pure argon
- the volume occupied by the liquid argon forces the use of a large coil, this can be reduced a lot using liquid Kripton instead.

We are confident that R&D on doping mixtures, readout techniques and the use of liquid Kripton can solve these problems, but the time needed for that makes impossible the installation of this device for the first data taking period foreseen at DaΦne.

5 Results of simulation.

The simulation of the detector has been implemented with the aid of the Geant 3.14 detector simulation package[12]. The $K_L K_S$ pair was produced with the right angular distribution and a beam spread of 2 mm $\times$ 20 μm $\times$ 3 cm in x, y and z respectively. Then the kaons where tracked in the detector, described as a number of cylinders as in Fig. 5 , taking into account possible interactions and decays. For the K_{l3} decay the proper $V-A$ matrix element, which plays an important role in background rejection, was inserted.

For the tracking devices, the charged particle crossing point on the cylinder, properly smeared with the detector resolution, was retained and afterwards used in the helix fit of the track. The helix parameters and their covariance matrix are then used to fit charged vertices or make invariant masses, as explained in the following. Electromagnetic and hadronic showers are generated and followed in the calorimeter with the EGS and Gheisha packages respectively. The energy depositions are summed up taking into account the electronics noise of the readout and the recombination in the liquid. For the moment only integral information are retained from the liquid argon calorimeter and its tracking capabilities are not exploited in the simulation.

We have tackled the problem of $K_{\mu 3}$ background rejection. As discussed in section 3 this is the most serious source of background for the ϵ'/ϵ measurement. To estimate the rejection power of kinematical cuts in a realistic way, we have inserted in Geant the correct $V-A$ matrix element for the $K \rightarrow \pi\mu\nu$ decay [2]. The matrix element effect is to somewhat deplete the $\pi - \mu$ angle distribution at 180° in the center of mass with respect to pure phase space, thus enhancing the kinematical difference from the signal $\pi^+\pi^-$. In the analysis we considered the case where no hardware muon identification was available, attributing the π mass to muons. The variables used for background

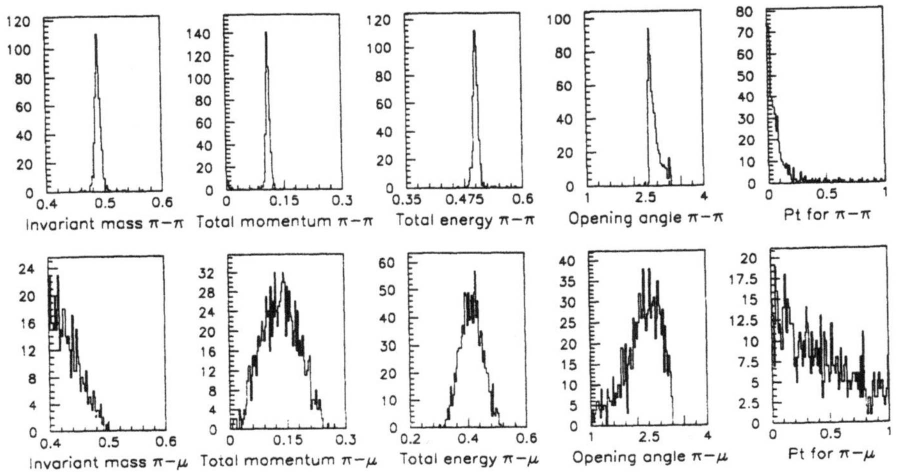

Figure 6: Distribution of the kinematical variables used for $K_{\mu 3}$ background rejection.

discrimination are based on the kinematical properties of the two particles in the final state ($\pi^+\pi^-$ or $\pi\mu$) and are defined as follows:

M_{12}: invariant mass of the two particles.

E_{12}: sum of the energies of the two particles.

θ_{12}: opening angle in laboratory system.

$p_{12} \equiv |\vec{p_1} + \vec{p_2}|$: module of the sum of the momenta of the two particles.

$p_{12}^T \equiv (p_1^T - p_2^T)/(p_1^L + p_2^L)$: difference of the transverse momenta of the two particles with respect to the K_L direction normalised to the sum of the longitudinal momenta. The direction of the K_L is the vector sum of the two 3-momenta.

Of course the rejection power of these variables depends strongly on the momentum resolution of the tracking system. In Fig. 7 the distribution of these quantities for $\pi^+\pi^-$ and $\pi\mu$ is shown in the case of $\Delta p/p = 2\%$. The applied cuts, chosen to keep 99.7% of the signal (3σ), are superimposed on the figure. The total contamination factor, defined as

$$CF = \frac{(\text{number of } \pi\mu\nu \text{ left})}{(\text{total number of } \pi\mu\nu)}$$

is shown in Fig. 8 for different momentum resolution in the range $0.2 \div 5\%$; the best performance that can be reasonably obtained, with $\Delta p/p = 0.2\%$, is 0.8×10^{-3}, which is still far from the 10^{-4} factor required to keep systematics under control. For our detector, that has a momentum resolution of about $1.5 \div 2\%$ the rejection factor lies in the range $(4 \div 8) \times 10^{-3}$. It becomes evident that an uncorrelated and powerful hardware μ identification is crucial for the detector performance and for the overall accuracy of the measurement.

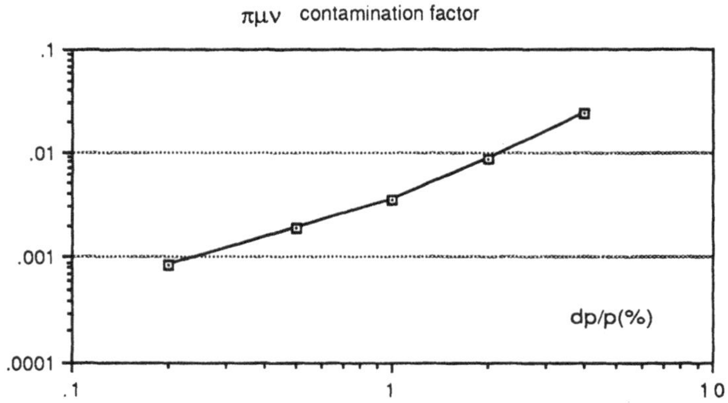

Figure 7: $\pi\mu\nu$ contamination factor for different momentum resolution.

6 Conclusions.

We have presented the Phebo general purpose detector for DaΦne. Phebo employs some of the most advanced detection techinques and, although some R&D is certainly needed before the design can be finalised, we believe that such a detector can cope with the challenging measurements to be performed at a Φ-factory.

References

1. C.Buchanan, R.Cousins, C. Dib, R.D. Peccei anb J.Quackenbush, "Testing CP and CPT Violation in the Neutral Kaon System at a Φ-Factory.", Preprint UCLA/91/TEP/44.
2. Particle data group, Phys. Lett. **239B** (1990) 1.
3. NA31 Results presented by G. Barr at the 1991 Symp. on Lepton and Photon Int. at High En., Geneva, Switzerland. July 23 - August 1, 1991.
4. E731 Results presented by B. Winstein at the 1991 Symp. on Lepton and Photon Int. at High En., Geneva, Switzerland. July 23 - August 1, 1991.
5. Proceedings of the "Workshop on physics and detectors for DaΦne", edited by G.Pancheri, INFN-LNF 1991.
6. A.Nappi, "ϵ'/ϵ Measurement at hadron machines" in "Proposal for a Phi Factory", LNF 90/031/R.
7. J.Bernabeu, F.J. Botella and J. Roldán, "ϵ'/ϵ and coherent decay of the K^0–$\bar{K}^0$ system" Phys. Lett. **211B** (1988) 226.
8. G.Batignani et al., "ϵ'/ϵ measurement in the process $\phi \to K_L K_S \to \pi^+\pi^-\pi^0\pi^0$", in [5], p. 505.
9. W.B.Atwood et al. "Performance of the Aleph Time Projection Chamber", N.I.M. **A306** (1991) 446.

10. G.Batignani et al., "Preliminary study for the design of a tracking system at DaΦne", in [5], p. 525.

11. "The Icarus R&D program", talk presented by S.Suzuki at the International Conference on liquid radiation detectors, Tokio, Apr. 92.

12. CERN report DD/EE/84-1

V. EXPERIMENTAL TECHNIQUES

Common Problems and Trends of Modern Physics, pp. 435–448
ed. T. Bressani, S. Marcello and A. Zenoni
©1992 World Scientific Publishing Company

GENERAL ASPECTS OF CALORIMETRY AT LOW ENERGIES

M. BRUSCHI and A. ZOCCOLI
Dipartimento di fisica dell'Università di Bologna, and
Istituto Nazionale di Fisica Nucleare , Sezione di Bologna, Italy

ABSTRACT

We describe some general aspects of electromagnetic calorimetry at low energies to show the importance of electromagnetic calorimeters for precision measurements and we discuss the physics processes, the designing parameters and the performances that are relevant for such detectors. Moreover, we analyse the foreseen behavior of the electromagnetic calorimeter of the Obelix experiment by means of extensive MonteCarlo simulations.

1. Introduction.

Calorimeters in elementary particle physics are devices in which the interacting particles deposit almost completely their energy. This happens in a cascade process or "shower" which produces particles of increasingly lower energy. In general a very little amount of the deposited energy is released as detectable signals [1]. These detectors are based on stocastic processes, hence it's possible to predict only the average properties of the shower by means of suitable parameters.

Calorimeters are used in a wide range of experiments because of their interesting characteristics:

1) They are sensitive both to neutral and charged particles;

2) The longitudinal dimension of the detector increases as $\ln E$ (where E is the particle energy), and their energy resolution scales as $1/\sqrt{E}$

3) They can measure the position and the direction of the incoming particles;

4) They can provide different responses depending on the type of particle interacting;

5) They can provide fast time signals for trigger purposes.

Calorimeters can be subdivided into two classes : ELECTROMAGNETIC and HADRONIC. The first ones are employed both in low and high energy physics with the main purposes of detecting gamma rays and of discriminating electrons from pions. The second-ones are used mainly in high energy physics, where the magnetic spectrometers cannot work properly because of their dimensions.

In this paper we point out our attention to the electromagnetic calorimetry in the low energy region. In section 2 we discuss the purposes of electromagnetic calorimeters, in sections 3 and 4 the physics processes and the shower development, in section 5 the spatial and energy resolution for such detectors and, finally, in section 6 we describe the performances of the electromagnetic calorimeter of the Obelix experiment (HARGD) as foreseen by extensive MonteCarlo simulations.

2. Calorimeter applications.

In low energy physics (for example in Meson Spectroscopy) there are many reactions involving the production of neutral particles, like Pions (π^0) and Etas (η^0), that decay electromagnetically into photons. For a precise analysis of these reactions it is necessary to recognize the neutral particles and measure their physics characteristics by detecting and measuring gamma rays with a good precision.

As an example we can consider the π^0 [2], the lightest particle that decays into two photons, that is produced in the 85% of the $P\overline{P}$ annichilation at rest.

The decay angle between the two gamma rays from a π^0 is strongly correlated with the pion momentum through the Jacobian peak distribution (see FIG. 1). This distribution increases sharply near the minimum decay angle given by the relation (c = 1) :

$$\cos \phi_{min} = 2\beta^2 - 1 \tag{1}$$

with

$$\beta = \frac{p}{\sqrt{m^2 + p^2}} \tag{2}$$

where m is the rest mass of the π^0 and P is the module of its momentum. The pion mass and momentum are obtained by measuring the two decay photons. The three observed quantities, the energy E_1 and E_2 of the photons, and the angle ϕ between them determine the invariant mass m_o of the two gamma rays:

$$m_o^2 = 2\, E_1\, E_2 \,(\, 1 - \cos\phi \,) \tag{3}$$

The mass resolution is given by the relation:

$$\frac{\Delta m_o}{m_o} = \frac{1}{2}\sqrt{\left(\frac{\Delta E_1}{E_1}\right)^2 + \left(\frac{\Delta E_2}{E_2}\right)^2 + \left(\frac{\Delta\phi}{\operatorname{tang}\frac{\phi}{2}}\right)^2} \tag{4}$$

At very low momenta the invariant mass resolution is dominated by energy resolution, whereas at higher momenta the angular resolution becomes important. Therefore to achieve a satisfactory reconstruction on a wide class of reactions it is necessary to use calorimeters with the following characteristics:

1) High solid angle covered (to achieve a good global efficiency).

2) Good detection efficiency (to measure the photons with a satisfactory precision).

3) Good spatial resolution (to measure the photon direction with a small error).

4) Good energy resolution (to measure the photon energy with a small error).

5) Good trigger capability to select with topological or energetic constraints a given reaction [3].

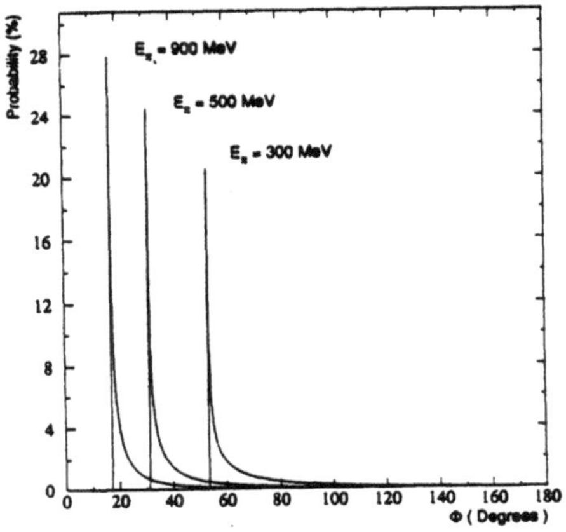

FIG. 1 Distribution of the decay angle between the two photons from three different π^0 total energies.

From the structural point of view calorimeters can be subdivided into two classes: HOMOGENEOUS calorimeters and SAMPLING calorimeters. The first ones are devices built enterely of active materials, ensuring, at low energies, optimal energy resolution from statistical point of view. To obtain the best performances they must contain electromagnetic showers as completely as possible, then it is necessary to use relatively large detector volumes placed near the interaction point. The best materials are heavy scintillators such as BGO ($Bi_4Ge_3O_{12}$),NaI,CsI, and Lead Glass (55% PbO, 45% SiO_2). The second-ones are employed when a good containment requires a too high volume, inacceptable both from a mechanical and a cost point of view. In these detectors converter materials (Lead, Iron, Uranium) are alternated with active materials (as liquid and solid scintillators or ionizing gases). In general they are placed far from the interaction point and can achieve good angular resolutions with a detriment to the energy resolution.

3. Physics Processes

Electromagnetic showers can be fully described by Quantum Electrodynamics (QED) [4,5] and evaluated by Feynman diagrams.

The most important reactions which undergo gamma rays (γ) and electrons / positrons ($e^{\pm}$) in a calorimeter are the following:

1) Pair production
2) Compton scattering
3) Photoelectric effect
4) Bremsstrahlung
5) Ionization
6) Bhabha scattering
7) Møller scattering
8) Pair annichilation

The first three points are referred to γ, the others to $e^{\pm}$. The dependence of these effects on the energy is shown in FIG. 2 [2]. As one can see, the cross section for pair production become predominant for photon energies greater than 10 MeV and saturates at about 100 MeV. For electrons and positrons the two predominant effects are Bremmsstrahlung at energies greater than $\approx$ 10 MeV and Coulomb scattering at lower energies.

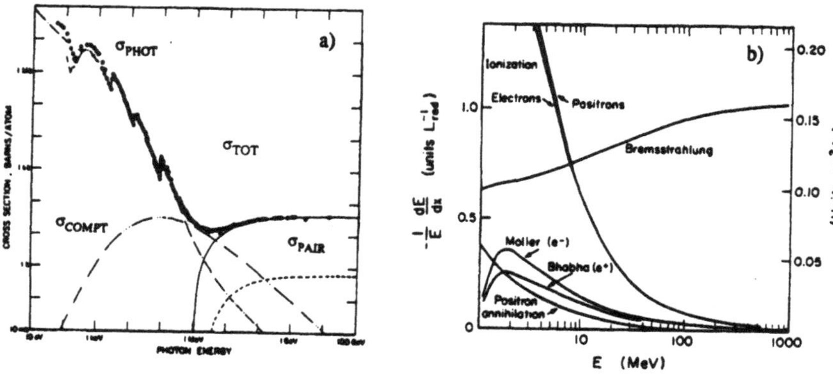

FIG. 2 a) Photon cross section σ in lead as a function of the photon energy [2].
b) Fractional energy loss per radiation length (left ordinate) and per gr/cm^2 (right ordinate) in lead as a function of electron and positron energies.

In the following paragraphs we will introduce some useful quantities for the description of calorimeters by studying the most important physics processes.

3.1 The Radiation Length.

The energy loss by electrons and positrons in Bremsstrahlung processes is given by the equation :

$$-\frac{dE}{dx} = 4\alpha N_o r_e^2 \frac{Z^2}{A} E \ln\left(\frac{183}{Z^{1/3}}\right) = \frac{E}{X_o} \qquad (5)$$

where α is the fine structure constant, N_o is the Avogadro number; Z,A are the atomic number and weigth , and $r_e = e^2/mc^2 \approx 2.8$ fm is the classical radius of the electron. Thus the energy loss is proportional to the energy through the constant X_o, called RADIATION LENGTH (X_o) of the material (depending only on the characteristics of the medium), defined as :

$$X_o = \frac{1}{4\alpha N_o r_e^2 \ln\left(\frac{183}{Z^{1/3}}\right)} \cdot \frac{A}{Z^2} \cong 180 \frac{A}{Z^2} \qquad (gr/cm^2) \qquad (6)$$

and corresponding to the thickness of medium which reduces the mean energy of an electron (positron) by a factor " e " for radiative processes.This relation is approximated

and shows an inverse dependence of the radiation length on the square of the atomic number.
This parameter is very useful in the description of calorimeters because it is also related to the cross-section for pair-production. In fact in the saturation region the cross-section can be written as:

$$\sigma_p \cong 4\,\alpha\, r_e^2\, Z^2\, \frac{7}{9}\, \ln \frac{183}{Z^{1/3}} \cong \frac{A}{N_o}\, \frac{7}{9}\, \frac{1}{X_o} \qquad (7)$$

Hence we can write the photon mean free path for pair production as:

$$\lambda = \frac{1}{N\,\sigma_p} = \frac{A}{N_o\,\sigma_p} \cong \frac{9}{7}\, X_o \qquad (\mathrm{gr/cm^2}) \qquad (8)$$

and it is possible to define the Radiation Length also as the thickness of material where a pair creation happens with a probability $P = 1 - e^{-7/9} = 54\%$ at high photon energies.

3.2 The Critical Energy.

The development of an electromagnetic shower stops when the energy lost by electrons in bremsstrahlung processes becomes comparable with the energy lost by ionization processes. The energy of an electron that looses as much energy in collision as in radiation is called CRITICAL ENERGY (ε).

Let us recall now the Bethe-Bloch formula for electrons:

$$-\frac{dE}{dx} = 4\,\pi\, N_o\, r_e^2\, m_e\, c^2\, \frac{Z}{A} \left(\ln\left(\frac{2 m_e v^2 \gamma^2}{I}\right) - 1 \right) \qquad (9)$$

where m_e is the electron mass, v is the electron velocity and I is the ionization potential of the material. By equalizing the relations (9) and (5) we obtain the approximated formula for the critical energy:

$$\varepsilon \cong \frac{550\ \mathrm{MeV}}{Z} \qquad (10)$$

A good conversion material should have a little ε to maximize the number of electrons and positrons produced in the shower. Therefore, high atomic number materials should be preferred.

3.3 The Molière radius.

The lateral spread of the showers is caused both by the average angle for Bremsstrahlung emission ($\theta_{Brem} \approx m_e/E_e$), and by the Coulombian multiple scattering in the converter. This effect becomes relevant with the decrease of the energy of the shower particles, when the radiative processes are negligible but the $e^{\pm}$ have enough energy to move away from the shower axis.

A useful parameter to evaluate this effect is the MOLIERE RADIUS defined as:

$$R_M = \frac{E_s}{\varepsilon} X_o \cong 7\frac{A}{Z} \qquad (gr/cm^2) \tag{11}$$

where E_S = 21 MeV is determined by the Molière theory of multiple scattering. This parameter is related to the average lateral deflection of $e^{\pm}$ of energy ε after one radiation length, and it is possible to show that the 99% of the energy of a shower is contained in a cylinder having a radius of $3R_M$.

4. Shower developement.

Now we can describe the developement of an electromagnetic shower as the result of the combined phenomena of Bremsstrahlung, pair production and energy loss by ionization.

In a very simplified model we can consider a primary gamma ray of energy E_o traveling in a material. After one radiation length (R.L.) it produces a pair of $e^{\pm}$ each-one receiving half energy ($E_o/2$). Assume that in the next R.L. each $e^{\pm}$ radiates half of its energy ($E_o/4$) as one photon. Thus after two R.L. we will have two photons of energy $E_o/4$, one electron and one positron each of $E_o/4$. Proceeding in this way it is possible to see that after n R.L. there will be $N = 2^n$ particles of energy $E_N = E_o/n$. This process continues until E_N become smaller than the critical energy ε ,after this it finishes sharply. Thus the shower maximum will be reached at a depth (in units of X_o) :

$$n = \ln\left(\frac{E_o}{\varepsilon}\right)/\ln 2 \tag{12}$$

with a number of particles: $N_P = 2^n = E_o/\varepsilon$.[4] We can now write the total track-length (S) of the charged particles of the shower as:

$$S = \frac{2}{3}X_o\sum_{i=1}^{n} 2^i + \frac{2}{3}S_o N_P \cong \left(\frac{4}{3}X_o + \frac{2}{3}S_o\right)\frac{E_o}{\varepsilon} \tag{13}$$

where X_o is the radiation length of the material and S_o is the path-length of $e^{\pm}$ below ε . The factor 2/3 is the percentage of charged particles in the shower.

As one can see, the total track-length is directly proportional to the initial energy E_o, if the charged particles of the shower can be detected until they stop. In a practical detector there is a minimum kinetic energy required for the detection (cut-off energy E_C). The consequence of this fact is that the real path-length becomes [6]:

$$S = F(x)\, X_o \frac{E_o}{\varepsilon} \tag{14}$$

with: $F(x) = e^x\left(1 + x\ln\frac{x}{1.526}\right)$ and $x = 4.58\,\frac{Z E_c}{A\,\varepsilon}$

where the cut-off energy E_c for electrons and positrons in gaseous detectors is about 0.1+1.MeV In this relation the total path-length is still proportional to the incident energy E_0 through the function F(x) that depends on the characteristics of the medium.

As a conclusion of this section we can state that the best converter materials for electromagnetic calorimeters should have a high atomic number. This means:

- Little radiation length
- Little critical energy
- Little Molière radius

Moreover, we outline the following important characteristics of such calorimeters:

1) The depth of the shower maximum scales logaritmically with the incident energy E_0.
2) The number N_P of the particles at the shower maximum is proportional to E_0 .
3) The total track-length (S) of the charged particles of the shower is still proportional to E_0 , hence the incident energy can be measured also with digital read-out by detecting and counting the number of charged particles of the shower.

5. Calorimeter Performances.

In the previous sections we have described the physical processes in the shower developement. Let us consider now the calorimeter performances in the gamma detection.

5.1 Spatial Resolution.

The impact point of photons and $e^{\pm}$ on an electromagnetic calorimeter can be obtained by measuring the lateral energy distribution of the showers. The precision of the position information (spatial resolution) increases with the number of the elementary detection cell hit by the shower particles, and decreases with the cell size (granularity). Moreover the precision (σ) of this measurement depends on the signal/noise ratio and hence improves by increasing the photon energy as $\sigma \approx 1/\sqrt{E}$.

Thus it is important to optimize the spatial resolution with respect to the requirements of the experiment , for the following motivations:

1) It is possible to achieve a good angular resolution in the determination of the direction of the incoming particles.

2) It is possible to optimize the capability of separating two or more close showers.

3) It is possible to minimize the saturation effects specially for digital calorimeters. This means that the probability that two charged particles cross the same elementary detection cell is minimized.

5.2 Energy Resolution.

For a calorimeter it is necessary to determine a correspondance (calibration) between the detected energy (corresponding to digital signals or scintillation light) and the incident energy (see FIG. 3a). This relation should be linear as we have written in equation (14). Howewer there can be intrinsic or electronic effects that keep this

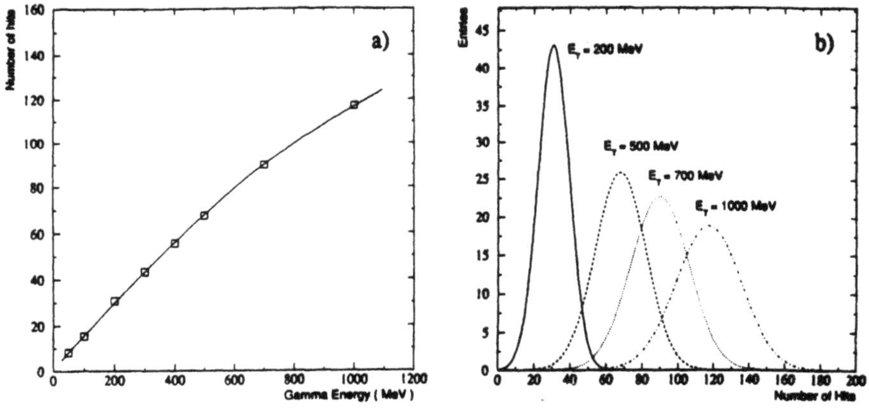

FIG. 3 a) Correspondance between the photon energy and the number of digital signals (hits) detected in the HARGD calorimeter (for more details see section 6).
b) Distribution of the hits number for monochromatic photons on the HARGD calorimeter.

correspondance off linearity. Moreover one has to determine, for a given incident energy, the spread of the detected energy around its avarage value (see FIG. 3b). It is possible to define the energy resolution of the detector as:

$$\frac{\Delta E}{E} = \frac{\sigma(E)}{E} \qquad (15)$$

$\sigma(E)$ being the overall statistical error (standard deviation) including all error sources. In this way it is possible to determine the energy of the incident photons with its error. We can state that a good calorimeter should have a calibration as linear as possible, and a narrow energy spread.

The energy resolution of an ideal homogeneous detector of infinite dimensions is limited only by statistical fluctuations, on the contrary for real calorimeters the contribution of many different effects must be summed quadratically. Let us analyze these effects that contribute to the whole energy resolution [6]:

1) STATISTICAL FLUCTUATIONS. They are fluctuations in the elementary processes of the shower, and limit the intrinsic resolution of a detector of infinite dimensions as [7]:

$$\frac{\Delta E}{E} < \frac{1.0\ \%}{\sqrt{E}} \qquad (\text{ E in GeV }) \qquad (16)$$

2) CONTAINEMENT FLUCTUATIONS. Any lack in the containement of the incident energy degrades the resolution. It can be demonstrated that, in general, the longitudinal losses are much more critical than the lateral ones, for this reason in the following we consider only the first ones. These losses depend on the dimensions of the calorimeter and increase by increasing the incident energy. One can estimate the degraded resolution as [8]:

$$\frac{\Delta E}{E} \cong A e^{-\frac{L}{\lambda_o}} \tag{17}$$

where L is the longitudinal dimension of the calorimeter, A and λ_o are parameters smoothly dependent on the energy.

3) LANDAU AND PATH-LENGTH FLUCTUATIONS. These effects are present if the sensitive layers of the calorimeter consist of gas or of a very thin layer of high density material, used as proportional counters.

Minimum ionizing particles traversing a material give an asymmetric distribution (Landau distribution) of the deposited energy because of the emission of δ rays at high angles. In this case the energy resolution can be expressed as [9]:

$$\frac{\Delta E}{E} \cong \frac{1}{\sqrt{N}} \frac{2}{\ln (10^4 x (gr / cm^2))} \tag{18}$$

where N is the number of crossing particles and x is the thickness of the material. For a gas proportional counter, with little x, Eq.(18) would give a worsening of the resolution .

The low energy electrons, due to δ rays emission, moving at large angles from the shower axis increase the fluctuations in the total path length of the shower. This effect is crucial in gas, where the cut-off energy E_c is little and the Coulombian multiple scattering become negligible, so that the electrons tend to remain in the active part of the calorimeter traveling for long path. This problem can be overcomed by putting in the gas transverse high-density walls in order to stop the low energy electrons. The worsening in the resolution, due to this effect, is of the same order of the previous one.

4) SAMPLING FLUCTUATIONS. They are to be taken into account when an electromagnetic calorimeter consists of a series of inactive converters layers of thickness δz interspaced with active detector layers. If the detector has a digital read-out and counts only the number (N_p) of particles traversing the active layers, the statistical fluctuation in N_p determines the contribution to the energy resolution. N_p depends on the total track length as:

$$N_p = \frac{S}{\delta z} = F(x) \frac{E_o}{\varepsilon} \frac{X_o}{\delta z} \tag{19}$$

where S is the total track length (given by Eq.(14)) and δz is the depth of the converter layers. We obtain for the energy resolution:

$$\frac{\Delta E}{E} = \frac{1}{\sqrt{N_p}} = 3.2\,\% \sqrt{\frac{\varepsilon\,(\text{MeV})}{F(x) \cos(\frac{E_s}{\pi\varepsilon})}} \sqrt{\frac{\delta z}{X_o\, E(\text{GeV})}} \tag{20}$$

where 3.2% is value of the intrinsic sampling resolution, E_s comes from the theory of Coulombian multiple scattering (see section 3.3), and $\cos(E_s/\pi\varepsilon)$ is the cosine of the average incident angle of $e^{\pm}$ on the converter layers. The two correcting factors F(x) and

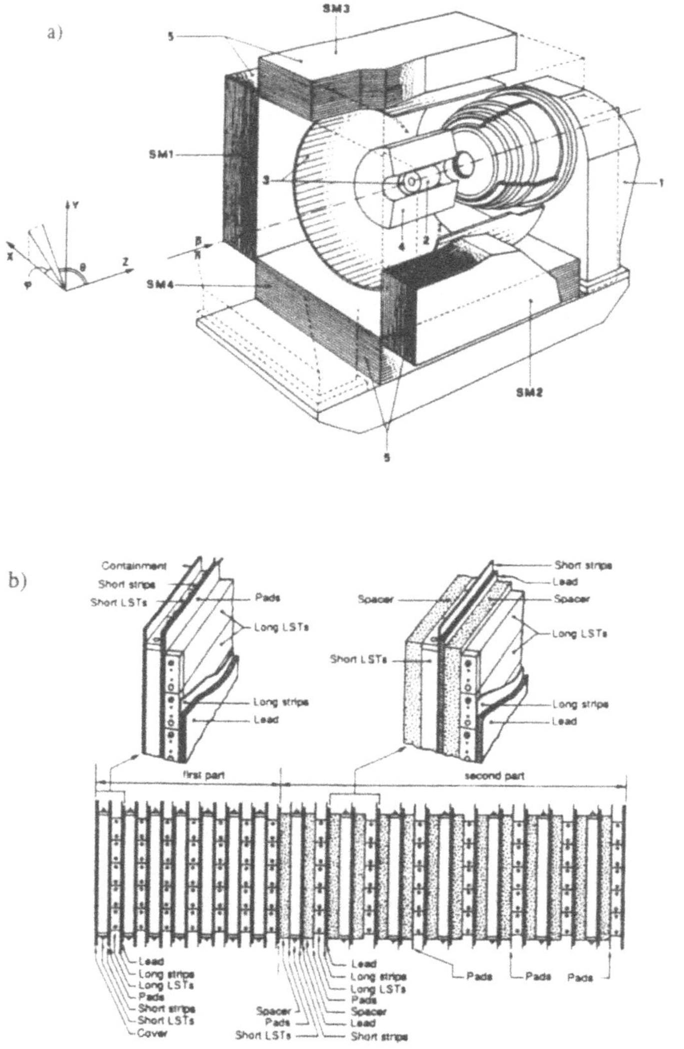

FIG. 4 a) Schematic view of the OBELIX apparatus: 1 Open Axial Field Magnet. 2) Spiral Projection chamber. 3) Internal and external Time of Flight barrel. 4) jet Drift Chamber. 5) Electromagnetic calorimeter (HARGD).
b) The calorimeter stratigraphy.

cos($E_S/\pi\varepsilon$) take into account the dependence of the resolution upon the cut-off energy E_k of the active part of the detector and upon the incident angle of $e^{\pm}$ on the converter layers.

We want to remark that the energy resolution is still proportional to $1/\sqrt{E}$, it increases by using high Z absorber materials and by lowering the thickness of the converter layers.

From these considerations we can state that the best energy resolution for electromagnetic calorimeters can be reached by homogeneous detectors with high Z and high density materials (as for exampole Lead Glass or BGO). While the best spatial resolution is achieved by sampling detectors with a high lateral segmentation of the active part, and placed at a big distance from the interaction point.

6 The electromagnetic calorimeter of the OBELIX detector

Let us consider briefly the behavior and the performances of the electromagnetic calorimeter of the Obelix experiment to better understand the result obtained in the previous sections.

Obelix [10,11] is an experiment installed on the LEAR accelerator at CERN (FIG. 4a), to study exclusive final states in the nucleon-antinucleon annichilation. HARGD, the electromagnetic calorimeter of the experiment, has been designed to detect gamma rays with high angular resolution in the energetic range between 0.1 and 1.0 GeV. It is a sampling calorimeter consisting of four indipendent supermodules. The converter is composed by 20 lead layers of half radiation length thickness (≈ 3 mm) . The active part consists of Limited Streamer Tubes (LST) [12] that are coupled to pick-up electrodes (strips and pads). The read-out of these electrodes is performed with digital electronic. In FIG. 4b is shown the stratigraphy of the calorimeter (for a more detailed description see ref.13).

FIG. 5 shows the MonteCarlo distribution of photon conversion points in the HARGD calorimeter for 500 MeV gammas. The Eq (8) of section 3.1 predicts an exponential distribution with a slope: $\lambda = \frac{1}{2}\frac{7}{9} \cong 0.4$

(the factor 1/2 is due to the thickness of the lead layers ≈ 0.5 X_0). The MonteCarlo results show a good agreement with this prevision, in fact the best fit of the data gives a slope of 0.42. The first bin of the picture contains the photons that have converted in the first Aluminium layer of the calorimeter.

In FIG. 6 we show the angular resolution of Hargd in function of the photon energy. The lateral segmentation of the detector is given by the pitch of the strips (1 cm), thus the spatial resolution should be $\sigma \approx 0.5$ cm. The distance of each supermodule from the interaction point is about $D \approx 150$ cm , therefore the angular resolution should be :

$$\Delta\vartheta \approx \frac{\sigma}{D} \approx \frac{0.5}{150} \cong 3.5 \text{ mrad} \qquad (21)$$

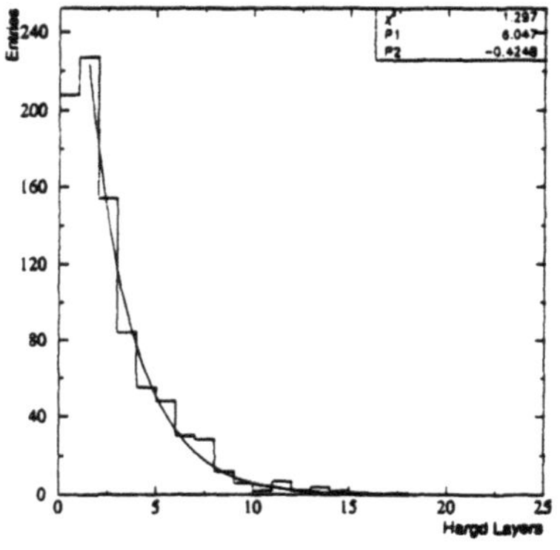

FIG. 5 Distribution of the photon conversion layers in the HARGD calorimeter for monochromatic photons of 500 MeV energy. The distribution is well fitted by the exponential form $e^{P1} \cdot e^{P2x}$ where P1 and P2 are written on the picture and show a good agreement with the previsions.

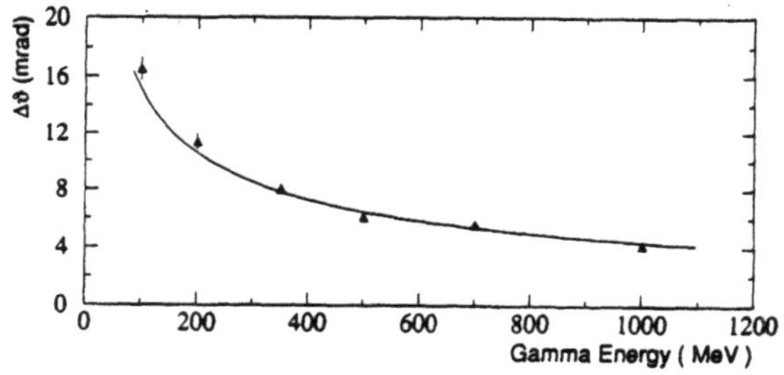

FIG. 6 Angular resolution of the HARGD calorimeter.

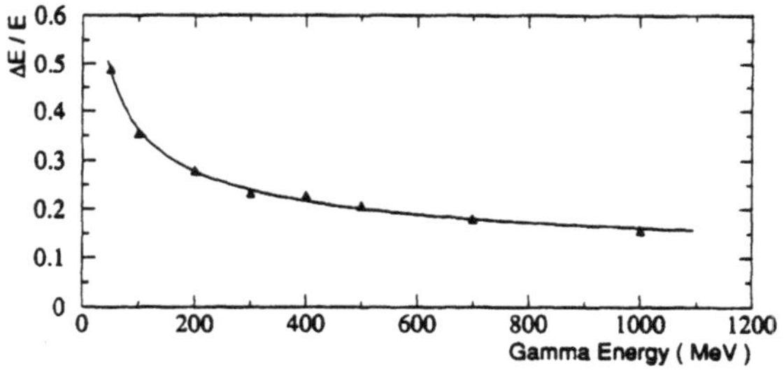

FIG. 7 Energy resolution of the HARGD calorimeter.

that is in good agreement with the MonteCarlo result at high photon energies. We want also to outline that the angular resolution increases with the photon energy proportionally to $1/\sqrt{E}$ in full agreement with the previsions of section 5.1 .

In FIG. 7 it is shown the energy resolution of the HARGD calorimeter in function of the photon energy. From the approximated relations (16 , 17 , 20) of the section 5.2 we can obtain the estimated contributions to the full resolution as :

$$\left(\frac{\Delta E}{E}\right)_I \cong \frac{0.007}{\sqrt{E}} \quad \left(\frac{\Delta E}{E}\right)_C \cong 0.07 \quad \left(\frac{\Delta E}{E}\right)_S \cong \frac{0.10}{\sqrt{E}}$$
$$\left(\frac{\Delta E}{E}\right)_L = \left(\frac{\Delta E}{E}\right)_P = 0 \qquad (22)$$

where I,C,S,L,P mean respectively intrinsic, containement,sampling,Landau and path-length fluctuations and E is expressed in GeV. In this case the Landau and the path-length fluctuations are negligible, in fact the calorimeter read-out is digital and the pick-up electrodes are displaced in order to not to be affected by these effects. These estimations give a resolution of $\approx$ 13% for E = 1 GeV and a behavior proportional to $1/\sqrt{E}$, quite in agreement with the MonteCarlo results.

7. Conclusions.

In this work we have analysed the physiscs processes, the parameters, and the elements of detector design to be taken into account to build a high performing calorimeter.

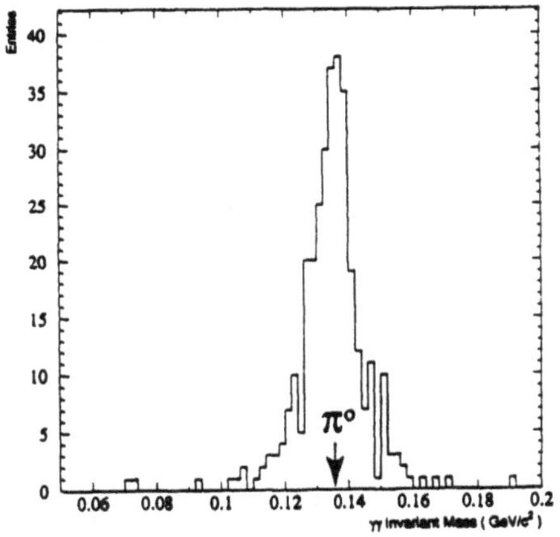

FIG. 8 Distribution of the $\gamma\gamma$ invariant mass from the π^0 decay in the reaction (23) as reconstructed by the OBELIX detector. The result is obtained by taking into account only the γ directions as reconstructed by HARGD, and the charged pion momenta reconstructed by JDC.

As we said in section 2 the fullfillment of these requirements permits to measure with high precision a wide range of exclusive reactions. As an example we show in FIG.8 the results of the study of the annichilation reaction :

$$P\overline{P} \rightarrow \pi^+\pi^-\pi^0 \rightarrow \pi^+\pi^-\gamma\gamma \qquad (23)$$

with the Obelix spectrometer as simulated by means of a MonteCarlo code. With the Angular and Energy resolutions described before it is possible to recognize the neutral pion by constructing the invariant mass of the two photon system. The precision achieved is ≈ 10% and permits a good measurement of this particle, and a high precision analysis of the reaction.

Acknowledgements

We are indebted to professors A. Bertin, M. Piccinini and A. Vitale for their continuous support and for stimulating discussions.

References

1. For a detailed description see for example: C. W. Fabjan and T. Ludlam *Calorimetry in High Energy Physics, Ann. Rev. Nucl. Part. Sci.* **32**(1982)335.
2. Review of Particles properties, *Physics Letters*. **B239**(1990).
3. See for example: M. Bruschi, I. D'Antone and A. Zoccoli, *The Trigger on Neutral Particles of the Obelix Experiment,* in this book..
4. For a detailed discussion see for example: B. Rossi , *High Energy Particles,* Prentice-Hall New York (1964).
5. Y. S. Tsai, *Rev. Mod. Phys.* **46**(1974)815.
6. For a detailed description see for example: U. Amaldi, *Fluctuations in Calorimetry Measurements, Physica Scripta.* **23**(1981)409.
7. E. Longo and I. Sestili , *Nucl. Instr. and Meth.* **128**(1975)283.
8. S. L. Stone et al. , *Nucl. Instr. and Meth.* **151**(1978)387.
9. T. Katsura et al. , *Nucl. Instr. and Meth.* **105**(1972)245.
10. R. Armenteros et al., *Study of Antinucleon Annihilations at LEAR with OBELIX, a Large - Acceptance and High - Resolution detector, based on the Open Axial Field Spectrometer,* CERN/PSCC/86-4, PSCC/P85, 21 Jan.1986.
11. A. Bertin et al., *Preliminary Study on a New version of the HARGD Calorimeter,* OX/BO/88/01, Bologna 1988.
12. E. Iarocci , *Nucl. Instr. and Meth.* **217**(1983)30
    G. Battistoni et al. , *Nucl. Instr. and Meth.* **A245**(1986)277
13. S. Affatato et al. , *Construction of a New Electromagnetic Calorimeter for the OBELIX Experiment ,* to be published .

Common Problems and Trends of Modern Physics, pp. 449–459
ed. T. Bressani, S. Marcello and A. Zenoni
©1992 World Scientific Publishing Company

The Trigger on Neutral Particles of the Obelix Experiment

M. BRUSCHI, I. D'ANTONE and A. ZOCCOLI

Dipartimento di Fisica and I.N.F.N. Sezione di Bologna, Via Irnerio 46, 40137 Bologna, Italy

Abstract

We describe the trigger system of the electromagnetic calorimeter of the Obelix experiment.

This system, completely designed built and tested by the Bologna Group of the collaboration, is capable of first level trigger on energy, on multiplicity and on identification of neutral particles in the final state of the reactions under study; higher level trigger is also possible.

Designing criteria and results on electronics and trigger efficiency on a sample reaction are here reported.

1. Introduction

The neutral particle detector of the Obelix[1] experiment (High Angular Resolution Gamma Detector, HARGD[2]) is designed to detect gammas having energies in the range 0.1 ÷ 1.0 GeV.

HARGD is a sampling electromagnetic calorimeter having 20 lead planes each 3 mm thick (half radiation length -r.l.) as converter; it is constituted by four supermodules (SM) which cover a solid angle of about $0.7 \cdot 4\pi$ around the interaction target.

The active detectors are coverless[3,4] Limited Streamer Tubes; the detection of the streamers is achieved by pick-up electrodes of two kinds:

- aluminium strips of 0.4 cm width and 1.0 cm pitch disposed parallel to LST;
- copper pads of about 30 x 30 cm^2 that divide the calorimeter into projective towers pointing toward the center of the experiment.

The sampling has a finer granularity in the conversion region of the gammas, for the energy range of interest (≈ first 5 r.l.).

In this region a pair of x-y LST planes interleaves two planes of converter material; both LST planes have readout strips, and only one of them has a plane of readout pads. The LST facing the pads have a high resistance (> 100 kΩ/square) cathode to pick up streamer signals[5].

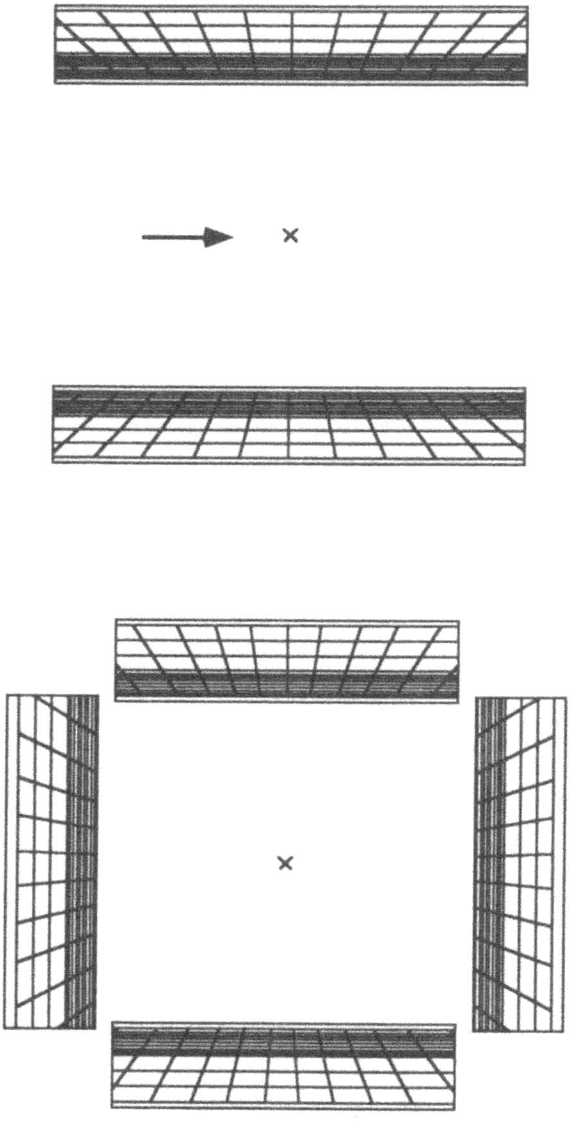

FIG. 1 . Segmentation of HARGD calorimeter in tower of pads:
a) view parallel to the beam;
b) view perpendicular to the beam.

The pair of LST planes closest to the center of the target does not have lead as converter material; this provides a mean to distinguish between a neutral or a charged particle.

The calorimeter is segmented into 120 main towers having 12 pads each (see FIG.1).

2. Data Readout

Strips have a digital readout by means of SGS electronic cards based on the D779 hybrid circuit[6]; the information one can extract is strip hit/ not hit. In this way, with the off-line data analysis we have information on hit strips and their coordinates; these quantities allow to reconstruct the energy and direction of the gamma. At trigger level every pair of LST planes (for a total of 14 couples of planes) gives a current proportional to the number of hit strips grouped four by four. We will refer to this signal as DIGOR (DIGital -OR) signal (one DIGOR = 30 mV on a 50 Ω load).

The pad readout is digital too. Hit pad configuration is passed to the off-line analysis; this allows to resolve ambiguities on clusters of hit detected by strips. We have at trigger level the same information; this means the readout of 1440 pads for each supermodule, corresponding to the 120 main towers.

Pad towers are dimensioned to contain, with high probability, e.m. showers of gammas in the range of energies 0.1 ÷ 1.0 GeV; this allows to get multiplicity in the final state by counting the number of hit towers.

3. Trigger Electronics

It has been designed and built the trigger electronics, fed by front end signals, that gives fast information about the following quantities:

- TRIGGER WITH STRIPS: energy; particle identification (charged or neutral or π/e);
- TRIGGER WITH PADS: particle identification; rough selection on neutrals energy; neutrals multiplicity in the final state.

We have designed a VME module that handles the 14 DIGOR signals of each SM, and which will be referred to as the Energy Trigger module from now on.

Input signals are added together by an analog adder and then fed into a window comparator whose threshold spans in the range 20 mV÷4 V, which translated in DIGOR units means 1÷133; this allows to make energy selections on the events of each supermodule.

Input signals are also fed into two series of 14 discriminators with two independent thresholds; depending on the threshold value set, the configurations of fired discriminators are compared with trigger configurations programmed in an EPROM. There is the possibility to select between 8 programmed triggers for each EPROM. In this way this module gives the possibility of triggering on patterns of hit planes to achieve particle identification; for example the request that the couple of LST planes closest to the beam target is not fired gives a method to distinguish between neutrals (if they do not convert in these two planes) or charged particles (a part inefficiencies of the same two planes).The trigger responses on energy and on hit planes can be further mixed to give an overall trigger response.The response time of this module is about 300 ns, due to the rise time of the DIGOR signal and to the EPROM access time.

A pad readout chain is characterized by a serial readout of daisy chained boards that transfers data for off-line reconstruction. A parallel bus allows to handle at trigger level hit pad signals; every HARGD supermodule contributes with 1440 pad signals to trigger.

Three VME modules have been designed to achieve a multiplicity trigger with pad towers.

The first module allows to scale from hit pads to hit towers; it verifies whether the pattern of hit pads of a tower corresponding to a certain event coincides with the one programmed for the trigger. The configuration of these towers is then fed to a second stage in which the number of hit towers, for each HARGD module, is counted to give tower multiplicity; then the four HARGD module multiplicities are fed into a further VME module that gives the overall one.

The first module, which will be referred to as RAMBOX, is essentially a memory look up table. It consists of ten 8k RAMs programmed so as to acknowledge input logic signal configurations that match the ones for the trigger. A single module accepts 120 input signals and therefore can deal with 10 towers (one for each RAM) of 12 pads each. An external STROBE signal is necessary to load the RAM. The RAM have an access time of about 60 ns that coincides with the module response time. The module is accessible from VME bus; each RAM can be written or read by a CPU placed in the same bus. Concerning this first module, it has also been developed a VME module that loads configurations on RAM without a CPU; it consists of a master VME module that performs the loading of programs contained on an EPROM on external RAMBOX modules.

The second module, which will be referred to as the Multiplicity Trigger, accepts as inputs 120 tower signals (the outputs of first stage RAMBOX). As output it furnishes a 7-bit word representing hit tower multiplicity for one HARGD module. This 7-bit sum is fed to :

i) VME bus for an off-line analysis;

ii) the following trigger module;
iii) a logic comparator programmable via VME to give fast information at trigger level on the multiplicity of that supermodule.

Besides, the Multiplicity Trigger module transfers also 120 hit towers data to VME bus where they can be read by a program running on a fast CPU to give a further level trigger. The response time of this module is about 300 ns.

The third and last module designed for multiplicity trigger , the Overall Multiplicity Trigger, performs a 7-bit summation starting with four 7-bit signals from the Multiplicity Trigger modules. These signals are first subdivided into two groups related to the two back to back configurations possible with the HARGD SM; the partial sums are fed to two logic comparators programmable via VME which give a fast response. Then the overall multiplicity is fed to another logic comparator programmable via VME. In this way we have information about the overall (or the back to back) multiplicities, equal or less to a prefixed one. The response time of this module is about 300 ns.

To summarize, with the designed electronics it is possible to have, at trigger level, (about 1.5 μs after the formation of an event in target) the following informations for each supermodule:

- the released energy;
- the identification of particle and pattern of hit planes;
- the multiplicity of neutrals.

All these informations for each supermodule can be further analyzed in the global HARGD event (by means of a RAMBOX module) to give the desired trigger.

4. Trigger Electronics Efficiency

The efficiency of the Trigger electronics has been tested with measurements performed with the calorimeter in conjunction with Obelix inner time of flight system, on cosmic rays. The inner time of flight system gave a trigger on cosmic rays when two of thirty scintillators slab in back to back configuration were hit.

In FIG. 2 is shown the response of Energy Trigger module on a summation over the DIGOR signals of a supermodule and a discrimination with threshold on number of DIGOR greater than 5; in abscissa we have the number of DIGOR and in ordinate the number of events having a certain number of DIGOR. For the histograms that will follow in the next figures, we will use this convention: the bars (or the area) filled with crossing lines refer to the response reconstructed on an off-line data analysis ("No HARGD Trigger" analysis); the shaded bars (or area) refer to the on-line response flagged, by HARGD trigger, on data acquired ("HARGD Trigger" analysis).

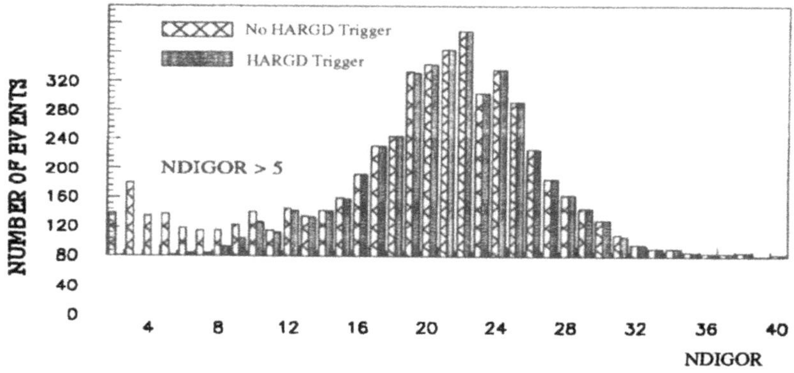

FIG. 2 Response of the Energy Trigger module (see text) on a summation over the DIGOR signals of a HARGD SM and a pulse height discrimination with a threshold on number of DIGOR greater than 5. Data taken with cosmic rays.

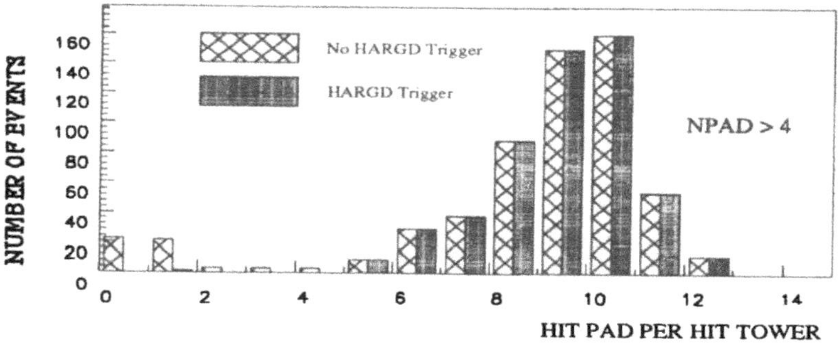

FIG. 3 Response of the RAMBOX module (see text) with the condition that a tower of pads is hit with at least 5 pads of it fired. Data taken with cosmic rays.

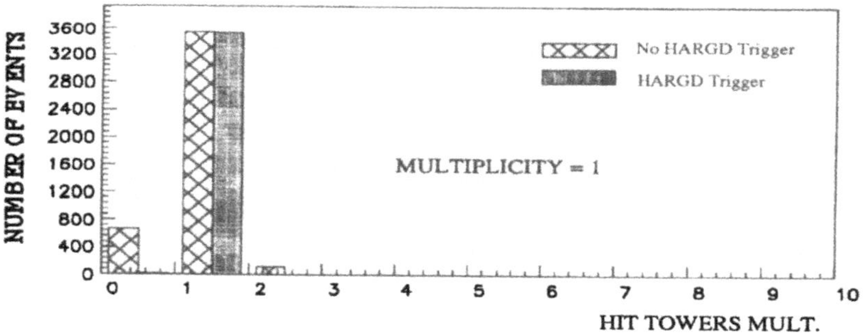

FIG. 4 Response of the Multiplicity trigger module (see text) when we put the condition that the multiplicity of hit towers on a HARGD SM is equal to 1. Data taken with cosmic rays.

We can see that the efficiency of trigger both to cut data with number of DIGOR less than 5, and to reproduce data with number of DIGOR greater than 5 is fairly good.

In order to evaluate trigger electronics efficiency we define the following quantities:

- ε_{00}: module efficiency to give a zero response when trigger conditions are not satisfied;
- ε_{11}: module efficiency to give a non zero response when trigger conditions are satisfied.

For the Energy Trigger module we obtain $\varepsilon_{00} \approx \varepsilon_{11} \approx 0.95$.

In FIG. 3 is shown the response of the RAMBOX module when we put the condition that a tower is hit with at least 5 pads fired. For the RAMBOX module we obtain $\varepsilon_{00} \approx \varepsilon_{11} \approx 0.95$.

In FIG. 4 is shown the response of the Multiplicity Trigger module when we put the condition that the multiplicity of hit towers on a supermodule is equal to 1. We obtain in this case $\varepsilon_{00} \approx \varepsilon_{11} \approx 0.97$.

5. Examples of Application of HARGD Trigger

We want now to discuss two examples of application of the designed trigger system, starting with data taken from the Obelix apparatus.

The data refer to $\bar{p}p$ annihilations at rest; the time of flight system gave a zero level "two prong" trigger (i.e. on events showing two inner scintillator barrel slabs hit and two outer scintillator barrel slabs hit); the purpose was to select events with two charged particles in the final state.

The goal of the following analysis is to inspect trigger HARGD capabilities in identifying $\bar{p}p \rightarrow \pi^+ \pi^- \pi^0$ reaction; to evaluate this we assume a 100 % HARGD trigger electronics efficiency that is rather close to the real case.

We start with a sample of about 10^5 two prong events; after a 3C kinematic fit event reconstruction we get about 400 π^0 candidates, i.e. events with a reconstructed $\gamma\gamma$ invariant mass 135 ± 40 MeV/c^2. So, without HARGD trigger we have that the percentage of events with one π^0 in the final state is about $4 \cdot 10^{-3}$.

In FIG. 5 we see three scatter plots on overall tower multiplicity vs. overall DIGOR signals (TM vs. NDIG); the threshold for a hit tower is a number of pads greater than 4. In FIG. 5a we see what we get for the original 10^5 events sample; we can see that the great majority of events lies in the bounds $0<NDIG<80$ and $0<TM<8$.

In FIG. 5b it is shown the same scatter plot for the about 400 π^0 candidates after 3C kinematic fit data analysis; we can see that events lie in the same boundaries as for part

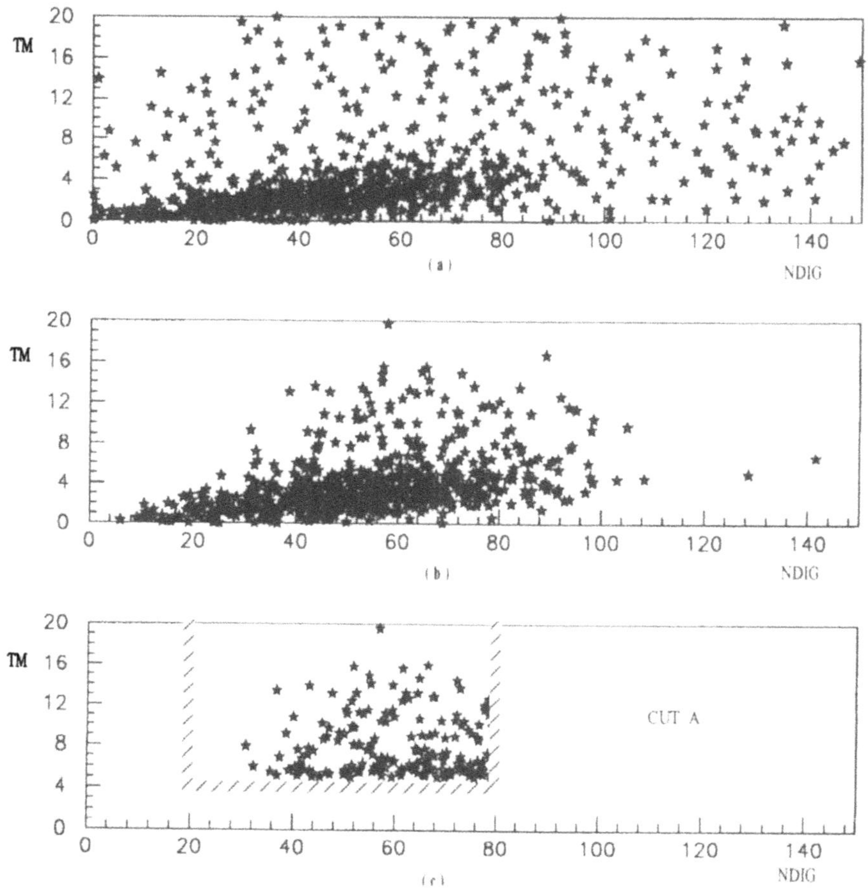

FIG. 5 Scatter plots on overall tower multiplicity (TM) vs. overall DIGOR (NDIG) signals with data taken on $p\bar{p} \dashrightarrow \pi^{+}\pi^{-}\pi^{0}$ annihilatons at rest.

a) original data sample of about 10^5 events taken with no HARGD trigger.

b) π^0 candidates (about 400 events) extracted from original sample after 3C cinematic fit data analysis.

c) Example of selection (CUT A in the text) on original data sample by means of HARGD trigger.

(a); so, at a first look, it seems rather difficult to separate one π^o final state reactions with a trigger that only makes cuts on NDIG and TM.

Anyway, we have looked for the best cuts (optimizing the number of π^o candidates in respect to overall data acquired) and we have found 20<NDIG<80 and TM>4 (cut "A", see FIG. 5c).

In FIG. 6 we have the $\gamma\gamma$ invariant mass spectrum for π^o candidates with no cut applied, and we see how the same spectrum modifies after the application of an HARGD trigger requiring cut A (dashed histogram).

The ratio of events in a window 40 MeV/c^2 around π^o mass to the total number of events is 0.32 ± 0.02 for the "No Trigger" case and 0.41±0.06 for the "Trigger" case; on the other side it must be outlined that the percent of events with one π^o in the final state with respect to the number of events acquired with an HARGD trigger is about $1.1 \cdot 10^{-2}$, showing an improvement factor of about 2.7 if compared to the case with no HARGD trigger; this means a gain of the same factor on the time needed to collect a given sample of $\bar{p}p \rightarrow \pi^+ \pi^- \pi^o$ events.

We can nevertheless note that apart from these improvements a trigger on TM>4 decreases the content of the low energy part of π^o energy spectrum.

In order to find a cut that preserves the low energy part of the π^o energy spectrum we impose the following requirements to the HARGD trigger (cut "B"): to select events with one supermodule hit by a neutral particle, or one supermodule hit by a neutral particle and another supermodule (at most) hit by a charged particle, with the condition 2<TM<6 on the overall hit tower multiplicity; also in this case the threshold for a hit tower is a number of pads greater than 4. We check the condition that a supermodule is hit by a charged or neutral particle inspecting whether the couple of LST planes closest to the experiment target is hit or not.

As for the previous case, in FIG. 7 we report a $\gamma\gamma$ invariant mass spectrum for π^o candidates with no cut applied, and how the same spectrum is modified after the application of a HARGD trigger requiring cut B(dashed histogram).

The ratio of events in a window 40 MeV/c^2 around π^o mass to the total number of events is 0.50±0.09 for this "Trigger" case; the percentage of events with one π^o in the final state with respect to number of events acquired with an HARGD trigger is also in this case about $1.1 \cdot 10^{-2}$ .One can therefore conclude that with trigger cut "B" the same factor 2.7 can be achieved with a reduced bias concerning the low energy part of the π^o spectrum.

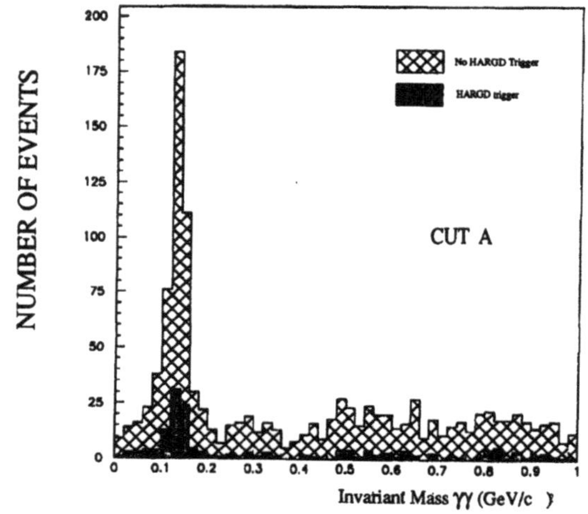

FIG. 6 Comparison of γγ Invariant Mass spectra for π^0 candidates in the case with no HARGD trigger, and with HARGD trigger CUT A (see text) applied.

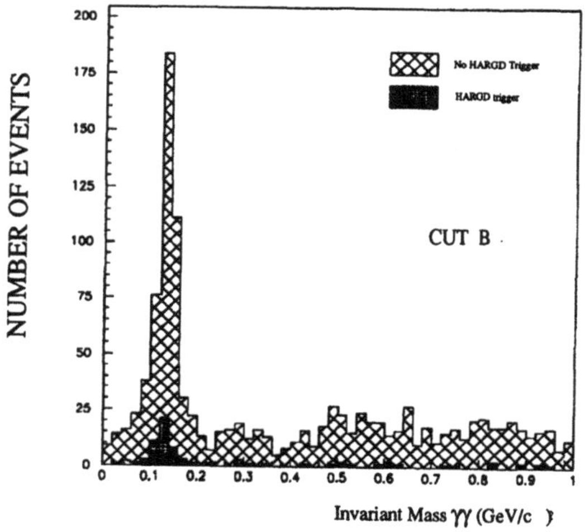

FIG. 7 Comparison of γγ Invariant Mass spectra for π^0 candidates in the case with no HARGD trigger, and with HARGD trigger CUT B (see text) applied.

6. References.

1. R. Armenteros et al., *Study of Antinucleon Annihilations at LEAR with Obelix, a Large - Acceptance and High - Resolution detector, based on the Open Axial Field Spectrometer*, CERN/PSCC/86-4, PSCC/P85, 21 Jan. 1986.
2. A. Bertin et al., *Preliminary Study on a New Version of the HARGD calorimeter*, OX/BO/88/01, Bologna 1988.
3. E. Iarocci, *Nucl. Instr. and Meth.* **A217**(1983)30.
4. G. Battistoni et al., *Nucl. Instr. and Meth.* **A245**(1986)277.
5. G. Battistoni et al., *Nucl. Instr. and Meth.* **152**(1978)423.
6. F. Beconcini et al., *IEEE Trans. Nucl. Sci.* **NS-35**(1988)311.

Common Problems and Trends of Modern Physics, pp. 461–468
ed. T. Bressani, S. Marcello and A. Zenoni
©1992 World Scientific Publishing Company

Proton extracted beam from high energy accelerators by channeling effect

M.P.Bussa, L.Busso and F.Tosello
Istituto di Fisica Generale dell'Universita' di Torino, Italy
Istituto Nazionale di Fisica Nucleare, Sezione di Torino, Italy

ABSTRACT

The availability of proton extracted beam out of the TeV accelerators would open very interesting possibilities for B-physics, in particular for the study of CP violation. Bent crystals offer an interesting alternative for beam extraction without disturbing collider mode operation of the accelerator.

1. Introduction

The availability of a 8 TeV/c proton beam, parasitically extracted out of the LHC accelerator, would open up very interesting physics possibilities.

Channeling in bent crystals appears to be the most promising method to produce an extracted beam of intensity 10^8 p/s without disturbing collider mode operation of the accelerator.

The physics of B mesons could be studied in a fixed-target experiments[1] reaching the sensitivity for a detailed investigation of CP violation. Measuring CP violation in Beauty decays is the ideal solution to test the validity of the Standard Model, since with a proper choice of the initial state (B_d or B_s) and of the decay mode, one can in principle measure all the three angles α, β, γ (fig.1) expressing the unitarity of the Cabibbo-Kobayashi-Maskawa (CKM) quark mixing matrix[2]. Since the b-quark mass is large the theoretical uncertainty is smaller than for Kaons, particularly when transitions of neutral B-mesons to final states f of defined CP-parity are involved. In that case the asymmetry:

$$\Lambda = \frac{N(B^0 \rightarrow f) - N(\overline{B^0} \rightarrow f)}{N(B^0 \rightarrow f) + N(\overline{B^0} \rightarrow f)}$$

assumes the form $\Lambda = \mathrm{sen}\, 2\Phi$ where Φ is a function of α, β, γ of the triangle. The asymmetry is expected to be large: numerical calculations predict $\Lambda = 0.30 \pm 12$ [3,4].

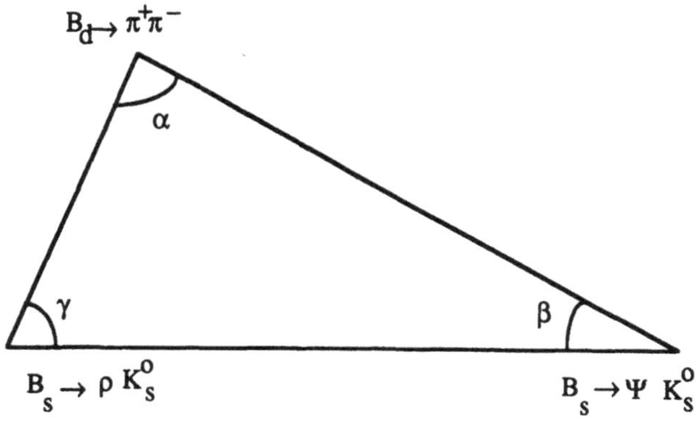

Fig. 1 – The Unitarity Triangle.

A problem with B decay is that the branching ratios of the interesting channels are small, typically of order 10^{-4}-10^{-5} : this requires a large initial $\overline{B}B$ sample. If we consider $B_d \rightarrow J/\psi + K_s$ decay and $\Lambda=0.3$ more than 10^8 $\overline{B}B$ per year are needed for collecting a significative number of events.

2. Beam extraction from LHC

To obtain more than 10^{10} $\overline{B}B$ pairs in 10^7 s in a fixed tungsten target at LHC proton momentum (8 TeV/c) an interaction rate $N \geq 1.6*10^7$ int/s is needed, considering $\sigma_{\overline{b}b}$ (p,p) = 0.6*A μb. This should happen without interfering with collider operation, the p beam for fixed target must not adsorb more protons than an intersection region of the beam, i.e. the fixed target operation should be completely transparent to the experiments working in the collider mode. Because classical resonant extraction schemes would in the best case demand a dedicated mode of operation, they would conflict with the request of a completely parasitic Fixed Target operation. Also using standard electrostatic septa to separate the extracted beam from the circulating one and to deflect it into the extraction line would be problematic. Given the modest transverse kick, they should be so long as to be incompatible with the length of the available accelerator straight sections.

Considering the conditions before exposed, the most promising methods for the beam extraction are based on crystal channeling [5].

Beam extraction from high energy accelerators using channeling in bent crystals was first demonstrated in the mid-eighties at the 8 GeV Dubna synchrophasotron [6]. The Dubna group used an 11 mm long silicon crystal bent through an angle of 35 mrad. The crystal transmission was 10^{-4}, limited by the area of the crystal transverse to the beam.

In the 1989 a beam of $5*10^6$ particles/s was extracted from the 76 GeV Serpukhov accelerator using a 7 cm long crystal bent through 80 mrad [7]. Recent results from the H8 micro-beam at CERN SPS have demonstrated high efficiency bending a 450 GeV/c protons beam by means of a 5 cm silicon crystal [8]. The beam was bent by 7 mrad with an efficiency larger than 10%.

Theoretical previsions about LHC (8 TeV/c) are to bend the beam of 1 mrad angle with 10-15% efficiencies.

3. Feeding the crystal

In the high luminosity colliders, like LHC, it is expected that elastic collisions between the two beams will form a flux of deflected particles called "halo". If the flux is grater than 10^7 protons/s, a set of collimators is required to capture them before they are lost in the superconducting magnets. It is possible to think to deflect these protons with a bent crystal before they are lost in the collimators. The crystal would be placed at the edge of the circulating beam [9].

Two parameters are important for bent crystal extraction in order to be totally parasitic:

-the impact parameter of the halo on the crystal, once the equilibrium is established,

-the flux of particles in the halo which in first approximation depends on the elastic cross-section p-p and on the beam optics.

It is reasonable to suppose that the flux in the beam halo is proportional to the luminosity because the expected RMS divergence between elastic collisions is 20 μrad, compared to the angular divergence of 30 μrad at the interaction point. For this reason the flux in the beam halo is roughly proportional to the luminosity of the machine in the collider intersections.The beam halo should be more intense, and then the crystal extraction more advantageous, from LHC over SSC since the luminosity is larger by a factor 40 and LHC has lower momentum.

When all these effects are considered, the halo is expected to be of the order of 10^9 protons/s in LHC and a factor 100 or 1000 less in SSC [10].

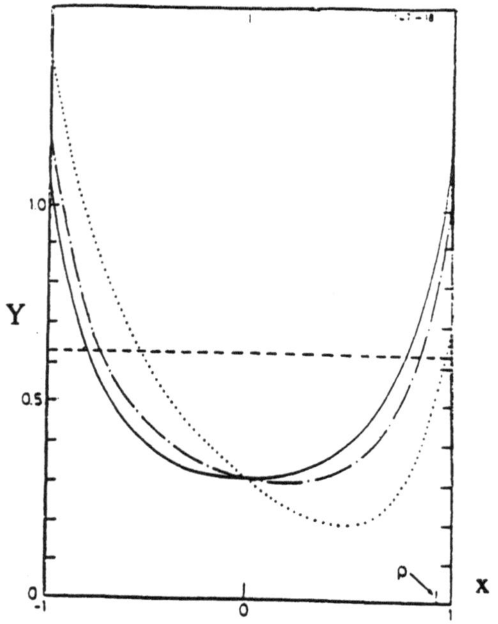

Fig. 2 – Effective planar potential for 12 GeV/c protons between two (110)-planes in silicon. x is the distance from the midplane between two atomic planes. The full line represents the potential in a straight crystal, the dashed-dotted curve the potential in a bent crystal with a radius of curvature of 40 cm., and the dotted curve the potential for a radius of curvature of 10 cm.The potential is in arbitrary units (Ref.15).

4. Channeling factors

Planar channeling is a phenomenon occurring when particles enter in a crystal nearly parallel to a planar direction [11]. These particles will be trapped and will oscillate in very steep potential wells with field gradient of 10^{11}V/cm. Even in a bent crystal the particles remain trapped between the planes unless the bending radius becomes very small or the beam momentum very high. Using the Linard potential [12] and taking into account the contribution from the two planes forming the planar channel, the effective potential may be plotted for various values of curvature (fig. 2)

The basic quantity determining the conditions under which planar channeling effect occurs is the critical angle:

$$\Psi_p = \sqrt{\frac{4\ Z_1\ Z_2\ e^2\ N\ d_p\ C\ a}{p\ v}}$$

where Z_1, Z_2 are the atomic numbers of projectile and target, p is the momentum of the particle, v its velocity, Nd_p is the areal density of atoms, C is a constant (roughly $\sqrt{3}$) and a $\approx a_0\ Z_2^{-1/3}$ is a screening length in the ion-atom potential, a_0 being the Bohr radius. Only for particles incident on a crystal with an angle to a plane less than or of the order of Ψ_p the channeling effect will be important, so in practice $\pm\Psi_p$ represents the angular acceptance of the crystal (note that Ψ_p is proportional to $1/\sqrt{p}$).

For example 12 GeV/c particles along {110} planes, Ψ_p =0.04 mrad; for 450 GeV/c protons in silicon Ψ_p =7 µrad.

In an unbent crystal there are three ways in which a channeled particle can change its transverse momentum, this effect is referred as dechanneling:

- scattering by electrons,
- scattering by nuclei in the lattice
- scattering by impurities or other crystal defects.

In the high quality crystals used in high-energy channeling, the latter possibility may be neglected. Scattering by nuclei is only effective for particles having a rather high transverse energy and therefore able to closely approach the atomic planes .

A dechanneling length L_d can be defined, which is proportional to the momentum p. For 450 GeV/c protons, L_d = 350 mm in silicon crystal. A reduction of L_d occurs when the crystal is bent.

5. Effect of bending

Particles channeled in a bent crystal will follow its curvature, but additional losses are introduced. This effect increases with the curvature of the crystal. The phenomenon can be understood as a distortion of the transverse potential generated by centrifugal term :

$$Y_{eff}(r) = Y(r) + \frac{L^2}{2\ m\ r^2}$$

here r is the distance of the particle from the center of curvature and L is the angular momentum. This expression implies that the particles are subjected to an additional constant force perpendicular to the bent planes:

$$F_{eff} = F(x) + pv/r$$

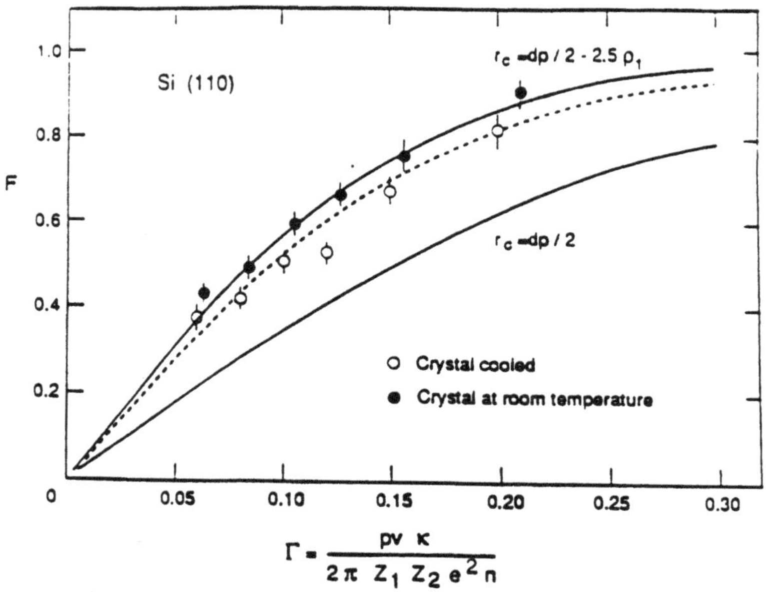

Fig. 3 – The channeled fraction F as a function of the dimensionless parameter Γ . Data are from ref. 13.

where F(x) is the force from the atomic planes and the constant term is the centrifugal force.

For strong bendings, corresponding to radii of curvature of 1÷10 cm, the minimum of the potential is shifted considerably towards the outer plane and few particles are able to follow the curvature (fig. 2)

Introducing a critical radius for bending, R_c, by demanding that the effective potential be shifted to a order of thermal vibrational amplitude, we obtain:

$$R_c = \frac{p\ v}{\pi\ Z_1\ Z_2\ e^2\ N\ d_p}$$

below which all particles in practice are lost. For 12 GeV/c particles in silicon, R_c is of the order of 1 cm. Considering R_c proportional to p, and v ≈ constant, for 100 GeV/c protons in silicon R_c ≈ 10 cm.

More detailed calculations on dechanneling in curved crystals have been performed by Kudo [13] and Ellison [14] who calculated the fraction of channeled particles which became dechanneled by entering in a region where the crystal is curved.
Fig. 3 presents the fraction F of particles being dechanneled upon entering in a region of constant curvature R as a function of the dimensionless parameter Γ:

$$\Gamma = \frac{p}{R} \frac{v}{2\pi Z_1 Z_2 e^2 n}$$

in bent crystals a reduction of dechanneling length $L_{d,b}$ is expected. A rough approximation gives:

$$L_{d,b} = L_d (1-F)^2$$

6. Conclusion

Bent crystal channeling offers an interesting alternative for beam extraction from TeV accelerators. Crystal channeling requires much less space than conventional extraction methods. More information is needed on dechanneling in bent crystals including the effects of the dislocations at TeV energies, a more thorough understanding of the theory of crystal extraction is also desirable.

7. References

1) B.N. Jensen et al.CERN/DRDC 91-25, p. 29 (1991).

2) C.O. Dib. et al., *Phys. Rev.* **D** **41** (1990) 1522.

3) (T. Nakada editor) CERN 90-92 (1990).

4) M. Lusignoli et al., Preprint Univ. Roma "La Sapienza" n. 792 (1991).

5) A. Carigan te al., *Nucl. Instr. Meth..* **B** **48** (1990) 167.

6) V.V. Audeichiov et al., *JINR Communication* **1**- 84 Dubna (1984).

7) M. Bavizhev UNK Workshop (1989).

8) S.P. Moller et al., *Phys. Lett.* **B** **256** (1991) 91.

9) W. Scandale; *Proceeding of the LHC Workshop* (G. Jarlskog and D. Rein Editors, Aachen 1990), Vol. 3 p. 760.

10) N. Mokhov, *SSC Note* SSCL- 49 (1990).

11) "Relativistic channeling" ASI Series, Vol 165 (R. A. Carrigan Jr. and J.A. Ellison Editors, Plenum Press, New York 1986).

12) J. Lindhard, *Mat. Fys. Medd. Dan. Vid. Selsk.* **34** (1965) 14.

13) H. Kudo, *Nucl. Instr. Meth.* **189** (1981) 609.

14) J.A. Ellison, *Nucl. Phys.* **B** **206** (1982) 205.

15) J.F.Bak et al., *Nuc. Phys.* **B** **242** (1984) 1.

Common Problems and Trends of Modern Physics, pp. 469–475
ed. T. Bressani, S. Marcello and A. Zenoni
©1992 World Scientific Publishing Company

SPIN PARITY METHODS IN MESON SPECTROSCOPY

VALERIO FILIPPINI

Istituto Nazionale di Fisica Nucleare
Sezione di Pavia
Via Bassi 6, 27100 Pavia

ABSTRACT

The spin parity method used to study meson spectroscopy is described. The covariant formalism is assumed, and the Zemach formalism is derived as a special case.

1 Covariant spin tensors

A covariant spin tensor is a 4tensor in the ordinary space-time with the following properties[1]:

- It is simmetric in every pair of indices:

$$S_{\mu\nu..} = S_{\nu\mu..} \tag{1}$$

- It is traceless in every pair of indeces:

$$S_{\mu..}^{..\mu} = 0 \tag{2}$$

- It is orthogonal to the 4momentum p_μ:

$$S_{\mu..}p^\mu = 0 \tag{3}$$

A 4tensor of rank s with the previous properties has $2s+1$ independent components.

If the rank is equal to 1 we are dealing with a 4vector describing a spin 1 particle (for example a ρ meson or a ${}^3S_1, {}^1P_1, {}^3P_1$ $p\overline{p}$ state). Only the last property is meaningful because there is one index only.

If the rank is equal to 2 we are dealing with a 4tensor describing a spin 2 particle (for example a f_2 meson or a ${}^3P_2$ $p\overline{p}$ state). Now all the three properties are meaningful.

If the particle decays into two spin zero particles, the spin tensor can be built from the momenta of the two particles $p_{1\mu}, p_{2\mu}$ using the difference $q_\mu = p_{1\mu} - p_{2\mu}$, and the 4velocity $u_\mu = (p_{1\mu} + p_{2\mu})/(p_{1\mu} + p_{2\mu})^2$:

Spin 1:

$$S_\mu = q_\mu - (q \cdot u)u_\mu \tag{4}$$

Spin 2:

$$S_{\mu\nu} = (q_\mu - (q \cdot u)u_\mu)(q_\nu - (q \cdot u)u_\nu) + \frac{1}{3}(q^2 - (q \cdot u)^2)(u_\mu u_\nu - g_{\mu\nu}) \tag{5}$$

Spin 3:

$$S_{\mu\nu\rho} = (q_\mu - (q \cdot u)u_\mu)(q_\nu - (q \cdot u)u_\nu)(q_\rho - (q \cdot u)u_\rho)$$
$$+ \frac{1}{5}(q^2 - (q \cdot u)^2)(q_\mu(u_\nu u_\rho - g_{\nu\rho}) + q_\nu(u_\rho u_\mu - g_{\rho\mu}) + q_\rho(u_\mu u_\nu - g_{\mu\nu})) \tag{6}$$

One can easily prove that these tensors satisfy the conditions (1,2,3).

In the rest frame of the resonance $u_\mu = (1,0,0,0)$, all the time components of these tensors are zero. This means that these 4tensors are 3tensors inserted in a 4 dimension space. In fact the spin describes how the particle wave function behaves under spatial (3 dimensional) rotation.

2 Three dimension formalism

If the particle decays into two particles of the same mass (for examples $\rho^0 \to \pi^-\pi^+$) then $q \cdot u = 0$ and the previous tensors can be written in the following way:

Spin 1:

$$S_\mu = q_\mu \tag{7}$$

Spin 2:

$$S_{\mu\nu} = q_\mu q_\nu + \frac{1}{3}q^2\delta_{\mu\nu} \tag{8}$$

Spin 3:

$$S_{\mu\nu\rho} = q_\mu q_\nu q_\rho - \frac{1}{5}q^2(q_\mu\delta_{\nu\rho} + q_\nu\delta_{\rho\mu} + q_\rho\delta_{\mu\nu}) \tag{9}$$

where:

$$\delta_{\mu\nu} = u_\mu u_\nu - g_{\mu\nu}$$

In the rest frame of the resonance:

$$\delta_{\mu\nu} = \begin{pmatrix} 0 & 0 & 0 & 0 \\ 0 & 1 & 0 & 0 \\ 0 & 0 & 1 & 0 \\ 0 & 0 & 0 & 1 \end{pmatrix}$$

and the spin tensors (7,8,9) can be thought as 3tensors built from the 3vector $\vec{q} = \vec{p}_1 - \vec{p}_2$ as follows:

Spin 1:

$$S_i = q_i \tag{10}$$

Spin 2:

$$S_{ij} = q_i q_j - \frac{1}{3} q^2 \delta_{ij} \tag{11}$$

Spin 3:

$$S_{ijk} = q_i q_j q_k - \frac{1}{5} q^2 (q_i \delta_{jk} + q_j \delta_{ki} + q_k \delta_{ij}) \tag{12}$$

That is the reason why very often in literature one finds the use of 3tensors (called Zemach tensors)[2] to describe the spin of the resonaces. One can do it only if the resonance decays in two particles of the same mass, and always using the resonance system (i.e. the system where the 4velocity of the resonace is (1,0,0,0)).

Using the covariant formalism has the great advantage that every reference system is good and that the masses of decay products can be different. The formalism becomes heavier and heavier when then spin increases, but all the formulae can be written in a computer program and easily used[3].

3 Tensor combination

Two tensors can be coupled to form a new tensor, following the quantum mechanics rules of angular momentum coupling: two angular momenta j_1, j_2 form an angular momentum J that ranges between $|j_1 - j_2|$ and $j_1 + j_2$.

In tensor algebra one has to combine tensors of different rank, using the rules for tensor combination and projection. Here we give only the results for the $1 \oplus 1$ case:

$1 \oplus 1 = 0$

$$S = A_\mu B^\mu \tag{13}$$

$1 \oplus 1 = 1$

$$S_\mu = \epsilon_{\mu\nu\rho\sigma} A^\nu B^\rho u^\sigma \tag{14}$$

$1 \oplus 1 = 2$

$$C_{\mu\nu} = \frac{1}{2}(A_\mu B_\nu + A_\nu B_\mu)$$

$$S_{\mu\nu} = C_{\mu\nu} - C_{\mu\rho} u^\rho u_\nu - C_{\nu\rho} u^\rho u_\mu + \frac{1}{3} C_{\rho\sigma} u^\rho u^\sigma (g_{\mu\nu} + 2u_\mu u_\nu) - \frac{1}{3} C^\rho_\rho (g_{\mu\nu} - u_\mu u_\nu) \tag{15}$$

In the following examples these formulae will be used.

4 Examples

4.1 $p\bar{p}_{^3S_1} \longrightarrow \pi^0\rho^0, \rho^0 \longrightarrow \pi^+\pi^-$

The $p\bar{p}_{^3S_1}$ initial state has quantum number $0^-(1^{--})$, the ρ meson $1^+(1^{--})$ and the π meson $1^-(0^{-+})$. The parity of $\pi^0\rho^0$ is $(-1)^L$ where L is the $\pi^0\rho^0$ angular momentum. To match the initial $p\bar{p}$ negative parity, L must be odd. The easiest solution is to set $L = 1$

The angular momentum between $\pi^+\pi^-$ is 1 coming from ρ^0 spin.

Let us define:

p_0 = 4momentum of π^0

p_+ = 4momentum of π^+

p_- = 4momentum of π^-

p_ρ = 4momentum of ρ^0 ($p_\rho = p_+ + p_-$)

The spin of ρ^0 can be built from p_- and p_+, using (4)

$$R_\mu = q_\mu - (q \cdot u)u_\mu$$

where $q = p_+ - p_-$ and $u = (p_+ + p_-)/(p_+ + p_-)^2$

A 4vector L_μ describing the $\pi^0\rho^0$ angular momentum is built with the same formula as before where now $q = p_0 - p_\rho$ and $u = (p_0 + p_\rho)/\sqrt{s}$.

$$L_\mu = q_\mu - (q \cdot u)u_\mu$$

These two 4vectors R_μ and L_μ must now be coupled to form a 4vector that gives the spin amplitude of the reaction. This can be done using formula (14) with u_μ equal to the 4velocity of $p\bar{p}$.

Now let us compute the same amplitude using the formulae of section 2

The two charged pions have the same mass, so we can write:

$$R_\mu = p_{+\mu} - p_{-\mu}$$

In the $p\bar{p}$ center of mass $u_\mu = (1, 0, 0, 0)$, giving:

$$L_\mu = (0, \vec{p}_0 - \vec{p}_\rho)$$

Formula (14) is nothing but the vector product between R and L, so the spatial components of the amplitude can be written:

$$\vec{A} = (\vec{p}_+ - \vec{p}_-) \times (\vec{p}_0 - \vec{p}_\rho)$$

Being $\vec{p}_\rho = \vec{p_+} + \vec{p_-}$ we have:

$$\vec{A} = (\vec{p}_+ - \vec{p}_-) \times \vec{p}_0$$

This result, very often quoted in literature, means that the signature for the reaction $p\overline{p}_{^3S_1} \rightarrow \pi^0\rho^0, \rho^0 \rightarrow \pi^+\pi^-$ is a $\sin^2\theta$ distribution for the angle between $\vec{p}_+ - \vec{p}_-$ and $\vec{p}_0$ (see Fig. 1).

In this case the result could be achieved also with the 3 dimensional formalism of section 2. In the next example it is not possible.

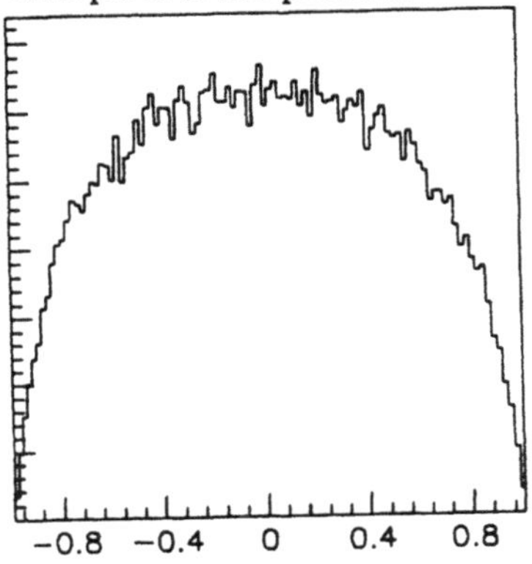

Fig.1

4.2 $p\overline{p}_{^1S_0} \rightarrow \overline{k^0}k^{*0}, k^{*0} \rightarrow k^+\pi^-$

This reaction is very similiar to the previous one, with a big important difference: k^{*0} does not decay into particles with equal mass. This means that we can not use the 3 dimension formulae of section 2, but we have to use the full covariant formalism.

The $p\overline{p}_{^1S_0}$ initial state has quantum number $1^-(0^{-+})$, the k^* meson $\frac{1}{2}(1^-)$, the k meson $\frac{1}{2}(0^-)$ and the π meson $1^-(0^{-+})$.

The parity of k^0k^{*0} is $(-1)^L$. To match the initial $p\overline{p}$ negative parity, we set $L = 1$

Let us define:

p_0 = 4momentum of $\overline{k^0}$

p_+ = 4momentum of k^+

p_- = 4momentum of π^-

p_* = 4momentum of k^{*0} $(p_* = p_+ + p_-)$

The spin of k^{*0} can be built from p_- and p_+, using (4):

$$K_\mu = q_\mu - (q \cdot u)u_\mu$$

where $q = p_+ - p_-$ and $u = (p_+ + p_-)/(p_+ + p_-)^2$

A 4vector L_μ describing the $L = 1$ $\overline{k^0}k^{*0}$ angular momentum is built with the same formula as before where now $q = p_0 - p_*$ and $u = (p_0 + p_*)/\sqrt{s}$.

These two 4vectors K_μ and L_μ must now be coupled with the help of (13)

The resulting angular distribution for the angle between $\vec{p}_+ - \vec{p}_-$ and $\vec{p}_0$ is plotted in Fig 2.

In Fig. 3 the $\cos^2\theta$ distribution of the $p\bar{p}_{^1S_0} \rightarrow \pi^0\rho^0, \rho^0 \rightarrow \pi^+\pi^-$ reaction is plotted.

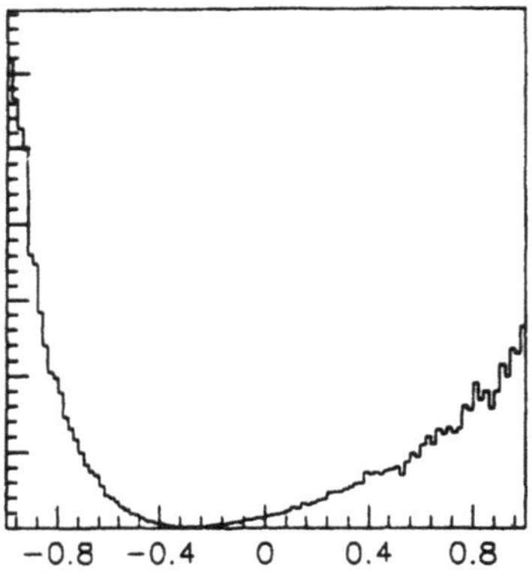

Fig.2

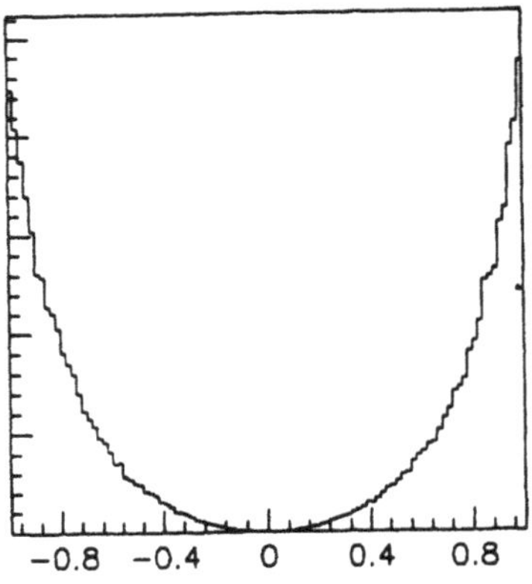

Fig.3

4.3 $p\bar{p}_{^3S_1} \longrightarrow \pi\rho, \rho \longrightarrow \pi\pi$

The only difference with the first example is that now we take into account all the three possible charged states of ρ meson. We are working with three interfering amplitudes:

$p\bar{p}_{^3S_1} \rightarrow \pi^0\rho^0, \rho^0 \rightarrow \pi^+\pi^-$ amplitude: $\vec{A} = (\vec{p}_+ - \vec{p}_-) \times \vec{p}_0$
$p\bar{p}_{^3S_1} \rightarrow \pi^+\rho^-, \rho^- \rightarrow \pi^0\pi^-$ amplitude: $\vec{A} = (\vec{p}_0 - \vec{p}_-) \times \vec{p}_+$
$p\bar{p}_{^3S_1} \rightarrow \pi^-\rho^+, \rho^+ \rightarrow \pi^0\pi^+$ amplitude: $\vec{A} = (\vec{p}_0 - \vec{p}_+) \times \vec{p}_-$
Each amplitude must now be multiplied by a Breit-Wigner[4] function like

$$BW(m) = \sqrt{\frac{m}{q}} \frac{\Gamma(m)^{\frac{1}{2}}}{m_\rho^2 - m^2 - im_\rho\Gamma(m)}$$

($m_\rho = 0.777$ GeV is the mass of ρ meson), and summed together.
The total amplitude is:

$$((\vec{p}_+ - \vec{p}_-)\times\vec{p}_0)BW(m_{\pi^+\pi^-}) + ((\vec{p}_- - \vec{p}_0)\times\vec{p}_+)BW(m_{\pi^-\pi^0}) + ((\vec{p}_0 - \vec{p}_+)\times\vec{p}_-)BW(m_{\pi^0\pi^+})$$

The sum of complex quantities will produce an interference among the different phases

References

[1] C.Zemach, *Phys. Rev.* 1B **140**, (1963) 97
[2] C.Zemach, *Phys. Rev.* 5B **133**, (1963) 201
[3] V.Filippini, A.Fontana, A.Rotondi, N.Semprini, *Obelix internal report* OX/PV/01-92
[4] A.Rotondi, *this proc.*

Common Problems and Trends of Modern Physics, pp. 477–486
ed. T. Bressani, S. Marcello and A. Zenoni
©1992 World Scientific Publishing Company

ANGULAR MOMENTUM ANALYSIS WITH ZEMACH TENSORS

N. SEMPRINI CESARI
Dipartimento di Fisica and INFN, Bologna (Italy)

ABSTRACT

Angular momentum analysis is a good tool in the determination of the spin-parity of a resonance. In the analysis of meson spectroscopy data the angular momentum tensors in the form developed by Zemach have been extensively used. In this paper we want to discuss some aspects of the tensorial formalism in the case of integer spin resonances decaying into spinless particles. The relativistic extension of the formalism is also discussed.

1. Introduction

During the sixties the discovery of resonant states decaying into barions and mesons has stimulated the search for spin-parity tests for the determination of the quantum numbers of the unstable particles[1,2]. One of the aims of these tests is to achieve the result with the minimum number of hypotheses on the mechanisms of production and decay.

In the case of nucleon-antinucleon annihilations, apart from a few tentatives, most of the data has been analysed using Watson's theory of the final state interaction[3]. In this model the dynamics is simplified by assuming that the interactions which occur among the final particles before getting outside the range of their mutual forces are strong compared to any other interaction which could be present in the reaction. In these cases it is useful to describe the angular part of the transition matrix elements by means of the angular momentum tensors following the prescription given by C. Zemach[4,5].

In this paper we want to discuss some aspects and applications of angular momentum tensors in the case of an integer spin particle decaying into spinless particles. The outline of the paper is the following: in the first part we recall some properties of cartesian tensors, in the second we show some examples regarding the three pion decay and in the third one we discuss the relativistic extension of tensorial formalism.

2. Non relativistic tensors

2.1 Transition amplitude

Let us consider a particle of spin J which decays in the center of mass system into n spinless particles with momenta $\mathbf{p}_1$, $\mathbf{p}_2, \ldots \mathbf{p}_n$, the transition amplitude is written

$$\langle \mathrm{Fin} \,|\, F \,|\, \mathrm{In} \rangle = f(\varepsilon_1, \varepsilon_2, \ldots \varepsilon_n)\, T_{k1,k2,\ \ldots\ kn}\, S_{k1,k2,\ \ldots\ kn} \tag{1}$$

where summation over repeated indices is assumed. The function $f(\varepsilon_1,\varepsilon_2,\ldots\varepsilon_n)$ called form factor depends upon particle energies (this is due to Watson theorem). The tensor $T_{k1,k2,\ldots kn}$ represents the angular momentum of the final state and is constructed with a sufficient number of components of the momenta of the final particles. $S_{k1,k2,\ldots kn}$ is the spin function of the initial state.

From a formal point of view these tensors realize the irreducible representations of the rotation group. They have rank J in the three space, symmetric and traceless in every pair of indices.

2.2 Vectors and Tensors

Let us consider a tridimensional rotation of the reference frame; the relation between the old and the new coordinates of a vector is

$$x'_j = R_{jk}\, x_k \qquad j, \kappa = 1, 2, 3 \tag{2}$$

the transformation preserves the scalar product

$$x'_j\, y'_k\, \delta_{jk} = x_p\, y_q\, \delta_{pq} \tag{3}$$

δ_{kl} is the Kroneker tensor. This equation imposes the following conditions on the matrix R

$$R_{jk}\, R_{jl} = \delta_{kl} \tag{4}$$

Relations 4) represent a set of 6 linear equations on the matrix R which has 9 independent parameters; it follows that a tridimensional rotation has 9 - 6 = 3 independent parameters. It is easy to show that the relations 4) imply

$$\varepsilon_{jkl}\, R_{pj}\, R_{qk}\, R_{sl} = \varepsilon_{pqs} \tag{5}$$

where ε_{jkl} is the Ricci tensor. In this equation we have taken into account the fact that the unit matrix which represents the identity transformation of the rotation group has unimodular determinant.

We define tensor of rank n (or with n indices) a set of 3^n quantities which transform themselves according to

$$w'_{jk\,\ldots\,z} = R_{jp}\; R_{kq}\; \ldots\; R_{zt}\; w_{pq\,\ldots\,t} \tag{6}$$

This transformation law is a natural generalization of the vector ones.

It is possible to use the equations 4) and 5) to reduce the rank of a tensor. Consider, for example, a tensor $W_{p,q,r\ldots t}$ of rank n and let us perform a contraction by the Kroneker delta

$$\delta_{jk}\; w'_{jkl\,\ldots\,z} = \delta_{jk}\; R_{jp}\; R_{kq}\; R_{lr}\; \ldots\; R_{zt}\; w_{pqr\,\ldots\,t} \tag{7}$$

Using equation 4) we obtain

$$w'_{jjl\,\dots\,z} = R_{lr} \,\dots\, R_{zt}\; w_{ppr\,\dots\,t} \tag{8}$$

which is the transformation law of a tensor with a rank lowered by two units. Analogously using equation 5) it is possible to show that the Ricci tensor lowers the rank by one or three units. We have no other way to lower the rank of a tensor because the only constraints imposed on the rotation matrix R are those expressed by the equations 4) and 5).

2.3 Irreducible Tensors

In the preceding paragraph we have defined a set of 3^n independent components (tensor of rank n) which transform themselves under rotation in a well defined manner. Now the question is the following: is it possible to reduce the number of independent components of a tensor? In general the answer is affermative. When this is impossible without zeroing all the components we call the tensor irreducible. We will show that the irreducible tensors realize the irreducible representations of the rotation group that we associate to the different states of angular momentum.

It is easy to show that the permutation operators commute with the rotation ones so that if a tensor has a defined symmetry with respect to the permutation of certain indices the rotated tensor $W'_{j,k,l,\dots z}$ has the same symmetry: the rotation preserves the symmetry properties. In this way the symmetry properties of a tensor have an absolute character and we can impose a well defined symmetry to reduce the number of independent components. Formally we have to construct the irreducible representations of permutation group (the group related to the permutations of the n indices of a tensor). The method of Young tables is a suitable way to build these representations.

Given a tensor $W_{j,k,l,\dots z}$ of rank n we have the minimum number of independent components when the tensor is symmetric with respect to the permutation of every pair of indices. The number of independent components is

$$\frac{n^2 + 3n + 2}{2} \tag{9}$$

The contraction with the Kroneker and Ricci tensors are the only invariant operations so we can impose on $W_{j,k,l,...z}$ only the following conditions

$$\delta_{pq} W_{jk \,...\, pq \,...\, vz} = 0 \tag{10}$$

$$\varepsilon_{pqr} W_{jk \,...\, pqr \,...\, vz} = 0 \tag{11}$$

The equations 11) are always fulfilled because of the symmetry of the tensor $W_{j,k,l,...z}$, the equations 10) instead represent a set of new $n(n-1)/2$ constraints (we have considered all the different pair of indices). The independent components of a completely symmetric and traceless tensor are

$$\frac{n^2 + 3n + 2}{2} - \frac{n(n-1)}{2} = 2n+1 \tag{12}$$

It is impossible to reduce further the number of independent components because we have no other constraints to impose than the total symmetry and the equations 10) and 11) ; these tensors are called irreducible.

2.4 Irreducible tensors and angular momentum

The irreducible tensors of rank n realize the irreducible representations of the rotation group that correspond to an angular momentum equal to n. It is possible to give a general proof but we will show its validity only in one particular case.

Consider the expression of the operator $\mathbf{L}^2$ (square of angular momentum operator) in cartesian coordinates

$$\mathbf{L}^2 = 2\, x_q \partial_q + x_p\, x_q\, \partial_p \partial_q - x_p\, x_p\, \partial_q \partial_q \tag{13}$$

and applying this operator on $T_{j,k}$ (tensor of rank 2 symmetric and traceless) we obtain

$$\mathbf{L}^2\, T_{jk} = \mathbf{L}^2 \left(x_j\, x_k - \frac{\delta_{jk}}{3}\, x_p\, x_p \right) = 6\, T_{jk} \tag{14}$$

that is the eigenvalue of angular momentum L = 2. This result holds also if the set of functions x_j are normalized to unity.

An irreducible tensor of rank n is an eigenstate of $\mathbf{L}^2$ but not of $\mathbf{L}_z$; in fact it is possible to show that the components of a cartesian tensor mix the components corresponding to different values of L_z. In all the cases where the third component of angular momentum is not essential the use of the angular momentum tensors is equivalent to that of spherical armonics.

3. Three pions decay

3.1 The kinematical variables

Following the prescription given by Zemach it is possible to construct the tensor of the angular momentum of the final state with the components of the momenta of the final particle.

Let us consider the case of a system which decays in three pions with momenta $\mathbf{p}_1$, $\mathbf{p}_2$ and $\mathbf{p}_3$ and suppose that two of them are produced by a resonance

$$A \Rightarrow \Omega + \pi_1 \Rightarrow \pi_2 + \pi_3 + \pi_1 \tag{1}$$

We can use the momentum $\mathbf{p}_1$ to construct the tensor which represents the angular momentum of the spectator and the momenta $\mathbf{p}_2$ and $\mathbf{p}_3$ to construct the tensor of the total angular momentum of the resonance. If we want to put in evidence the dependence from the spin we have to consider in the frame of the resonance the relative momentum **q** of π_2 and π_3. The other two variables may be $\mathbf{p}_r$ and $\mathbf{p}_1$, the momenta of the resonance and of the spectator respectively. In this way with $\mathbf{p}_1$ we can construct the tensor of the angular momentum of the spectator, with $\mathbf{p}_r$ the tensor of the angular momentum of the resonance and with **q** the tensor of the spin of the resonance. These variables are not indipentent because of the conservation of momentum $\mathbf{p}_r + \mathbf{p}_1 = 0$, so that we can choose only two independent momenta. If we choose $\mathbf{p}_1$ and **q** we can construct the tensors of the angular momentum of the system resonance-spectator and the spin of the resonance.

3.2 Examples and applications

Let us consider the reaction

$$p\,\bar{p} \Rightarrow \Omega + \pi^0 \Rightarrow \pi^+ + \pi^- + \pi^0 \tag{2}$$

$$L_{in} = 1\,,\ L_s = 1\,,\ S_r = 2$$

where L_{in}, L_s and S_r are respectively the initial angular momentum, the orbital angular momentum of the system resonance-spectator and the spin of the resonance. We describe the angular momentum of final state by using the tensors

$$T^s_j(\vec{p}) = p_j \tag{3}$$

$$T^r_{kl}(\vec{q}) = q_k\, q_l - \frac{1}{3}\,\delta_{kl}\, q^2$$

where **p** and **q** are respectively the relative momentum of two pions in the frame of the resonance and the momentum of the spectator in the center of mass frame (see par. 3.1). We have used an irreducible tensor of rank 1 to describe the angular momentum of the resonance-spectator system (L_s=1) and an irreducible tensor of rank 2 for the resonance spin (S_r=2).

The initial angular momentum is L_{in}=1, so we must form an irreducible tensor of rank 1 starting from the tensors of equations 3). We have

$$M_l \propto \delta_{jk}\, T^s_j(\vec{p})\, T^r_{kl}(\vec{q}) = (\vec{p}\,\vec{q}\,)\, q_l - \frac{1}{3} q^2\, p_l \tag{4}$$

$$M_l\, M_l = |M|^2 \propto q^2 (\vec{q}\,\vec{p}\,)^2 + \frac{1}{9} q^4 p^2 - \frac{2}{3} q^2 (\vec{q}\,\vec{p})^2 = \tag{5}$$

$$= \frac{1}{3} q^4 p^2 \,(\cos^2(\theta) + \frac{1}{3})$$

To reduce the rank of the tensor product we have used the Kroneker tensor; it is possible also to use two times the Ricci tensor, the result is obviously the same. The ambiguity in the choice of the indices to contract is not essential. In equation 5) we have performed a summation over the polarizations of the initial state. It is easy to obtain the same result by using the Clebsh - Gordan coefficients.

We consider the same reaction as 2) with the angular momentum $L_{in}=1$, $L_s=2$ and $S_r=2$, we have the following tensors

$$T^s_{jk}(\vec{p}) = p_j\, p_k - \frac{1}{3}\, \delta_{jk}\, p^2 \tag{6}$$

$$S^r_{lm}(\vec{q}) = q_l\, q_m - \frac{1}{3}\, \delta_{lm}\, q^2 \tag{7}$$

$$M_i \propto \varepsilon_{ijl}\, \delta_{km}\, T^s_{jk}(\vec{p})\, S^r_{lm}(\vec{q}) = (\vec{p}\times\vec{q})_i (\vec{p}\,\vec{q}) \tag{8}$$

$$M_i\, M_i = |M|^2 \propto p^4\, q^4\, \sin^2(\theta)\, \cos^2(\vartheta) \tag{9}$$

We have performed the reduction of the tensor product of rank 4 with the Kroneker and Ricci tensors.

4. Relativistic tensors

The relativistic extension of the formalism may be realized considering the Lorentz transformed of particles momenta of the non relativistic description.

Let us consider a particle of mass m which decays in n final particles. Let us introduce now the following four-vectors :

$$X = (m, \vec{0}) \tag{1}$$

$$P = (p_0, \vec{p}) \tag{2}$$

$$P' = (0, \vec{p}) \tag{3}$$

The first is simply the four-momentum of the decaying particle, the second is the four-momentum of a generic produced particle, the third one is an opportune four-vector whose temporal part is zero and spatial part is the momentum of the particle. All the four-vectors are written in the system of the decaying particle.

The relativistic invariant associated to P' is equal to the tridimensional invariant of the momentum **p**

$$p_j \, p_j = - P'_\mu \, P'^\mu \qquad j = 1, 2, 3 \qquad \mu = 0, 1, 2, 3 \tag{4}$$

In this way is possible to write in a covariant form a non relativistic expression simply by replacing the vector **p** with the four-vector P'. We need only the relation between the four-vector P' and the four-vectors X and P associated to the particles. Multiplying equation 1) by P_0 and equation 2) by m we obtain easily

$$P' = P - \frac{P_0}{m} X \tag{5}$$

Considering the relativistic invariant

$$X_\mu \, X^\mu = m^2 \qquad P_\mu \, X^\mu = m \, P_0 \tag{6}$$

we obtain

$$P' = P - X \frac{P_\nu X^\nu}{X_\mu X^\mu} \quad (7)$$

So we can obtain the relativistic expression with the following substitution

$$p_j \Rightarrow P'_\mu - X_\mu \frac{P_\eta X^\eta}{X_\rho X^\rho} \quad (8)$$

With similar considerations we can calculate the substitution for the Kroneker and Ricci tensors

$$\delta_{jk} \Rightarrow g_{\mu\nu} - \frac{X_\mu X_\nu}{X_\rho X^\rho} \quad (9)$$

$$\varepsilon_{ijk} \Rightarrow \varepsilon_{ijkl} \frac{X^l}{\sqrt{X_\rho X^\rho}} \quad (10)$$

With the previous equations it is possible to obtain the expression of a relativistic tensor starting from the non relativistic one. When the non relativistic formulation is correctly used the relativistic one adds nothing except being more compact and independent from the frame of reference.

5. References

1. R. Gatto and H. P. Stapp, *Phys. Rev.* **121**, (1961) 1553.
2. M. Ademollo, R. Gatto and G. Preparata, *Phys. Rev.* **139B**, (1965) 1608.
3. K. M. Watson, *Phys. Rev.* **88**, (1952) 1163.
4. C. Zemach, *Phys. Rev.* **140**, (1965) B97, *Phys. Rev.* **140**, (1965) B109.

LIST OF PARTICIPANTS

Alessandra ADAMO
I.N.F.N. - sez. di Cagliari
via Ada Negri n. 18
I09127 - CAGLIARI

Oscar ADRIANI
I.N.F.N. - sez. di Firenze
largo Enrico Fermi n. 2
I50125 - FIRENZE

Michelangelo AGNELLO
Dipartimento di Fisica del
Politecnico di Torino
I.N.F.N. - sez. di Torino
via Pietro Giuria n. 1
I10125 - TORINO

Raffaella ARANI
I.N.F.N. - sez. di Milano
via Celoria n. 16
I20133 - MILANO

Rinaldo BALDINI FERROLI
L.N.F. dell'I.N.F.N.
via Enrico Fermi n. 40
I00044 - FRASCATI (RM)

Ruggero BARNI
I.N.F.N. - sez. di Milano
via Celoria n. 16
I20133 - MILANO

Francesco BECATTINI
I.N.F.N. - sez. di Firenze
largo Enrico Fermi n. 2
I50125 - FIRENZE

Giorgio BENDISCIOLI
Dipartimento di Fisica
Nucleare Teorica
dell'Universita'di Pavia
via Ugo Bassi n. 6
I27100 - Pavia

Antonio BERTIN
Dipartimento di Fisica
dell' Universita' di Bologna
I.N.F.N. - sez. di Bologna
via Irnerio n. 46
I40126 - BOLOGNA

Cesare BINI
Universita' "La Sapienza"
Dipartimento di Fisica e
I.N.F.N.
piazzale Aldo Moro n. 2
I00185 - ROMA

Aldo BONASERA
Laboratori Nazionali
del Sud dell' I.N.F.N.
viale Andrea Doria
I95125 - CATANIA

Ivan BONO
I.N.F.N. - sez. di Milano
via Celoria n. 16
I20133 - MILANO

Elena BOTTA
I.N.F.N. - sez. di Torino
sede di via Sette Comuni n. 56
I10127 - TORINO

Tullio BRESSANI
Dipartimento di
Fisica Sperimentale
dell'Universita' di Torino
I.N.F.N. - sez. di Torino
via Pietro Giuria n. 1
I10125 - TORINO

Marco BRUSCHI
I.N.F.N. - sez. di Bologna
via Irnerio n. 46
I40126 - BOLOGNA

Luigi BUSSO
Dipartimento di
Fisica Sperimentale
dell'Universita' di Torino
I.N.F.N. - sez. di Torino
via Pietro Giuria n. 1
I10125 - TORINO

Francesco Saverio CAFAGNA
Dipartimento di Fisica
I.N.F.N. - sez. di Bari
via Giovanni Amendola n. 173
I70126 - BARI

Daniela CALVO
I.N.F.N. - sez. di Torino
via Pietro Giuria n. 1
I10125 - TORINO

Barbara CEREDA
I.N.F.N. - sez. di Bologna
via Irnerio n. 46
I40126 - BOLOGNA

Piergiorgio CERELLO
I.N.F.N. - sez. di Torino
via Pietro Giuria n. 1
I10125 - TORINO

Emilio CHIAVASSA
Dipartimento di
Fisica Sperimentale
dell' Universita' di Torino
I.N.F.N. - sez. di Torino
via Pietro Giuria n. 1
I10125 - TORINO

Stefano DE CASTRO
I.N.F.N. - sez. di Bologna
via Irnerio n. 46
I40126 - BOLOGNA

Emilio DEL GIUDICE
I.N.F.N. - sez. di Milano
via Celoria n. 16
I20133 - MILANO

Louis DICK
I.N.F.N. - sez. di Milano
via Celoria n. 16
I20133 - MILANO

Ferdinando D'ISEP
I.N.F.N. - sez. di Torino
via Pietro Giuria n. 1
I10125 - TORINO

Antonietta DONZELLA
Dipartimento di Elettronica
per l'Automazione
dell'Universita' di Brescia
via Valotti n. 9
I25060 - MOMPIANO (BR)

Luciano FAVA
I.N.F.N. - sez. di Torino
via Pietro Giuria n. 1
I10125 - TORINO

Alessandro FELICIELLO
I.N.F.N. - sez. di Torino
via Pietro Giuria n. 1
I10125 - TORINO

Alessandra FILIPPI
I.N.F.N. - sez. di Torino
via Pietro Giuria n. 1
I10125 - TORINO

Valerio FILIPPINI
I.N.F.N. - sez. di Pavia
via Ugo Bassi n. 6
I27100 - PAVIA

Francesco FORTI
I.N.F.N. - sez. di Pisa
via Livornese n. 582/A
I56010 - PISA

Domenico GALLI
I.N.F.N. - sez. di Bologna
via Irnerio n. 46
I40126 - BOLOGNA

Luca GAMBERALE
I.N.F.N. - sez. di Milano
via Celoria n. 16
I20133 - MILANO

Remo GARATTINI
I.N.F.N. - sez. di Milano
via Celoria n. 16
I20133 - MILANO

Paolo GENSINI
Dipartimento di Fisica
dell'Universita' di Perugia
via Elce di Sotto n. 10
I27100 - PERUGIA

Paola GIANOTTI
I.N.F.N. - sez. di Torino
via Pietro Giuria n. 1
I10125 - TORINO

Marina GIBILISCO
I.N.F.N. - sez. di Milano
via Celoria n. 16
I20133 - MILANO

Carlo GUARALDO
L.N.F. dell'I.N.F.N.
via Enrico Fermi n. 4
I00044 - FRASCATI (RM)

Felice IAZZI
Dipartimento di Fisica del
Politecnico di Torino
corso Duca degli Abruzzi n. 24
I10125 - TORINO

Carlo LAMBERTI
I.N.F.N. - sez. di Torino
via P. Giuria n. 1
I10125 - TORINO

Armando LANARO
L.N.F. dell'I.N.F.N.
via Enrico Fermi n. 4
I00044 - FRASCATI (RM)

Massimo LENTI
I.N.F.N. - sez. di Firenze
largo Enrico Fermi n. 2
I50125 - FIRENZE

Giuseppe LIBERTI
Universita' della Calabria
I87053 - ARCAVACATA DI RENDE (CS)

Evandro LODI RIZZINI
Dipartimento di Elettronica
per l'Automazione
dell'Universota' di Brescia
via Valotti n. 9
I25060 - MOMPIANO (BR)

Vincenzo LUCHERINI
L.N.F. dell'I.N.F.N.
via Enrico Fermi n. 40
I00044 - FRASCATI (RM)

Luciano MAIANI
Dipartimento di Fisica
dell'Universita' - Roma I
piazzale Aldo Moro n. 2
I00185 - ROMA

Gianpiero MANGANO
Dipartimento di Scienze Fisiche
dell'Universita' di Napoli
I.N.F.N. - sez. di Napoli
Mostra d' Oltremare
Padiglione 19
I80125 - NAPOLI

Simonetta MARCELLO
I.N.F.N. - sez. di Torino
via Pietro Giuria n. 1
I10125 - TORINO

Stefano MARCHI
I.N.F.N. - sez. di Milano
via Celoria n. 16
I20133 - MILANO

Umberto MARCONI
I.N.F.N. - sez. di Bologna
via Irnerio n. 46
I40126 - BOLOGNA

Giacomo MARGAGLIOTTI
Dipartimento di Fisica
dell'Universita' di Trieste
I.N.F.N. - sez. di Trieste
via Valerio n. 2
I34127 - TRIESTE

Alberta MARZARI CHIESA
Dipartimento di
Fisica Sperimentale
dell' Universita' di Torino
I.N.F.N. - sez. di Torino
via Pietro Giuria n. 1
I10125 - TORINO

Alberto MASONI
I.N.F.N. - sez. di Cagliari
via Ada Negri n. 18
I09127 - CAGLIARI

Ignazio MASSA
Dipartimento di Fisica
dell' Universita' di Bologna
I.N.F.N. - sez. di Bologna
via Irnerio n. 46
I40126 - BOLOGNA

Renata MELE
I.N.F.N. - sez. di Milano
via Celoria n. 16
I20133 - MILANO

Enrica MEZZETTI
Dipartimento di Fisica del
Politecnico di Torino
corso Duca degli Abruzzi n. 24
I10125 - TORINO

Emilio MIGNECO
Laboratori Nazionali
del Sud dell' I.N.F.N.
viale Andrea Doria
I95125 - CATANIA

Bruno MINETTI
Dipartimento di Fisica
del Politecnico di Torino
corso Duca degli Abruzzi n. 24
I10125 - TORINO

Paolo MONTAGNA
I.N.F.N. - sez. di Pavia
via Ugo Bassi n. 6
I27100 - PAVIA

Giacomo MORPURGO
Dipartimento di Fisica
dell'Universita' di Genova
via Dodecaneso n.33
I16146 - GENOVA

Aldo MORSELLI
I.N.F.N. - sez. di Roma II
via Raimondi
I00100 - TOR VERGATA (RM)

Alessandro MUGGIA
I.N.F.N. - sez. di Milano
via Celoria n. 16
I20133 - MILANO

Matteo PARIS
I.N.F.N. - sez. di Milano
via Celoria n. 16
I20133 - MILANO

Giovanni PASSALEVA
I.N.F.N. - sez. di Firenze
largo Enrico Fermi n. 2
I50125 - FIRENZE

Maurizio PICCININI
I.N.F.N. - sez. di Bologna
via Irnerio n. 46
I40126 - BOLOGNA

Luca PIROVANO
I.N.F.N. - sez. di Milano
via Celoria n. 16
I20133 - MILANO

Mario POLI
Dipartimento di Energetica
dell'Universita' di Firenze
via santa Marta n. 3
I50131 - FIRENZE

Giuliano PREPARATA
Dipartimento di Fisica
dell' Universita' di Milano
I.N.F.N. - sez. di Milano
via Celoria n. 16
I20133 - MILANO

Stefano RAGAZZI
I.N.F.N. - sez. di Milano
via Celoria n. 16
I20133 - MILANO

Philip RATCLIFFE
I.N.F.N. - sez. di Milano
via Celoria n. 16
I20133 - MILANO

Renato Angelo RICCI
L.N.L. dell'I.N.F.N.
via Romea n. 4
I35020 - Legnaro (PD)

Elena ROSSETTO
I.N.F.N. - sez. di Torino
via Pietro Giuria n. 1
I10125 - TORINO

Alberto ROTONDI
Dipartimento di Fisica
dell'Universita' di Pavia
via Ugo Bassi n. 6
I27100 - PAVIA

Paola SALVINI
I.N.F.N. - sez. di Pavia
via Ugo Bassi n. 6
I27100 - PAVIA

Enrico SCOMPARIN
I.N.F.N. - sez. di Torino
via Pietro Giuria n. 1
I10125 - TORINO

Massimo SCORLETTI
I.N.F.N. - sez. di Milano
via Celoria n. 16
I20133 - MILANO

Nicola SEMPRINI CESARI
I.N.F.N. - sez. di Bologna
via Irnerio n. 46
I40126 - BOLOGNA

Sergio SERCI
Dipartimento di Fisica
dell'Universita' di Cagliari
I.N.F.N. - sez. di Cagliari
via Ada Negri n. 18
I09127 - CAGLIARI

Maria Teresa SERGI
Dipartimento di Fisica
dell'Universita' di Cagliari
via Ospedale n. 72
I09127 - CAGLIARI

Susanna TESSARO
Dipartimento di Fisica
dell'Universita' di Trieste
I.N.F.N. - sez. di Trieste
via Valerio n. 2
I34127 - TRIESTE

Victor TRETIAK
I.N.F.N. - sez. di Pavia
via Ugo Bassi n. 6
I27100 - PAVIA

Arnaldo UGUZZONI
Dipartimento di Fisica
dell' Universita' di Bologna
I.N.F.N. - sez. di Bologna
via Irnerio n. 46
I40126 - BOLOGNA

Stefania VECCHI
I.N.F.N. - sez. di Bologna
via Irnerio n. 46
I40126 - BOLOGNA

Marco VERPELLI
I.N.F.N. - sez. di Milano
via Celoria n. 16
I20133 - MILANO

Antonio VITALE
Dipartimento di Fisica
dell' Universita' di Bologna
I.N.F.N. - sez. di Bologna
via Irnerio n. 46
I40126 - BOLOGNA

Joseph WEBER
Department of Physics and
Astronomy
University of Maryland
College Park
MD - 20742 U.S.A.

She-sheng XUE
L.N.F. dell'I.N.F.N.
via E. Fermi n. 40
I00044 - FRASCATI (RM)

Guido ZAVATTINI
I.N.F.N. - sez. di Bologna
via Irnerio n. 46
I40126 - BOLOGNA

Aldo ZENONI
I.N.F.N. - sez. di Pavia
via Ugo Bassi n. 6
I27100 - PAVIA

Antonio ZOCCOLI
I.N.F.N. - sez. di Bologna
via Irnerio n. 46
I40126 - BOLOGNA

SUBJECT INDEX

AUTHOR INDEX

www.ingramcontent.com/pod-product-compliance
Lightning Source LLC
LaVergne TN
LVHW020834261026
850861LV00047B/41
9789810212032